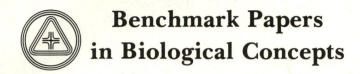

Benchmark Papers
in Biological Concepts

Series Editor: Peter Gray
University of Pittsburgh

Published Volumes and Volumes in Preparation

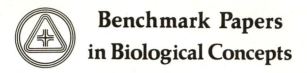

Benchmark Papers
in Biological Concepts

—— A *BENCHMARK* ᴛᴍ Books Series ——

VERTEBRATE
REGENERATION

Edited by
C. S. THORNTON and S. C. BROMLEY
Michigan State University

Dowden, Hutchinson & Ross, Inc.
Stroudsburg, Pennsylvania

Copyright © 1973 by **Dowden, Hutchinson & Ross, Inc.**
Benchmark Papers in Biological Concepts, Volume 2
Library of Congress Catalog Card Number: 73–13830
ISBN: 0–87933–057–0

Library of Congress Cataloging in Publication Data

Thornton, Charles Stead, 1910-   comp.
    Vertebrate regeneration.

    (Benchmark papers in biological concepts, v. 2)
    Bibliography: p.
    1. Regeneration (Biology) 2. Vertebrates--
Physiology. I. Bromley, Stephen C., 1938-
joint comp. II. Title. [DNLM: 1. Regeneration--
Collected works. 2. Vertebrates--Collected works.
W1 BE515 v. 2 1973 / QH499 V567 1973]
QH499.T52   596'.03'1   73-13830
ISBN 0-87933-057-0

Manufactured in the United States of America.

Exclusive distributor outside the United States and
Canada: John Wiley & Sons, Inc.

Permissions

The following papers have been reprinted with the permission of the authors and copyright owners.

Academic Press, Inc.—*Developmental Biology*
"Electron Microscopic Observations of Muscle Dedifferentiation in Regenerating *Amblystoma* Limbs"
"Origin of the Blastema in Regenerating Limbs of the Newt *Triturus viridescens*. An Autoradiographic Study Using Tritiated Thymidine to Follow Cell Proliferation and Migration"
"Localization of Newly Synthesized Proteins in Regenerating Newt Limbs as Determined by Radio-autographic Localization of Injected Methionine-S^{35}"
"Collagenolytic Activity in Regenerating Forelimbs of the Adult Newt (*Triturus viridescens*)"
"The Urodele Limb Regeneration Blastema: A Self-organizing System. II. Morphogenesis and Differentiation of Autografted Whole and Fractional Blastemas"
"Lens Antigens in a Lens-regenerating System Studied by the Immunofluorescent Technique"
"The Influence of Neural Retina and Lens on Lens Regeneration from Dorsal Iris Implants in *Triturus viridescens* Larvae"
"Nerve Fiber Requirements for Regeneration in Forelimb Transplants of the Newt *Triturus*"
"Denervation Effects on Newt Limb Regeneration: DNA, RNA, and Protein Synthesis"

American Association for the Advancement of Science—*Science*
"Limb Regeneration: Induction in the Newborn Opossum"

American Society of Zoologists—*American Zoologist*
"Relationship Between the Tissue and Epimorphic Regeneration of Muscles"

Cambridge University Press—*Biological Reviews*
"The Loss and Restoration of Regenerative Capacity in the Limbs of Tailless Amphibia"

The Company of Biologists Limited—*Journal of Embryology and Experimental Morphology*
"Effects of Adrenal Transplants upon Forelimb Regeneration in Normal and in Hypophysectomized Adult Frogs"
"Experimental Investigations of Morphogenesis in the Growing Antler"

Masson et Cie, Paris—*Comptes Rendus Société de Biologie*
"Démonstration de l'existence de territoires spécifiques de régéneration par la méthode de la déviation des troncs nerveux"

North-Holland Publishing Company—*Proceedings, Regeneration in Animals*
"Autonomous Morphogenetic Activities of the Amphibian Regeneration Blastema"

The University of Chicago Press—*Regeneration in Vertebrates*
"The Cellular Basis of Limb Regeneration"

The Wistar Institute Press
Anatomical Record
"Studies on Limb Regeneration in X-rayed *Amblystoma* Larvae"
The Journal of Experimental Zoology
"Stability of Chondrocyte Differentiation and Contribution of Muscle to Cartilage During Limb Regeneration in the Axolotl (*Siredon mexicanum*)"
"Preblastemic Changes of Intramuscular Glycogen in Forelimb Regeneration of the Adult Newt (*Triturus viridescens*)"
"Methods of Initiating Limb Regeneration in Adult Anura"
"The Nervous System and Regeneration of the Forelimb of Adult *Triturus*. VII. The Relation Between Number of Nerve Fibers and Surface Area of Amputation"

"The Relation Between the Caliber of the Axon and the Trophic Activity of Nerves in Limb Regeneration"

"Blastema Formation in Sparsely Innervated and Aneurogenic Forelimbs of *Amblystoma* Larvae"

"Tissue Interaction in Amputated Aneurogenic Limbs of *Amblystoma* Larvae"

"Recuperation of Regeneration in Denervated Limbs of *Ambystoma* Larvae"

"Hormonal and Nutritional Requirements for Limb Regeneration and Survival of Adult Newts"

Journal of Morphology

"The Inductive Activity of the Spinal Cord in Urodele Tail Regeneration"

"Analysis of Tail Regeneration in the Lizard *Lygosoma laterale*. I. Initiation of Regeneration and Cartilage Differentiation: The Role of Ependyma"

Series Editor's Preface

"Biological Concepts," the title of this series in the Benchmark Program, is a difficult term to define; and the monographs in this series do not fall into any easily delineated category. For example, the present volume, *Vertebrate Regeneration,* is not a central element of the topic "animal behavior," although there is a vast difference between the ethology of a lizard that sheds its tail and one that does not. Nor is regeneration a subtopic of ecology, although it in part conditions an organism's response to its environment. Genetics certainly enters the picture, but it is not the overriding consideration. Nonetheless, because of its relation to so many aspects of biology, it would be difficult to find a better example of a basic biological concept than regeneration.

The approach used by Doctors Thornton and Bromley is as basic as their statement of the topic: "Why can a newt regenerate a limb, while a mammal cannot?" This question has puzzled observers from both the practical and theoretical points of view since the phenomenon was first observed, but the answer is not yet available. If some adult mammals can annually regenerate an antler and some embryonic mammals a limb, why is an adult limb amputation irreversible? Why, in the course of evolution, should an adaptive character of such obvious survival value have been lost?

There is an impressive mass of information, admirably represented in this Benchmark volume, on how regeneration takes place. There are some fairly clear indications as to the control of the process by nervous and hormonal mediation, but the basic question—"Why?"—cannot be answered from the existing literature. One of the greatest values of a Benchmark book, however, is that the reader can study, in their original form, papers that ask the central questions and try to answer them. No other approach, certainly not that of a textbook, could so arouse one's curiosity—which is, after all, the beginning of scholarship.

Peter Gray

Contents

III. SYSTEMIC FACTORS IN REGENERATION

IV. MAMMALIAN REGENERATION

Contents by Author

Introduction

A particular occupational pleasure of biologists comes with witnessing complex and highly orchestrated responses of organisms to simple stimuli. One of the most spectacular of such displays occurs in tailed Amphibia following the loss of a limb. It is especially impressive that the new limb that slowly appears, literally before one's eyes, is not less perfectly formed than the one that was lost. Almost anyone who observes such an event, biologist or not, seems compelled to begin asking questions about it.

Why can a newt regenerate a limb while a mammal cannot? What is the adaptive significance of regeneration and, conversely, of its absence? How does this developmental process, occurring as it does in an adult, compare with embryonic developmental processes? But, more practically, what is the mechanism of regeneration? For years, scores of people who are particularly fascinated by regenerative processes have posed these questions, or parts of them, in experimental terms. The rules of the game are, of course, that the questions will always be answered with complete honesty; but the questions actually answered will be those actually asked, which may or may not be the same ones that the investigator thought had been asked. For this reason, if one genuinely wishes to know what some of the answers are likely to be, one must know precisely what questions were asked. Also, it is not permissible in this context to paraphrase answers, lest summaries intended to clarify should have the opposite effect.

It is thus most useful in examining regenerative phenomena, where the questions come so easily and the answers frequently much less so, to examine the matter as near to firsthand as can be arranged.

Early students of growth confined themselves to first principles: the description of regeneration in as precise terms as possible. They simply made careful descriptions of the gross events of regeneration in an organ such as the limb or tail, often that of a newt, since these amphibians possess most remarkably the powers of regeneration.

1

Microscopic observations of the process soon disclosed that under the wound epidermis of a limb or tail stump there gradually accumulated a mass of mesenchymatous cells termed the regeneration blastema. It was recognized early that an injury is normally the stimulus which sets in train those processes of repair we recognize as regenerative. Although an injury appears to be an essential prerequisite to the onset of regeneration of tail and limb, it is *not* needed for the regeneration of a new lens from the dorsal margin of the iris. All that is needed here (although this applies only in the Salamandridae) is the *absence* of the old lens!

Following wound healing there is a phase—the *regressive phase*—which lasts for several days, its duration depending upon the age of the animal and the nature of the organ injured. In limb and tail this period begins with phagocytic removal of tissue and cellular debris resulting from the trauma of amputation. Then the injured stump tissues begin a limited phase of dissociation (often called "dedifferentiation") in which approximately the distal third of the stump tissues (muscle, skeleton, connective tissues) break up into mononucleate cells that soon become indistinguishable from each other and form the mesenchymatous cells of the blastema. In the iris, however, there is first a loss (extrusion) of pigment in the dorsal iris cells, with phagocytosis. The depigmented cells begin to divide and round up into a vesicle which progressively differentiates lens protein. Unlike the limb or tail blastema, the regenerating lens is composed of only a single type of cell, of ectodermal origin.

During the Regressive Phase of regeneration there is an increase in hydrolytic enzymes (i.e., cathepsins, acid phosphatase, peptidases, and collagenase) as numerous investigators have described (reviewed by Flickinger, 1967; Schmidt, 1968). Mesenchymatous cells derived from the dissociating tips of the stump tissues show DNA synthesis within 4 days after amputation (Hay and Fischman, this volume). As a result of distal migration and proliferation, a population of mesenchymatous cells accumulates beneath a thickened wound epithelium where, by continued proliferation, the cells increase in number and, reminiscent of developing limb and tail buds in the embryo, differentiate into the missing tissues of the organ. This constitutes the *progressive phase*.

With a detailed description of the morphological events of regeneration, a number of more specific questions present themselves. First, where do the cells of the blastema come from? Before the advent of the electron microscope, many investigators were skeptical of the histological evidence which seemed to point to a tissue-specific origin of blastema cells. Stump tissues could be seen to undergo dissociation, but the viability of the released cells was questioned. To some, the invasion of the wound area by many leucocytes was impressive and the supposed transformation of these blood cells to blastema cells was described. The possibilities of an hematogenetic contribution to the regenerate was finally and conclusively eliminated when Butler (this volume) grafted unirradiated limbs to fully x-irradiated salamander larvae with no subsequent loss in the ability of the graft limb to regenerate. These and other studies focused the search for the source of the blastema cells to the immediate neighborhood of the amputation surface.

There has accumulated, during the past 20 years, an impressive body of evidence which implicates the organ stump tissues as the sources of the blastema cells. The early description of cartilage and muscle dissociation (Butler, 1933; Thornton, 1938) has been followed by the careful studies of Chalkley (this volume) and Hay (this volume) which have confirmed and extended the earlier observations. The mitotic index studies by Chalkley disclosed an astonishing correlation between specific mitoses of dedifferentiating stump tissues and the later, redifferentiating tissues of the regenerate. The cellular details of stump tissue contributions to the blastema, as followed by Hay in her elegant ultrastructural studies, clearly established the viability of the released cells, particularly when DNA synthesis could be observed in the nuclei of these cells 4 days after amputation (Hay and Fischman, this volume). These cells also incorporate amino acids (Bodemer and Everett, this volume).

Another question arises when one attempts to relate the actions of the cells thus far discussed to what is known about cells generally. Is morphological dedifferentiation equated with genetic dedifferentiation? Do blastema cells become truly multipotential? In the excitement that permeated developmental biology in the decade following Spemann's demonstration of embryonic induction, many experiments were conducted to test whether similar mechanisms were operative in regenerating systems. Early limb and tail blastema were interchanged with the apparent result that subsequent specific morphogenesis depended on the age of the blastema, a situation seeming parallel to that in the embryo. Unfortunately no critical controls were used and the identity of graft and host tissues was never clear. More recent work indicates that young blastemata are simply absorbed by the host stump, which then regenerates normally (Polezhayev, 1936).

Other possible examples of "multipotency" in regeneration have not been confirmed. Thus, an apparent epidermal contribution to the mesenchymatous blastema (Godlewski, 1928; Rose, 1948) has been neither confirmed nor denied by the careful experiments of Riddiford (1960), O'Steen and Walker (1961), and Hay and Fischman (this volume) who, all found independently, that labeled epidermal cells could not be followed into the blastema. Steen (this volume), using a double-label technique, elegantly showed that cartilage cells of the limb stump give rise only to cartilage cells in the regenerate. Limbs from which bones have been extirpated, however, give rise to regenerates with well-formed bones (Weiss, 1925). Holtzer (this volume) also describes the differentiation of vertebrae in larval salamander tail regenerates where the stump normally possesses only notochord—no vertebrae at all! The way in which all of this information relates to the manner in which genetic information that directs regenerative processes is stored or retrieved is no less enigmatic than fascinating. One would like to know the relation of genetic information directing limb ontogeny to that directing limb regeneration. Are they largely the same, or does the ability to regenerate imply a genetic endowment completely apart from that used in ontogeny? If information used in ontogeny and in regeneration does come largely from the same source, is loss of the ability to regenerate simply a matter of suppression of a retrieval mechanism?

The formation of a blastema on a limb or tail stump has been found to be dependent upon a number of factors. Wound healing, for example, is of critical importance in regeneration. This has been shown well by the experiments of Rose (this volume), who was able to induce regeneration in the forelimb of adult frogs by maintaining dermis-free contact between wound epidermis and internal stump

3

tissues. Normally, when the frog limb stump heals, whole skin (dermis and epidermis) slides over the wound to seal it. By using salt treatments, Rose was able to induce a more larval type epidermal migration over the wound surface in which dermis did not participate. Further evidence of a need for epidermal–mesodermal contact in successful regeneration has been provided by Steen and Thornton (this volume) and Stocum and Dearlove (1972). How the wound epidermis stimulates outgrowth is still not understood.

Blastema differentiation may also be in part dependent upon a wound epidermis–blastemal cell interaction. Faber (this volume) interpreted his data from blastemal transplants to mean that distal outgrowth and morphogenesis of the blastema is a function of an "apical organization center" in the blastema. Stocum (this volume) extended this analysis and has shown clearly that whenever blastemal cells are in contact with wound epidermis, outgrowth occurs at this point, resulting eventually in the differentiation of distal structures. Without such epidermal–mesenchymal contact, outgrowth fails, as does differentiation of distal structures.

Systemic influences are also important in regeneration. In a long series of elegant papers, Singer has demonstrated a neurotrophic function of nerves in newt limb regeneration. A quantitative relation exists between amputation wound area and neuronal volume (Singer, this volume) so that regeneration occurs when the nerve/amputation surface ratio is above threshold but fails when the ratio is subthreshold. Singer theorizes that a neurotrophic substance passes down the axon and "leaks" to stump tissues. At the axon terminus, it is responsible for stimulating limb regeneration. Increasing forelimb nerve quantity in adult frogs by sciatic nerve deviation results in the stimulation of regeneration in these normally nonregenerating limbs (Singer, 1954). In the lizard tail (Simpson, this volume), the ependyma of the spinal cord actively controls regeneration and is also responsible for inducing cartilage around the regenerated nerve tube. Dresden (this volume) has found that denervation markedly decreases the synthesis of RNA, DNA, and protein in newt limb blastemata. Since these effects are also obtained in denervated, but not in innervated, blastemata cultured for 20 hours *in vitro*, he suggests that they indicate a direct control by the nerve on synthesis of DNA, RNA, and protein in the blastema. Carlson (this volume), whose interesting finding that minced rat leg muscle will reconstitute itself, also finds this phenomenon nerve-dependent. The well-known annual replacement of deer antlers (Goss, this volume) seems, however, to be nerve independent.

It is, therefore, most interesting to learn that if nerves have never grown into a limb ontogenetically (Yntema, this volume), the regeneration of such an "aneurogenic" limb will proceed normally. Perhaps, as the work of Thornton and Thornton (this volume) indicates, there is merit in the idea (Singer, 1965) that all embryonic tissues synthesize the neurotrophic factor but that this function is suppressed in them during development and is taken over exclusively by nerves.

The importance of hormones in controlling regeneration has been extensively studied since the pioneering experiments of Schotté (1926). In general, pituitary hormones have been considered essential for limb regeneration. Schotté has developed the theory that the trauma of amputation stimulates production of pituitary adrenocorticotrophic hormone whose stimulation of adrenal steroid synthesis is necessary for regeneration to proceed (Schotté and Wilber, this volume). There is also evidence that adrenocorticosteroids accumulate preferentially near an amputation surface (Bromley, 1971). Whatever the role of adrenocorticosteroids, it is fairly clear that ungulate prolactin and growth hormone are more effective in promoting survival and regeneration in hypophysectomized newts than is ACTH (Tassava, this volume).

At present, the central problem of the entire field seems to be the nature of the neurotrophic factor. In view of the possibility that this factor is shared by embryonic tissues (Singer, 1965), it is interesting that aneurogenic limbs of chick and hamster embryos seem to initiate regenerative responses after amputation (Tassava, unpublished). It is important that the mechanism of the interaction between wound epidermis and mesodermal blastema be carefully investigated. The many resemblances between this interaction in regenerating limbs and in developing salivary glands in mice (Bernfield and Banerjee, 1972) is most intriguing and requires mention. Although regeneration is a developmental process occurring in a postembryonic situation, it would be very satisfying to discover basic underlying similarities in the processes of embryonic and of regenerating systems.

In any event, we do not really know how organisms can make the elegant restitutions of large losses as they do. The papers that follow, comprising as they do an extended interview with the subject matter itself, contain some very astute questions and interpretations of answers, and some which are perhaps not so astute. The measure of skill and success of future workers in the field depends in large part on their abiity to tell one from the other.

References

Bernfield, M. R., and S. D. Banerjee, 1972. Acid mucopolysaccharide (Glycosaminoglycan) at the epithelial–mesenchymal interface of mouse embryo salivary glands. J. Cell Biol., *52*: 664.

Bromley, S. C., 1971. The distribution of H^3 cortisol in adult regenerating newt limbs. Anat. Rec., *169*: 284.

Butler, E. G., 1933. the effects of x-radiation on the regeneration of the forelimb of *Amblystoma* larvae. J. Exptl. Zool., *65*: 271.

Flickinger, R. A., 1967. Biochemical aspects of regeneration. In *The Biochemistry of Animal Development* (R. Weber, ed.), Vol. 2, p. 303. Academic Press, New York.

Godlewski, E., 1928. Untersuchungen über Auslösung und Hemmung der Regeneration beim Axolotl. Arch. Entw.-mech., *114*: 108.

O'Steen, W. K. and B. E. Walker, 1961. Radioautographic studies of regeneration in the common newt: II. Regeneration of the forelimb. Anat. Rec., *139*: 547.

Polezhayev, L. 1936. La Valeur de la structure de l'organe et les capacités du blastème régénératif dans le procès de la détermination du régénérat. Bull. Biol. France et Belgique, *70*: 54.

Riddiford, L. M., 1960. Autoradiographic studies of tritiated thymidine infused into the blastema of the early regenerate in the adult newt *Triturus viridescens*. J. Expl. Zool., *145*: 61.

Rose, S. M., 1948. Epidermal dedifferentiation during blastema formation in regenerating limbs of *Triturus viridescens*. J. Expl. Zool., *108*: 337.

Schmidt, A. J., 1968. *Cellular Biology of Vertebrate Regeneration and Repair*. University of Chicago Press, Chicago.

Schotté, O. E., 1926. Hypophysectomie et régénération chez les Batraciens urodèles. C. R. Soc. Phys. Hist. Nat. (Genève), *43*: 67.

Singer, M., 1954. Induction of regeneration of the forelimb of the post-metamorphic frog by augmentation of the nerve supply. J. Expl. Zool., *126*: 419.

———, 1965. A theory of the trophic nervous control of amphibian limb regeneration, including a re-evaluation of quantitative nerve requirements. In *Regeneration in Animals and Related Problems* (V. Kiortsis and H. Trampusch, eds.). North-Holland, Amsterdam.

Stocum, D. L., and G. E. Dearlove, 1972. Epidermal–mesodermal interaction during morphogenesis of the limb regeneration blastema in larval salamanders. J. Expl. Zool., *181*: 49.

Thornton, C. S., 1938. The histogenesis of muscle in the regenerating forelimb of larval *Amblystom punctatum*. J. Morph., *62*: 17.

Weiss, P., 1925. Unabhängigkeit der Extremitätenregeneration vom Skelett (bie *Triton cristatus*). Arch. Mikr. Anat. u. Entw.-mech., *104*: 359.

Origin and Fate of Blastemal Cells

I

Editors' Comments on Papers 1 Through 8

Regeneration of complex structures is a localized affair. When parts of eyes or limbs are being replaced, animals show no clearly marked general changes whatsoever. Specialized cells appear in localized places and undergo highly organized transformations. Where do these cells, possessing such enormous and diverse potential, come from? And what, exactly, are their various potentials for transformation?

These are primary questions, and much ingenuity and talent has been expended in trying to answer them. Obviously, the question of the origin of the cells must be answered in either of two ways. Either the cells migrate from distant places, congregate appropriately, and do not undergo particularly impressive changes in order to attain the appearance and morphogenetic capacities that they possess, or cells already in the area, engaged in the performance of tasks not immediately related to regenerative processes, respond to a local emergency by shedding their duties (and the cellular equipment that goes with them) and becoming something quite new.

The origin of the blastema has been examined in some of the investigations that follow in an attempt to distinguish between these two possibilities and to render more precise the understanding of the specifics involved. Earlier speculations about the presence of a scattered population of specially competent cells, after the manner of coelenterates, have received very little support. Decades ago, Hertwig (1927) grafted haploid limb buds onto diploid hosts; after subsequent amputation, the internal parts of the regenerated tissue were apparently haploid. Also, during the first part of this century, it was discovered that x-radiation was able to divest cells of regenerative competence. Elmer G. Butler, in addition to collaborating with or training an uncommonly large number of workers now active in the field, utilized this new tool as a means of asking the same question. Again, all indications are that competent cells do not make relatively long migrations to the amputation area. And while the mode of action of x-rays on tissue in this respect is not much better understood now than it was then, there have been few serious inquiries raised about generalized migration of blastemal cells since the appearance of that paper.

Asumming that blastemal cells do originate close to the amputation plane, the magnitude of the contribution of the various tissue types in the area is of immediate interest. Donald Chalkley met the requirements of a Ph.D. thesis with an elegant and painstaking investigation designed to elucidate these various roles. As a result of measuring changes in cell type, number, and volume with respect to various locations in regenerating limbs, he postulated that the bulk of the new tissue that appears is constituted of cells derived from associated muscle and nerve sheath. Epidermis apparently makes little contribution. The general methodology used by Chalkley could undoubtedly be more widely utilized for various inquiries, although the care and persistence necessary to make it work are not easily summoned.

Elizabeth D. Hay and Donald Fischman examined the same general questions as Chalkley, although with a completely different methodology. Hay applied considerable electron micrographic skill to elucidation of muscle dedifferentiation, and produced clear and unmistakable photographic evidence that muscles do indeed transform into blastemal cells. Soon thereafter, in collaboration with Fischman, she allowed cells to incorporate tritiated thymidine at various times during the regenerative process and thus was able to follow, autoradiographically, their subsequent movements. As with Chalkley's data, it appears that blastema cells originate from subjacent tissue, not epidermis. Anthony J. Schmidt has correlated histochemical evidence of glycogen depletion in subjacent muscles with their dissolution and the subsequent appearance of blastemal cells.

Once cells transform into recognizable blastemal cells, must their fates reflect their origin, or is the blastemal mass essentially homogeneous with regard to subsequent developmental potentiality? Trygve Steen used a combination of techniques and concluded that blastemal cells that begin as chondrocytes also end as chondrocytes. He also postulated that muscle cells operate similarly, although their initial state is more difficult to categorize. Earlier, Paul Weiss (1925), Charles Thornton (1938), and others showed that limbs from which bones were extirpated yield regenerates with bones in the right places. Evidently these bones arise from cells associated with muscle, but not bearing identical commitments.

As the general trends of the transformations become more clear, one becomes concerned with the molecular events associated with morphological change. Unraveling this tangle of reactions is a Gordian undertaking indeed, except that in this case the knot is composed of hundreds of different materials, many of which are ephemeral, and the whole mass is constantly changing, which is what aroused the interest in the first place. In this case, the solution will doubtless be piecemeal; the paper by Charles Bodemer and Newton Everett reflects current efforts with respect to general patterns of protein synthesis. This paper is a start, but interest attends the specific properties of those proteins. And in view of the vast disparity in the properties of newly synthesized proteins and the limitations of current methodological approaches, it seems unlikely that comprehensive answers will be immediately forthcoming.

Adult cells are often essentially immobilized in a collagenous network, whereas cells that demonstrate properties associated with more embryonic states seem much more free to move. Turning that statement around, it is certainly worthwhile to ask whether or not blastemal cells, which demonstrate some embryonic properties, are able to become so as a consequence of changes in the collagenous network in which the preblastemal cells reside. Workers in Jerome Gross's laboratory have been interested in this line of investigation; the report by Grillo, LaPière, Dresden, and Gross details changing patterns of collagenase activity in regenerating newt limbs. The degree to which collagenase activity is causal to the formation of blastemal cells is an intriguing question indeed that is probably susceptible to experimental approach.

References

Hertwig, G., 1927. Beiträge zum Determinations-und Regenerationsproblem mittels der Transplantation haploidkerniger Zellen. Wilhelm Roux Arch. Entw.-mech. Organ., *III* 292–316.

Thornton, C. S., 1938. The histogenesis of the regenerating fore limb of the larval *Amblystoma* after exarticulation of the humerus. J. Morph., *62*: 219–241.

Weiss, P., 1925. Unabhängigkeit der Extremitäten-regeneration vom Skelett (bei *Triton cristatus*). Wilhelm Roux Arch. Entw.-mech. Organ., *104*: 359–394.

Reprinted from *Anat. Rec.*, **62**, 295–307 (1935)

1

STUDIES ON LIMB REGENERATION IN X-RAYED AMBLYSTOMA LARVAE

ELMER G. BUTLER

Department of Biology, Princeton University

EIGHT FIGURES

INTRODUCTION

Within recent years the effects of x-rays on regeneration have been studied by several investigators. An historical résumé of the principal work done in this field has been included in a previous paper (Butler, '33), and will not be considered further at this time. The experiments to be reported in the present paper deal with the effects of x-rays on growth and regeneration of limbs of Amblystoma punctatum larvae, and are a continuation of the author's earlier work. The primary purpose of the experiments to be discussed in the following pages has been to determine whether the effect of x-rays in suppressing limb growth and regeneration is due to local action of the radiation on cells in the limb region, or to general action of the radiation on the body of the animal as a whole. To a great extent the experiments deal, also, with the fundamental problem of the source of the cells on which regeneration of the Amblystoma limb depends.

Up to the present time studies on the effect of x-rays on growth and regeneration of the amphibian limb, except for the work of Brunst and Scheremetjewa ('33), have been carried out on animals in which the entire body has been radiated. This was the method used in my first experiments (Butler '31 and '33), and also was the method employed by Pucket ('34) for his studies on limb growth and regeneration in Amblystoma. In the series of experiments to be dealt with

295

in the present paper other methods have been used. In some
cases Amblystoma larvae have been shielded with lead in
such a manner that only a portion of the body of each individual was subjected to the action of x-rays. In other cases a
normal limb from an unradiated larva has been transplanted
to the body of a radiated individual. By the use of these
methods it has been possible to determine: 1) whether limb
regeneration will be prevented when the limb region alone is
radiated, 2) whether limb regeneration will be prevented when
a large part of the body is radiated with either fore or hind
limbs shielded, and 3) whether an unradiated limb is capable
of growth and regeneration when transplanted to a radiated
animal, which in itself completely lacks the capacity for limb
regeneration.

Larvae of Amblystoma punctatum, which were reared in the
laboratory, were used for all experiments. The source of
x-rays was a Coolidge medium focus tube. Factors governing
the radiation which were kept constant for all experiments are
as follows: 60 kv., 6 ma., distance from target to Amblystoma,
25 cm. No filter was used. Dosages in Roentgen units, as
stated in following pages, were determined by the use of a
standard ionization chamber.

RADIATION OF THE LIMB REGION ONLY

In order to study the effects of radiating the limb region
alone of Amblystoma larvae an apparatus, such as shown
diagrammatically in figure 1, was used. By means of this
apparatus it was possible to administer to one limb of an individual any desired dosage of x-rays, while the remainder of
the body was completely shielded from the x-rays by a lead
plate. The method of procedure was as follows: An anesthetized larva (exp.) was placed with its ventral side uppermost
on moist cotton in a Petri dish (p.d.). The larva was then
covered with a plate of lead (l.p) ¼ inch in thickness, which
was supported by the edges of the Petri dish. In one edge of
the lead plate was a vertical groove (g.) about 2 mm. deep.
In covering the larva with the lead plate the groove in the

plate was placed directly above the limb to be radiated, the orientation of groove and limb being made with care under a dissecting microscope, so that the entire limb as far as the head of the humerus and including also part of the shoulder girdle was exposed to the x-rays. With the exception of the radiated limb, the body was completely shielded from the x-rays by the lead plate. A second larva (con.), which was entirely shielded by the lead plate, served as a control. After the two larvae had been properly oriented, the Petri dish was

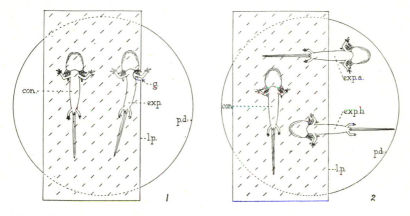

Fig. 1   Diagram of apparatus used for radiation of a single limb.   con., control larva; exp., experimental larva; g., vertical groove in edge of lead plate; l.p., lead plate (cross hatched) ¼ inch in thickness; p.d., edge of Petri dish.

Fig. 2   Diagram of apparatus used for partial radiation of the body.   exp.a., larva with anterior half of body exposed to x-rays; exp. b., larva with posterior half of body exposed to x-rays; other legends as in figure 1.

placed beneath the x-ray tube and the radiation was made from above.

Using the method just described, a series of experiments were performed in which the limbs of Amblystoma larvae were given various dosages of x-rays. The description of a single experiment will serve to demonstrate the results which have been obtained in all experiments in which effective dosages have been employed. The larvae used for the experiment to be described measured 15 mm. in total length at the time of the radiation, and possessed fore limbs which had developed to

THE ANATOMICAL RECORD, VOL. 62, NO. 3

the four-digit condition. On May 21st one fore limb on each
of four larvae was given 900 Roentgen units in a single exposure (exp., fig. 1). In some cases the right, and in other
cases the left fore limb was radiated. Immediately after the
radiation of one fore limb, both right and left fore limbs on
each of the larvae were amputated through the mid-region of
the shaft of the humerus. At the same time both fore limbs
were similarly amputated on the control larvae (con., fig. 1),
which had been completely shielded from the radiation by the
lead plate.

In figure 3 is shown in ventral view one of the larvae (AF-4)
of this experiment in which the right fore limb was radiated,
while the remainder of the body was shielded. The photograph shows the larva on June 21st, 31 days after the right
fore limb was radiated, and 31 days after both fore limbs were
amputated. As is evident from the photograph, the radiated
right limb has failed completely to regenerate, while the left
fore limb has regenerated in a normal manner and again
possesses four digits. The regeneration of the left fore limb
on this larva took place as rapidly and as completely as on the
control larva (con., fig. 1), which was completely shielded
from the x-rays.

Furthermore, not only has the right fore limb on the larva
shown in figure 3 failed completely to regenerate, but, in addition, there has been considerable degeneration and resorption
of tissue, so that the small stub which remains on the limb site
31 days after amputation is much smaller than the limb stump,
which was left at the time of amputation. The resorption,
which is always evident in the stumps of amputated limbs of
radiated larvae, is due to an extensive dedifferentiation of
tissues within the limb stump. The histology of dedifferentiated limb stumps has been extensively dealt with elsewhere
by the author, and will not again be considered at this time.
It is important to point out, however, that local radiation of
the limb in the experiment under consideration was apparently
as effective in preventing limb regeneration and producing
tissue dedifferentiation as was radiation of the entire animal,

as reported in previous papers (Butler, '33 and '34; Puckett, '34).

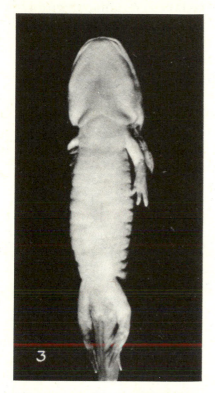 

Fig. 3   Ventral view of Amblystoma larva (AF-4) showing failure of right fore limb to regenerate after local radiation.   Larva photographed 31 days after amputation of both fore limbs.

Fig. 4   Ventral view of Amblystoma larva (AB-3a) showing failure of hind limbs to regenerate after radiation of the posterior half of the body.   Fore limbs have regenerated completely.   The tail has atrophied as a result of the radiation. Larva photographed 43 days after amputation of both fore and hind limbs.

RADIATION OF A LARGE PORTION OF THE BODY WITH EITHER FORE OR HIND LIMBS SHIELDED

The type of experiment just described demonstrates conclusively that local radiation of a limb alone will prevent its regeneration, and will cause tissue dedifferentiation within the limb stump.   On the other hand, it does not prove that failure

15

of a limb to regenerate may not be due, in other circumstances, to a general effect which the x-rays exert on the body as a whole. In other words, if an entire larva be radiated with the exception of one limb, will the regenerative capacity of this unradiated limb be affected? Experiments were first performed, therefore, in which an attempt was made to shield only a single limb of a larva with lead, leaving the remainder of the body entirely exposed to x-rays. These experiments, however, did not lead to clear-cut results, for the reason that in shielding a single limb it was necessary to use so small a piece of lead that often there was considerable leakage of x-rays around the edges of the small lead plate. Because of this leakage of x-rays, the shielding of the limb in most instances was only partial instead of complete, and the results were not conclusive. In order to eliminate this difficulty a slightly different procedure was substituted, with which conclusive results were obtained. Instead of shielding with lead a single limb only, one-third or one-half of the body of a larva was shielded, leaving the remaining one-half or two-thirds of the body exposed to x-rays. By this method it has been possible to determine, for example, whether radiation of the anterior half of the body will affect hind limb regeneration, and, conversely, whether radiation of the posterior half of the body will affect fore limb regeneration. The apparatus used for this type of experiment is shown diagrammatically in figure 2. Anesthetized larvae were placed, as in the previous experiment, on moist cotton in a Petri dish. A $\frac{1}{4}$-inch lead plate was then placed over the portion of the larva to be shieded from the x-rays. In a single exposure to x-rays either the anterior half (exp. a.), or the posterior half (exp. b) of several larvae could be radiated at one time. One larva (con.), completely shielded by the lead plate, served as a control. After the radiation, fore and hind limbs of radiated and control larvae were amputated in order to test their regenerative capacity.

Results of all experiments show that the failure to regenerate occurs only in the region of the body actually exposed to x-rays. Limbs situated in the shielded region of a radiated larva still retain the regenerative capacity. In figure 4, for example, is shown a larva (AB-3a), the posterior half of which was radiated, while the anterior half of the body was shielded (exp. b., fig. 2). Both fore and hind limbs were amputated through the mid-regions of humerus and femur, respectively. The photograph in figure 4 was taken 43 days after radiation and limb amputation. Fore limbs have completely regenerated. Hind limbs, on the contrary, have failed completely to regenerate, and in the hind limb region there remain simply two degenerate limb stumps. The animal shown in figure 4 was given 900 Roentgen units in one exposure, which is a large dosage for an Amblystoma larva of this age. As a result of this large dosage the entire posterior half of the body was severely affected. The tail atrophied, and the pelvic region of the body failed to develop normally. Nevertheless, regeneration of the fore limbs was unaffected, and took place as completely and as rapidly as in control unradiated larvae.

In an experiment similar to that described above, the anterior halves of other larvae (exp. a., fig. 2) were given the same large dosage of 900 Roentgen units. As a result of the radiation of the anterior half of the body the fore limbs lost the capacity to regenerate, and, in addition, the entire anterior half of each larva exhibited serious effects of the radiation. The gills atrophied, the animals ceased to eat, and became very emaciated. Nevertheless, the hind limbs, which were shielded from the x-rays, regenerated as well as in unradiated control larvae.

Other experiments have been conducted using younger larvae and lighter dosages of x-rays. For example, by radiating the posterior half of the body, using larvae in which hind limbs have not yet appeared, one can obtain animals on which the fore limbs are entirely normal, but which are completely limbless in the posterior region. In these cases x-radiation completely suppresses hind limb development in the manner described by Puckett ('34), without interfering with fore limb development. Moreover, fore limbs on such animals retain the capacity for regeneration.

17

REGENERATION OF UNRADIATED LIMBS TRANSPLANTED TO X-RAYED LARVAE

Experiments have been made to determine whether normal limbs taken from unradiated larvae are capable of growth and regeneration when transplanted to x-rayed larvae. Two ages of larvae have been used for these experiments: a) In some experiments larvae of approximately 20 mm. total length with well-developed four-digit fore limbs were used as donor and host. Transplantation was made to the normal limb site of the radiated host. b) In other experiments younger larvae (stage 43 of Harrison), in which the fore limbs were in the two-digit stage, served as donor and host. In some cases the two-digit limb was transplanted to the normal limb site on the radiated host, and in other cases it was placed in a heterotopic position on the lateral body wall. In all experiments, the transplanted limbs, after they had become established on the radiated host, were then amputated in order to test their regenerative capacity.

The method of limb transplantation was as follows: A larva to be used as a host was first given in one exposure a dosage of x-rays which totally incapacitated it for limb regeneration; the entire body of the larva was completely exposed, no shielding being used. After the radiation, transplantation of a limb from an unradiated donor was made to the radiated host. In cases in which the normal limb site of the radiated host was used, one fore limb including the scapula was first removed from the host. To the wound on the host was then transplanted the corresponding fore limb from an unradiated donor. With a little care the transplant could be adjusted so that it occupied about the same position as the original limb, and it was then held in place for several hours by small glass rods. In cases in which the limb was transplanted to a heterotopic position on the lateral body wall, a short slit was first made in the integument of the host at the site for transplantation. The scapula of the transplant was then slipped through the slit. The integument at the edge of the slit closed over the scapula, and served to hold the transplanted limb quite firmly in place. During the operation each larva was anesthetized in a 1:500 solution of M.S. 222.[1] After the transplantation had been completed, the larva was allowed to remain in a weaker solution of the anesthetic to keep it immobile for from 3 to 4 hours. During this time the wound healed sufficiently, so that when the larva was released from the anesthetic the transplanted limb ordinarily remained in place.

The success of tranplantation appears to depend to a great extent on the rapidity and completeness with which the transplanted limb becomes vascularized. This is particularly true when fully developed limbs of older larvae are transplanted. In the most successful cases circulation of blood can be observed from 3 to 5 days after transplantation. Circulation first appears in the proximal region of the limb, and then gradually progresses distad until finally, in many cases, the limb appears to be as completely vascularized as before transplantation. When vascularization of the limb fails to take place, then the limb undergoes a gradual degeneration and finally drops off.

a) Regeneration of unradiated four-digit limbs transplanted to normal limb site of x-rayed hosts

The description of a single experiment will serve to illustrate the results which have been obtained in this type of transplantation. Twelve Amblystoma larvae (Exp. AM) about 20 mm. total length were given, in a single exposure, 450 Roentgen units. After the radiation, the left fore limb was removed from each larva and to the limb site was transplanted a left fore limb from an unradiated larva of the same age as the host. In seven cases the transplantation was successful, two larvae died as a result of the operation, and in three cases the transplanted limbs failed to become vascularized and consequently degenerated. After the seven successful transplants had become well vascularized, each of the transplanted limbs was amputated through the mid-region of

---

[1] I am indebted to the Sandoz Chemical Works for supplying this anesthetic.

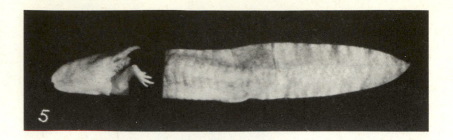

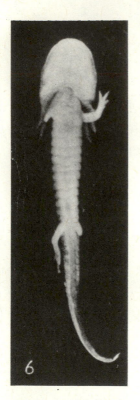

Figs. 5, 6 and 7   Radiated Amblystoma larvae (AM-4, AM-9, AM-3), to which unradiated left fore limbs were transplanted and subsequently amputated.   In each case the transplanted left fore limb has regenerated completely.   Larvae photographed 33 days after limb amputation.   For further details see text.

Fig. 8   Ventral view of radiated Amblystoma larva showing a regenerated heterotopic transplant.   Larva photographed 21 days after amputation of the left fore limb of host and of the left heterotopic transplant.   The unradiated heterotopic transplant has regenerated; the left fore limb of the radiated host has failed completely to regenerate.   Radiation has also prevented the development of hind limbs.

304

the humerus. At the same time the right fore limb was similarly amputated, and also either one or both of the hind limbs was amputated through the mid-region of the femur. In six of the seven cases regeneration of the transplanted left fore limb took place. In one case no regeneration was evident. The failure to regenerate in this single case was probably due to faulty circulation in the limb at the time of amputation.

Figure 5 shows in lateral view one of the larvae (AM-4) of this series. An unradiated left fore limb was transplanted to the radiated host on May 28th. On June 4th, 7 days after the transplantation, both fore limbs and the left hind limb were amputated. The right fore limb and the left hind limb of the host failed to regenerate, because of the radiation. The transplanted left fore limb alone regenerated, and on July 7th, 33 days after the amputation (fig. 5), the regenerated limb possessed four digits, and, to all external appearances, was normal. Moreover, the regenerated limb was capable of well-coordinated movement, indicating that motor innervation had been established.

Figure 6 shows in ventral view another larva (AM-9) of this series. In this case the transplanted unradiated left fore limb was amputated 6 days after transplantation, and at the same time the right fore limb and the left hind limb of the host were also amputated. The photograph in figure 6 shows the larva 33 days after the amputations. The transplanted left fore limb regenerated nicely, while the right fore limb and left hind limb of the radiated host failed completely to regenerate. In this case, as in the previous one, the regenerated left fore limb was capable of well-coordinated movement.

In figure 7 is shown another larva (AM-3), in which all limbs failed to regenerate except the transplanted unradiated left fore limb. Amputation of all limbs on this larva was made 7 days after the left fore limb had been transplanted. The photograph was taken 33 days after limb amputation. The regenerated left fore limb was capable of movement, but in this case the movement was not so well coordinated as in the cases previously described.

b) Regeneration of unradiated limbs transplanted to a hetero-topic position on the lateral body wall of x-rayed hosts

Unradiated fore limbs transplanted to a heterotopic position on the lateral body wall of x-rayed hosts will retain the capacity for regeneration, although the fore limbs of the host are entirely incapable of regeneration. For heterotopic transplantation larvae in stage 43 of Harrison, which possess fore limbs in the early two-digit stage, were used. Hind limbs have not yet appeared at this stage.

Figure 8 shows a larva (AL-8), which possesses a regenerated heterotopic transplant. An unradiated left fore limb in the two-digit stage was transplanted to the lateral body wall of the host, which had previously been given 450 Roentgen units. Fourteen days after the transplantation the heterotopic limb, which at that time had reached the three-digit stage of development, was amputated through the proximal end of the radius and ulna. At the same time the left fore limb of the radiated host was amputated at the same level. The photograph in figure 8 shows the larva 21 days after limb amputations. The heterotopic transplant has regenerated so that it is again in the three-digit stage. The left fore limb of the host has failed completely to regenerate and some resorption has taken place. No hind limbs have ever developed on this larva as a result of the action of x-rays on limb development as described by Puckett ('34).

DISCUSSION AND SUMMARY

The experiments described in foregoing pages demonstrate conclusively that the effect of x-rays in preventing limb regeneration in Amblystoma larvae is due to local action of the radiation on cells in the limb, and is not the result of general action of the radiation on the body of the animal as a whole. This conclusion is fully supported by the following evidence: 1) Radiation of the limb alone will prevent its regeneration. 2) An unradiated limb will still retain the capacity for regeneration, although a large part of the body of the animal has been radiated. 3) An unradiated limb transplanted to a radiated host will retain the ability to regenerate, although the limbs of the host itself are entirely incapable of regenerating.

The observations reported in preceding pages are of significance, furthermore, in regard to the problem of the source of the cells on which regeneration of the Amblystoma limb depends. The experiments, both with partial radiation and with transplantation of unradiated limbs to radiated hosts, show clearly that the cells used in regenerative processes must arise in loco, and are not brought to the site of regeneration from other parts of the body. The evidence from these experiments is contrary to the suggestion of Hellmich ('29 and '31), that some of the cells utilized in the regeneration of the amphibian limb are of hematogenetic origin. On the other hand, all evidence demonstrates that the cells which are concerned with regeneration are of histogenetic origin, and that regeneration results from changes which these cells undergo at the site of regeneration. There appears to be no conclusion other than this to be drawn from the transplantation experiments in particular, for in these cases the circulating blood in the limb is from a host which has lost completely the capacity for limb regeneration. All materials for regeneration must arise, therefore, in the transplanted limb itself.

LITERATURE CITED

BRUNST, V. V. AND E. A. SCHEREMETJEWA 1933 Untersuchung des Einflusses von Röntgenstrahlen auf die Regeneration der Extremitäten beim Triton. Arch. f. Entwmech., Bd. 128, S. 181.

BUTLER, E. G. 1931 X-radiation and regeneration in Amblystoma. Science, vol. 74, p. 100.

———— 1933 The effects of x-radiation on the regeneration of the fore limb of Amblystoma larvae. J. Exp. Zoöl., vol. 65, p. 271.

———— 1934 The effect of x-rays on tissue regeneration. Scientific Monthly, vol. 39, p. 511.

HELLMICH, W. 1929 Untersuchungen über Herkunft und Determination des regenerativen Materials bei Amphibien. Arch. f. Entwmech., Bd. 121, S. 135.

———— 1931 Histology of regeneration in different species of adult and larval urodeles. Anat. Rec., vol. 48, p. 303.

PUCKETT, W. O. 1934 The effects of x-radiation on limb development in Amblystoma. Anat. Rec., vol. 58, supplement, p. 32.

THE ANATOMICAL RECORD, VOL. 62, NO. 3

Reprinted from *Regeneration in Vertebrates*, C. S. Thornton (ed.),
University of Chicago Press, 34–56, 102–104 (1956)

2

· III ·

The Cellular Basis of Limb Regeneration

D. T. CHALKLEY[1]

Department of Biology, University of Notre Dame

Regeneration, like other biologic processes, is susceptible of analysis at various levels of organization. While an attack on the fundamentals of the process should unquestionably begin at or near the molecular level and progress to more complex supramolecular situations, the absence of adequate methods compels us, instead, to move down the historical avenue of approach dealing with superficially more simple natural units—the cells and tissues. The knowledge gained in such studies is basic only to an understanding of tissue and organ development and cannot provide information as to the more fundamental processes of regeneration. However, any study of the physiology and biochemistry of regeneration will benefit from a clear statement of the morphologic events of the process. Physiologists, particularly those who are much impressed by the amazing uniformity of intermediary metabolic processes in various cell processes, frequently need to be reminded of the complexity of morphogenetic processes. A brei presents a very satisfyingly homogeneous appearance, even when prepared from an extremely non-homogeneous tissue source such as an embryo. It is desirable to emphasize that in morphogenetically active tissues this morphologic non-homogeneity may be paralleled by physiologic non-homogeneity.

This chapter of the symposium reviews the present state of our knowledge of the cellular basis of regeneration and its possible interpretation in terms of cell migration and of the possible changes in the states of differentiation of the participant cells at various stages of the process.

Except for differences in extent, the events which follow amputation of the amphibian limb appear to be identical with those following amputation of the tail. The soft tissues contract over a somewhat greater distance in the limb because of the increased length of the muscle fibers. A

---

[1] Present address: Division of Research Grants, National Institutes of Health, Bethesda, Maryland.

34

clot is rapidly formed over the amputation surface, and the adjacent epithelium begins to migrate over the wound within 15–45 minutes (Lash, 1955). The process appears to be complete within 24 hours or less. This migration is not associated with local epidermal cell division, nor, for that matter, are mitoses found at any level of the stump during the first 12 hours (Inoue, 1954). A similar mesenchymal migration occurs beneath the wound surface as dissociated cells, emerging from the stump tissues, accumulate between the wound epithelium and the subjacent sub-epithelial formed tissues. This immigration continues for at least 72 hours, even in the presence of antimitotic agents. Colchicine (Luscher, 1946), nitrogen mustard (Skowron and Roguski, 1953), and aminoketones (Dettelbach, 1952) have no effect on this process. Further accumulation of cells is blocked by these inhibitors, but, interestingly enough, there is no detectable mitotic activity in the tissues adjacent to the amputation surface on which these inhibitors might be presumed to act. Thus Litwiller (1939) reported an accumulation of the order of 9,000 cells during the first 15 days of regeneration (Fig. 1) in the complete absence of mitosis (Fig. 2). Bassina (1940) reported blastemata averaging 15,000 cells which were formed in less than 6 days (Fig. 1) and also in the complete absence of mitoses (Fig. 2). Rose (1948) found an accumulation of only 375 cells (Fig. 1) in the complete absence of mitoses but noted the addition of another 18,300 cells between his "epithelial mound" and "early blastema" stages accompanied by a mitotic index of 1.4 per cent. Assuming that the interval between these stages is not more than 48 hours, as indicated by the chronological ages of the examples cited by Rose, and applying the compound-interest formula to the initial accumulation, we find that a total of 164 division intervals would be necessary to produce the required number of cells. This allows only 18 minutes per division, while observed times are appreciably higher. The only observations on amphibian cell division in vivo are those of Clark (1912), who found frog epidermal mitoses to last 263 minutes. Observations in vitro on frog and newt fibroblasts indicate times of from 83 to 93 minutes. Computations by Luscher (1946), employing the colchicine method, indicate an in vivo division time of 99–104 minutes for tail mesenchyme cells.

The inadequacy of the observed mitotic activity in the early blastema to account for the growth of that tissue had been generally recognized before Rose's documentation of the phenomenon. Various theories have been put forward to explain it. The oldest of these theories is that of a hematogenic origin of the blastema cells, presumably from extravasated leukocytes. First proposed by Colucci (1884), contributions of varying

35

magnitude from this source have been described by Hellmich (1930), Kazancev (1934), and Ide-Rozas (1936). Any idea that such a contribution was in any way critical was disposed of by a series of experiments by Butler (1935), Brunst and Chérémetiéva (1936), and Butler and O'Brien (1942). These authors alternately shielded and exposed restricted sites on a large number of amphibians to X-rays and demonstrated that regeneration invariably followed amputation through a

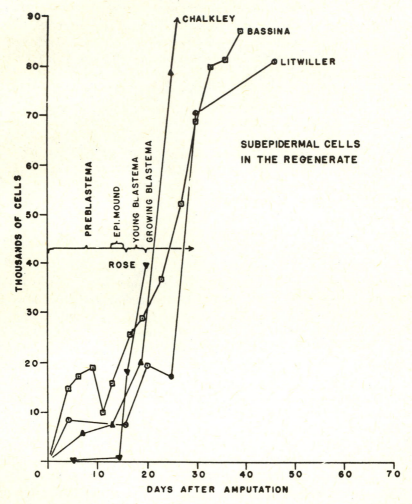

Fig. 1.—Numbers of cells in the regenerate at various dates after amputation (Bassina, Chalkley, Litwiller) or during various stages of regeneration (Rose). The chronological spread and degrees of overlap of the stages are not necessarily accurate but reflect the ages of the specimens assigned to the various stages by Rose (1948).

36

shielded zone, even if all other portions of the organism had been exposed. Regeneration failed with equal invariability if the amputation plane passed through even a small irradiated zone in an otherwise intact animal. Butler and O'Brien concluded that the cells of the blastema "have an exclusive and restricted origin at the site of the amputation."

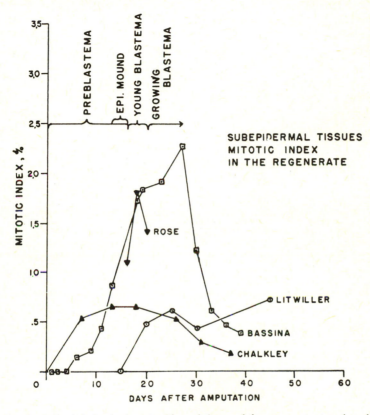

Fig. 2.—Mitotic index in the subepidermal tissues of the regenerate at various dates after amputation (Bassina, Chalkley, Litwiller) or during various stages of regeneration (Rose). The chronological spread and degrees of overlap of the stages are not necessarily accurate but reflect the ages of the specimens assigned to the various stages by Rose (1948).

An amitotic origin of the early blastema was first proposed by Towle (1901) and later by Bassina (1940). The evidence in both instances was circumstantial. No descriptive or quantitative studies of the distribution of amitotic figures were undertaken. The mechanism was simply proposed as a last resort after all other possibilities had been considered and rejected. Several authors, including Kazancev (1934), Ide-Rozas

37

(1936), Mettetal (1939), and Polezhayev and Ginsburg (1942), have specifically denied the occurrence of amitoses.

Figures closely resembling the amitotic figures of certain protozoan macronuclei can be found in the blastema and stump tissues of *Triturus*. Their distribution does not show any correlation with increase in cell number in any part of the limb. The total number of such figures found at any one time does not exceed 0.6 per cent of the mitoses found at the same time in the same tissue. It seems highly probable that these are "pseudo-amitotic figures" produced by the constriction of the nuclei of cells or by the fragmentation of large polymorphic nuclei.

It has also been suggested that the critical mass of the definitive blastema is provided by the immigration of epidermal cells from the thickened late wound epithelium. This hypothesis was first advanced by Godlewski (1928) and has recently been revived by Rose (1948). There can be no question that the epidermal and subepidermal tissues are closely associated during the early period of blastema formation and become clearly separate only after the formation of a basement membrane between the two tissues. Frequently, large masses of epithelial cells protrude into the underlying blastema, and the boundary between the two tissues becomes progressively less distinct near the tip of the protrusion.

While Rose has interpreted these areas of intrusion and contact as indicative of the ingression of epidermal cells, Ide-Rozas (1936) has described them as areas of egression of cells from blastema to epidermis. I interpret them as areas of chance apposition of tissues of approximately equal cellular density along the borders of epidermal infoldings produced at the time of amputation. As a result of the rapid proliferative growth of the young blastema, the average size of the subepidermal cells decreases, the intercellular matrix formation fails to keep pace with cell division, and the number of subepidermal cells per unit volume of tissue rises to a value comparable with that of normal epidermis. At the same time, the epidermal cells lose their close-packed arrangement and become separated, bound together only by tenuous and poorly resolved intercellular bridges. The loss of these and other characteristic properties of the two tissues renders the line of demarcation between them indistinct. The impression of exchange between the tissues in such areas is easily obtained.

Rose has supported his interpretation of this phenomenon with measurements of cell numbers which show an abrupt drop in the number of epidermal cells from 43,900 in his "epidermal mound" stage to 27,600 in the "young blastema" stage. This is shown in Figure 3. Simultaneously, as shown in Figure 1, his counts of subepidermal cells show an increase

38

of from 1,100 to 19,300 cells. The changes are almost compensatory, and the data, as presented, are statistically significant. Unfortunately, the corresponding data of Litwiller (1939), Bassina (1940), and Chalkley (1954) fail to indicate any such abrupt rise in the number of subepidermal cells or, with the exception of Bassina's data, any comparable drop in the number of cells in the overlying epidermis.

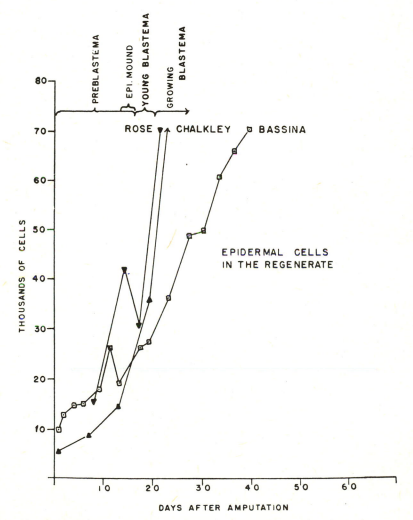

FIG. 3.—Numbers of epidermal cells in the regenerate at various dates after amputation (Bassina, Chalkley) or during various stages of regeneration (Rose). The chronological spread and degrees of overlap of the stages are not necessarily accurate but reflect the ages of the specimens assigned to the various stages by Rose (1948).

39

Strict comparison with Rose's data is hampered by his failure to refer them to a truly independent variable. He separated his limbs on the basis of histologic stages, which, he states, cannot be identified with distinct chronologic intervals. This is unfortunate because the identifications of the several stages are made in terms of histologic characteristics which conform to a previously conceived hypothesis. It is not surprising that quantification of these characteristics should produce data which support the hypothesis.

Fortunately, experimental studies also bear on this problem. Rose, Quastler, and Rose (1953) irradiated forelimbs with doses of as much as 10,000 r and then stripped the irradiated epidermis from the limbs, permitting the normal, unirradiated epidermis of the shoulder to migrate distally and re-cover the exposed subepidermal tissues. Following amputation, these limbs regenerated, though those receiving the highest dosages showed considerable delay. On the other hand, Fimian (1951) shielded the distal millimeter of recently amputated limbs and irradiated the shoulder region with 5,000 r. Regenerates appeared but failed to develop normally and regressed after a few days. Histologic examination of the limb tissues showed that epidermal mitoses were reduced even in the shielded epidermis but that a normal mitotic index, for that stage of the regenerative process, was attained by the eleventh day following amputation. Mitotic activity in the subepidermal tissues was completely suppressed in the exposed region until the eleventh day and, though it reappeared at that date, never reached a normal level thereafter. Thus in the presence of normal mitotic activity in a histologically normal wound epidermis, but of markedly reduced mitotic activity in the subepidermal tissues, regeneration proceeded aberrantly and ultimately halted. The observations of Rose *et al.* may be explained by the phenomenon described by Luther (1948) following distal irradiation of intact tails. He found that the replacement of the irradiated epidermal cells is followed by a similar replacement of the immediately subepidermal cells. It would appear highly probable that some viable subepidermal cells may have accompanied the replacement epithelium during the recovery of the stripped limbs. In Fimian's cases the replacement was necessarily slower, since the original epidermis remained in place. In such a case the accumulation of a sufficient mass of blastema cells was long delayed, and the inadequate accumulation blastema, formed only by dissociated and mitotically inactive cells, regressed.

The hypothesis of a definitive epidermal origin of the regeneration blastema is also not supported by the work of Heath (1953). He created tissue chimeras by exchanging epidermal and subepidermal limb-bud tis-

40

sues between embryos with markedly different growth potentials. Both
the chimeras and the regenerates formed by chimeras consisting of
grafted epidermis and host subepidermis showed size characteristics of
the host tissue. Regenerates formed by chimeras consisting of grafted
subepidermal tissues and host epidermis occasionally showed some characteristics of the host limb.

The possibility of an epidermal contribution to the blastema remains,
but the extent of such a contribution in normally developing regenerates
must be slight and of no significance as far as the developmental characteristics of the subepidermal blastema are concerned.

The sole remaining possible source of blastema cells is the old stump.
This is the source indicated by the great majority of descriptive studies.
Reviews of the earlier literature have been presented by Mettetal
(1939), Schotté (1939), and Weiss (1939). The former author attributes blastema formation solely to the proliferation of loose connective tissue elements. He specifically denies contributions by the two
other distinct subepidermal tissues—muscle and cartilage. This view of
a limited fibroblastic origin of the blastema has been supported recently
by Manner (1953). Schotté, on the other hand, concluded that the
blastema took origin from all the tissues of the limb—muscle, cartilage,
and the ubiquitous connective tissues. Weiss attributes blastema origin
to the local proliferation of still unidentified reserve cells, presumably
associated with connective tissue elements.

In 1947 I began a quantitative histologic analysis of the histologic
changes occurring during forelimb regeneration in *Triturus viridescens*.
The object of these studies was to determine whether any correlation
existed between the proliferation rates of the tissues of the old stump
and the rates of appearance of new cells of the same tissues in the regenerate. This would provide evidence as to the identity and degree of contribution of all elements of the limb to the regenerate. As a matter of
fact, a very close correlation of the type anticipated can be shown to
exist. But it became quite evident during the course of the study that
the difficulties inherent in cell identification during morphologic dedifferentiation and dissociation of the stump tissues and the impossibility of
determining the physiologic state of a cell from the study of fixed tissues
make it impossible to distinguish clearly between the three hypotheses
mentioned above.

The data presented are drawn largely from Chalkley (1954) and from
unpublished material collected in the same study. The animals used are
not a haphazard sample. Three animals were sacrificed from the experimental group on each of seven dates. The over-all changes in stump and

41

regenerate volumes of the three limbs, determined on the day prior to sacrifice, placed one of these three animals near the group mean and the other two near the values indicated by the group standard deviations.

As shown in Figure 4, the rate of accumulation of cells in the regenerating limbs (graph *L*) is essentially constant during the first 30 days after amputation. Initially, the bulk of this accumulation is confined to the stump (graph *S*), and it is not until the twenty-second day that it is transferred to the regenerate proper (graph *R*). The decrease in the stump accumulation, which is concomitant with an abrupt rise in the

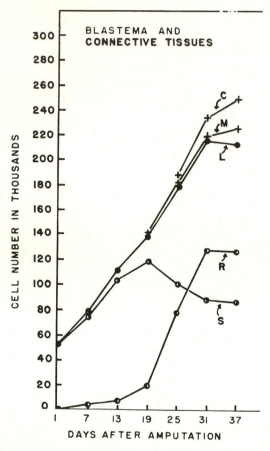

FIG. 4.—Changes in the numbers of blastema and connective tissue cells combined in the regenerating stump (*S*), in the regenerate proper (*R*), in the entire regenerating limb (*L*), and in the entire limb plus the redifferentiating muscle cells (*M*) and plus the redifferentiating cartilage cells (*C*). (From Chalkley, 1954.)

42

number of cells in the regenerate, strongly suggests a distal migration of cells from stump to regenerate.

The pattern of this migration is clearly indicated by the change in cell distribution along the limb axis. This is shown in Figure 5. Accumulation of cells occurs along the entire length of the stump and regenerate, though this accumulation centers about a peak which, as of the seventh day after amputation, is located some 0.56 mm. proximad from the amputation plane (indicated by the arrows). By the nineteenth day the accumulation in the stump is at a maximum. The increase has continued

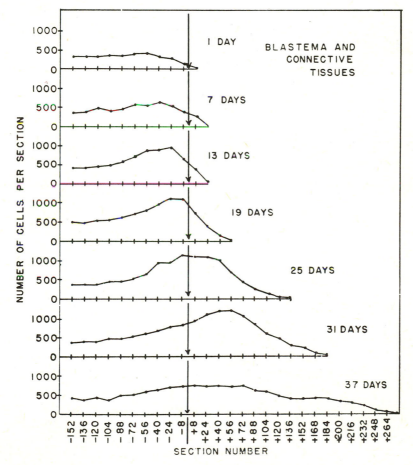

Fig. 5.—Changes in the distribution of blastema and connective tissue cells along the axis of the regenerating limb. Arrows indicate the amputation plane. (From Chalkley, 1954.)

43

along the entire limb axis, but the level of peak accumulation has moved distally and is now only 0.24 mm. proximad from the amputation plane. The peak then shifts into the regenerate, reaching a level 0.56 mm. from the amputation plane on the thirty-first day. The shift is real. Notice that the number of cells in the stump decreases along the entire length of the stump axis. There is no indication of a proximad migration and little or no necrosis.

Distal migration of a mitotically active zone, from stump to regenerate, accompanies the shift in numbers of cells. This is clearly shown in Figure 6. Only 2 per cent of the mitoses are found in the regenerate on the seventh day, though it contains 17 per cent of the blastema cells. By the nineteenth day the regenerate contains 29 per cent of the mitoses and 23 per cent of the blastema cells. Its continued growth is then essentially independent of mitotic activity and cell migration from the stump, though both these processes continue in diminishing degree for some time. Notice that the distribution of the mitotic index in Figure 6 suggests an "apical growth cone" such as that described by Holtzer, Holtzer, and Avery (1953) for the tail. A similar distribution of mitoses was described in *T. pyrrhogaster* by Litwiller (1939). The distribution of total mitoses is more or less normal and extends along much of the length of the blastema, with the largest number of the cells being added well back from the limb tip. Also notice that there is no interruption in any of the distributions at or near the amputation plane.

This evidence of early cellular migration is of particular interest in view of the demonstration by Singer (1955) of the existence of a barrier, or barriers, between the stump and regenerate as of the nineteenth day after amputation. This barrier resists the passage of micropipettes from stump to regenerate and the passage of infusates from regenerate to stump. Singer has suggested the existence of a coarse barrier, possibly consisting of a loose network of collagenous fibrils, occupying roughly the position of the amputation plane. Such a network could be felt but would be difficult, if not impossible, to detect microscopically with the ordinary histologic techniques used in most studies, including the present one. The barrier to infusates, Singer suggested, may be little more than a change in the amount of intercellular space available for rapid perfusion, or it may be no barrier at all in the physical sense but simply an increased vascular bed which rapidly removes perfusates from the distal stump. Measurements of available intercellular space and of vascular space, as shown in Figure 7, indicate no abrupt changes at or near the amputation plane. This leaves untested a third hypothetical barrier mechanism suggested by Singer—that the movement of edema fluids

44

from stump to regenerate interferes with the basipetal movements of perfusates from regenerate to stump.

It seems reasonably well demonstrated that the major source of the cells of the forelimb regenerate in *T. viridescens* is the old stump and that there are three overlapping stages in the formation of the regenerate proper: (1) an initial distal migration of cells unaccompanied by cell

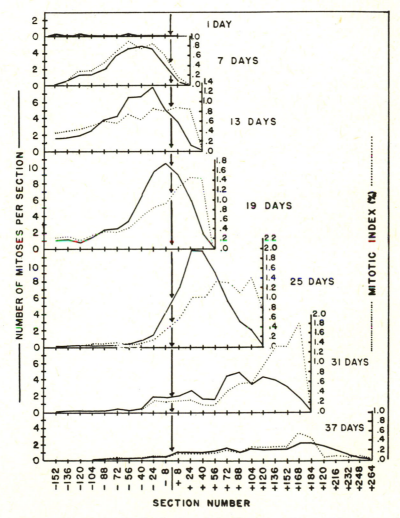

FIG. 6.—Changes in the distribution of the blastema and connective tissue cell mitoses and mitotic index along the axis of the regenerating limb. The index is expressed as per cent cells per section. Arrows indicate the amputation plane. (From Chalkley, 1954.)

45

division, a process which contributes, at most, a few thousand cells and which terminates within the first 3 or 4 days; (2) migration of mitotically produced cells from stump to regenerate, a process which begins some 4 days after amputation and produces a minimum of 60,000 cells; (3) continued division of these migratory cells in the regenerate proper. This produces a minimum of 100,000 additional cells.

I have repeatedly called attention to the absence of breaks in the distribution of cell counts or space measurements at or near the amputation plane. It is evident that a distinction between stump and regenerate is

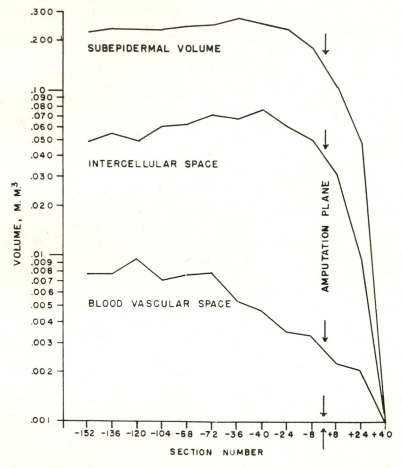

Fig. 7.—Changes in the distribution of total subepidermal volume, intercellular space and blood vascular space along the axis of the regenerating limb on the nineteenth day after amputation. Arrows indicate the amputation plane. Note that the volume scale is logarithmic.

46

completely arbitrary insofar as blastema cell production and migration are concerned. Identification of the blastema with the young regenerate is an anachronism when, as in *T. viridescens,* it can be shown that blastema cell production begins well back in the old stump. In this species the forelimb blastema can be conceived as a mass of cells occupying not only the region distal to the amputation plane but also the space previously taken up by the now dissociated soft tissues of the subjacent stump. Only by the twenty-fifth day is the mitotic activity of the regenerating limb so clearly concentrated in the regenerate proper that it can be reasonably identified with the blastema. Unfortunately, this is also the date at which redifferentiation begins in the regenerate, and the number of blastema cells rapidly decreases.

During the early stages of blastema accumulation it is possible to identify a large number of mitoses with specific stump tissues. On the seventh day after amputation, approximately 68 per cent of all mitoses can be identified with residual stump tissues, while the remaining 32 per cent of the cell divisions are classed as blastema mitoses (Fig. 8). During the next 18 days the percentage of cell divisions identifiable only as blastema cell mitoses, since they show neither clearly identifiable morphologic characteristics nor close association with differentiated tissues, increases to a maximum of 89.9 per cent. All the identifiable mitoses are confined to the stump, but only a few perichondrial mitoses are associated with newly differentiating tissues. On the thirty-first day many new mitoses are identifiable in differentiating tissues, and by the thirty-seventh day only 30.8 per cent of the mitoses are still recognizable only as blastema cell divisions. By this latter date all the mitoses in differentiating cells are found in the regenerate. A comparison of the distribution of these mitoses to the various categories on the seventh day with that found on the thirty-first day (see Fig. 8), when plotted on a scatter diagram as in Figure 9, shows that a high degree of correlation exists between them. If correlation were perfect, all points would lie on the interrupted line, and the correlation coefficient, *r,* would have a value of 1.00. The value of *r* found for the observed scatter from perfect correlation is 0.965. The data indicate a very broad origin of the blastema cells from all the stump tissues and strongly suggest a possible tissue-specific origin—muscle contributing to muscle, connective tissue to connective tissue, and so on.

If this latter suggestion were true, a similarly close correlation should exist between the distribution of cells among the tissues of the unamputated limb (see Fig. 10, *A*) and the distribution of mitotic activity among the tissues of the regenerating limb (see Fig. 10, *D*). These are

47

37

obviously not directly comparable, since there is no equivalent of the blastema cell in the unamputated limb. If, however, we distribute the blastema cell mitoses among the other categories in the proportions found among those tissues on the seventh day after regeneration (Fig. 8), we get the distribution shown in Figure 10, *B*. If this distribution of the total subepidermal mitoses is plotted against the distribution of subepidermal cells in unamputated limbs, as in the scatter diagram of Figure 11, a close correlation, in which the value of *r* is 0.960, is obtained. The only striking deviations from more perfect correlation are the low number of mitoses in the blood vessels (*BV*) and the high num-

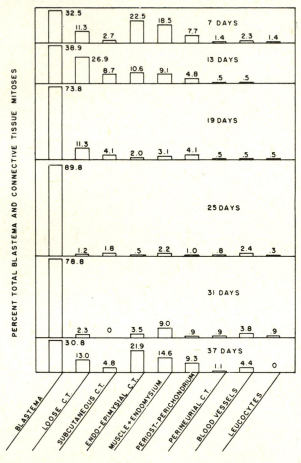

FIG. 8.—Percentages of the total blastema and connective tissue mitoses at 7, 13, 19, 25, 31, and 37 days after regeneration, attributable to various tissue components. (From Chalkley, 1954.)

48

ber of mitoses in the epimysium-endomysium (*EE*) categories. Since the small regenerate does not require the large, heavily walled arteries present in the normal limb, the first deviation is readily understandable. The high count in the epimysium-endomysium category is not so easily explained. Since most of the other counts are low, it may reflect a predilec-

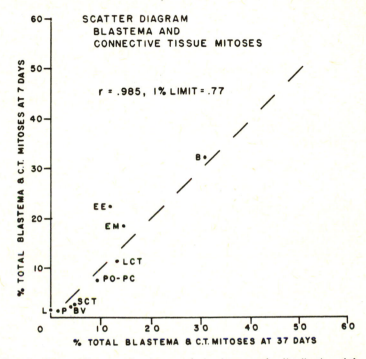

FIG. 9.—Scatter diagram, showing correlation between the distribution of the total blastema and connective tissue mitoses at 7 days as compared with their distribution at 37 days. The correlation coefficient, *r*, for this scatter is 0.985. The abbreviations are: *B*, blastema cells; *BV*, blood vessel walls; *EE*, epimysium and endomysium; *EM*, endomysium and muscle (assignment to *EE* or to muscle uncertain); *L*, leukocytes; *LCT*, loose connective tissue; *P*, perineurium; *PO-PC*, periosteum-perichondrium; *SCT*, subcutaneous connective tissue.

tion on the part of the investigator to place doubtful counts in this category.

We have now established correlation between mitotic activity in the stump tissues and in the redifferentiating regenerate tissues and between the distribution of cells in the normal limb and the mitotic activity in the regenerating limb. It now remains to establish the degree of correlation between mitotic activity in the regenerating limb (see Fig. 10, *B*) and the numbers of new cells produced (see Fig. 10, *C*). Only the first five

49

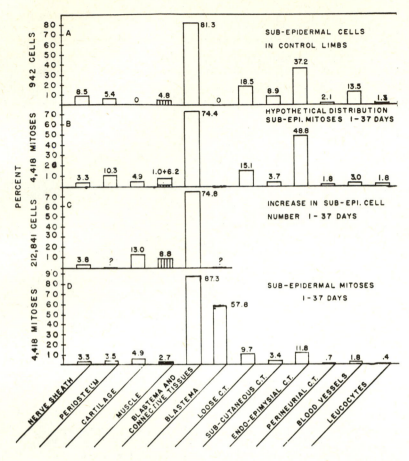

Fig. 10.—A, Distribution of subepidermal cells in the control limbs. B, Hypothetical distribution of subepidermal mitoses through the thirty-seventh day. These data were obtained by distributing the blastema mitoses among the possible connective tissue contributors on the basis of the distribution shown in Fig. 8 on the seventh day (see detailed explanation in text). C, Increase in cell number in the regenerating limbs through the thirty-seventh day. Periosteal cells are inseparable from cartilage. D, Distribution of subepidermal mitoses through the thirty-seventh day if blastema mitoses are counted as a distinct type (possible reserve cells?).

In all graphs the column "blastema and connective tissues" is the sum of all columns graphed to its right. Hatched portions of the muscle columns represent counts on intact muscle; clear portions represent counts in dissociated or redifferentiating muscle. (From Chalkley, 1954.)

40

categories—nerve sheath cells, periosteal cells, cartilage cells, muscle cells, and the combined blastema and connective tissue cells—can be compared in the scatter diagram in Figure 12. Routine identification of mitoses is possible because of the relatively small number of cells involved. Similar precision in identification of the various types of con-

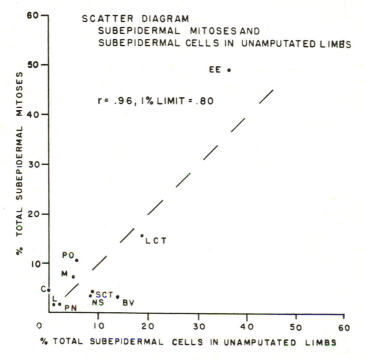

Fig. 11.—Scatter diagram, showing correlation between the distribution of the total subepidermal mitoses and the distribution of the subepidermal cells in the unamputated limb. The dashed line indicates the line of perfect correlation. The correlation coefficient, *r*, for this scatter is 0.96. The abbreviations are: *BV*, blood vessel walls; *C*, cartilage; *EE*, epimysium and endomysium (includes 75 per cent of *EM*, endomysium and muscle, mitoses); *L*, leukocytes; *LCT*, loose connective tissue; *M*, muscle (includes 25 per cent of *EM*, endomysium and muscle, mitoses); *NS*, nerve sheath; *PN*, perineurium; *PO*, periosteum; *SCT*, subcutaneous connective tissue.

nective tissue cells is impracticable, if not impossible. The correlation obtained with the raw data gives a high *r* value of 0.973, but, because of the small number of points, this is only slightly higher than the 1 per cent confidence limit of 0.96. The reason for this is readily evident. Periosteal divisions account for 10.3 per cent of the total mitoses, but there are no periosteal cells in the regenerate and no detectable new periosteal cells in the old stump. On the other hand, cartilage accounts for

51

13.0 per cent of the new cells but only 4.9 per cent of the mitoses. If we recognize the probability that periosteal and perichondrial cells represent different functional states of the same cell and that periosteal mitoses contribute to the formation of perichondrial, and thus cartilage, cells, then the number of cartilage-producing mitoses ($C + PO$ in Fig. 12) amounts to 15.2 per cent of the total number of cell divisions. This compares favorably with the 13.0 per cent of the new cells which are cartilage.

A similar adjustment can be made with respect to the formation of new muscle nuclei. These account for 8.8 per cent of the added cells and

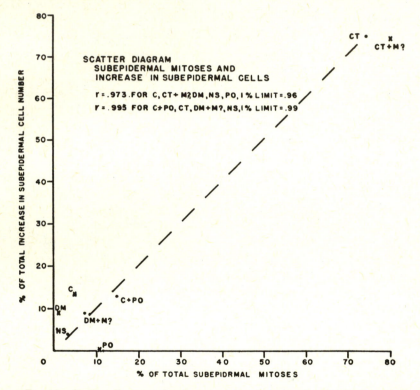

Fig. 12.—Scatter diagram, showing correlation between the distribution of the increase in subepidermal cell numbers in various categories and the distribution of mitoses in various similar or identical categories during the same period. The dashes indicate the line of perfect correlation. The correlation coefficient, *r*, for this scatter is 0.973 or 0.995, depending upon the selection of points (see text for details). The abbreviations are: *C*, cartilage; *PO*, periosteum; *C + PO*, cartilage plus periosteum; *CT*, connective tissue; *CT + M?*, connective tissue plus some probable muscle mitoses; *DM*, differentiated muscle tissue; *DM + M?*, cells and mitoses in differentiated muscle plus probable muscle mitoses; *NS*, nerve sheath cells and mitoses.

52

nuclei but for only 1.0 per cent of the contributory mitoses. However, certain of the mitoses occurring in the empty endomysial sheaths of the old stump cannot be clearly identified with either connective tissue or muscle (*CT* + *M?* in Fig. 12). Assuming that approximately 25 per cent of these mitoses are muscle, since 25 per cent of the nuclei in the normal muscle-enodomysium complex are in that category, they account for 6.2 per cent of the total mitoses. Adding the questionable muscle mitoses to mitoses in differentiated muscle tissue (*DM* + *M?* in Fig. 12), a close fit to perfect correlation is obtained. The fit of connective tissue mitoses and increase in connective tissue cell number is also improved. The value of *r* obtained with these modified data is appreciably higher, 0.995, though the elimination of a point by the combination of cartilage and periosteal cells and mitoses increases the value of the 1 per cent confidence limit to 0.99.

Correlation exists, then, between the distribution of available cells in the stump, the mitotic activity in the stump, the continued mitotic activity in the redifferentiating cells of the regenerate, the increase in cell number, and, since it is identical with availability, the ultimate demand for cells in the restored limbs. The correlation coefficients obtained range between 0.96 and 0.995 and are in excess of the 1 per cent confidence limits for each correlation.

Allowing for the necessary recapitulation of embryonic processes in the formation of bone, there is nothing in these data that would preclude their interpretation as a description of tissue-specific regeneration.

Such a conclusion is incompatible with pertinent experimental observations, particularly the X-radiation experiments of Thornton (1942), Luther (1948), Trampusch (1951), Rose, Quastler, and Rose (1953), and Ichikawa and Okada (1954). In all these experiments limbs have been exposed to X-ray doses sufficient to inhibit regeneration, and then various tissues have been replaced or supplemented with normal unirradiated tissues. The "radiation chimeras" thus produced formed regenerates with fair frequency and normality. Dr. Trampusch describes certain of these results elsewhere in this volume. There is no point in repeating them here. It is interesting to note that chimeras frequently develop normally, though the subcutaneous connective tissue is supposedly the only mitotically competent tissue present. Normally, it provides only 3.7 per cent of the mitoses. Where the mitotically competent tissue has been muscle, accompanied inevitably by the epimysium and endomysium, which normally provide 56 per cent of the mitoses, regenerates are infrequent and differentiate poorly.

It would be interesting to know the distribution of mitotic activity in

53

43

these chimeras. Are the tissues of an irradiated limb equally inhibited? Does the addition of mitotically competent tissue to an irradiated limb in any way affect the mitotic activity of the irradiated tissues? Brunst and Chérémetiéva (1935) have found partial regeneration in 75 per cent of *Triturus* treated with 15,000 r. Fimian (1951) found near-normal mitotic activity in two out of fifteen limbs exposed to doses of 5,000 r. Litschko (1934) noted accelerated regeneration following removal of the dense wound epidermis formed by regenerates exposed to 300 r. Ichikawa and Okada (1954) added unirradiated hind-limb tissues to forelimbs which had been exposed to 1,000 r. Of the normal limbs regenerated, 62 per cent were forelimb in type. They suggest that irradiated tissues are somehow stimulated to recover their regenerative and proliferative capacities by the presence of normal tissue.

Lacking quantitative histologic studies on these radiation chimeras, we are forced to admit that a blastema can be formed in the absence of the usual contributing tissues. This is not a difficult admission to make where such tissues as the endomysium, endoneurium, and subcutaneous connective tissues are concerned, since there is little difference in the functions of their component cells. Cartilage and muscle present different problems. In the absence of unirradiated muscle or of unirradiated cartilage, perichondrium, or periosteum, what is the source of these tissues? It would seem necessary to propose the presence in the limb of indifferent reserve cells or of physiologically dedifferentiated cells. Brunst and Chérémtiéva (1936) suggested that the variability in the response of regenerating limbs to regeneration could be explained on the assumption that the limb contained a relatively small number of reserve cells. The widespread nature of the mitotic pool that produces the blastema would seem to rule out this particular hypothesis. Weiss (1923, 1939) has also proposed a reserve-cell origin of the blastema. He has pointed out that morphological dedifferentiation may occur with ease under tissue culture conditions, while the occurrence of physiologic dedifferentiation is extremely infrequent. In the absence of a mechanism for physiologic dedifferentiation or for the orderly migrations or reassociations required by tissue-specific regeneration, Weiss prefers to conceive of blastema formations as the result of the proliferation of reserve cells. To what extent is this a necessary device? Where are the reserve cells? Again we do not need to provide a reserve-cell mechanism for the connective tissues. They are ubiquitous and presumably easily mobilized into whatever categories are needed. The close association of connective tissues and the perichondrium and periosteum might form a basis for an argument that a residual ability to form cartilage and bone was retained by

54

a large number of connective tissue elements. On the other hand, the distinctive behavior and appearance of the periosteal and perichondrial cells in the blastema suggest that they are truly distinct from the connective tissues. A reserve cell is therefore required which can differentiate into cartilage and muscle cells and possibly into connective tissue cells. No distinctive reserve cell appears to be present, and it seems surprising, if such cells are present, "disguised" as normal tissue cells, that they are so uniformly distributed throughout all the normal tissues of the limb and are not limited to a single mitotically competent cell type, such as the fibroblast, as inferred by Manner (1953). One could, of course, argue that some of the mitoses observed in the limb are those of differentiated cells dividing cell-specifically and that only a small fraction of the divisions are those of reserve cells. At the moment this possibility would have to be admitted, though the distinct periosteal divisions would argue against it.

The physiologic conservatism of cultivated tissues constitutes an argument against the reserve-cell hypothesis, as well as against the hypothesis of dedifferentiation. The persistent retention of physiologic characteristics by gonadotrophin-secreting chorionic cells (Waltz *et al.*, 1954), of glycogen retention by cultivated liver cells (Westfall *et al.*, 1953), and the reversion to their characteristic epithelial patterns of fibroblast-like skin cells following their return to an in vivo environment (Bassett *et al.*, 1955) suggest that functional dedifferentiation of adult tissues may be a myth. But tissue culture has also failed to disclose any reserve-cell potentials in the variety of cells included in the culture inoculum, even though they have been exposed to a variety of culture environments. Cultivation in vitro of blastema cells produces cultures that are indistinguishable from connective tissue cultures obtained from non-regenerating parts of the organism (Lecamp, 1947, 1949).

We are faced with a dilemma. Radiation chimeras and extirpation experiments clearly suggest the existence of some *doppelte Sicherung* mechanism in the regenerating limb. Other experiments fail to identify either a dedifferentiating mechanism or a reserve-cell system that could explain the ability of the limb to regenerate normally when certain characteristic tissues are absent from the amputation plane. Clearly, we need to know more about the behavior of irradiated cells and tissues. Does irradiation block mitosis completely and permanently? What tissues in a radiation chimera are dividing? Following extirpation of limb parts, are there any changes in the rates of division of the cells near the ampu-

55

tation plane? Over what distances can mitotically competent cells migrate to reach the blastema?

Quantitative histologic techniques can help to answer some of these questions, albeit in a slow and cumbersome fashion. The use of tagged cell-specific or even tissue-specific antibodies as tracers for morphologically dedifferentiated cells would be of considerable help. The development of "tissue pure" radiation chimeras which avoid the complications of previous experiments which always involved the transfer of some mitotically competent connective tissue would present a surgical challenge of no small magnitude. Only with a considerable advance in our present knowledge of the histology, histochemistry, and physiology of the blastema, can we hope to be in a position to place any reliance on any of the currently available hypotheses as to the nature of the regeneration process.

REFERENCES FOR PAPER III (CHALKLEY)

BASSETT, C. A. L., EVANS, V. J., CAMPBELL, D. H., and EARLE, W. R. 1955. Characteristics and potentials of long-term cultures of human skin. Nav. M. Res. Inst., **13**:509.

BASSINA, J. A. 1940. An inquiry into regeneration by the method of calculating mitotic coefficients. Bull. Exper. Biol. Meditsny., **10**:389 (in Russian).

BRUNST, V. V., and CHÉRÉMETIÉVA, E. A. 1935. Recherches sur l'influence des

102

rayons X sur la régénération de la queue du Triton après une seule irradiation à doses variées. Arch. biol., **46**:357.

———. 1936. Sur la perte locale du pouvoir régénérateur chez le Triton. Arch. zool. exper. génét., **78**:56.

Butler, E. G. 1933. The effects of X-radiation on the regeneration of the forelimb of *Amblystoma* larvae. J. Exper. Zoöl., **65**:271.

———. 1935. Studies on limb regeneration in x-rayed *Amblystoma* larvae. Anat. Rec., **62**:295.

Butler, E. G., and O'Brien, J. P. 1942. Effects of localized X-irradiation on regeneration of the urodele limb. Anat. Rec., **84**:407.

Chalkley, D. T. 1954. A quantitative histological analysis of forelimb regeneration in *Triturus viridescens*. J. Morphol., **94**:21.

Clark, E. L. 1912. Further observations on living, growing lymphatics. Their relations to the mesenchyme cells. Am. J. Anat., **13**:351.

Colucci, Vincenso, 1884. Introno alla rigenerazione degli arti e della coda nei Tritoni. Studio sperimentale. Mem. ric. Accad. sc. Inst. Bologna, Ser. 4, **6**:501.

Dettelbach, H. R. 1952. Histostatic and cytostatic effects of some amino-ketones upon tail regeneration in *Xenopus* larvae. Rev. suisse zool., **59**:339.

Fimian, Walter J. 1951. Determination of the distribution of mitotic activity in the regenerating urodele limb after partial X-irradiation. Thesis, University of Notre Dame.

Godlewski, E. 1928. Untersuchungen über die Auslösung und Hemmung der Regeneration beim Axolotl. Arch. f. Entw.-Mech., **114**:108.

Hellmich, W. 1930. Untersuchung über Herkunft und Determination des regenerative Materials bei Amphibien. Arch. Entw.-Mech., **121**:135.

Holtzer, H., Holtzer, S., and Avery, G. 1955. Morphogenesis of tail vertebrae during regeneration. J. Morphol., **96**:145.

Ichikawa, M., and Okada, K. 1954. Studies on the determination factors in amphibian limb regeneration. Mem. Col. Sc. Univ. Kyoto, Ser. B, **21**:23.

Ide-Rozas, Alberto. 1936. Die cytologische Verhältnisse bei der Regeneration Kaulquappenextremität. Arch. Entw.-Mech., **135**:552.

Inoue, Sakae. 1954. The mitotic rate in the regenerating epithelium of the forelimb and the dorsal skin of the adult newt. Sci. Rep., Tohoku Univ., Ser. 4, **20**:182.

Kazancev, W. 1934. Histologische Untersuchungen der Regenerationsprozesse an amputierten Extremitäten beim Axolotl, hauptsächlich zwecks Klärung der Frage nach der Herkunft der Zellen des Regenerats. Acad. Sc. U.R.S.S. trav. lab. zool. exper. morphol. animaux, **3**:23.

Lash, James W. 1955. Studies on wound closure in urodeles. J. Exper. Zoöl., **128**:13.

Lecamp, Marcel. 1947. Les Tissus de régénération en culture in vitro. Compt. rend. Acad. sc. Paris, **224**:674.

———. 1949. Régénération chez le Triton in vitro et in vivo. *Ibid.*, **226**:695.

Litschko, E. J. 1934. Einwirkung der Röntgenstrahlen auf die Regeneration der Extremitäten des Schwanzes und eines Teiles der Dorsalflosses beim Axolotl. Akad. Nauk. U.R.S.S. trav. lab. zool. exper. morphol. animaux, **3**:101.

Litwiller, Raymond. 1939. Mitotic index and size in regenerating amphibian limbs. J. Exper. Zoöl., **82**:280.

103

Lüscher, Martin. 1946. Die Hemmung der Regeneration durch Colchicin beim Schwanz der Xenopus-Larve und ihre entwicklungsphysiologische Wirkungsanalyse. Helvet. physiol. et pharmacol. acta, **4**:465.

Luther, W. 1948. Regenerationsversuche mit Hilfe von Röntgenstrahlen. Verhandl. deutsch. zool. Gesellsch., Kiel, p. 66.

Manner, H. W. 1953. The origin of the blastema and of new tissues in regenerating forelimbs of the adult *Triturus viridescens*. J. Exper. Zoöl., **122**:229.

Mettetal, Christian. 1939. La Régénération des membres chez la Salamandra et le Triton. Histologie et détermination. Arch. anat. hist. embryol., **28**:1.

Polezajew, L. W., and Ginsburg, G. I. 1944. Investigation of ways of formation of regeneration blastema based on calculation of mitotic coefficient. Compt. rend. (Doklady) Acad. sc. U.R.S.S., **43**:315.

Rose, F. C., Quastler, H., and Rose, S. M. 1953. Regeneration of x-rayed limbs of adult *Triturus* provided with unirradiated epidermis. Anat. Rec., **117**:619.

Rose, S. M. 1948. Epidermal dedifferentiation during blastema formation in regenerating limbs of *Triturus viridescens*. J. Exper. Zoöl., **108**:337.

Schotté, O. E. 1939. The origin and morphogenetic potencies of regenerates. Growth Suppl. 59.

Singer, Marcus. 1955. A study of limb regeneration in the adult newt, *Triturus*, by infusion of solutions of dye and other substances directly into the growth. J. Exper. Zoöl., **128**:185.

Skowron, S., and Roguski, H. 1953. Tail regeneration in *Xenopus laevis* tadpoles. Bull. Acad. polon. Sc. Cl. II, **1**:15.

Thornton, C. S. 1942. Studies on the origin of the regeneration blastema in *Triturus viridescens*. J. Exper. Zoöl., **89**:375.

Towle, E. W. 1901. On muscle regeneration in the limbs of *Plethodon*. Biol. Bull., **2**:289.

Trampusch, H. A. L. 1951. Regeneration inhibited by X-rays and its recovery. Proc. k. nederl. Akad. wetensch., Ser. C, **54**:373.

Waltz, Helen K., Tullner, W. W., Evans, V. J., Hertz, Roy, and Earle, W. R. 1954. Gonadotrophic hormone secretions from hydatid mole grown in tissue culture. J. Nat. Cancer Inst., **14**:1173.

Weiss, Paul. 1939. Principles of development. Chicago: University of Chicago Press.

Westfall, B. B., Evans, V. J., Shannon, J. E., and Earle, W. R. 1953. The glycogen content of cell suspensions prepared from massive tissue culture. J. Nat. Cancer Inst., **14**:655.

Reprinted from *Develop. Biol.*, **1**, 555–585 (1959)

3

Electron Microscopic Observations of Muscle Dedifferentiation in Regenerating *Amblystoma* Limbs[1]

ELIZABETH D. HAY

Department of Anatomy, Cornell University Medical College, New York, New York

Accepted October 27, 1959

INTRODUCTION

The regenerating salamander limb has a great capacity for dedifferentiation of the formed tissues of the stump following amputation. Cells having an undifferentiated appearance arise from the muscle, cartilage, Schwann cells, and fibroblasts of the stump and accumulate to form a blastema under the tip epithelium. The blastema then proliferates rapidly and later redifferentiates to restore the complete limb. Although earlier workers, beginning with Towle in 1901, had described changes in the muscles of regenerating limbs, the most careful study of muscle dedifferentiation in the limb was made by Thornton in 1938 with the light microscope and regenerating limbs of *Amblystoma* larvae (see Thornton, 1938a, b and 1942, for review of the earlier literature). There remains, nevertheless, a tendency to confuse muscle dedifferentiation with muscle degeneration in regenerating limbs (see Manner, 1953; Bodemer, 1958). The present study with the electron microscope confirms Thornton's excellent description and extends it to include observations on the disappearance of myofibrils, alterations in cell organelles, acquisition of cytoplasmic ribonucleoprotein, and formation of cell boundaries during dissociation of the syncytial muscle fiber into mononucleate blastema cells.

[1] Supported by grant C-3708 from the National Institutes of Health, United States Public Health Service.

555

49

There has been some controversy over the use of the term "dedifferentiation" to describe the changes which occur in the amphibian limb tissues during blastema formation. Therefore it may be pertinent to justify our retention of this term before embarking upon a detailed description of the process. There can be no reasonable doubt that the differentiated tissues of the stump do contribute morphologically undifferentiated cells to the blastema. Some authors have, however, insisted that the term "dedifferentiated" be used only for those transformations in which cells can be shown to have reverted to a more generalized embryonic state characterized by pluripotentiality as well as structural simplicity (Bloom, 1937) Experimental conditions can seldom be devised that will demonstrate the pluripotentiality required to satisfy this exacting definition. The plea for a rigid definition of dedifferentiation was a reaction to the indiscriminate use of the term to describe any simplification of cell structure observed in tissues *in vitro* or in neoplasia *in vivo*. The term "modulation" was suggested by Weiss for those reversible cellular transformations in which there is loss of structural complexity with no associated gain in developmental potentialities (see Weiss, 1949, for review and discussion).

Experience with tissues *in vitro* has led some investigators to conclude that cell morphology is of little help in determining whether changes observed represent true dedifferentiation or merely "loss of some visible equipment" (Godman, 1958) with no fundamental change in the cell's identity. This pessimistic view may be justified where observations are restricted to the relatively gross changes detectable with the light microscope. However, examination of developing tissues with the electron microscope has provided us with a number of additional criteria for judging the state of differentiation of cells, and our study of the fine structure of dissociating muscle fibers has convinced us that this remarkable transformation is more than a mere "loss of some visible equipment." The dedifferentiating cells of the stump do not simply lose the distinctive structural components that facilitate their identification, but they also *acquire* numerous embryonic features such as rounded nuclei with prominent nucleoli, increased nucleocytoplasmic ratio, and increased numbers of free cytoplasmic ribonucleoprotein granules. Although it has yet to be proved that any given blastema cell can differentiate into cell types other than the one from which it took origin, nonetheless, the complete reversion of both cartilage (Hay, 1958) and muscle cells to a common basophilic cell type with all the ultrastructural characteristics of embryonic cells must be considered a change of more far-reaching significance than is implied by the term "modulation."

The changes in the fine structure of the muscle of regenerating salamander limbs to be described in this paper seem to us to constitute as good an example of *dedifferentiation* as can be found among the vertebrates. The observations recorded here also bear upon such cytological problems as the mode of formation of cell membranes, the structure of the myofibril, and the behavior of the endoplasmic reticulum and Golgi apparatus under conditions of functional dedifferentiation.

MATERIALS AND METHODS

The *Amblystoma punctatum* larvae used in this study were raised from eggs collected in the vicinity of Cornwall, New York, and Baltimore, Maryland. The animals were kept in individual fingerbowls at 20° C in a constant temperature room during regeneration. Preliminary studies were made of the regenerating limbs of larvae 18 mm in length, but the limbs of older larvae which were 30–40 mm in length proved to be considerably better for the study of muscle and were utilized for most of the observations recorded here. Limbs were amputated through the distal third of the humerus, and regenerating limbs were fixed at daily intervals after amputation. The fixative consisted of 1% OsO_4 buffered to pH 7.8–8.0 (Palade, 1952) to which sucrose (0.05 gm/ml) was added. After a 1–2-hour period in the fixative, limbs were transferred directly to 50% alcohol and dehydrated rapidly in graded alcohols (total time in alcohols, 1–1½ hours). They were then embedded in N-butyl methacrylate in capsules with the longitudinal axis of the limb perpendicular to the long axis of the capsule. This orientation facilitated sectioning the limb in the longitudinal plane. The present study deals with the period 2–6 days after amputation, during which time dedifferentiation of the stump tissues is occurring. Approximately sixty limbs in this age group were sectioned with a glass knife in a Servall microtome (Porter and Blum, 1953). Thin sections 0.05–0.1 μ in thickness were mounted on copper grids and examined with an RCA EMU 3B or 3D electron microscope. Thicker methacrylate sections, 2–5 μ in thickness, were mounted on glass slides in balsam and examined unstained with a Zeiss phase

51

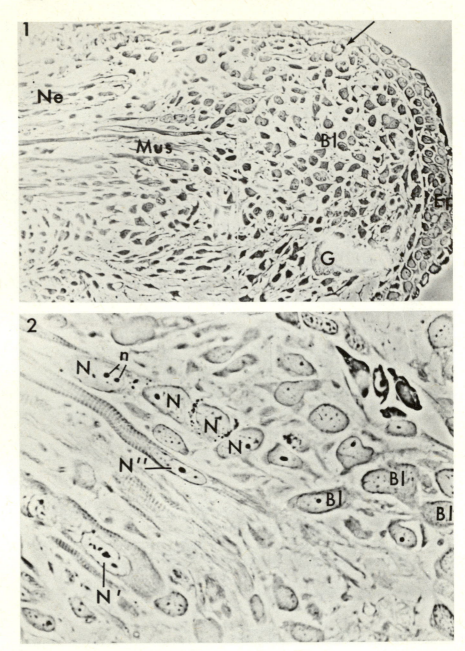

contrast microscope. Other thick sections were stained with thionine or Wright's stain and studied without phase contrast, using the light microscope.

OBSERVATIONS

General Characteristics of Dedifferentiating Muscle Cells

During the initial 2 days following amputation of the forelimb of larval *Amblystoma*, cellular activity in the limb stump is primarily concerned with wound healing. The amputation surface is covered by epithelium migrating from the sides of the limb, and the cellular debris resulting from hemorrhage and tissue damage at the site of amputation is removed by phagocytosis. Dedifferentiation of muscle, cartilage, and other formed tissues of the stump begins about the third day after amputation and continues until the sixth or seventh day. The blastema cells derived from these dedifferentiating tissues accumulate under the thickened apical epithelium of the amputated limb. After a period of cellular proliferation, redifferentiation begins in the blastema on the tenth or eleventh day after amputation and a fully formed new limb is produced by about the sixteenth day.

The general features of the process of muscle dedifferentiation are illustrated in Figs. 1–3, which are phase contrast photomicrographs of

FIG. 1.   Phase contrast photomicrograph of an unstained section of a regenerating limb 5 days after amputation. The limb was fixed in osmic acid and embedded in methacrylate for sectioning. This section passes lateral to the humerus and illustrates a bundle of dedifferentiating muscle (*Mus*). Numerous blastema cells (*Bl*), most of which derived from muscle, appear distal to the intact muscle fibers. The tip epidermis (*Ep*) is slightly thickened. The fibrous basement membrane underlying the proximal epithelium ends at the arrow. A multinucleate giant cell (*G*) is associated with a remnant of cartilage and part of a nerve (*Ne*) appears in the proximal portion of the section: Magnification: × 140.

FIG. 2.   Enlarged photomicrograph of the tips of the dedifferentiating muscle fibers illustrated in Fig. 1 (*Mus*). Typical blastema cells with relatively little cytoplasm appear on the right (*Bl*). A series of nuclei (*N*) belonging to muscle units that have almost completed the transformation to blastema cells appears on the left. During the transformation, nuclei change from the elongate shape typical of differentiated muscle (*N''*) to a more oval shape (*N'*) and finally become rounded (*N*). Nucleoli (*n*) become prominent. Myofibrils have partly disappeared from the cytoplasm of the cell the nucleus of which is labeled *N'* and are no longer detectable with the light microscope in the cytoplasm of the cells labeled *N*. Magnification: × 500.

unstained sections of regenerating limbs fixed in osmic acid. Although the process of muscle dedifferentiation, as seen with the light microscope, has been illustrated with camera lucida drawings by Thornton (1938a), it seems worth while to include these photomicrographs for general orientation and to illustrate the excellent fixation achieved with osmic acid. Features which were not seen previously, such as myofibrils in mononucleate cells derived from dedifferentiating muscle, can be detected with the light microscope in methacrylate sections of this material.

Figure 1 is a longitudinal section of a 5-day regenerate. The plane of sectioning is lateral to the cartilaginous humerus. The amputation level is indicated by the cut end (at the arrow) of the fibrous basement membrane that underlies the stump epithelium. This basement membrane does not extend under the thickened tip epithelium of the regenerating limb. The blastema is loosely organized initially, but later the cells composing it come to lie closer together as they accumulate in greater number under the apical epithelium. In the particular region included in the section illustrated in Fig. 1, the blastema (*Bl*) is composed of dedifferentiated cells derived largely from the skeletal muscle and fibroblasts previously present in the area. A multinucleate giant cell (*G*) can be seen associated with a remnant of cartilage matrix which is being resorbed. More proximally, a portion of a nerve (*Ne*) and the tips of a bundle of dedifferentiating muscle fibers (*Mus*) appear. The ends of the dedifferentiating muscle fibers are shown at higher magnification in Fig. 2.

In Fig. 2, typical blastema cells with round nuclei and relatively little cytoplasm appear on the right (*Bl*). An intact muscle fiber with an elongated nucleus (*N″*) and more abundant cytoplasm can be seen on the left. During the transformation of muscle cell to blastema cell, myofibrils are lost from the cytoplasm and the nucleus assumes a more oval shape (*N′*). Further rounding of nuclei then occurs, as is well illustrated by the row of nuclei (*N*) in the dedifferentiating muscle fiber near the upper portion of the photomicrograph. During this process nucleoli (*n*) become more prominent and later the cytoplasm will become basophilic in stained histological sections. A considerable amount of cytoplasm disappears. This apparently occurs by the pinching off of anucleate fragments of muscle cytoplasm which are subsequently lysed (Thornton, 1938a). Electron microscopy demonstrates that new cell boundaries have already formed between muscle nuclei such as those illustrated at N in Fig. 2, dividing them into mononucleate cells which then migrate away from the intact muscle and become blastema cells.

Figure 3 is a phase contrast photomicrograph illustrating the presence of striations (at the arrows) in mononucleate cells which have separated from the mass of muscle (Mus) in the upper right. Observations such as these, together with the electron microscopic observations of muscle fiber fragmentation, leave little doubt that mononucleate cells in the vicinity of intact muscle arise from dedifferentiating muscle, as Thornton originally claimed. By the time these cells leave the vicinity of the muscle to migrate distad, all trace of myofibrillar material has disappeared and they can no longer be distinguished from blastema cells of other origin.

Figure 4 is a low magnification electron micrograph of an early blastema cell similar to those illustrated in Fig. 2 (Bl). This cell appears to have been migrating from the main muscle mass toward the tip of the regenerate (arrow) and has cytoplasmic processes extending behind it such as were described by Thornton. The mesenchyma-like blastema cells, although diverse in origin, are similar in their fine structural features. Their characteristics have been described in detail in a previous report (Hay, 1958) and will only briefly be summarized here. The nuclei are diffusely granular and the chromatin pattern is less pronounced than in differentiated muscle cells. The cytoplasm may be relatively scanty, as illustrated in Fig. 4, or somewhat more abundant, as illustrated in our earlier report. The most notable characteristic of the cytoplasm is the presence of numerous small granules ~ 100 Å in diameter of the type shown by Palade and Siekevitz (1956) to be composed of ribonucleoprotein. These granules (inset, Fig. 4) are predominantly free in the cytoplasm, although a few are attached to the membranes of the endoplasmic reticulum. They presumably account for the diffuse cytoplasmic basophilia of the blastema cell. Mitochondria (m, Fig. 4) have a light matrix and irregular cristae. The endoplasmic reticulum is considerably reduced in amount as compared with the differentiated muscle or cartilage cell and occurs mainly in the form of vesicles, but occasional cisternae are also observed (er, Fig. 4). The layer of amorphous material investing differentiated muscle cells, commonly referred to as a "basement membrane" (bm, Fig. 11), disappears during dedifferentiation and the plasma membranes of adjacent blastema cells may be in close

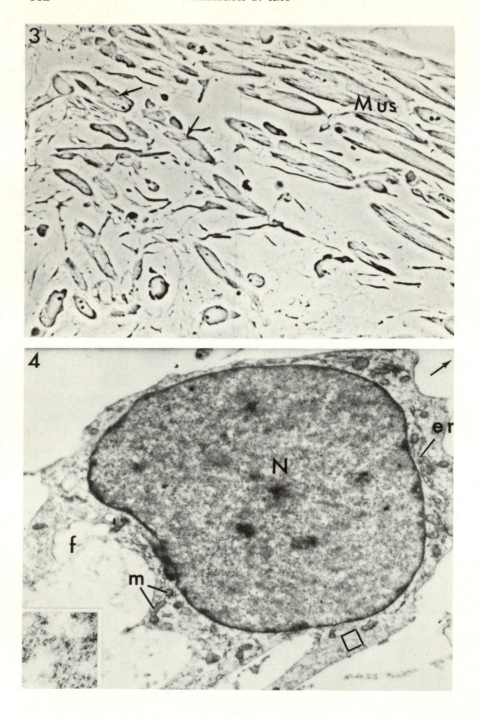

apposition later in regeneration. The extracellular ground substance contains scattered collagenous fibrils (f, Fig. 4) which are not striated and are not organized into fibers.

Separation of the Muscle Fiber into Mononucleated Cells

Figure 5 is an electron micrograph illustrating an area of dedifferentiating muscle similar to that pictured in the upper portion of Fig. 3. The plane of sectioning coincides with the long axis of the muscle. A muscle unit containing a single nucleus appears at N''' and below it, a fragment with part of its nucleus (N'') showing. Each fragment contains myofibrils in various stages of disintegration (my). A variable amount of nuclear material will be included in any given section. It is possible, however, to ascertain from serial sections and from light microscopy that many of the muscle fragments do not actually contain a nucleus. A muscle fragment which apparently has no nucleus appears in the center of Fig. 5 (F). Only the nucleated units complete the transformation to blastema cell and enter the regenerate. In addition to longitudinal fragmentation of muscle fibers illustrated here, we have observed portions of cytoplasm apparently in the process of being pinched off the fiber laterally.

The separation of the syncytial muscle fibers of the limb stump into mononucleate cells requires formation of new plasma membranes between fragments of muscle. We have the impression from the study of osmic acid-fixed material that the delineation of new cell boundaries

FIG. 3. Phase contrast photomicrograph of an unstained plastic section of a regenerating limb fixed in osmic acid 4 days after amputation. The distal tip of the limb is to the left and a bundle of dissociating muscle fibers appears in the upper right (Mus). Below and to the left of these fibers are two mononucleate cells which have separated from the main mass of muscle and which still contain striations in their cytoplasm (at the arrows). Magnification: × 450.

FIG. 4. Low magnification electron micrograph (× 7000) of an early blastema cell in a regenerating limb 6 days after amputation. The tip of the limb is toward the upper right in the direction indicated by the small arrow. This cell is similar to those appearing on the right of Fig. 2. The nucleus (N) is diffusely granular and the cytoplasm is rather sparse. The cytoplasm contains a small amount of endoplasmic reticulum (er) and scattered mitochondria (m). Its dense appearance is due to numerous free ribonucleoprotein granules which are shown at higher magnification (× 44,800) in the inset. The area from which the inset was taken is indicated by the small square in the lower right of the picture. A few nonstriated collagenous fibrils (f) appear in the extracellular ground substance.

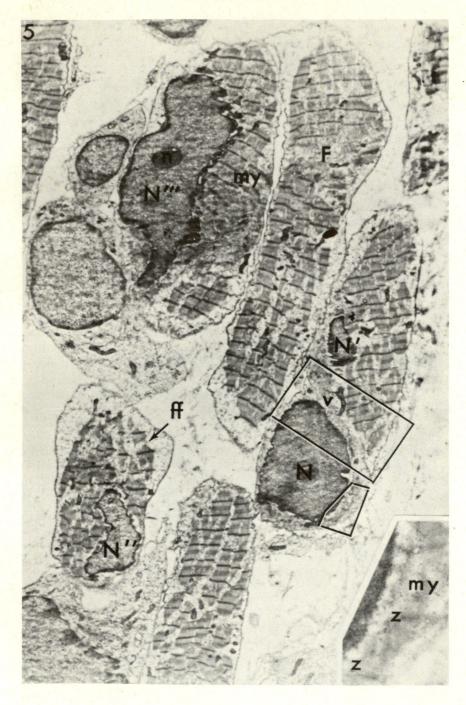

begins while the muscle fiber is still relatively intact, that is, while it still contains myofibrils and still appears to be part of the main muscle mass. We have observed with the electron microscope a series of changes in areas of cell separation which we consider to be stages in the production of new plasma membrane. In regions of future cell boundaries, a dense material can be seen to extend across the fiber (v, Fig. 5). Higher magnification of an area such as that enclosed in the rectangle in Fig. 5 reveals that this dense material consists of numerous small vesicles clustered together to form an irregular line across the cell (arrows, Fig. 6). In Fig. 6 a portion of the nucleus (N) shown in Fig. 5 appears to the right and cytoplasm containing myofibrils to the left (my). We believe that the small membrane-bounded vesicles extending across the fiber (arrows) are the formative material for the new cell membranes that would ultimately demarcate the muscle fragment on the left from that on the right in Fig. 6. The vesicles are 300–700 Å in diameter and the membrane bounding them is of approximately the same thickness (~ 70 Å) as the cell membrane (cm, Fig. 6). They appear to contain a material denser than the surrounding ground substance of the cytoplasm (arrow, inset Fig. 6). The similarity of these vesicles to the "pinocytosis vesicles" normally found next to the plasma membrane of muscle cells (pv, Fig. 6) does, however, raise the question of whether or not the row of vesicles illustrated in Fig. 6 is not merely a row of "pinocytosis vesicles" along obliquely cut plasma membranes. We do not believe this to be the case because similar rows of small vesicles are observed consistently

FIG. 5. Low magnification electron micrograph ($\times 2800$) of a longitudinal section through an area of dedifferentiating muscle in a regenerating limb 6 days after amputation. The micrograph is oriented in such a way that the distal tip of the limb would be below the picture. A muscle unit containing one nucleus (N''') and prominent nucleolus (n) appears near the top of the picture, and below it, a mononucleate unit with only part of its nucleus (N'') showing. The muscle fragment in the center (F) apparently has no nucleus. Myofibrils (my) are losing their longitudinal orientation and are fragmenting (ff). A dense line (v) demarcates the future cell boundary which will separate the two muscle units whose nuclei are labeled N and N'. The myofibrils in the fragment whose nucleus is labeled N have partly disintegrated and are less apparent at this magnification. Part of the cytoplasm of this fragment is shown at higher magnification ($\times 19,600$) in the inset. Two Z bands (z) of a disintegrating myofibril (my) are labeled. The region enclosed in the rectangle is shown at higher magnification in Fig. 6.

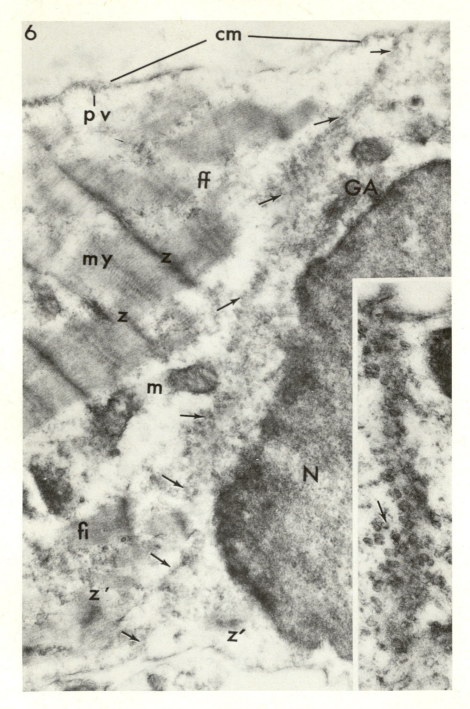

in areas of future cell boundaries in material which is sectioned longitudinally and are not observed to the same extent along the lateral surfaces of the muscle fibers. Serial sections lend further support to the idea that these vesicles appear before the definitive cell membranes. Moreover, instances in which the accumulation of vesicles does not yet extend completely across the cell may be found.

Figure 7 illustrates what we interpret to be a more advanced stage in the formation of new plasma membranes between muscle fragments. In this figure, two portions (F, F') of a muscle fiber are separated by a row of vesicles (v) centrally, whereas laterally new cell membranes have formed and extracellular space can be seen between the fragments. This space (s') is continuous with the surrounding extracellular space (s). In the area of junction between vesicles and plasma membranes (see rectangle, Fig. 7) the vesicles appear to be fusing with the new membrane (arrow, Fig. 8). In the lower left corner of Fig. 7, vesicles have apparently fused to produce a vacuole in the interior of the cell, and it seems likely that the space (s'') contained in this vacuole is also future extracellular space. We have the impression that the fusion of vesicles to produce new plasma membranes usually begins on the surface of the cell and proceeds inward. Figure 10 illustrates a final stage in the process of cell separa-

FIG. 6. Higher magnification (× 28,000) electron micrograph of the region enclosed in the rectangle in Fig. 5. A part of the nucleus labeled N in Fig. 5 appears on the right (N). The dense line (v) in Fig. 5 is seen at higher magnification to consist of a row of vesicles extending across the fiber (arrows). These vesicles are considered to be the formative material for the new cell membranes which will demarcate the muscle unit on the left from that on the right. Myofibrils (my) in the cell on the left are fragmenting (ff). The Z band (z) will become smaller and more condensed (z') and utimately disappear. Myofilaments (fi) will also completely disappear.

Several short lamellae and associated vesicles which resemble the Golgi apparatus (GA) appear next to the nucleus in the muscle unit on the right. The vesicles associated with the Golgi cannot be distinguished, except by their location, from the vesicles that demarcate the future cell boundaries (at the arrows). "Pinocytosis vesicles" (pv) of the endoplasmic reticulum are closely associated with the cell membrane (cm) of the fiber. These vesicles (pv) are also similar in size to the vesicles at the arrows. A mitochondrion is labeled m.

The inset shows at higher magnification (× 44,800) part of a row of vesicles extending across another dissociating muscle fiber. It can be seen that the vesicles are bounded by a membrane and contain a material (arrow) which is denser than the surrounding ground substance.

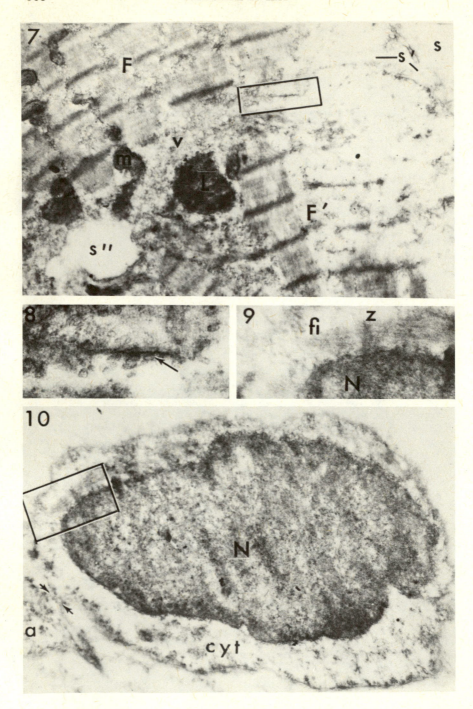

tion. Only a small tag of tissue (arrows) is left separating the mono-nucleated cell pictured from its neighbor (*a*).

Dedifferentiation of Cytoplasmic Organelles

Differentiated amphibian skeletal muscle is characterized by a highly organized, smooth-surfaced endoplasmic reticulum consisting of interconnected longitudinal and horizontal tubular elements with specialized differentiations in the region of the Z band (Porter and Palade, 1957). During dedifferentiation the endoplasmic reticulum breaks up into vesicles and all vestiges of its previous high degree of organization are lost. The fate of the vesicles is unknown. Some may be utilized in plasma membrane formation (Fig. 6). Some of the reticulum does persist in the blastema cell in the form of vesicles and occasional cisternae, but it is reduced considerably in amount as compared with the differentiated muscle cell.

Figure 11 illustrates a portion of the cytoplasm of a muscle cell which has almost completed its dedifferentiation. The acquisition of ribonucleoprotein granules (*g*) by the cytoplasm is a conspicuous

---

Fɪɢ. 7. Electron micrograph (× 16,800) showing an intermediate stage in the formation of cell boundaries between two fragments (*F* and *F′*) of a dissociating muscle fiber. Extracellular space appears at *s* and is continuous with the space (*s′*) between the lateral portions of the two separating muscle units. A row of vesicles (*v*) extends across the central portion of the fiber. These vesicles are apparently fusing in the region of the rectangle to produce the new cell membranes of the two fragments. In the left-hand portion of the micrograph, vesicles have apparently fused to produce a larger vacuole containing a space (*s″*) that would probably become continuous with the extracellular space later. A dense lipid droplet (*L*) and mitochondrion (*m*) in the center of the fiber are labeled. 6 days.

Fɪɢ. 8. Higher magnification electron micrograph (× 56,000) of the area enclosed in the rectangle in Fig. 7. One of the vesicles which we interpret to be fusing with other vesicular components to produce cell membranes is labeled by an arrow.

Fɪɢ. 9. Electron micrograph (× 33,600) of the portion of the dedifferentiating muscle cell indicated by the rectangle in Fig. 10. Part of a disintegrating myofibril is shown. The myofilaments (*fi*) are disappearing, but the Z band (*z*) is still relatively intact. Part of the nucleus appears at *N*.

Fɪɢ. 10. Electron micrograph (× 16,800) of a mononucleate cell which has almost completely dissociated from a dedifferentiating muscle fiber. The nucleus is labeled *N*. The cytoplasm (*cyt*) is relatively scanty and contains a few disintegrating myofibrils (rectangle and Fig. 9). This cell is still attached by a small tag of cytoplasm (arrows) to the fiber from which it originated (*a*). 6 days.

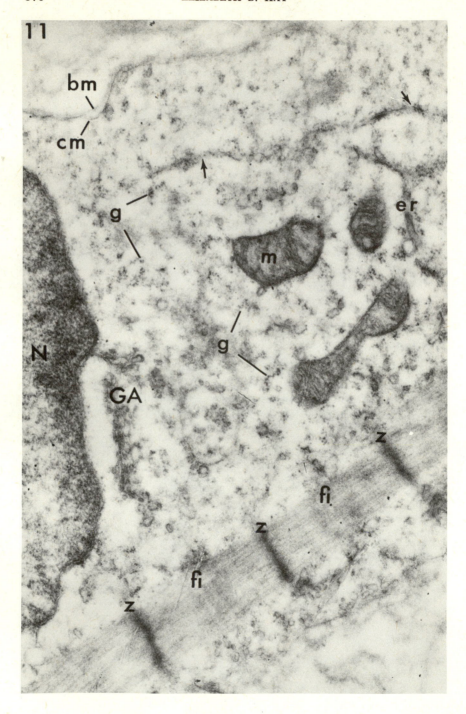

and important feature at this stage. Only one myofibril (*fi*) remains in the cytoplasm and the endoplasmic reticulum (*er*) is disorganized. Some of the ribonucleoprotein granules appear to be attaching to the membrane of the endoplasmic reticulum (arrows), but most of them (*g*) are free in the cytoplasm, the condition which is typical of undifferentiated cells. An appreciation of the change in the character of the cytoplasm during dedifferentiation can be obtained by comparing Fig. 11 with Fig. 6. The cell pictured in Fig. 11 would subsequently lose its remaining myofibrils and investing "basement membrane" (*bm*) and would then no longer be distinguishable as a cell of muscle origin.

The Golgi apparatus is not highly differentiated in muscle. In Fig. 11, it can be identified next to the nucleus (*N*) as a parallel array of several short lamellae surrounded by small vesicles (*GA*). The lamellae are about 300 Å in diameter and are composed of a pair of membranes enclosing a space with a content which is denser than the surrounding ground substance of the cytoplasm. The membrane-bounded vesicles associated with the lamellae enclose a cavity containing a material of similar electron density. The Golgi apparatus, like the endoplasmic reticulum, breaks up into vesicles during dedifferentiation.

Figure 12 serves to emphasize the similarities between Golgi apparatus and the smooth-surfaced endoplasmic reticulum of muscle cells. Several short lamellae with associated vesicles which resemble the Golgi apparatus can be seen at *L*, and tubules and vesicles which

---

FIG. 11. Electron micrograph (× 28,000) of a portion of the cytoplasm of a dedifferentiating muscle fiber to show the acquisition of free cytoplasmic ribonucleoprotein granules (*g*) during the transformation of muscle to blastema cells. A part of the nucleus (*N*) appears on the left and next to it an aggregate of short lamellae and vesicles thought to be the Golgi apparatus (*GA*). The endoplasmic reticulum has been reduced to a few vesicles and short cisternae (*er*), the membranes of which are becoming covered with ribonucleoprotein granules (arrows). Most of the ganules are free in the cytoplasm and may be arranged in small clusters and short rows. Mitochondria (one of which is labeled, *m*) have inner cristae and a moderately dense matrix.

One myofibril still remains in the cytoplasm of this mononucleate muscle cell. The Z bands (*z*) are intact and the myofilaments (*fi*) have not yet fragmented. A and I band striations are not apparent. The cell membrane (*cm*) is still surrounded by the layer of amorphous material termed a "basement membrane" (*bm*). 6 days.

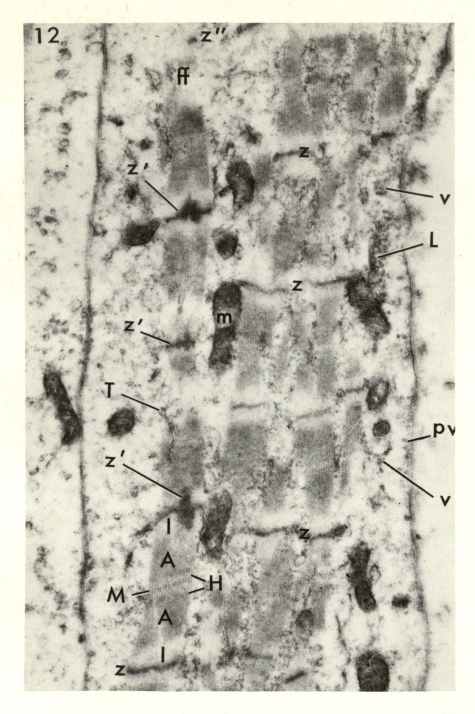

we would classify as endoplasmic reticulum appear at T, v, and pv. The elements of the reticulum and the Golgi apparatus are commonly identified by the organized form in which they usually occur, the parallel array of closely packed, smooth-surfaced lamellae with associated vesicles serving to distinguish the Golgi. When this distinctive organization is lost, as in dedifferentiation, the identification is no longer clear. Thus it is impossible to make a clear distinction between the fates of these two organelles in muscle dedifferentiation. For example, the vesicles (arrows) extending across the fiber in Fig. 6 could have arisen from the nearby Golgi apparatus (GA). In Fig. 7, however, there is no associated Golgi apparatus and many of the vesicles pictured extending across that fiber may have arisen from the fragmenting endoplasmic reticulum, which they so closely resemble.

The appearance of mitochondria does not change markedly during dedifferentiation. Mitochondria in muscle cells (m, Figs. 11 and 12) are bounded by an outer pair of closely apposed membranes and contain cristae (Palade, 1953) surrounded by a dense matrix. The matrix may lose some of its original density and the cristae become somewhat more irregular by the time dedifferentiation is completed (Hay, 1958).

Disappearance of Myofibrils

The intact sarcomere of the myofibril extends from Z band to Z band and contains half of an I band at each end (Fig. 12, lower left). The A band is transected in the middle by a lighter H band, which is, in turn, transected by the M band. The distinction between A and I band depends on the state of contraction of the muscle and is not

---

FIG. 12. Electron micrograph ($\times$ 28,000) of a portion of the cytoplasm of a dedifferentiating muscle fiber showing the agglutination of the Z substance into an amorphous material (z') during fragmentation of myofibrils. An intact sarcomere with A, M, H, I, and Z bands appears in the lower part of the picture. In the upper portion of the picture, a sarcomere has fragmented in the region of the A and I band (ff). Part of the Z band (z'') is still apparent.

Tubules of endoplasmic reticulum (T) are associated with some of the myofibrils. Vesicles (v) derived from the endoplasmic reticulum are labeled in the right-hand side of the picture. Several short lamellae (L) appear which resemble lamellae of the Golgi apparatus. Other vesicles appear near the cell membrane and probably represent the "pinocytosis vesicles" characteristic of muscle (pv). A mitochondrion is labeled m. 6 days.

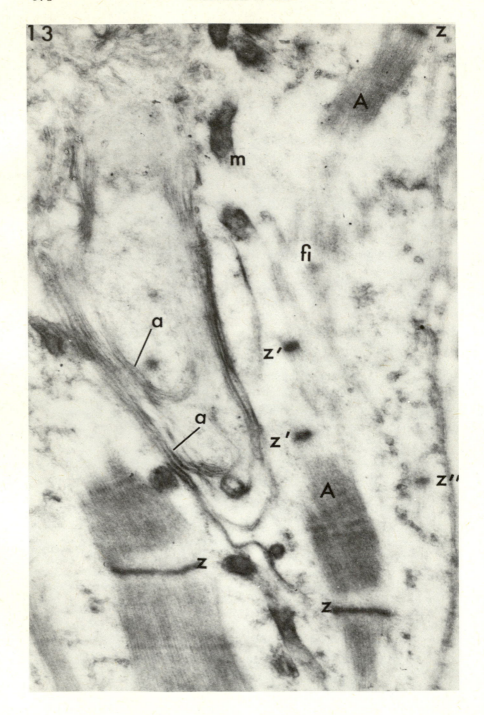

apparent when the myofibril is fully contracted (*my*, Fig. 6). In differentiated *Amblystoma* skeletal muscle, myofibrils are arranged parallel to one another with the Z bands of adjacent myofibrils closely aligned (Porter and Palade, 1957).

In the early stages of dedifferentiation, the myofibrils lose this parallel, longitudinal arrangement and begin to fragment (Fig. 5). The fragmentation of the myofibrils is random with respect to location of the I, A, H, and M bands. The sarcomeres never fragment along an intact Z band. It is possible to observe in some preparations of dedifferentiating muscle an agglutination of Z material into an amorphous mass prior to the disappearance of the myofilaments (*z'*, Fig. 12). We believe this to be the first step in the degeneration of the unknown osmiophilic material composing the Z band.

The disrupted myofilaments and Z bands disappear in some unknown manner into the ground substance to the cytoplasm (*z'*, *fi*, Fig. 6). In some cases, the myofibril may partly disintegrate without losing its original orientation, as for example, in the region *A-A* in Fig. 13. In such instances, it is quite clear that the Z material is a distinct extrafilamentous component of the myofibril. The Z band remains distinguishable (*z'*, Fig. 13) as a contracted dense material at a time when the myofilaments have almost completely disappeared (*fi*, Fig. 13). Some of the persisting myofilaments are apparently still attached to the Z material. Less dense osmiophilic regions such as *z''* in Figs. 12 and 13 probably represent the final stage in disappearance of Z bands.

The loss of myofibrils occurs at the same time as organelle dedifferentiation, acquisition of cytoplasmic basophilia, rounding of nuclei, and formation of new cell boundaries. The degree of overlap of these processes varies from fiber to fiber. In Fig. 11, an intact myofibril

---

FIG. 13. Electron micrograph ($\times$ 28,000) of a portion of the cytoplasm of a dedifferentiating muscle fragment showing condensation of the Z band into an amorphous mass (*z'*). The A bands labeled are presumed to have been part of a single myofibril (*A-A*) which has broken up in the region indicated at *z'*. A few disorganized myofilaments (*fi*) can also be seen in this region of the cytoplasm. Later, these filaments and the Z substance will disappear. A dense mass which probably represents the last stage in the disappearance of a Z band can be seen on the right (*z''*). A mitochondrion is labeled *m*. Several whorls of membranes which probably belong to a myelin figure appear at *a*. Such myelin figures are rare in dedifferentiating muscle and will be described in a separate report. 6 days.

with myofilaments (fi) joined at Z bands (z) persists in the cytoplasm of a muscle cell which is almost completely dedifferentiated. Figure 10 illustrates a mononucleate cell which has almost completely lost its myofibrils (fi, z, Fig. 9) before acquiring significant cytoplasmic ribonucleoprotein. This cell is still attached to its neighbor by a small strand of cytoplasm (arrows). On the other hand, the mononucleate cell (N'''') pictured in Fig. 5 has apparently completely separated from the original fiber but still contains a considerable amount of myofibrillar material. These minor variations permit us to establish with reasonable certainty the origin of mononucleate blastema cells from muscle fibers. If all the myofibrils were lost from the distal cytoplasm before fragmentation of the syncytial muscle fiber, as is suggested by Thornton's (1938a) observations, it would be more difficult to prove that mononucleate cells in the vicinity of muscle do arise from the muscle. Finally, it should be emphasized again that myofibrils do disappear completely from the cytoplasm of these mononucleate cells when they enter the blastema. The end result of the processes of dedifferentiation which have been described above is an undifferentiated-appearing cell which cannot be distinguished from blastema cells of other origin.

DISCUSSION

In spite of the careful description by Thornton in 1938 of the transformation of muscle into blastema cells, some investigators have accepted later work by Manner (1953) which suggested that the amphibian limb blastema originates entirely from fibroblasts. Manner dismisses the changes which occur in the muscle of the amputated *Triturus* limb as degenerative. The present study has confirmed Thornton's original description of muscle dedifferentiation. The cells originating from dedifferentiating muscle fibers are certainly not degenerating. Nuclei are acquiring the characteristics of nuclei of actively growing cells and the cytoplasm is acquiring the basophilia and abundant free ribonucleoprotein granules typical of undifferentiated cells. Examination of Fig. 1 of this paper is sufficient to dispel the notion that there is extensive cellular degeneration at the time of blastema formation. Although some cells are injured by amputation, they are phagocytosed in the first 2–3 days after amputation and there is no significant cell death during the major phase of blastema formation in muscle or other stump tissues. The dedifferentiating muscle fibers illustrated here 4–6 days after amputation are predominantly shoulder muscles which were not even cut at amputation but have become detached from the humerus in the manner described by Thornton. The observation, with the electron microscope, of a continuous series of changes in the transformation of the syncytial muscle fibers of the stump into mononucleate blastema cells leaves little doubt as to the validity of Thornton's original observations on the dedifferentiation of muscle in *Amblystoma* limb regeneration.

Manner (1953) also erroneously concluded that there is no mitosis in the differentiated stump tissues of the amputated limb. Chalkley's subsequent painstaking analysis (Chalkley, 1954) has established beyond doubt the fact that mitosis does indeed occur in all the differentiated stump tissues as they undergo dedifferentiation. Chalkley calculated that in the *Triturus* limb regenerate, dedifferentiating muscle accounted for 8% of the blastema cells, and he points out that this figure might be too small because the onset of mitotic activity is delayed in dedifferentiating muscle until cell dissociation is completed. It would therefore be difficult to recognize by mitotic activity all the cells originating from muscle. Our observations substantiate this suggestion and lead us to believe that in older *Amblystoma* limbs perhaps as much as a quarter of the blastema originates from dedifferentiating muscle.

Actively growing, undifferentiated cells in general are characterized by a cytoplasm rich in free ribonucleoprotein granules (Porter, 1954; Palade, 1955; Howatson and Ham, 1955; Slautterback and Fawcett, 1959). On the other hand, the endoplasmic reticulum of such cells is usually quite poorly developed. The diffuse basophilia imparted to the cytoplasm by numerous free ribonucleoprotein granules is to be contrasted with the more localized basophilia seen in such protein-secreting cells as the pancreatic acinar cell. In the latter, basophilia is attributable to ribonucleoprotein granules that are, for the most part, attached to the surface of the endoplasmic reticulum which is organized into parallel cisternae (see review by Palade, 1956). The authors quoted above have generalized that the numerous free ribonucleoprotein granules in the cytoplasm of rapidly dividing cells are related to the synthesis of cytoplasmic protein for growth. Bodemer and Everett (1959) have presented autoradiographic evidence for the beginning of protein synthesis in the cytoplasm of dedifferentiating cells in amphibian limb regeneration. The acquisition of numerous free cytoplasmic granules by blastema cells observed here, could be interpreted as further evidence that these dedifferentiating cells are beginning the synthesis of intracellular protein for cell growth. The rounded nucleus and prominent nucleoli in blastema cells are other cytological features which accompany cytoplasmic basophilia and active protein synthesis (see Brachet, 1950).

During the dedifferentiation of cartilage, as well as muscle, endoplasmic reticulum breaks up into vesicles and decreases in amount. This disorganization could be spoken of as dedifferentiation of the organelle. The cartilage cell actively producing cartilage matrix has a highly elaborate rough-surfaced endoplasmic reticulum which resembles the ergastoplasm of the pancreatic acinar cell. The function of this specialized reticulum most likely has to do with production of the new cartilage matrix (Hay, 1958). The endoplasmic reticulum of the differentiated muscle cell, on the other hand, is smooth surfaced. There are no ribonucleoprotein granules attached to its membrane. Its structure is highly organized, with longitudinal tubules running the length of sarcomeres and terminating at Z bands in vacuoles which have smaller tubules interposed between them. Horizontal continuity is provided by connecting tubules in the region of the A band. This elaborate reticulum is believed to play a role in the conduction of the impulse from the cell surface to the myofibrils (Bennett, 1956; Porter and Palade, 1957). It seems likely that with the cell dedifferentiation that occurs in the transformation of muscle and cartilage to blastema cells, such specialized functions of the reticulum disappear. It is tempting to speculate that the dispersed reticulum in the undifferentiated cell plays some fundamental role in the intracellular circulation of metabolites and in the turnover of cytoplasmic membrane, whereas the more highly differentiated reticulum seen in muscle and cartilage cells has acquired additional functions concerned with conduction and secretion, respectively.

The Golgi apparatus is never very highly developed in muscle; this is not surprising in view of the probable role of the Golgi in secretion (see review by Palay, 1958). In the transformation of muscle cell to blastema cell, the Golgi apparatus and endoplasmic reticulum behave as similar intracellular membrane systems. Both become disorganized during dedifferentiation, in contrast to the mitochondria, for example, which are relatively unchanged structurally in the dedifferentiated cell.

It seems likely that some of the vesicles derived from the endoplasmic reticulum, and possibly also from the Golgi apparatus, enter into the formation of new cell membranes during the dissociation of syncytial fibers into individual cells. Clusters of vesicles have been observed extending in a line across the fiber in regions of future separation of cells. Our observations indicate that these vesicles then fuse to form the definitive plasma membranes. Such a fusion of vesicles to produce opposing membranes would mean that their contents now become continuous with the contents of the extracellular space. This would be entirely in keeping with the current concept that the interior of the endoplasmic reticulum does in some instances represent extracellular space from a functional viewpoint. It is of considerable interest in this regard that in the formation of platelets from the megakaryocyte, a string of vesicles delineating the new platelet appears first and then the vesicles apparently fuse to form the plasma membranes separating platelet from the megakaryocyte (Yamada, 1957). The formation of new plasma membrane in other cases of cell separation, such as in cell division, has not yet been clarified. In dividing cells of the plant, *Allium cepa*, development of the new cell wall is preceded by the appearance of vesicles (Porter and Caulfield, 1958). The formation of the cell membrane in these plants may, however, involve different mechanisms than those discussed here.

It is apparent that in the dedifferentiating muscle, there is considerable variety in membrane behavior. Specialized membrane systems of the cytoplasm, such as the highly organized endoplasmic reticulum, are fragmenting while vesicles are fusing to form new plasma membranes. We know nothing of the conditions influencing these processes and it is to be hoped that studies of membrane formation in isolated cells, such as those by Fawcett and Ito (1958) of spermatids and spermatocytes *in vitro*, may provide us with a solution to this problem. In the dedifferentiating muscle cell there are simultaneous events, such as the breaking up of myofibrils, which might influence the structure of the endoplasmic reticulum. We have the impression, however, that fragmentation of the reticulum and disappearance of myofibrils are not necessarily correlated events. The sarcoplasmic reticulum is sensitive to changes in the intracellular metabolism. It breaks up into vesicles in vitamin E deficiency (Rumery and Hampton, 1959) and in dystrophic muscle (van Breeman, 1957; Ross *et al*, 1958).

73

The changes observed in dedifferentiating muscle could represent a response to changes in intracellular metabolism induced by the stimulus of the wound conditions.

The disappearance of the myofibrils from the cytoplasm usually begins with fragmentation of sarcomeres in the region of the A or I band. The Z band agglutinates into a dense, amorphous mass and persists with a few broken myofilaments attached to it. Fragmentation never occurs along an intact Z band. These observations are compatible with the concept that the Z substance acts as an extrafilamentous adhesive bonding together the myofilaments. Fawcett and Selby (1958) postulated that the Z substance may be the same as the dense cytoplasmic components of the desmosome, terminal bar, and intercalated disk. Pease and Baker (1949) noted resistance to fracture at Z bands in electron micrographs of specimens disrupted by shrinkage. It is of interest in this regard that sarcomeres may be isolated intact by trypsin digestion of the Z band (Ashley *et al*, 1951) and that when isolated sarcomeres occur in nature, as in the Purkinje fiber (Muir, 1957b), no Z band material is associated with them. On the other hand, in some types of muscle degeneration the Z band may disappear first without the myofibril breaking up (Moore *et al*, 1956).

The ultimate fate of the Z substance and myofilaments after they disappear into the cytoplasm is unknown. We have not been able to distinguish between primary and secondary myofilaments (Huxley, 1957) during the breaking up and disappearance of myofilaments. Ferris (1959), however, has reported the appearance of two types of myofilaments in differentiating muscle. It would be of interest to compare the disappearance of the myofibril with stages in its formation, but electron microscopic studies of differentiating muscle (e.g., van Breeman, 1952; Hibbs, 1956; Muir, 1957a; Ruska and Edwards, 1957; Gilev, 1958; Ferris, 1959) have not yet given a complete enough picture of the development of myofibrils.

The major cytological changes occurring during the process of cartilage and muscle dedifferentiation in the amputated salamander limb may be summarized as follows: (1) disappearance of differentiated intracellular and extracellular products; (2) dedifferentiation of cytoplasmic organelles; (3) acquisition of cytoplasmic ribonucleoprotein granules and other features characteristic of active cytoplasmic protein synthesis, such as enlargement of nucleoli and rounding-up and enlargement of nuclei; (4) loss of cytoplasm by fragmentation and dissociation of syncytial fibers into mononucleate cells in the case of muscle, with an increased nucleocytoplasmic ratio resulting which is characteristic of the embryonic cell. These remarkable changes lead to the production of a group of blastema cells which are completely undifferentiated in appearance and similar in their fine structure. In mammalian muscle regeneration where tissue replacement alone is involved, there may be some fragmentation of muscle fibers, sarcoplasmic budding, and possibly formation of mononucleate cells from the injured muscle (see reviews by Godman, 1958; Lash *et al.*, 1957). Sarcoplasmic budding and formation of mononucleate cells from muscle in tissue culture has been observed (Lewis and Lewis, 1917) and terminal sarcoplasmic buds have been described as playing a role in muscle regeneration in amphibian tails (Naville, 1922; and others). In amphibian limb regeneration, however, terminal sarcoplasmic budding plays a minor role in muscle regeneration (Thornton, 1938a) and the major part of the muscle dedifferentiates, dissociating by fragmentation rather than budding. The new muscle which is formed later in the regenerating blastema differentiates from embryonic-appearing blastema cells. The regenerating salamander limb is unique in that all the various types of tissues in the distal region of the stump transform into undifferentiated cells which accumulate under the tip epithelium to form a blastema of closely aggregated cells. It seems reasonable to conclude that part of the secret of regeneration of a whole appendage may lie in the remarkable ability of the formed elements of the limb stump to acquire the embryonic characteristics which allow the cells to proliferate rapidly to form a large mass of undifferentiated cellular material upon which the new limb pattern may be imposed without restriction.

SUMMARY

Regenerating limbs of *Amblystoma punctatum* larvae were fixed in osmic acid, sectioned, and studied with the electron microscope. The present report describes the changes which occur in the syncytial muscle fibers of the stump as they transform into undifferentiated, mononucleate blastema cells during the 2–6-day period after amputation. Only the nucleated cell units derived from the dissociating muscle fiber complete the process of dedifferentiation. Anucleate fragments apparently are lysed. Nuclei become rounded in shape, the chromatin more diffusely granular, and nucleoli more prominent. The nucleocytoplasmic ratio is increased by loss of cytoplasm by fragmentation during dedifferentiation.

Both the intracellular and extracellular products of cell differentiation are lost during dedifferentiation. Myofibrils fragment along the A and I bands and then disintegrate. The Z substance behaves as a distinct extrafilamentous component during disorganization of the myofibril, and its behavior is compatible with the concept that it is an adhesive material serving normally to bind sarcomeres together. Individual myofibrils, undetected heretofore, persist in the cytoplasm of some of the mononucleate cells for a short period after their separation from the muscle fiber. Thus there remains no doubt that mononucleate cells in the vicinity of intact stump muscle do arise from multinucleate muscle fibers. By the time the transformation to blastema cell is completed, all the myofibrils have disappeared from the cytoplasm.

The new plasma membranes separating muscle fragments appear to be formed by fusion of small vesicles which are probably derived largely from the endoplasmic reticulum. The endoplasmic reticulum, which is highly developed in differentiated muscle, breaks up into small vesicles in dedifferentiating muscle. The Golgi apparatus also becomes unorganized in appearance, whereas the mitochondria show relatively little change. One of the most striking changes in the cytoplasm is its acquisition of numerous free ribonucleoprotein granules, a feature characteristic of undifferentiated cells in general. These features together with the nuclear changes described above indicate that the dedifferentiating muscle cell is probably beginning active cytoplasmic protein synthesis. They dispel the idea that the muscle might be degenerating.

The end result of the process of muscle dedifferentiation is an undifferentiated-appearing cell which cannot be distinguished from the blastema cells derived from cartilage and other tissues of the stump. By virtue of its acquisition of such embryonic features as free cytoplasmic ribonucleoprotein granules and prominent nucleoli, as well as loss of differentiated structure, the dedifferentiated blastema cell seems to be remarkably well equipped for the ensuing rapid proliferation of a large cellular mass capable of redifferentiation into a complete limb.

REFERENCES

ASHLEY, C. A., PORTER, K. R., PHILPOTT, D. E., and HAAS, G. M. (1951). Observations by electron microscopy on contraction of skeletal myofibrils induced with adenosinetriphosphate. *J. Exptl. Med.* **94:** 9–22.

BENNETT, H. S. (1956). The sarcoplasmic reticulum of striped muscle. *J. 'Biophysic. Biochem. Cytol.* **2**, Suppl. 4, 171–174.

BLOOM, W. (1937). Cellular differentiation and tissue culture. *Physiol. Revs.* **17**, 589–617.

BODEMER, C. W. (1958). The development of nerve-induced supernumerary limbs in the adult newt, *Triturus viridescens. J. Morphol.* **102**, 555–582.

BODEMER, C. W., and EVERETT, N. B. (1959). Localization of newly synthesized proteins in regenerating newt limbs as determined by radioautographic localization of injected methionine-S^{35}. *Developmental Biol.* **1**, 327–342.

BRACHET, J. (1950). Synthesis, localization, and physiological role of the nucleic acids. "Chemical Embryology," pp. 189–250. Interscience, New York.

CHALKLEY, D. T. (1954). A quantitative histological analysis of forelimb regeneration in *Triturus viridescens. J. Morphol.* **94**, 21–70.

FAWCETT, D. W., and ITO, S. (1958). Observations on the cytoplasmic membranes of testicular cells, examined by phase contrast and electron microscopy. *J. Biophysic. Biochem. Cytol.* **4**, 135–142.

FAWCETT, D. W., and SELBY, C. C. (1958). Observations on the fine structure of the turtle atrium. *J. Biophysic. Biochem. Cytol.* **4**, 63–72.

FERRIS, W. (1959). Electron microscope observations of the histogenesis of striated muscle. *Anat. Record* **133**, 275.

GILEV, V. (1958). Studien über einige Elemente des Muskelgewebes in seiner Histogenesis und Regeneration. *4th Intern. Conf. Electron Microscopy, Berlin, 1958*, p. 57.

GODMAN, G. C. (1958). Cell transformation and differentiation in regenerating striated muscle. *In* "Frontiers in Cytology" (S. L. Palay, ed.), pp. 381–416. Yale Univ. Press, New Haven, Connecticut.

HAY, E. D. (1958). The fine structure of blastema cells and differentiating cartilage in regenerating limbs of *Amblystoma* larvae. *J. Biophysic. Biochem. Cytol.* **4**, 583–592.

HIBBS, R. G. (1956). Electron microscopy of developing cardiac muscle in chick embryos. *Am. J. Anat.* **99**, 17–52.

HOWATSON, A. F., and HAM, A. W. (1955). Electron microscope study of sections of two rat liver tumors. *Cancer Research* **15**, 62–69.

HUXLEY, H. E. (1957). The double array of filaments in cross-striated muscle. *J Biophysic. Biochem. Cytol.* **3**, 631–648.

LASH, J. W., HOLTZER, H., and SWIFT, H. (1957). Regeneration of mature skeletal muscle. *Anat. Record* **128**, 679–698.

LEWIS, W. H., and LEWIS, M. R. (1917). Behavior of cross striated muscle in tissue cultures. *Am. J. Anat.* **22**, 169–194.

MANNER, H. W. (1953). The origin of the blastema and of new tissues in regenerating forelimbs of adult *Triturus viridescens viridescens. J. Exptl. Zool.* **122**, 229–257.

MOORE, D. H., RUSKA, H., and COPENHAVER, W. M. (1956). Electron microscopic and histochemical observations of muscle degeneration after tourniquet. *J. Biophysic. Biochem. Cytol.* **2**, 755–764.

MUIR, A. R. (1957a). An electron microscopic study of the embryology of the intercalated disc in the heart of the rabbit. *J. Biophysic. Biochem. Cytol.* **3**, 193–202.

MUIR, A. R. (1957b). Observations of the fine structure of the Purkinje fibres in the ventricle of the sheep's heart. *J. Anat.* **91**, 251–258.

NAVILLE, A. (1922). Histogenèse et régénération du muscle chez les Anoures. *Arch. Biol.* **32**, 37–171.

PALADE, G. E. (1952). A study of fixation for electron microscopy. *J. Exptl. Med.* **95**, 285–298.

PALADE, G. E. (1953). An electron microscope study of the mitochondrial structure. *J. Histochem. and Cytochem.* **1**, 188–211.

PALADE, G. E. (1955). A small particulate component of the cytoplasm. *J. Biophysic. Biochem. Cytol.* **1**, 59–68.

PALADE, G. E. (1956). The endoplasmic reticulum. *J. Biophysic. Biochem. Cytol.* **2**, Suppl. **4**, 85–98.

PALADE, G. E., and SIEKEVITZ, P. (1956). Liver microsomes, an integrated morphological and biochemical study. *J. Biophysic. Biochem. Cytol.* **2**, 171–200.

PALAY, S. L. (1958). The morphology of secretion. *In* "Frontiers in Cytology" (S. L. Palay, ed.), pp. 305–342. Yale Univ. Press, New Haven, Connecticut.

PEASE, D. C., and BAKER, R. F. (1949). The fine structure of mammalian skeletal muscle. *Am. J. Anat.* **84**, 175–200.

PORTER, K. R. (1954). Electron microscopy of basophilic components of cytoplasm. *J. Histochem. and Cytochem.* **2**, 346–373.

PORTER, K. R., and BLUM, J. (1953). A study of microtomy for electron microscopy. *Anat. Record* **117**, 685–712.

PORTER, K. R., and CAULFIELD, J. B. (1958). The formation of the cell plate during cytokinesis in *Allium cepa*. *4th Intern. Conf. Electron Microscopy, Berlin, 1958*, p. 141.

PORTER, K. R., and PALADE, G. E. (1957). Studies on the endoplasmic reticulum. III. Its form and distribution in striated muscle cells. *J. Biophysic. Biochem. Cytol.* **3**, 269–300.

ROSS, M. H., PAPPAS, G. D., and HARMAN, P. J. (1958). A comparison of normal and dystrophic muscle in the mouse by means of electron microscopy. *Anat. Record* **130**, 365–366.

RUMERY, R. E., and HAMPTON, J. C. (1959). Microscopic and submicroscopic observations on skeletal muscle from vitamin E-deficient rats. *Anat. Record* **133**, 1–12.

RUSKA, H., and EDWARDS, G. A. (1957). A new cytoplasmic pattern in striated muscle fibers and its possible relation to growth. *Growth* **21**, 73–88.

SLAUTTERBACK, D. B., and FAWCETT, D. W. (1959). The development of the cnidoblasts of *Hydra*. An electron microscope study of cell differentiation. *J. Biophysic. Biochem. Cytol.* **5**, 441–452.

THORNTON, C. R. (1938a). The histogenesis of muscle in the regenerating forelimb of larval *Amblystoma punctatum*. *J. Morphol.* **62**, 17–47.

THORNTON, C. R. (1938b). The histogenesis of the regenerating forelimb of larval *Amblystoma* after exarticulation of the humerus. *J. Morphol.* **62**, 219–242.

THORNTON, C. R. (1942). Studies on the origin of the regeneration blastema in *Triturus viridescens*. *J. Exptl. Zool.* **89**, 375–390.

TOWLE, E. W. (1901). On muscle regeneration in the limbs of *Plethedon*. *Biol. Bull.* **2**, 289–299.

VAN BREEMEN, V. L. (1952). Myofibril development observed with the electron microscope. *Anat. Record* **113**, 179–196.

VAN BREEMEN, V. L. (1957). Endoplasmic reticulum and other structural details of dystrophic muscle. *Anat. Record* **127**, 486.

WEISS, P. (1949). Differential growth. *In* "The Chemistry and Physiology of Growth" (A. K. Parpart, ed.), pp. 135–186. Princeton Univ. Press, Princeton, New Jersey.

YAMADA, E. (1957). The fine structure of the megakaryocyte in the mouse spleen. *Acta Anat.* **29**, 267–290.

Copyright © 1968 by The Wistar Institute Press

Reprinted from the *J. Exptl. Zool.*, **167**, 49–77 (1968)

Stability of Chondrocyte Differentiation and Contribution of Muscle to Cartilage during Limb Regeneration in the Axolotl (*Siredon mexicanum*) [1]

4

TRYGVE P. STEEN [2]

Department of Biology, Yale University, New Haven, Connecticut

ABSTRACT   In order to determine whether chondrocytes are stable as to type during limb regeneration and whether any cells from limb muscle become chondrocytes in the regenerate, labeled cartilage and muscle were transplanted into limbs which were then allowed to regenerate. Transplanted cells were identified after redifferentiation in the regenerate by three kinds of label: H³-thymidine detected by radioautography; triploidy as judged by the presence of three nucleoli per nucleus; or a combination of the two.

The results of this investigation indicate that limb chondrocytes are intrinsically stable with respect to cell type, for grafts of both triploid and H³-thymidine labeled chondrocytes gave rise to morphologically dedifferentiated blastema cells which almost exclusively differentiated into chondrocytes. The small percentage of labeled non-cartilage cells observed could have been the result of metaplasia, although it is more likely that all or most of them arose as spontaneous polyploids or as contaminants of the original grafts. Isotope dilution in H³-thymidine labeled chondrocytes indicated that they had usually divided at least five times. Hence, the basis for the intrinsic stability of chondrocytes must be heritable.

Chondrocytes from the coracoid, scapula, and visceral arch cartilages also contributed cells to cartilage of limb regenerates, and the available evidence indicates that cells released from these cartilages are similarly stable as to type. These cartilages did not dedifferentiate and participate in limb regeneration to the same extent as did limb cartilage.

Results from transplanted limb muscle, labeled by all three methods used in this study, indicate that certain cells of muscle can become chondrocytes during limb regeneration. The exact cellular origin of these chondrocytes, however, is uncertain, for muscle is a mixture of cell types.

The study of cell differentiation during limb regeneration has fundamental implications for understanding both the process of regeneration and the properties of differentiated cells. Early in the regeneration of a urodele limb, mesenchymatous cells derived from the remaining limb tissues accumulate at the tip of the stump and form the regeneration blastema (Butler, '33; Hay, '62; Hay and Fischman, '61; O'Steen and Walker, '61; Thornton, '38a). The loss of differentiated properties by cartilage and muscle cells, as they form mesenchymatous blastema cells, has been observed in histological studies (Butler, '33; Thornton, '38a,b; Trampusch and Harrebomée, '65) and in cytological studies (Hay, '58, '59, '62; Salpeter and Singer, '62). The loss of differentiated properties has also been observed in immunochemical studies of muscle (DeHaan, '56; Laufer, '59). Since blastema cells originating from such diverse sources as cartilage, connective tissue, muscle, or dermis are morphologically indistinguishable (Hay, '58, '62), two different pathways for their redifferentiation are possible. A blastema cell could redifferentiate into its former type, exhibiting an intrinsic stability of cell differentiation; or a blastema cell could redifferentiate into a different cell type, undergoing metaplasia in response to local environmental influences.

From what we know of the differentiative stability of various cells, it is clear that either alternative for the redifferentiation of the blastema cells would be possible. In fact, following morphological dedifferentiation the same cell type can exhibit either metaplasia or stability in different test sys-

[1] A dissertation presented to the faculty of the graduate school of Yale University in candidacy for the degree of Doctor of Philosophy.
[2] Present address: Department of Biology, St. Olaf College, Northfield, Minnesota 55057.

tems. In urodele amphibians, retinal pigment cells will form a neural retina *in situ,* if the normal one is removed or degenerates (Stone, '50a,b; see also review by Reyer, '62). In contrast, chick retinal pigment cells, which have dedifferentiated and proliferated in mass culture, will still re-express their two major differentiated properties, melanin pigment formation and epithelial morphology, during proliferation in clonal culture (Cahn and Cahn, '66). Redifferentiation of pigment cells in culture has also been observed under conditions of reduced growth rate (Doljanski, '30; Whittaker, '63). The behavior of chick cartilage cells also varies with the extrinsic conditions. Dedifferentiation has been observed in mass cultures, where chondrocytes apparently lose the ability to redifferentiate, even when formed into a pellet and subsequently organ cultured (Holtzer et al., '60; Coon, '66). However, after chondrocytes have apparently lost the ability to redifferentiate in organ culture, many cells retain the ability to give rise to cartilage making colonies when cultured as clones (Coon, '66). Heritable stability of myoblasts from chick skeletal muscle has also been observed under conditions of clonal culture (Konigsberg, '63). In contrast, differentiated cells of the crayfish hepatopancreas have been traced directly from one cell type into another, going through a sequential series of different cell properties (Davis and Burnett, '64). The differentiated state is therefore clearly not a fixed, permanent cell property in all cell types. Stability varies with cell type, environmental conditions, and developmental history. Consequently, when considering the problem of the stability of cell differentiation, a strictly operational outlook is necessary (see reviews by Grobstein, '59; Trinkaus, '56).

Many different labeling methods have been used to investigate cell and tissue participation in regeneration. The difference in nuclear size between diploid and haploid cells enabled G. Hertwig ('27) to show that the cells of a limb regenerate are local in origin; however, the criteria used to identify cells on the basis of nuclear size were not adequate to permit an unequivocal conclusion. The work of Blaikher and Kraskina ('49)[3] has been cited by Vorontsova and Liosner ('60) as indicating that muscle contributes cells to all major parts of a limb regenerate; when triploid muscle was transplanted into a diploid limb, which then regenerated, labeled cells were found in all major parts of the regenerate: skin, skeleton, and muscle. The cell types being considered and their exact locations are uncertain, however, because the location of the triploid cells was determined by an indirect method: each of the above limb parts was transplanted into another limb, and during the regeneration of that limb, the blastema was examined for triploid mitotic figures. In the case of skeleton, for example, it is impossible to know whether the labeled cells from muscle had become cartilage, perichondrium, joint connective tissue, or merely loose connective tissue associated with the skeleton. Thus, these observations could be simply explained by a movement of connective tissue cells from one part of the limb to another. Hay ('52) was able to trace polyploid graft cells into the blastema on the basis of nucleolar number, but after differentiation, the graft cells could not be identified with certainty. In a study published during the course of this investigation, Patrick and Briggs ('64) used triploid cartilage grafts to determine the fate of cartilage cells participating in limb regeneration. The equivocal nature of their results, however, indicates a need for further study (see discussion below, page 19). Nucleolar number was also used as a cell marker by Barr ('64) to study the participation of epidermal cells in *Xenopus* limb regeneration; the reduced nucleolar number caused by the Oxford nucleolar mutation made it possible to use normal cells with two nucleoli as the labeled cells. H[3]-thymidine has been injected into animals before or during the early stages of limb regeneration in order to label differentiated stump tissues, which could then be followed *in situ* and shown to contribute to the blastema (Hay and Fischman, '61; O'Steen and Walker, '61). In other studies, H[3]-thymidine labeled cells were transplanted and subsequently located in order to determine whether they participated in blastema formation (Riddiford, '60; Steen and Thornton, '63). The work of Riddiford

---

[3] I have been unable to examine the original paper of Blaikher and Kraskina.

indicated that epidermal cells do not contribute to the blastema, while that of Steen and Thornton indicated that internal stump tissues from an innervated limb will contribute to the blastema formed in an aneurogenic limb. It is clear that for one reason or another, these studies on cell participation in regeneration have left obscure the ultimate fate of particular cell types as they differentiate in the regenerate (with the possible exception of epidermis).

In the present study, cells from transplanted cartilage and muscle were followed through limb regeneration and identified after redifferentiation by means of three kinds of label: H³-thymidine, ploidy as judged by nucleolar number, or a combination of the two. The use of these different labeling methods minimized the limitations of the individual methods and reduced the possibility of error due to the possible influence of a particular labeling method on cell behavior. In addition, the availability of two independent labeling methods permitted an analysis of the properties of each method.

The primary purpose of this study has been to determine whether blastema cells originating from cartilage redifferentiate only into cartilage, thereby exhibiting an intrinsic stability of differentiation, or whether they redifferentiate into other cell types, undergoing metaplasia. Cartilage cells originate from a homogeneous, avascular tissue, containing only one cell type. Therefore, labeled cartilage cells transplanted into regenerating limbs can provide information about the participation of chondrocytes in regeneration and an unequivocal answer to the question of chondrocyte stability of differentiation. In addition, labeled limb muscle has been used to determine whether other cell types contribute to the cartilage of the regenerate, and triploid non-limb cartilages have been transplanted into limbs in order to ascertain whether their cells are able to participate in the formation of cartilage in normal limb regenerates.

MATERIAL AND METHODS

Animals and their maintenance

Young axolotls, *Siredon mexicanum,* ranging in length from 50 to 85 mm were used in all experiments. The animals in any given experiment were within 5 mm of being the same length. Most experimental animals were raised from spawnings in which one parent was heterozygous and the other homozygous for the recessive white (d) gene. This permitted transplantations to be done between phenotypically white (dd) and phenotypically black (Dd) sibling animals, which would facilitate the detection of any pigment cells that might arise from grafted cells. Homozygous white and homozygous black animals were also used, and since no differences were observed, the genotype of the experimental animals will not be reported. Animals were maintained in individual bowls from the time of hatching, in order to prevent siblings from amputating each others' limbs. They were fed strips of beef liver at least three times a week, with additional feedings of tubifex worms on two or three of the other days, except for those which were being injected with thymidine. Two days before and on the days of thymidine injections, the animals involved were not fed. Animals were maintained at a temperature of $20.5 \pm 0.5°C$, in aged tap water changed three times a week.

Triploidy as a cell label

In one of the labeling methods used for the present study, triploid cells were identified by the presence of three nucleoli. Since nucleolar number is an intrinsic cellular property, this cell label has certain advantages. It cannot be diluted by proliferation, will not transfer to other cells, and can cause no radiation damage. There are some disadvantages to the use of triploidy, however: the occurrence of spontaneous polyploidy, the impossibility of detecting more than about half of the triploid cells, difficulties in finding the nucleoli at some stages of differentiation, and the extremely slow rate at which data can be collected, since the number of nucleoli in every nucleus in the limbs being studied must be observed.

Triploid graft cells were obtained from animals which had developed from heat treated eggs; this method of producing triploids is similar to that used for newts by Fankhauser and Watson ('42). In order to collect eggs for heat treatment without disturbing the spawning animals significantly, the spawning tank was equipped with several glass pyramids made of three pieces of 8 mm glass rod fused together at one end. Newly laid eggs could be seen on the glass rods by using very low levels of transmitted light directed through the spawning tank. Eggs were collected at intervals of no more than 30 minutes by removing the glass pyramids and transferring the eggs immediately into water at $35.5 \pm 0.5°C$ for seven minutes. At the end of this heat treatment, the eggs were transferred to water at $20.5°C$ and were allowed to develop without further interference.

In determining the conditions for heat treatment, the temperature was varied from 33.0 to 38.0°C and the duration from 6 to 30 minutes. The parameters found most effective for inducing triploidy in the axolotl were $35.5 \pm 0.5°C$ for seven minutes. Under these conditions, 17% of all treated eggs developed into young animals, and 48% of these animals were triploid, which is an improvement over the 34% triploid animals obtained after cold treatment (Fankhauser, '45). The rate of triploidy varied widely from spawning to spawning, ranging from 33% to 100%. It should be noted that only 41% of the non-treated, control eggs developed; therefore, considering the eggs which would normally develop, 41% survived the heat treatment.

Triploid larvae were identified by counting nucleoli in cells of whole mounts of tail tips which had been fixed in Clarke's fixative and stained in Azure B (Flax and Himes, '52).

H^3-thymidine labeled cells

In contrast with triploid cells, H^3-thymidine labeled diploid ($2NH^3$) cells, detected by radioautography, are easier to find than triploid cells, can be detected at any stage of differentiation, can obviously not be confused with spontaneous polyploids, give the possibility of a higher detection rate for labeled cells, and permit conclusions to be made about the amount of cell division. However, the usefulness of H^3-thymidine as a label is limited in several ways: dilution of isotope during proliferation, the possibility of isotope transfer, the possibility of radiation damage, and the possibility of confusing the pigment granules normally present in host cells with radioautograph silver grains. All of these limitations have been considered in the course of this study, and methods have been developed to assure the effective labeling and appropriate interpretation of H^3-thymidine labeled cells.

H^3-thymidine labeled triploid cells ($3NH^3$ cells) utilize the advantages of both methods. When a cell contains both markers simultaneously, it can clearly be designated as labeled. In addition, both classes of single label are found, i.e. apparently non-radioactive cells with three nucleoli and radioactive cells with two visible nucleoli. The only disadvantage of this double label is that the detection rate is lower than with either method alone.

The procedure used for producing cells labeled with H^3-thymidine was the same for both diploid and triploid cells. Cellular DNA was labeled by administering H^3-thymidine in a series of four intraperitoneal injections while the cells proliferated during the cone blastema stage of regeneration. The process of regeneration was initiated by amputating both forelimbs of a given animal within the proximal third of the upper arm. The first injections were given when the regenerates had reached the early cone stage, typically on the tenth day after amputation, and the three subsequent injections were given at 24 hour intervals. This timing permitted the last injection to be given 24 to 48 hours before the regenerate flattened and entered the palette stage, when chondrogenesis begins in the interior of the blastema. Since the chondroblasts essentially stop dividing after chondrogenesis begins, they do not dilute their label by proliferating significantly in the absence of H^3-thymidine. Development of the regenerate was allowed to continue for seven to nine weeks after the time of amputation, before cells were used for grafts. This permitted advanced differentiation of the regenerated tissues, including the beginnings of periosteal ossification in the humerus.

Undiluted H^3-thymidine with a specific activity of 6.7 c/mM (New England Nuclear) was used for all injections. The total dose administered to a given animal was 5, 7, 10, or 15 μc/gm of animal weight.

The first injection contained 29.5% of the total dose; while injections 2, 3, and 4 each contained 23.5% of the total. The total dose of H^3-thymidine was divided into unequal parts in order to compensate for the significant amounts of isotope that remain in the blood of the axolotl at 24 hours postinjection (Steen, unpublished).

In order to evaluate possible deleterious effects of H^3-thymidine on labeled cells of limb regenerates, two different kinds of control experiments were carried out. Control observations were made on the rate of regeneration as well as on the structure of the regenerate, evaluated on the basis of both gross and histological study. In one experiment, cold thymidine was substituted for H^3-thymidine in the usual sequence of four injections. The total amount of cold thymidine injected was equivalent to the amount of thymidine in a 10 μc dose of H^3-thymidine. In a second control experiment, two groups of animals were used. Animals in one group received H^3-thymidine at total doses of 5, 7, or 15 μc/gm in a normal injection procedure, and the other animals received equal amounts of distilled water equal in volume to that injected with the 5 μc/gm dose of H^3-thymidine. These limbs remained intact, and the animals were maintained for an extended period of time, in order to let the cells in the regenerate be exposed to radiation from the incorporated H^3-thymidine. Two or four months after being injected, the limbs were reamputated in order to observe their ability to regenerate. The animals injected with distilled water were reamputated four months after injection and provided a set of control limbs with which to compare the rate of regeneration in the H^3-thymidine-containing limbs.

Preparation of graft cells and graft procedure

All transplantations were performed in the same manner. Labeled tissues were removed from one animal, cleaned, and placed into a sibling host limb. Steinberg's solution (Steinberg, '57) was used as the medium for all operations. Animals were anaesthetized in 1:2,000 MS 222, and during the operations they were maintained under anaesthesia by a 1:10,000 concentration.

Host animals were put into anaesthetic during the preparation of graft tissue, so they would be ready to operate on when the graft became available. Grafts were placed into either the upper arm or the forearm, depending on how the graft cells were labeled: triploid cells were placed into the upper arm, in order to give the labeled cells a maximal opportunity for proliferation and participation in regeneration; while cells labeled with H^3-thymidine were usually placed into the forearm, in order to minimize dilution of their H^3-thymidine label by proliferation. In all operations, care was taken to minimize the trauma to the host limb. For grafts into the upper arm, the host limb was amputated just distal to the elbow. The elbow, all internal forearm tissues, and the distal two-thirds of the humerus were removed. Graft cells were placed into the cavity left by the humerus, and the end of the stump was closed by tying with a cotton fiber. For grafts into the forearm, the host limb was amputated through the wrist. The wrist cartilages as well as the distal two-thirds of the ulna and radius were removed. The graft cells were placed into the cavity left by the forearm skeleton, and the end of the stump was closed by tying with a cotton fiber. These basic procedures were occasionally varied by removing most of the muscle too, in order to leave more room for a large graft (see table 6 for a definition of a large graft and for a case where one was used). Host limbs usually began to regenerate immediately. If there was no sign of regeneration 7–10 days after grafting, a piece of skin from the tip of the host limb was cut off in order to initiate regeneration. Immediate regeneration was desired in order to encourage graft cell dissemination and participation in the regenerate.

The cleaning procedure for a piece of limb cartilage involved several steps, with each step carried out in a separate petri dish of Steinberg's solution. Instruments were cleaned with new cotton between each stage of the cleaning operation. 1. When isolating the cartilage, care was taken to remove it intact, including the proximal and the distal epiphyses. 2. All muscle was removed. 3. All remnants of connective tissue and most of the perichondrium were cleaned from both the epiphyses and the diaphysis. 4a. If no periosteal ossification was present, the ends of the cartilage were trimmed off and the last bits of protruding material removed. 4b. When periosteal ossification was present, which was usually the case, the thin shell of bone was peeled off. Cases in which the ossification extended most of the length of the cartilage yielded particularly favorable material for grafts, because once the ossified shell was removed and the ends trimmed off, the enclosed cartilage was especially pure. Such grafts supplied most of the data used in this study. Sometimes, the ossified region comprised only a small part of the total length of the cartilage piece. In such cases, after the ossified part was removed, the rest was cleaned as in 4a. 5. The cartilage piece was squirted in and out of a pipette in order to wash off any loose cells and to raise contaminants from the surface of the cartilage, making it possible to recognize and remove them. These final steps of cleaning were carried out at a magnification of 50 ×. 6. The cleaned cartilage piece was put on a glass slide with ridges to support the coverslip and given a final check for purity under a Zeiss compound microscope at 100 ×. Photographs of the cleaned cartilage were usually taken (see fig. 1). 7. The cartilage was put back into a petri dish, and if any contaminating cells had been observed (they seldom were), they were removed. 8. The cartilage was either placed in a host limb or fixed as a control to check for contamination.

The non-limb cartilages grafted were coracoid, scapula, and visceral arch. They were cleaned by the same procedure as that used for limb cartilages. Step 4a was always used, since there was no periosteal ossification, and at that point the edges of the plate cartilages (coracoid and scapula) were trimmed off. In addition, before step 5 the plate cartilages were cut into narrow strips to facilitate their insertion into the host limb. Finally, the larger pieces of visceral arch cartilage were scored with shallow cuts, in order to give more surface area for the release of cells into the blastema.

All grafted muscle came from either the proximal forearm or the upper arm, in order to minimize the possibility of contamination by cartilage. Also, care was taken during the removal of the skin in order to prevent contamination by epidermal cells. Muscle grafts were cleaned in dishes different from those used for initial isolation. Cleaning consisted of removing major circulatory vessels, major nerves, and tendons.

Histological procedures

Host limb regenerates and triploid control regenerates were fixed following formation of the digital cartilages, between 30 and 42 days after the initiation of regeneration. Fixation was in Clarke's fixative (absolute ethanol-glacial acetic acid, 3:1) for 24 hours. Longitudinal sections were cut at 12 μ, and the sections were affixed to slides with fresh 1% egg albumin for radioautography or with Mayer's albumin for routine histology. Both routine slides and developed radioautographs were put into citrate buffer (pH 4.0) for one minute then stained for 80 minutes at 40°C in Azure B (0.25 mg/ml in pH 4.0 citrate buffer, 0.05 M), differentiated overnight in tertiary butyl alcohol, cleared in toluene, and mounted in permount. The use of this staining method (Flax and Himes, '52) gave metachromatically stained, purple nucleoli which contrasted with the orthochromatically stained blue-green DNA of the nucleus.

Slides of control cartilage pieces were prepared as discussed above and were stained according to Cason's one-step Mallory-Heidenhain technique (Cason, '50).

Radioautographic procedure

Since H[3]-thymidine labeled cells were identified in radioautographs by the presence of silver grains over their nuclei, great care was taken to minimize the number of background grains in these radioautographs, so that as few as seven grains would be sufficient to identify a labeled cell. Background levels after exposure usually did not exceed an average of 1.5 grains/1,000 μ² and often were much lower. The rare cases where local or general background was much higher were not utilized. All operations for dipping and development were carried out in total darkness. Emulsion was melted only once. After testing for the presence of activated silver grains by dipping, drying, and developing as usual (except with no peroxide treatment), it was then used for a group of slides if the initial background level was less than seven grains/1,000 μ^2. Control of drying conditions prevented the artifacts which can form at that stage (Kopriwa and Leblond, '62), and treatment with H_2O_2 oxidized any latent image present from previous background or mechanical activation of grains during dipping (Caro, '64). Careful control of temperature, time, and agitation during development minimized background and gave a standard grain size.

Slides with their affixed sections were prepared for radioautography by passing them through two changes of toluene, two changes of absolute alcohol, absolute alcohol–ether (50:50), and 100% diethyl ether, then allowing them to air dry. After warming at 42°C, slides were dipped into melted, 40°C Ilford K-5 emulsion which had been diluted 1:1 with doubly distilled water. Slides were dried in a vertical position at 28° to 29°C with 80% relative humidity (Kopriwa and Leblond, '62) and were then placed into a H_2O_2 atmosphere for five hours, according to the directions of Caro ('64). After airing, they were exposed for 7–41 weeks at 4°C in an atmosphere dried by silica gel and drierite. The longest exposures were required when 5 μc/gm of H^3-thymidine had been used for labeling and when the label had been diluted extensively by proliferation. Photographic processing was carried out at $17.8 \pm 0.2°C$ (with 5 seconds of agitation every 30 seconds) using Kodak D-19 developer for three minutes, a water rinse, Kodak fixer for four minutes, wash for one minute, Permawash solution (Heico) for one minute, and finally ten water washes of one minute each. Slides were then stained in Azure B (see histological procedures).

Grain size in these radioautographs was very important, because melanin granules in the axolotl are within the size range of the silver grains formed by many emulsions, especially Kodak NTB-2 and NTB-3. Even when routinely using white (dd) host limbs, with their minimum of pigment, confusion between melanin granules and silver grains could be a source of error in identifying labeled cells. In addition, a sensitive emulsion was desired so that exposure times could be minimized. In the light of available data on grain size and conflicting reports on the sensitivity of Ilford K-5 emulsion (Kopriwa and Leblond, '62; Caro, '64), Kodak NTB-2, Kodak NTB-3, and Ilford K-5 emulsions were all tested for grain size and sensitivity. Test slides were made of limbs fixed two days after the usual injection series where total doses of 15.3, 0.27, or 0.015 μc of H^3-thymidine were used. Dipping was carried out according to the procedure detailed above. This test demonstrated Ilford K-5 (diluted 1:1) to be best for these experiments. The silver grains formed in K-5 after one and one-half to three and one-half minutes of development in Kodak D-19 were clearly smaller than axolotl melanin granules, and this small grain size also made it easier to count the nucleoli in an H^3-thymidine labeled nucleus. In addition, the K-5 tested was approximately two and one-half to three times as sensitive as the NTB-2 and about one and one-half times as sensitive as the NTB-3. With peroxide treatment in the preparation procedure, background levels in radioautographs made with K-5 were as low as with NTB-2 and much lower than with NTB-3.

Identification of labeled cells

When triploid graft cells were used, nucleoli were counted in all possible nuclei in the regenerates. The presence of three nucleoli in a given nucleus was taken to indicate that the cell was labeled.

H^3-thymidine labeled cells were identified in radioautographs by the presence of silver grains over their nuclei. The minimum grain number used to identify a cell as labeled is defined and supported in the section on transfer of H^3-thymidine label (page 11).

When $3NH^3$ graft cells were used, a labeled cell was identified in radioautographs by the presence of both three nucleoli in the nucleus and silver grains over the nucleus. In addition to cells containing both labels, $3NH^3$ grafts gave rise to cells which contain three nucleoli but no silver grains (due to excessive cell division), as well as cells with silver grains in the radioautograph but fewer than three nucleoli (due to sectioning or nucleolar fusion). From a single transplant, therefore, data from three kinds of label were obtained.

Purity of cartilage grafts

In order to determine whether chondrocytes are stable as to type during limb regeneration, the labeled cartilage grafts that are transplanted should be composed exclusively of chondrocytes. To verify their purity, 11 pieces of limb cartilage were selected at random from those used in the grafting experiments, fixed, sectioned, and stained so that any contaminating cells could be detected with certainty (see table 1 and fig. 2). These control cartilage pieces contained a total of 20,463 cells, with a maximum of 99.9% and a minimum of 99.7% chondrocytes, the range being due to the presence of cells that could not be identified with certainty. The only contaminating cells were from the perichondrium or the connective tissue; muscle or other cell types were never observed.

The contamination found on piece number 1 requires comment, for this was the only case in which connective tissue was found. It, along with pieces 2 and 3, was cleaned by a procedure which was a little less rigorous than the one used for all later pieces. With subsequent pieces of limb cartilage, the ends were trimmed off after most of the cleaning was completed, as in step 4 of the graft preparation procedure.

This removal of the ends of cleaned cartilage pieces eliminated the possibility of connective tissue contamination such as that found on piece 1. Also, in contrast with pieces 1, 2, and 3, almost all subsequent cartilage was at least partially enclosed within periosteal ossification which facilitated the cleaning process. In addition, the connective tissue cells counted as contaminants showed signs of mechanical damage, so they probably would not have survived to participate in regeneration if they had been grafted. The only experiment which used cartilage cleaned like the first three control pieces in the table involved grafts of $2NH^3$ cartilage into triploid limbs. Since these limbs were used primarily to check for transfer of H^3-thymidine label and provided no critical information on stability of chondrocytes, this contamination does not seriously affect the conclusions drawn.

PROPERTIES OF LABELED CELLS

Detection of triploid cells

Since triploid cells contain a maximum of three nucleoli per nucleus and diploid cells a maximum of two, provided the cells are not aneuploid (Fankhauser and Humphrey, '43), a triploid graft cell in a diploid host limb can be unequivocally identified by its three nucleoli. However, all triploid cells can not be detected for

TABLE 1

Contamination in control cartilages

Piece	Amount of periosteal ossification [1]	Total number cells	Perichondrocytes	Questionable perichondrocytes [2]	Connective tissue cells	Questionable connective tissue cells [3]	Maximum % contamination	Minimum % contamination
1	None	2353	7	8	14	3	1.4	0.89
2	None	867	0	0	0	0	0	0
3	None	3041	3	3	0	0	0.20	0.10
4	Some	2323	1	4	0	0	0.22	0.04
5	Some	4964	1	5	0	0	0.12	0.02
6	Some	876	0	0	0	0	0	0
7	Some	667	0	0	0	0	0	0
8	Some	1621	1	7	0	0	0.49	0.06
9	Complete	690	0	0	0	0	0	0
10	None	1797	2	5	0	0	0.39	0.11
11	Some	1264	2	2	0	0	0.32	0.16
	Totals	20463	16	29	14	3 Mean	0.30	0.15

[1] "None" indicates no ossification present. "Complete" indicates whole length of cartilage piece came from within periosteal ossification.
[2] These cells were on the surface of the cartilage and could be either chondrocytes or perichondrocytes.
[3] These cells were on the surface of the cartilage and could be chondrocytes, perichondrocytes, or connective tissue.

several reasons. The number of nucleoli may be reduced by fusion (Fankhauser and Humphrey, '43; Wallace, '63) or by cutting through a nucleus so that its nucleoli are in different sections. Also, nucleoli may be obscured by heterochromatin.

In order to determine the proportion of triploid cells detected when using the presence of three nucleoli as the criterion, the percentage of cells containing three nucleoli was determined for five cell types in each of five triploid regenerates. In each regenerate, 500 cells were observed, as five groups of 100 each. Separate counts were made in areas of cells that were cut similarly. Nuclei which were pale, because they obviously had been sectioned through the edge, were not counted. Muscle and its associated connective tissue were not counted separately, because it was usually impossible to determine the cell type for a given nucleus. The data are summarized in table 2. Since the percentage of triploid cells detected is dependent upon the relationship between the plane of the section and the axes of the nuclei, maximum and minimum percentages are included, in order to indicate the range in detection rate that is primarily due to sectioning.

In these observations on the detection of triploid cells, it is of interest to note that both right and left limbs from one animal (31–3), contained a low percentage of very large, polyploid cells with 4, 5, or 6 nucleoli. Most of the polyploid cells were found in the muscle or the connective tissue, and a few were found in the cartilage. These observations imply that spontaneous doubling of chromosome number may occur in a triploid animal.

The participation of H^3-thymidine labeled cells in regeneration

In the experiments described in this paper regeneration after injections of H^3-thymidine appeared to take place more slowly than normal and to give rise to abnormal limbs more frequently. In earlier experiments, injection of 62.5 µc/gm of H^3-thymidine into *Ambystoma opacum* larvae caused the formation of abnormal regenerates containing cells which exhibited gross abnormalities, including anaphase bridges and acentric fragments (Steen, unpublished). Toxic effects of H^3-thymidine have also been observed in cell culture (Drew and Painter, '59) as well as in developing chick embryos (Sauer and Walker, '61), and thymidine alone has been observed to inhibit cell division (Xeros, '62). It was therefore considered necessary to determine the nature and degree of influence that H^3-thymidine has on the behavior of labeled cells in axolotl limb regenerates.

Following isotope administration, the first properties available for assessing the influence of H^3-thymidine were the rate and the quality of the development of the regenerate. The stage of regeneration attained at 23 days post-amputation (usually 10 days after the final injection) was used as an index for the rate of regeneration. Comparisons were made between pairs of limbs: one from an experimental, H^3-thymidine injected animal, and the other from a control animal injected with water or non-radioactive thymidine. Between 10 and 20 pairs of limbs were compared for each level of H^3-thymidine injected: 5, 7, 10 or 15 µc/gm. The significance of differences between experimentals and controls was

TABLE 2

Proportion of cells with three nucleoli detected in triploid axolotl limbs

Cell types counted	Mean percentage of cells with three nucleoli		
	2500 Cells	Two counts, 100 cells each, in five limbs	
	Overall	Maximum	Minimum
Cartilage	43	48	36
Epidermal cells (tangentially sectioned)	52	58	45 [1]
Leydig cells (unicellular glands)	41	44	36
Loose connective tissue	42	51	35
Muscle or muscle connective tissue	51	56	44

[1] In sections cut perpendicularly to the surface of the epidermis, as few as 23% of the cells had three nucleoli.

TRYGVE P. STEEN

determined using the sign test (Siegel, '56). Before the injections, no significant difference was ever detected between experimental and control groups. Cold thymidine injections appeared to inhibit the rate of regeneration slightly, but this inhibition was not significant at the 90% level. In the case of both diploid and triploid regenerates, the stage of regeneration reached at day 23 by regenerates on animals subjected to 5, 7, or 10 µc/gm of H³-thymidine was also not significantly different from the controls. A 15 µc/gm dose of H³-thymidine, however, clearly inhibited the rate of regeneration, at a confidence level of over 99%. In fact, even when 30 day regenerates from 15 µc injected animals were compared with 23 day control regenerates, the controls were still more advanced, at a confidence level of 91%.

Morphologically abnormal limbs occurred at a significant frequency with all doses of H³-thymidine used (see table 3), the gross morphology becoming increasingly abnormal with increasing doses of H³-thymidine. Abnormalities observed after 5, 7, or 10 µc doses included failure of digits to separate and the formation of three digits instead of the normal four. Limbs with three digits comprised about one-third of the regenerates formed after a 5 or 10 µc dose. Although the three digits were relatively normal in length, they often diverged from the hand at abnormal angles. At 15 µc, three digit limbs comprised about one-half of the regenerates; the central digit was usually abnormally long, and the two lateral digits were abnormally short or consisted simply of tiny bumps on the sides of major spike (see fig. 3). In addition, at 10 and 15 µc doses the forearm and hand were usually shorter than normal. Although these gross abnormalities were observed, it should be emphasized that histological structure was normal, at all levels of H³-thymidine used in this study.

The possibility of long term radiation damage to cells labeled with H³-thymidine was assessed by observing their ability to form a regenerate after either two or four months of exposure to radiation from their incorporated isotope. Two months after the final injection, two regenerates labeled with 7 µc/gm and two regenerates labeled with 15 µc/gm were amputated in the mid third of the forearm, and four months after the final injection six regenerates labeled with 5 µc/gm and six unlabeled, control regenerates were similarly amputated in the mid third of the forearm. Both the rate of regeneration and the quality of development of these second regenerates were observed. Since the results from all levels of labeling were the same, this discussion will consider only the general outcome. Comparisons of the rate of regeneration were made at day 23, after the second amputation in the same way as previously, and they showed that labeled cells formed regenerates at a rate indistinguishable from unlabeled, control cells. Since normal regenerates formed on eight of the ten experimental limbs and on nine of the ten control limbs, the quality of the labeled second regenerates clearly did not differ significantly from the controls. The two abnormal regenerates that formed on labeled limbs had been labeled with the dose of 5 µc/gm, and the abnormal control limb was on a sibling control for the 5 µc group. The formation of eight in ten normal second regenerates contrasts with the two in

TABLE 3

Occurrence of abnormal regenerates after H³-thymidine injections

Dose/gram	Ploidy	Number of control limbs		Number of experimental limbs		Confidence level for difference [1]
		Normal	Abnormal	Normal	Abnormal	
µc						
5	2N	12	0	5	7	99.2
5	3N	6	0	3	3	—
7	2N	20	0	6	14	99.9
10	2N	18	0	4	14	99.9
15	2N	16	0	1	15	99.6
15	3N	10	0	3	7	99.2

[1] Using the sign test (Siegel, '56).

ten normal regenerates formed by the same set of limbs immediately after the H³-thymidine injections. Labeled cells clearly participated in the formation of the second regenerates, because radioautographs of the second regenerates revealed large numbers of labeled cells. This participation of labeled cells in normal regeneration took place after exposure to either two or four times the amount of radiation encountered in grafting experiments. Thus, even though abnormalities occurred in regenerates that developed immediately after H³-thymidine injections, one can reasonably expect graft cells labeled with H³-thymidine to be normal in their behavior; for their participation in regeneration would be analogous to the formation of a second regenerate.

Transfer of H³-thymidine

When cells labeled with H³-thymidine are grafted into an unlabeled limb, cytolysis of labeled cells near unlabeled cells synthesizing DNA could result in transfer of the H³-thymidine label. In order to determine whether transfer occurs under the conditions of these experiments, H³-thymidine labeled diploid cells (2NH³ cells) were grafted into triploid limbs. In radioautographs of the regenerates, H³-thymidine labeled cells were located, and their nucleoli were counted. If there were no transfer, all radioactive cells would be expected to have a maximum of two nucleoli. However, if transfer did take place, some radioactive cells with three nucleoli should be found (see fig. 4). Twelve limbs were studied in detail, and nine limbs provided useful data. Three limbs, all containing 7 μc/gm labeled grafts, were eliminated from further consideratoin for a number of reasons: one contained a non-integrated graft, another contained no clearly labeled cells, and a third, 14–5R, contained polyploid cells (see next section for details).

When regenerates of triploid limbs, which had received transplants of 2NH³ cells were observed, cells with three nucleoli were found, either with some of the rare groups of three or four background grains [4] over them or with larger numbers of grains indicating the presence of radioactivity derived from transfer. The grain counts over cells which received label from transfer were never more than 5% of the grain count over the most heavily labeled cells in a given limb (see table 4). Accordingly, the following definition for an H³-thymidine labeled cell was formulated, primarily in order to eliminate from consideration in the results cells which had become radioactive from low level transfer. Cells with seven or more grains above their nuclei were considered to be labeled, as long as no nuclei with more than 75 grains were found. Where labeling was heavier than 75 grains, the minimum was set at 10% of the mean grain number over the five most heavily labeled cells, provided this new minimum was greater than seven. A new minimum was defined for each limb observed, and when radioautographs for the same limb were exposed for different lengths of time, a new minimum was defined for each exposure time. By using seven as the minimum grain number for a labeled cell, background grains were effectively eliminated as a source of apparently labeled cells, because any cell under the rare groups of three or four background grains that appear in these radioautographs would be considered not labeled by a safe margin. Cells which had received their label as a result of transfer of radioactivity were eliminated from consideration by both the simple minimum grain number of seven and the elevated minimum grain numbers generated by the definition. When this definition for an H³-thymidine labeled cell was applied, no cell considered to be labeled ever contained three nucleoli, indicating a total lack of transfer at a level sufficient to exceed the defined grain number (see table 4). Thus, the cells considered to be labeled under this definition were derived only from labeled graft cells.

The use of this definition of a labeled cell was particularly important for interpreting the results from large muscle grafts placed into limbs from which both cartilage and some muscle had been removed, for in such cases there was much low level transfer. The use of this definition provided a large margin of safety when small

---

[4] Although background levels were usually lower than an average of 1.5 grains/1,000 μ², groups of 3 or 4 background grains were occasionally observed within the area of a single nucleus. When such a group of grains happened to occur over a nucleus, a cell would appear to be lightly labeled.

TRYGVE P. STEEN

TABLE 4

Results from transplanting H³-thymidine labeled diploid cells into triploid limbs, to determine the extent of transfer

Triploid host limb	Diploid graft source	H³-thymidine dose/gm	Grain number limits for labeled cells		Number of radioactive, labeled cells with given no. of nucleoli[3]					Radioactive 3N cells from transfer[4]	Maximum grain number over transfer cells[5]
			Defined minimum[1]	Maximum grain no. observed[2]	4	3	2	1	0		
		μc									
14–2R	Humerus	15	7	70	2	0	130	188	85	0	—
14–8R	Humerus	15	9	—	0	0	36	43	16	0	—
21–2R	Humerus	10	33	500	0	0	4	7	5	0	—
21–2L	Humerus	10	7	300	0	0	7	46	14	0	—
21–3R	Humerus	10	7	71	0	0	14	40	7	0	—
21–10R	Humerus	10	32	500	0	0	6	30	14	16	17
9–5L	Muscle and cartilage	7	18	300	0	0	16	13	5	2	13
21–3L	Muscle	10	28	400	0	0	27	47	9	10	18
21–10L	Muscle	10	21	240	0	0	38	103	68	2	8

[1] Minimum numbers of grains listed were derived from the definition of a labeled cell used in this paper (see text, p. 11).

[2] Grain counts over 250 are ± 15%.

[3] All cells tabulated in this section of the table were identified as labeled by having above them more than the defined minimum number of silver grains. They all contained three nucleoli as well as having over their nuclei between seven and the tabulated maximum number of grains for labeled cells. They were considered not labeled on the basis of the definition for a labeled cell used in this paper.

[4] All cells tabulated in this column were considered not labeled over these cells unequivocally indicate the presence of radioactivity derived from transfer.

[5] Note that these maximum grain numbers over transfer cells are substantially less than the minimum grain numbers used to define a labeled cell.

cartilage grafts were placed into the cavity left after removal of the host limb cartilage, for in such cases transfer was almost never observed, even at a minimum grain count of seven.

The only disadvantage of using the elevated minimum grain numbers derived from the definition was the elimination of certain originally labeled cells that showed a small number of grains. However, the loss of these cells from the data was more than compensated for by the increased confidence with which a cell could be designated as labeled with H^3-thymidine.

Cells containing four or more nucleoli

In experiments where triploidy is used as a cell marker, a rare but very important event is the spontaneous generation of higher levels of ploidy in either host or graft cells, for it complicates the interpretation of the data. Polyploid mosaics have been reported in the axolotl (Fankhauser, '45), and polyploid amounts of DNA have been reported in a low percentage of somatic cells in *Taricha* (Truong and Dornfeld, '55). During the present study, spontaneous polyploidy also occurred in the somatic cells of both diploid and triploid axolotls. This is a matter of some importance, for if the cells in a diploid limb that is host to a triploid graft give rise to some tetraploid cells, these cells could easily have their maximum of four nucleoli reduced to three because of either sectioning or fusion. As a result, cells of host origin could be interpreted as originating from the graft.

Polyploid cells have been observed several times in these experiments (see fig. 5). In one case a triploid cartilage graft from a single humerus was divided into several pieces and placed into two different limbs on two different animals, 31–13R and 31–14L. Only one cell with four nucleoli was found in the entire limb of 31–14L, indicating that the graft probably did not contain any cells with four or more nucleoli. On the other hand, four cells containing four nucleoli were found in less than one-fifth of the sections of limb 31–13R, two in dermal connective tissue and two in cartilage. These polyploid cells in 31–13R were considered to be of host origin, and since their presence precluded

a clear interpretation of the data, this limb was eliminated from consideration.

In two other cases where diploid cells probably gave rise to cells with four nucleoli, diploid cartilage labeled with H^3-thymidine had been implanted into triploid limbs, in order to check for transfer of the H^3-thymidine label. In one particularly instructive limb, 14–2R, two cells with four nucleoli in each were very clearly labeled with H^3-thymidine; indeed, they were among the most radioactive cells in the limb. None of the other 402 H^3-thymidine labeled cells in this limb contained more than two nucleoli. Clearly, both cells were from the diploid graft. This limb was included in the results, because the presence of the polyploid cells generated no ambiguity. The other case (14–5R) contained a significant number of radioactive cells with three nucleoli as well as with four nucleoli. It was eliminated, because the presence of cells with four nucleoli made it impossible to account for the origin of the H^3-thymidine labeled cells with three nucleoli.

RESULTS

The fate of chondrocytes in limb regeneration

Evidence applicable to the question of chondrocyte stability during limb regeneration came from experiments using two different methods for labeling cells, from $2NH^3$ cartilage grafts transplanted into triploid limbs and from triploid cartilage grafts transplanted into diploid limbs. Labeled chondrocytes were found dispersed among unlabeled cells and integrated into all parts of the normal limb skeleton that formed in any given experimental regenerate. Moreover, when limbs were fixed during the mound blastema stage, the cone blastema stage, or the early digital plate stage, labeled mesenchymatous cells were found in the blastema. Therefore, since the labeled chondrocytes were clearly separated spatially from their grafts and were part of the regeneration blastema, their behavior can be considered representative of normal cell participation in regeneration.

In order to be certain that individual cell behavior was being studied in these experiments, care was taken to make sure that the chondrocytes found in the regenerates had been released from the grafts and had gone through the mesenchymatous blastema cell stage before being incorporated into the regenerate cartilage. This could be verified by noting the dispersion of labeled cells among unlabeled cells and by making sure that the extra-cellular matrix around labeled cells was identical with that around adjoining cells. A few limbs were found to contain small remnants of the graft, which usually had been incompletely integrated into the otherwise normal regenerate cartilage. The cells in such pieces were not included in these results, for clearly they had not participated in regeneration. Limbs containing such coherent pieces of graft cartilage are noted in table 6. Only cells which were individually integrated into the regenerate are recorded. Since these cells were intermixed with host cells, it is reasonable to assume that these cells came under the influence of the environment prevailing in the blastema during normal limb regeneration. The behavior of these cells therefore enables one to distinguish between an intrinsic stability of differentiation and differentiation in response to local environmental influences.

2NH³ cartilage grafts. Five regenerates containing grafts of 2NH³ cartilage cells yielded data consistent with chondrocyte stability (see table 5 and fig. 6); all labeled, differentiated cells found were either within the cartilage or at its surface. Cells at the surface of the developing cartilage were designated perichondrocytes, although they were probably destined to become chondrocytes. If one considers immature perichondrocytes to be contaminants, then chondrocytes comprise 96% of the differentiated, labeled cells in these limbs. These five regenerates with 2NH³ cartilage grafts also contained many undifferentiated, labeled mesenchymatous cells that could not be identified as to type, because these limbs were fixed before many of the cells in the distal parts of the regenerate had differentiated. For example, in limbs 14–2R and 14–8R, fixed at the earliest stage (early digital plate stage), 29% of the labeled cells were mesenchymatous, whereas in the three limbs fixed at later stages, only 15% of the labeled cells were mesenchymatous.

A comparison of grain counts over labeled chondrocytes located in these regenerates with grain counts over labeled cells in control cartilage pieces fixed before transplantation, indicates that the graft chondrocytes in these experiments usually divided at least five times before differentiating in the regenerates. For example, in cartilage labeled by a 15 μc/gm dose, the median number of grains originating from the initial level of label is between 36 and 48 grains per day per nucleus (a distribution-free 90% confidence interval for the median of 35 control nuclei, using Owen, '62, table 12.1), while the median number

TABLE 5

Cell types formed by 2NH³ limb chondrocytes that participated in regeneration of 3N limbs

Host limb	Stage when regenerate was fixed	H³-thymidine dose/gm	Total no. of cells	Chondrocytes	Perichondrocytes	Mesenchymatous cells
		μc				
14–2R	Digital plate	15	404	304	8	92
14–8R	Digital plate	15	92	30	11	51
		Subtotal	(496)	(334)	(19)	(143)
21–2R	Late two digit regenerate	10	26	26	0	0
21–2L	Late two digit regenerate	10	66	45	0	21
21–3R	Early two digit regenerate	10	71	68	0	3
		Subtotal	(163)	(139)	(0)	(24)
		Total	659	473		

of grains found over labeled cells in the experimental regenerates is between 0.57 and 0.71 grains per day per nucleus (a distribution-free 90% confidence interval for the median of 198 labeled nuclei, using Owen, '62, table 12.1). In this example, the dilution of isotope shown by the grain counts indicates six divisions took place, assuming reciprocity of grain number and exposure time (see Kopriwa and Leblond, '62). However, although the data indicates a 2^6 dilution of isotope, more than six divisions could be involved, because after six divisions one could expect to find only one radioactive chromosome in a given cell (note $2N = 28$ in the axolotl). Additional divisions after a sixth would result in no further exponential dilution, but would be limited to dilution resulting from an event such as somatic crossing over.

Triploid cartilage grafts. Transplantation of triploid cartilage cells into diploid limbs clearly demonstrated a high level of chondrocyte stability as to cell type (see table 6 and fig. 7). Experimental regenerates in this group were fixed at the four digit stage, and in every case the gross morphology was normal. Moreover, all limbs were histologically normal, with normal amounts of muscle and with normal skeletal elements in all but the three limbs which contained coherent remnants of graft cartilage (see table 6). Ten limbs were studied in detail, but these data come from eight limbs. Two limbs (24–24L and 31–13R, see p. 13) were eliminated from further consideration, because they contained significant numbers of cells with four or more nucleoli. In the results from these eight triploid grafts, 93.7% of all labeled cells were chondrocytes, while 2.9% of the labeled cells were undifferentiated, mesenchymatous cells, and 3.4% of the labeled cells differentiated into other cell types. It should be emphasized, however, that 96.5% of all differentiated, labeled cells were chondrocytes, and that an additional 2.0% of the apparently differentiated cells could have ultimately become chondrocytes in their respective regenerates. These potential chondrocytes were classified as connective tissue, as joint connective tissue, or as perichondrocytes, and they were all located in close proximity to the surface of the cartilage. The labeled, undifferentiated cells were located among other mesenchymatous cells in the wrist and digital parts of the regenerates. Most of the labeled connective tissue cells were associated with the muscle in the proximal part of the regenerate but a few were located in the loose dermal connective tissue. Labeled cells in the immature muscle usually could not be precisely identified, so they were grouped as muscle or muscle connective tissue. Epidermis and Leydig cells were never found to be labeled. Regardless of the origin of the labeled chondrocytes, whether from the ulna, radius, humerus, coracoid, or scapula, they were found in all parts of the limb skeleton. It is of interest to note the high degree of stability as to cell type exhibited by chondrocytes from the coracoid and scapula, when they contribute to a limb regenerate, for like chondrocytes originating from limb cartilages, they almost exclusively differentiated into cartilage cells in the limb regenerate skeleton.

Triploid grafts of various sizes were used in order to put the labeled chondrocytes into different environments. Cells originating from small grafts, containing approximately 200 to 700 chondrocytes, were so dispersed among cells of stump origin, that they could not have interacted with each other significantly. Labeled cells in this situation were therefore clearly acting as individuals and not as a population. In the case of large grafts, with an estimated size of 3,000 to 7,000 cells, cells of graft origin tended to remain together and so had the opportunity to interact with each other as well as with cells of stump origin. A large population of stable cartilage cells also could possibly provide all the cells necessary for formation of an entire regenerated skeletal element. In such a situation, graft cartilage cells unable to be included in limb cartilage could perhaps become something else. It is of interest to note, therefore, that chondrocyte stability as to cell type was not significantly different under the conditions created by differing graft sizes (see table 6).

Three alternative hypotheses could explain the 1.5% of the cells with three nucleoli found as muscle or as connective tissue in locations where they could not possibly have become cartilage. An addi-

94

TABLE 6

Cell types formed by triploid chondrocytes that participated in regeneration in diploid limbs

Host limb	Graft cell source	Graft size [2]	Label observed [3]	Total no. of cells [4]	Chondrocytes	Joint C.T. [5]	Perichondrocytes	Muscle	Muscle C.T. [5] or muscle	C. T. [5]	Mesenchymatous cells
24–12L	3N Humerus	small	3N	68	68	0	0	0	0	0	0
24–11R	3N Ulna, radius and perichondrium	small	3N	7	6	0	1	0	0	0	0
24–16R [1]	3N Humerus	medium	3N	7	6	0	0	0	1	0	0
31–14L	3N Humerus	large	3N	505	485	7	0	1	2	6	4
			4N	1	0	0	0	0	1	0	0
31–16R	3N Humerus	small	3N	59	58	0	0	0	0	1	0
31–21L	3N Humerus	medium	3N	85	75	1	1	0	0	8	0
21–11R [1]	3N Scapula	medium	3N	85	71	0	0	0	0	0	14
21–11L [1]	3N Coracoid	medium	3N	122	111	0	0	0	0	2	9
				939	880	8	2	1	4	17	27

[1] The graft in this limb was not completely integrated into the regenerate as individual cells. The only cells recorded were separate from the coherent graft piece and were integrated into normal limb tissue.

[2] A small graft contained between 200 and 700 cells. A medium sized graft contained between 700 and 1500 cells and was usually approximately the same size as the cartilage removed from the host limb. A large graft contained from 3000 to 7000 cells and was usually several times the size of the humerus removed from the host limb.

[3] 3N indicates that cells with three nucleoli were being considered and have their data tabulated. 4N indicates data from cells with four nucleoli are tabulated.

[4] All cell numbers tabulated are those actually observed; no correction has been applied for the fact that all triploid cells can not be detected.

[5] C.T. = connective tissue.

tional 2% of the cells with three nucleoli were found as connective tissue cells very near the cartilage and could be explained by these hypotheses too, although they were possibly destined to become cartilage. (1) These labeled non-cartilage cells could originate from contaminating cells introduced with the graft. This possibility is quite unlikely, however, because of the low contamination rate (0.3%) observed with the control cartilages, and the observed damage to almost all contaminating cells on the control cartilages. Also, since it is clear that all graft cells do not survive, it seems reasonable to consider the most exposed cells, the contaminants, the least likely of all to survive. (2) It is also possible that all or most of the non-cartilage cells with three nucleoli arose as spontaneous polyploids. Since most of the non-cartilage cells were found in limb 31–14L, which contained one muscle or muscle-connective tissue cell with four nucleoli, this is a very likely possibility. Moreover, almost all of the labeled muscle and connective tissue cells were found in the most proximal parts of the limbs, where it is least likely that a graft cell would have penetrated. For example, only two of the ten labeled non-cartilage cells found in limb 31–14L were located in a part of the regenerate that most probably was formed from the blastema, and eight cells were found in an even more proximal location than almost all of the labeled chondrocytes that had been found in the humerus. (3) The labeled muscle or connective tissue cells could have originated from the graft chondrocytes as a result of metaplasia. While this possibility can not be completely eliminated on the basis of the available evidence, it appears unlikely in the light of the possibility of spontaneous polyploid cells and of graft contamination. One can only conclude that whenever triploid labeled cells are used in the axolotl with grafts of the size used in this study, it will probably be impossible to distinguish low levels of metaplasia (2% or less) from spontaneous polyploidy in the host diploid limb.

The contribution of muscle to the cartilage of limb regenerates

Since it was clear that many cells in regenerated cartilage did not originate from labeled chondrocyte transplants, even with the largest grafts used, the contribution of labeled limb muscle to regenerates was observed in order to determine whether muscle supplies cells to the cartilage. Triploid, 2NH[3], and 3NH[3] muscle were transplanted and allowed to participate in regeneration. Regenerates were fixed at the four digit stage, and in every case both gross morphology and histology were nor-

TABLE 7

Contribution of labeled limb muscle to cartilage in limb regenerates

Host limb	Host ploidy	Graft label method	H[3]-thymidine dose/gm	Label observed [2]	Chondrocytes	Muscle or muscle C. T.[3]
			μc			
21–3L	3N	2NH[3]	10	H[3]	35	+
21–10L	3N	2NH[3]	10	H[3]	59	+
24–8L	2N	3N	0	3N	6	+
21–25R	2N	3NH[3]	5	3NH[3]	12	+
				H[3]	76	+
				3N	6	+
31–35L [1]	2N	3NH[3]	15	3NH[3]	5	+
				H[3]	40	+
				3N	4	+

[1] In this limb, graft muscle primarily replaced host muscle, and most of the host cartilage remained in the stump.
[2] 3N indicates that cells with three nucleoli were being observed and tabulated. H[3] indicates that H[3]-thymidine labeled cells were being considered. When 3NH[3] graft cells were used, H[3] alone indicates that the cells tabulated were labeled with H[3]-thymidine but had two or fewer nucleoli 3NH[3] indicates the presence of two labels in each cell tabulated, both three nucleoli and H[3]-thymidine.
[3] The presence of labeled muscle and muscle connective tissue is indicated by a "+". In all such cases, the number of labeled muscle or muscle connective tissue cells greatly outnumbered the labeled chondrocytes.

mal. In all cases, labeled chondrocytes were found, indicating that muscle consistently supplies cells to regenerated cartilage (see table 7 and fig. 8). Labeled cells were found throughout the regenerated muscle as well (fig. 9). Labeled cells were found in the dermis, but no label was ever found in the epidermis. The 17 3NH³ cells found in limbs 21–25R and 31–35L constitute particularly strong evidence for cells from muscle becoming chondrocytes, because the double label precludes any possibility of transfer or of spontaneous polyploidy. Unfortunately, it is impossible to state which cell type(s) within muscle became chondrocytes, because limb muscle is a mixture of several cell types: muscle cells, connective tissue cells, and endothelial cells (to name a few).

Participation of non-limb cartilage cells in limb regeneration

Triploid cartilages from the coracoid, scapula, and visceral arches were transplanted into limb stumps, in order to determine the ability of non-limb cartilage cells to participate in limb regeneration. In all cases, labeled cells were released from the non-limb cartilage transplants and were found within the regenerated limb skeleton. These non-limb cartilage cells that were incorporated into the limb regenerate cartilage can be considered to have been mesenchymatous blastema cells between the time of their release from the graft and their incorporation into the limb regenerate. This is supported by the following observations: labeled cells were dispersed among the cells in the regenerate cartilage; the matrix surrounding the differentiated, labeled cells was identical with that around adjoining cells; and some labeled mesenchymatous cells were present in these regenerates (see table 6). Cells released from the non-limb cartilages apparently were stable as cartilage cells, because when detailed observations were made (see table 6) and when about 10% of the cells in the regenerates were scanned, the labeled cells which had been released from the transplants had differentiated into chondrocytes almost exclusively. Coracoid and scapula cartilage transplants (8 and 7 cases respectively) supplied a greater number of cells to limb regenerates

than did visceral arch cartilages. In fact, in two of the better cases (table 6), the number of labeled cells contributed to the limb regenerate cartilage compared favorably with the number supplied by limb cartilage grafts. Visceral arch cartilage transplants (11 cases) contributed very few cells to the regenerated cartilage and in three cases probably none. In all regenerates containing non-limb cartilage, the number of labeled cells in the regenerated cartilage was insufficient to give any evidence about cell properties beyond stability as a chondrocyte, e.g. cell arrangement and gross morphology. In contrast with the results from limb cartilage transplants, in which only two regenerates in 11 contained an intact remnant of the original graft, significant remnants of non-limb cartilage transplants remained in all regenerates studied. This occurred in spite of the extra cuts which had been made in the surface of the transplants in order to increase the area available for matrix dissolution and cell release. Coracoid and scapula transplants left the smallest remnant pieces, while visceral arch cartilage transplants left such large remnants that they appeared to have been hardly reduced in size. Either the cells in these non-limb cartilages were not as active as limb cartilage cells in the process of matrix dissolution, or the matrix in the non-limb cartilages was more difficult to dissolve or to remove. It should be noted that the limb skeletons in these regenerates were quite abnormal in the proximal regions which contained the non-limb cartilage remnants. In contrast, the limb skeletons in the distal regions, which contained no transplant pieces and few or no labeled cells, were normal.

DISCUSSION

The observations presented in this paper indicate that chondrocytes are stable as to cell type during normal limb regeneration in the axolotl. Triploid cartilage graft cells participating in limb regeneration give rise almost exclusively to chondrocytes. The small percentage of labeled non-cartilage cells which were observed probably arose as spontaneous polyploids or as contaminants of the original grafts, although the possibility that some of them resulted from metaplasia can not be excluded. Results from cells labeled with H³-thymidine also support the conclusion of chondrocyte stability. Labeled chondrocytes no doubt lost their differentiated properties during their participation in blastema formation. During this period, they were also dispersed in a much larger population of other cells, thereby precluding any significant interaction among the labeled cells. In spite of this exposure to the influence of cells from other tissues, while in the blastema, the grafted chondrocytes did not transform. Consequently, it must be concluded that some mechanism intrinsic to the cells was responsible for their stability. Moreover, the basis for this intrinsic stability must be heritable, since observations on the dilution of isotope in H³-thymidine labeled cells showed that the chondrocytes in these regenerates had usually divided at least five times.

One striking feature of these results is that chondrocytes maintain their character during regeneration in spite of a loss of morphological features during blastema formation. Thus, for chondrocytes at least, the term dedifferentiation should be used in a morphological sense only, with no implication of increased competence or potential. The fact that mesenchymatous cells derived from muscle and cartilage are indistinguishable at the level of fine structure when in the blastema (Hay, '62) does not indicate that chondrocytes could freely become muscle cells. The intrinsic stability of the chondrocyte cell type is in significant contrast to the instability of morphology that is apparent during the dedifferentiation associated with blastema formation (Hay, '62), as well as during the regression that follows denervation (Butler and Schotté, '41). A particularly striking demonstration of the extrinsic influence of innervation on the stability of chondrocyte morphology occurs after local crushing or perforation of the skeleton of a denervated limb; first the limb skeleton and then all internal limb tissues lose their morphological properties as the limb regresses (Thornton and Kraemer, '51; Thornton, '53).

During the course of this investigation, a paper appeared which reported an investigation of the fate of cartilage cells during regeneration of the axolotl limb, by grafting triploid cartilage into diploid limbs (Patrick and Briggs, '64). Although the authors concluded that their results provide no evidence to indicate that cartilage cells change in type during limb regeneration, the conclusion appears not to be justified by their observations. Thirty-nine regenerates were studied. In 28 cases no triploid cartilage was found, and the authors state that "other tissues were also devoid of triploid cells." In four limbs the triploid graft was present but did not participate in regeneration, and no triploid cells were found in either the muscle or the epidermis. In the seven limbs where the triploid cartilage grafts did contribute cells to the cartilage, cells with three nucleoli were also found in the muscle (2 of the 7 limbs) as well as in the epidermis (6 of the 7 limbs). The authors ascribed these non-cartilage cells with three nucleoli to spontaneous polyploidy in host limb cells, because cells with four nucleoli were also found. It may be asked, however, why no polyploid cells were observed in the epidermis or muscle of the 32 host limbs where the graft did not contribute to the regenerate cartilage. Consequently, the origin of these cells with three nucleoli found in the muscle and the epidermis is ambiguous. They could have arisen from host cells as postulated, from the transformation of donor chondrocytes, or from contaminating cells brought in with the graft.

The heritable stability of chondrocytes in the urodele limb parallels the stability in clonal culture of chick cartilage cells (Coon, '66), chick skeletal muscle (Konigsberg, '63), and chick retinal pigment cells (Cahn and Cahn, '66). In all these cases, differentiated cells maintained their character in spite of morphological dedifferentiation and extensive proliferation. It should be noted, however, that this parallel behavior took place in radically different test situations, with the urodele cartilage *in vivo* and the chick cells *in vitro*. It would be interesting to see if urodele limb chondrocytes were also stable under conditions of clonal culture. The stability of chondrocyte differentiation in limb regeneration contrasts with the metaplasia observed during regeneration of the neural retina, iris, or lens from the pigmented retina (Stone, '50a,b, '55) and lens formation from the dorsal iris (see review by Reyer, '62).

Non-limb cartilages were transplanted into limbs in order to ascertain whether their cells were regionally determined, i.e. unable to participate in the formation of normal limb cartilage. In all cases, however, insufficient numbers of non-limb cartilage cells were present in the regenerate skeletons to furnish information on the question at issue. Also, the participation of non-limb cartilage cells in the formation of the limb regenerate skeleton was difficult to evaluate because of the skeletal abnormalities which occurred in association with remnants of the transplants. Although all non-limb cartilages did contribute cells to limb regenerate cartilage and were stable as chondrocytes, further work will be required to ascertain whether non-limb cartilage cells are determined with respect to their distinctive tissue properties. Since chondrocytes released from the scapula were found to be stable, the results of the present study support Eggert's ('66) hypothesis that scapula cartilage cells are stable as to type after forming a blastema in an x-rayed newt limb.

Many experiments involving the removal of cartilage from the limb stump have shown that the cartilage of the regenerate can originate from cells that were not previously chondrocytes (Bischler, '26; Bischler and Guyénot, '26; Goss, '58; Morrill, '18; Thornton, '38b; Weiss, '25). Histological studies of regeneration after exarticulation of skeletal elements indicated that the remaining stump tissues formed the blastema which gave rise to the new cartilage (Bischler and Guyénot, '26; Thornton, '38b). The following tissues were observed to contribute to such a blastema: muscle, muscle connective tissue, connective tissue of the brachial plexus, and some subcutaneous connective tissue (Thornton, '38b). By actually tracing cells from the stump muscle into the cartilage that regenerated after exarticulation, the present study clearly indicates that regenerate cartilage formed after exarticulation originates at least partially from one or more of the cell types found in skeletal muscle.

The contribution of cells from labeled muscle to limb regenerate cartilage is of considerable interest. Since this contribution takes place even in the presence of cartilage in the stump and since cartilage in the regenerate does not all originate from cartilage, even when using large labeled grafts, it seems clear that muscle must normally contribute cells to the cartilage of limb regenerates. Because muscle is a mixture of cell types, however, the cellular origin of the chondrocytes derived from muscle is uncertain. This ambiguity could be resolved by using the technique of cloning muscle cells pioneered by Konigsberg ('63). By labeling and grafting the pure cell populations thus obtained, it should be possible to determine not only which cells from muscle contribute to the regenerated cartilage but also whether muscle cells and connective tissue cells are stable in their differentiation.

The possible contribution of epidermal cells to the blastema and to the internal tissues of the regenerate, at first suggested by Godlewski ('28) and later revived by Rose ('48), has also attracted much attention. Recent investigations of this problem (Barr, '64; Hay and Fischman, '61; O'Steen and Walker, '61; Riddiford, '60; Rose and Rose, '65) have used both H^3-thymidine and differences in nucleolar number to follow epidermal participation in regeneration. Riddiford ('60) showed that when epidermis was labeled by H^3-thymidine during the period of blastema formation and then transplanted onto a non-radioactive blastema, no epidermal contribution to the blastema could be found. When H^3-thymidine was injected before amputation, the only labeled cells found in the limb were the epidermis and a few blood cells (Hay and Fischman, '61; O'Steen and Walker, '61). The epidermis could be followed *in situ*, and no labeled blastema cells were ever observed to be derived from this labeled epidermis. When animals were injected ten days before amputation, O'Steen and Walker ('61) noted that labeled cells were present in all layers of the epidermis during the period of maximal blastema accumulation in the regenerating forelimb and that labeled cells were present only in the epidermis, even at the early digit stage and what they designated as the "5 digit

stage"[5] of the regenerate. These results, therefore, present no evidence for an epidermal contribution to the blastema or to the regenerate. In Barr's study ('64), reported only in a preliminary note, epidermis was removed from *Xenopus* hind limbs which were composed of cells containing one nucleolus; these limbs were then used to replace hind limbs of recipient larvae composed of cells containing two nucleoli, so that epidermal cells labeled with two nucleoli would grow over transplanted limbs. When regenerates of these chimeric limbs were studied, no cells with two nucleoli were found among the inner tissues, again indicating an absence of an epidermal contribution to the regenerate. In contrast, Rose and Rose ('65) conclude, on the basis of an extensive investigation using H[3]-thymidine labeled cells, that epidermis contributes cells to the blastema and to the regenerate. This conclusion must be accepted with caution, however, for other directly labeled stump tissues could have given rise to the labeled blastema cells observed. The time of isotope administration used by Rose and Rose, i.e. within one day of amputation, could have directly labeled the stump tissues, because H[3]-thymidine circulates for an extended period after its injection, as demonstrated indirectly by O'Steen and Walker ('61) and directly in the present study. In addition, the very lightly labeled blastema cells (3–5 grains), thought by Rose and Rose to have come from the epidermis, could have been labeled by transfer. In the present study, H[3]-thymidine transfer was routinely detected, by the double label system employed, at a level sufficient to give 3 to 5 grains over a transfer cell. In some cases transfer occurred even when the H[3]-thymidine must have originated from fewer than 600 cells, which were labeled less heavily than the epidermal cells in the Rose's study. In the light of these potential problems with the H[3]-thymidine labeling method and of Barr's ('64) result with an intrinsic cell marker, the conclusion that the epidermis contributes to the blastema (Rose and Rose, '65) seems unjustified. Simultaneous use of more than one labeling method, as was done in the present study, should furnish a basis for unequivocal conclusions about an epidermal con-

tribution to the blastema and regenerate.

In work on limb regeneration, considerable attention has been focused on the cellular origin of the blastema and of the regenerate. While Hertwig's ('27) pioneering investigation of this problem utilized a potentially useful cell label, i.e. the nuclear size difference between diploid and haploid cells, his conclusion that the regenerate is local in origin does not necessarily follow from his data. His criteria for identifying a labeled cell were probably not reliable; nuclear size was determined only by the measurement of two diameters of each nucleus, and the range of nuclear sizes observed was never indicated, either for the haploid-diploid difference or for the different cases observed. X-irradiation of whole limbs or the whole body (Butler, '35) and especially of a small region of the limb (Butler and O'Brien, '42), demonstrated that the ability to regenerate is local in origin. Histological and cytological observations of blastema formation (Butler, '33; Hay, '58, '59, '62; Salpeter and Singer, '62; Thornton, '38a) as well as an extensive quantitative analysis of newt limb regeneration (Chalkley, '54) are also consistent with a local origin for the blastema. Additional evidence comes from H[3]-thymidine labeling of stump tissues *in situ*, for labeled cells were found to contribute to the blastema (Hay and Fischman, '61; O'Steen and Walker, '61). In the study by O'Steen and Walker some cells labeled after they were in the blastema were traced into the regenerates as well. In the present study, both labeled chondrocytes and labeled cells from muscle were followed from their location in the limb stump through blastema formation and finally into the regenerate. Important evidence has thus been added supporting a local origin of both the blastema and the regenerate.

ACKNOWLEDGMENTS

I wish to thank Dr. J. P. Trinkaus for his counsel during the course of this investigation and in the preparation of the manuscript. I am grateful to Drs. E. J. Boell and C. L. Markert for their advice and critical reading of the manuscript. I am

---

[5] This is obviously an error since newt forelimbs have only four digits.

indebted to Dr. R. R. Humphrey, of Indiana University, for a gift of axolotls to begin a colony at Yale. Finally, I wish to express my appreciation to Dr. C. S. Thornton, of Michigan State University, for introducing me to the study of limb regeneration and for stimulating my interest in the cellular aspects of regeneration.

This investigation was supported by fellowships from the Danforth Foundation and from the USPHS (Developmental Biology Training Grant T01 HD-00032), as well as by NSF grant GB 4265 to Dr. J. P. Trinkaus.

LITERATURE CITED

Barr, H. J.   1964   The fate of epidermal cells during limb regeneration in larval *Xenopus*. Anat. Rec., *148:* 358.

Bischler, V.   1926   L'influence du squellette dans la régénération, et les potentialités des divers territoires du membre chez *Triton cristatus*. Rev. Suisse Zool., *33:* 431–560.

Bischler, V., and E. Guyénot   1926   Les potentialités régénératives différentielles des divers segments du membre sont une expression de la masse du blastème squelettogène. Comptes Rendus Soc. Biol., *94:* 968–971.

Butler, E. G.   1933   The effects of x-radiation on the regeneration of the fore limb of *Amblystoma* larvae. J. Exp. Zool., *65:* 271–315.

——— 1935   Studies on limb regeneration in x-rayed *Amblystoma* larvae. Anat. Rec., *62:* 295–307.

Butler, E. G., and J. P. O'Brien 1942   Effects of localized x-radiation on regeneration of the urodele limb. Anat. Record, *84:* 407–413.

Butler, E. G., and O. E. Schotté   1941   Histological alterations in denervated non-regenerating limbs of urodele larvae. J. Exp. Zool., *88:* 307–341.

Cahn, R. D., and M. B. Cahn 1966   Heritability of cellular differentiation: clonal growth and expression of differentiation in retinal pigment cells *in vitro*. Proc. Natl. Acad. Sci., U. S., *55:* 106–114.

Caro, L. G.   1964   High-resolution autoradiography. In: Methods in Cell Physiology. D. M. Prescott, ed. Academic Press, New York, vol. *1:* 327–363.

Cason, J. E.   1950   A rapid one-step Mallory-Heidenhain stain for connective tissue. Stain Tech., *25:* 225–226.

Chalkley, D. T.   1954   A quantitative histological analysis of forelimb regeneration in *Triturus viridescens*. J. Morph., *94:* 21–70.

Coon, H. G.   1966   Clonal stability and phenotypic expression of chick cartilage cells in vitro. Proc. Natl. Acad. Sci., U. S., *55:* 66–73.

Davis, L. E., and A. L. Burnett 1964   A study of growth and cell differentiation in the hepatopancreas of the crayfish. Develop. Biol., *10:* 122–153.

DeHaan, R. L. 1956   The serological determination of developing muscle protein in the regen-erating limb of *Amblystoma mexicanum*. J. Exp. Zool., *133:* 73–85.

Doljanski, L.   1930   Sur le rapport entre la prolifération et l'activité pigmentogène dans les cultures d'épithélium de l'iris. Comptes Rendus Soc. Biol., *105:* 343–345.

Drew, R. M., and R. B. Painter 1959   Action of tritiated thymidine on the clonal growth of mammalian cells. Radiation Research, *11:* 535–544.

Eggert, R. C.   1966   The response of x-irradiated limbs of adult urodeles to autografts of normal cartilage. J. Exp. Zool., *161:* 369–390.

Fankhauser, G.   1945   The effects of changes in chromosome number on amphibian development. Quart. Rev. Biol., *20:* 20–78.

Fankhauser, G., and R. R. Humphrey   1943   The relation between number of nucleoli and number of chromosome sets in animal cells. Proc. Natl. Acad. Sci., U. S., *29:* 344–350.

Fankhauser, G., and R. C. Watson   1942   Heat induced triploidy in the newt, *Triturus viridescens*. Proc. Natl. Acad. Sci., U. S., *28:* 436–440.

Flax, M. H., and M. H. Himes   1952   Microspectrophotometric analysis of metachromatic staining of nucleic acids. Physiol. Zool., *25:* 297–311.

Godlewski, E.   1928   Untersuchungen über Auslösung und Hemmung der Regeneration beim Axolotl. Wilhelm Roux' Arch. Entwicklungsmech. Organ., *114:* 108–143.

Goss, R. J.   1958   Skeletal regeneration in amphibians. J. Embryol. Exp. Morphol., *6:* 638–644.

Grobstein, C.   1959   Differentiation of vertebrate cells. In: The Cell. J. Brachet and A. E. Mirsky, eds. Academic Press, New York, vol. *1:* 437–496.

Hay, E. D.   1952   The role of epithelium in amphibian limb regeneration, studied by haploid and triploid transplants. Am. J. Anat., *91:* 447–482.

——— 1958   The fine structure of blastema cells and differentiating cartilage cells in regenerating limbs of *Amblystoma* larvae. J. Biophys. Biochem. Cytol., *4:* 583–592.

——— 1959   Electron microscopic observations of muscle dedifferentiation in regenerating *Amblystoma* limbs. Develop. Biol., *1:* 555–585.

——— 1962   Cytological studies of dedifferentiation and differentiation in regenerating amphibian limbs. In: Regeneration. D. Rudnick, ed. Ronald Press, New York, pp. 177–210.

Hay, E. D., and D. A. Fischman   1961   Origin of the blastema in regenerating limbs of the newt *Triturus viridescens*. Develop. Biol., *3:* 26–59.

Hertwig, G. 1927   Beiträge zum Determinations- und Regenerationsproblem mittels der Transplantation haploidkerniger Zellen. Wilhelm Roux' Arch. Entwicklungsmech. Organ., *111:* 292–316.

Holtzer, H., J. Abbott, J. Lash and S. Holtzer 1960   The loss of phenotypic traits by differentiated cells in vitro. I. Dedifferentiation of cartilage cells. Proc. Natl. Acad. Sci., U. S., *46:* 1533–1542.

Konigsberg, I. R. 1963 Clonal analysis of myogenesis. Science, *140:* 1273–1284.

Kopriwa, B. M., and C. P. Leblond 1962 Improvements in the coating technique of radioautography. J. Histochem. Cytochem., *10:* 269–284.

Laufer, H. 1959 Immunochemical studies of muscle proteins in mature and regenerating limbs of the adult newt, *Triturus viridescens.* J. Embryol. Exp. Morphol., 7: 431–458.

Morrill, C. V. 1918 Some experiments on regeneration after exarticulation in *Diemyctylus viridescens.* J. Exp. Zool., 25: 107–133.

O'Steen, W. K., and B. E. Walker 1961 Radioautographic studies of regeneration in the common newt. Anat. Rec., *139:* 547–555.

Owen, D. B. 1962 Handbook of statistical tables. Addison-Wesley, Reading, Mass., pp. xii + 580.

Patrick, J., and R. Briggs 1964 Fate of cartilage cells in limb regeneration in the axolotl (*Ambystoma mexicanum*). Experientia, 20: 431–432.

Reyer, R. W. 1962 Regeneration in the amphibian eye. In: Regeneration. D. Rudnick, ed. Ronald Press, New York, pp. 211–265.

Riddiford, L. M. 1960 Autoradiographic studies of tritiated thymidine infused into the blastema of the early regenerate in the adult newt, *Triturus.* J. Exp. Zool., *144:* 25–32.

Rose, S. M. 1948 Epidermal dedifferentiation during blastema formation in regenerating limbs of *Triturus viridescens.* J. Exp. Zool., *108:* 337–361.

Rose, F. C., and S. M. Rose 1965 The role of normal epidermis in recovery of regenerative ability in x-rayed limbs of *Triturus.* Growth, *29:* 361–363.

Salpeter, M. M., and M. Singer 1962 The fine structure of mesenchymatous cells in regenerating limbs of larval and adult *Triturus.* In: Electron Microscopy. S. S. Breese, ed. Academic Press, New York, vol. 2: 00–12.

Sauer, M. E., and B. E. Walker 1961 Radiation injury resulting from nuclear labeling with tritiated thymidine in the chick embryo. Radiation Research, *14:* 633–642.

Siegel, S. 1956 Nonparametric Statistics for the Behavioral Sciences. McGraw-Hill, New York.

Steen, T. P., and C. S. Thornton 1963 Tissue interaction in amputated aneurogenic limbs of *Ambystoma* larvae. J. Exp. Zool., *154:* 207–222.

Steinberg, M. S. 1957 A nonnutrient culture medium for amphibian embryonic tissues. Carnegie Institution of Washington Year Book, 56: 347–348.

Stone, L. S. 1950a Neural retina degeneration followed by regeneration from surviving retina pigment cells in grafted adult salamander eyes. Anat. Rec., *106:* 89–110.

———— 1950b The role of retina pigment cells in regenerating neural retinae of adult salamander eyes. J. Exp. Zool., *113:* 9–32.

———— 1955 Regeneration of the iris and lens from retina pigment cells in adult newt eyes. J. Exp. Zool., *129:* 505–534.

Thornton, C. S. 1938a The histogenesis of muscle in the regenerating fore limb of larval *Amblystoma punctatum.* J. Morph., 62: 17–47.

———— 1938b The histogenesis of the regenerating fore limb of larval *Amblystoma* after exarticulation of the humerus. J. Morph., 62: 219–235.

———— 1953 Histological modifications in denervated injured fore limbs of *Amblystoma* larvae. J. Exp. Zool., *122:* 119–149.

Thornton, C. S., and D. W. Kraemer 1951 The effect of injury on denervated unamputated fore limbs of *Amblystoma* larvae. J. Exp. Zool., *117:* 415–437.

Trampusch, H. A. L., and A. E. Harrebomée 1965 Dedifferentiation a prerequisite of regeneration. In: Regeneration in animals and related problems. V. Kiortis and H. A. L. Trampusch, eds. North-Holland Publishing Company, Amsterdam, pp. 341–376.

Trinkaus, J. P. 1956 The differentiation of tissue cells. Am. Naturalist, *90:* 273–289.

Truong, S. T., and E. J. Dornfeld 1955 Desoxyribose nucleic acid content in the nuclei of salamander somatic tissues. Biol. Bull., *108:* 242–251.

Vorontsova, M. A., and L. D. Liosner 1960 Asexual Propagation and Regeneration. Pergamon Press, New York, pp. 432–433.

Wallace, H. 1963 Nucleolar growth and fusion during cellular differentiation. J. Morph., *112:* 261–278.

Weiss, P. 1925 Unabhängigkeit der Extremitätenregeneration vom Skelett (bei *Triton cristatus*). Wilhelm Roux' Arch. Entwicklungsmech. Organ., *104:* 359–394.

Whittaker, J. R. 1963 Changes in melanogenesis during the dedifferentiation of chick retinal pigment cells in cell culture. Develop. Biol., 8: 99–127.

Xeros, N. 1962 Deoxyriboside control and synchronization of mitosis. Nature, *194:* 682–683.

1 Living cartilage from a humerus, cleaned and ready for grafting. $\times$ 210.

2 Section of control limb cartilage piece 6, see table 1. $\times$ 160.

3 A 54 day limb regenerate representative of the abnormality observed in some limbs labeled with 15 μc of H^3-thymidine (see text, p. 10). $\times$ 13.

4 An early cartilage cell containing three nucleoli which became radioactive from transfer of isotope in limb 21–3L (see table 4). The actual grain count for this cell is 12, with eight grains visible in this photomicrograph. $\times$ 2800.

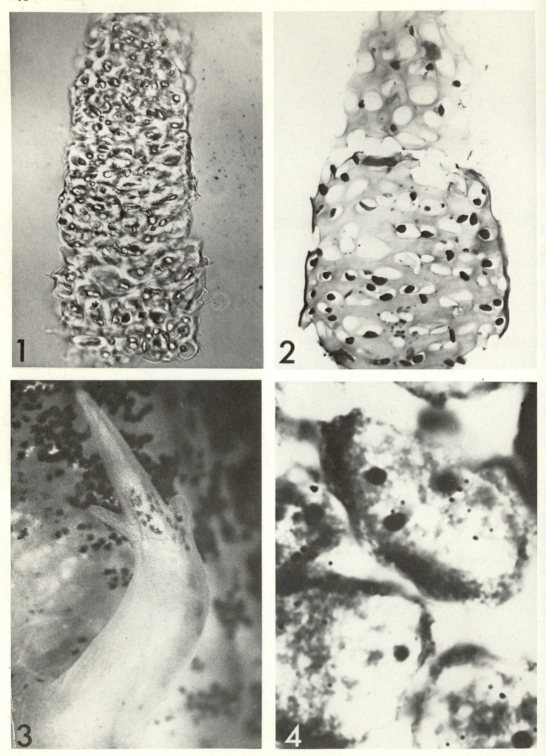

PLATE 2

5 The polyploid muscle connective tissue cell indicated by the arrow contains five nucleoli, with only four of them visible in this photomicrograph. The limb containing this cell (24–24L) was eliminated from consideration in the data (see p. 15), because of the ambiguity caused by the presence of polyploid cells. × 530.

6 An early H^3-thymidine labeled chondrocyte in the humerus of limb 21–2L. Since it originated from a limb cartilage graft, this cell illustrates stability as to cell type. × 3100.

7 A differentiating triploid chondrocyte that originated from a small limb cartilage graft into limb 24–12L. × 1600.

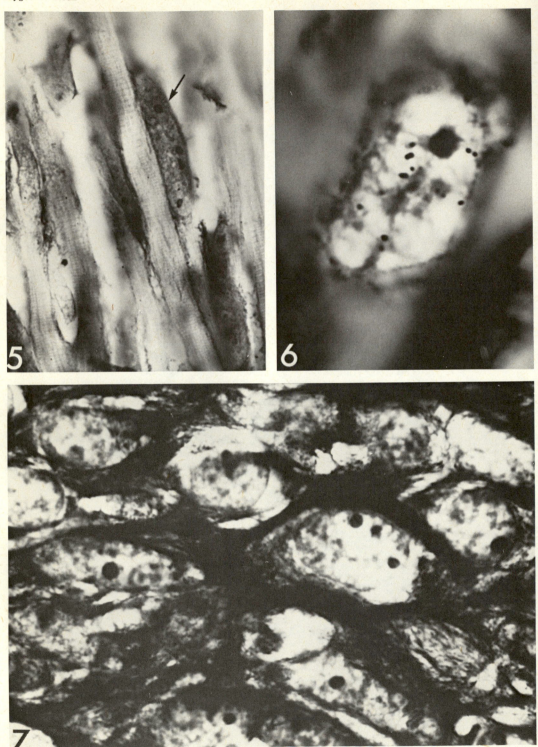

PLATE 3

8   Labeled chondrocytes that originated from limb muscle grafts, illustrating the incorporation of cells from muscle into limb regenerate cartilage.

   a $3NH^3$chondrocyte in the wrist of limb 21–25R. Plane of focus includes the three nucleoli (arrows) and some of the 32 silver grains found over this nucleus. $\times$ 3500.

   b   Labeled chondrocyte with three nucleoli, found in humerus of limb 24–8L. $\times$ 1600.

9   The differentiating muscle cell with three nucleoli originated from the same graft that gave rise to the cartilage cell in figure 8a. $\times$ 1300.

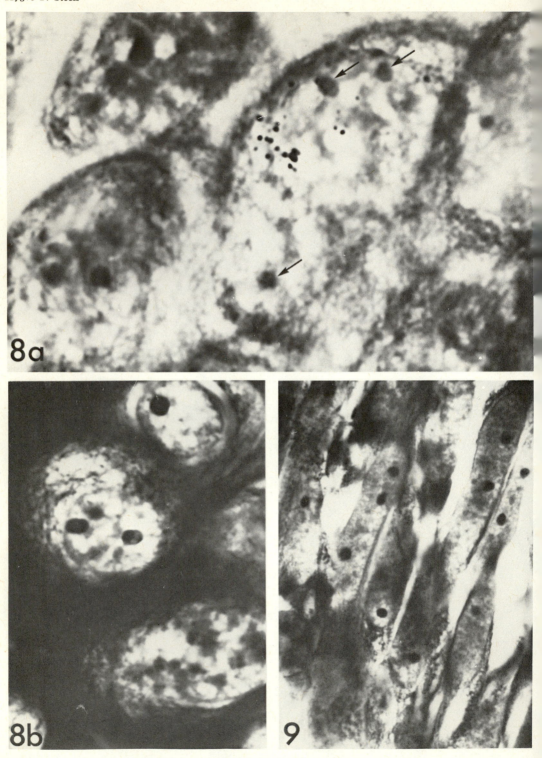

8a

8b

9

Reprinted from *Develop. Biol.,* **3,** 26–59 (1961)

5

Origin of the Blastema in Regenerating Limbs of the Newt *Triturus viridescens*

An Autoradiographic Study Using Tritiated Thymidine to Follow Cell Proliferation and Migration[1]

Elizabeth D. Hay[2] and Donald A. Fischman

*Departments of Anatomy, Cornell University Medical College,
New York, New York,
and Harvard Medical School, Boston, Massachusetts*

Accepted September 1, 1960

INTRODUCTION

Recent advances in autoradiography have led to the development of a technique by which newly synthesized nucleic acids can be located in the cell with great accuracy. A tritiated nucleoside has been produced (Taylor *et al.* 1957; Firket and Verly, 1958) which is incorporated into deoxyribonucleic acid (DNA) during its synthesis by the cell. Hughes *et al.* (1958) demonstrated the value of tritiated thymidine in determining sites of cell proliferation in tissues. This radioactive nucleoside can also be employed to follow chromosome duplication with considerable accuracy (Taylor *et al.,* 1957) and is useful in analyzing the temporal relation of DNA synthesis to cell division (Firket, 1958; Firket and Verly, 1958; Painter, *et al.,* 1958b; Painter and Drew, 1959). Moreover, by fixing tissues at longer intervals after treatment with tritiated thymidine, the subsequent fates of the cells labeled at the time of treatment can be studied (Leblond *et al.,* 1959; Cronkite *et al.,* 1959; Everett *et al.,* 1960; Sidman *et al.,* 1959; Sauer and Walker, 1959).

The present study is an application of this technique to an analysis of cell proliferation and cell migration in blastema formation during limb regeneration in *Triturus viridescens.* Most investigators now accept the theory that the blastema arises by a transformation of the

[1] Supported by Grant C-3708 from the United States Public Health Service.
[2] Present Address: Harvard Medical School, Boston, Massachusetts.

26

cells composing the muscle, cartilage, and other inner tissues of the limb stump into mesenchyma-like cells (see reviews by Thornton, 1938; Schotté, 1939). In 1948, however, Rose presented evidence reviving an older theory (Godlewski, 1928) that also limb epidermis transformed into blastema cells. The implications with respect to stability of cell differentiation that are raised by the possibility of epidermal dedifferentiation are important, and investigators have attempted to prove (Hay, 1952; Rose, Quastler and Rose, 1955) and disprove (Karczmar and Berg, 1951; Heath, 1953) the theory without convincing results. The best evidence against an epidermal transformation into blastema cells was presented by Chalkley in 1954. Chalkley showed that cell division begins in the inner limb tissues much earlier and more proximally than was previously realized, and could account for the appearance of the blastema without the necessity of postulating an epidermal origin of cells. The present study confirms Chalkley's determinations of sites of cell proliferation in the regenerate and provides the first direct evidence, by tracing labeled cells, that blastema cells do indeed originate from the dedifferentiating internal tissues and not from the apical limb epidermis.

In the first experimental series, the location and number of cells engaged in DNA synthesis during blastema formation was determined by fixing regenerates on the day they received tritiated thymidine. The information gained from this series was then utilized in planning two additional experiments in which the regenerating limbs were followed at daily intervals after treatment with the radioactive nucleoside. In the first group, the muscle and connective tissues of the stump were labeled and traced, whereas in the second group, epithelium and blood cells were followed. It is possible to obtain differential labeling of the tissues indicated above by injecting the animal with tritiated thymidine at a time when the inner stump tissues are synthesizing DNA but the apical epithelium is not (Series II), and at a time when epithelium and blood are synthesizing DNA but the inner limb tissues are not (Series III). An abstract of this work has been published (Hay and Fischman, 1960).

MATERIALS AND METHODS

Adult newts of the species *Triturus viridescens viridescens* were obtained in Massachusetts. Animals were kept in groups of six or more in small aquaria, rather than in individual fingerbowls, because they molt too often when kept in the smaller containers. The experiments were conducted at 20°C, a temperature which allowed us to compare our stages in regeneration with those observed by both Chalkley (1954) and Rose (1948). The forelimbs of each newt were amputated through the distal third of the radius and ulna either before or after administration of the isotope to the animal. Each newt was injected intraperitoneally with 1.25 μc of tritiated thymidine in 0.1 cc of sterile water at hourly intervals for a total of four injections in all the experiments described. In a preliminary experiment, regenerating limbs from animals receiving only one 1.25-μc dose of the isotope were fixed at daily intervals during regeneration. The results were similar to those reported in Series I. The more prolonged treatment (1.25 μc × 4) was chosen, however, with the hope that the total amount of labeling would be somewhat increased, thus permitting cells to be followed for a longer period of time in the cell migration studies. We hesitated to increase the dose further because of reports and/or warnings of a toxic effect of high levels of the isotope (Painter *et al.*, 1958a; Plaut, 1959; Greulich, 1960). The dosage employed in this study (1.25 μc × 4) does not interfere with the general growth and differentiation of the regenerate. We have followed a number of treated limbs for several months of regeneration and found that they resembled the controls in all important respects.

The tritiated thymidine, purchased from Schwarz Bioresearch, Inc., Mt. Vernon, New York, had a specific activity ranging from 0.36 to 3.0 curie per millimole. Determinations of the radioactivity of blood smears made in an ionization chamber with a Nuclear Corporation Scaling Unit (model 165) showed that circulating plasma thymidine had fallen to a nonmeasurable level 3 hours after the last intraperitoneal injection. It was assumed that the available thymidine had been utilized by this time, and therefore the limbs in Series I were fixed 3 hours after the last injection to analyze the immediate uptake of the isotope in the regenerating tissues. There was no evidence of isotope incorporation after the day of injection in any of the experiments.

Details on the treatment of the regenerating limbs in Series I, II, and III are presented in the observations. Two limbs were fixed on each of the dates stated there. This report describes 100 regenerating limbs from newts receiving four injections of 1.25 μc of tritiated thymidine. An additional 50 limbs were studied in preliminary experiments using lower doses of thymidine.

Regenerating limbs were fixed in Bouin's fluid, embedded in paraffin, and sectioned longitudinally at 10-μ thickness. The serial sections were floated on a mixture of gelatin and water at 45°C and picked up on slides which were then dried overnight in a 37°C oven. The paraffin was removed with xylene and the slides were passed through 100% alcohol into a mixture of ether and alcohol and then pure ether. From ether, they were dried at room temperature. No protective coating was placed over the sections.

The melted-emulsion technique refined by Messier and Leblond (1957) was used to coat the sections with emulsion. We are grateful to Eastman Kodak, Inc., Rochester, New York for giving us the NTB-3 bulk emulsion employed in this study. The coated slides were placed in light-proof boxes for 3 weeks to expose the emulsion and then developed in Kodak D-19 developer for 6 minutes at 18°C. After 10 minutes in acid fixer, they were washed 1 hour in running water and stained with the Cason technique (acid fuchsin, aniline blue, and orange G) recommended by Sidman *et al.* (1959). A few slides were stained with hematoxylin and eosin.

The report to follow is largely descriptive. A few calculations of percentages of epithelial cells incorporating thymidine are included in Series I because they are more useful than the terms "many" and "few" in comparing regenerating and nonregenerating epidermis. The percentage of proximal epithelial cells incorporating thymidine was estimated by counting the number of radioactive nuclei per 200 contiguous cells in three or more representative longitudinal sections through the midplane of the limb. The apical epithelial cap was defined as the thickened epidermis distal to a line drawn at the level of the cut ends of the bones in a plane perpendicular to the long axes of the bones. In estimating the percentage of cells synthesizing DNA in the apical cap, all the radioactive cells in the cap were counted in three or more representative sections of each limb, and this figure was divided by the total number of apical epidermal cells in these sections. To give the reader some idea of the relative changes in number of labeled internal cells at different stages in regeneration, the number of radioactive internal cells in three or more representative sections was counted. The average number of radioactive cells per section was multiplied by the number of 10-μ sections through the regenerate. The inner cells are approximately 10 μ in diameter, and the figure obtained is a rough estimate of the total number of labeled cells present in the regenerate.

These rough estimates are useful in describing the order of magnitude of the increase in labeling in the internal tissues as regeneration progresses.

OBSERVATIONS

Series I. Regenerating Limbs Fixed 3 Hours after Injection of Tritiated Thymidine

The object of the first experimental series was to determine locations of cells engaged in DNA synthesis during blastema formation. The nuclear uptake of tritiated thymidine would be expected to follow the general pattern of mitotic activity, and, indeed, the major sites of thymidine incorporation did correspond with areas previously determined (Chalkley, 1954) to have high mitotic indexes during regeneration of the limb. Since thymidine incorporation occurs before actual mitosis, the present method yielded new information on the temporal and spatial location of cells actively preparing for proliferation. The account to follow describes sites of thymidine incorporation in limbs fixed 1, 2, 3, 4, 5, 8, 10, 12, 15, 18, 20, 22, 24, 26, and 28 days after amputation. The animals were treated with thymidine on the day they were fixed (see Materials and Methods).

Days 1–4 of regeneration. During the first 2 days after amputation, the major cell activity in the limb stump is concerned with closure of the wound and removal of debris. Epidermis migrates over the ex-

FIG. 1. Thickened wound epidermis migrating over the amputation surface 1 day post amputation. Part of the radius appears in the lower right-hand portion of the photomicrograph, and the cut end of the bone is slightly farther to the right, not shown in the picture. The limb was treated with tritiated thymidine 3 hours before fixation. There are fewer exposed grains over the nuclei of these migrating epidermal cells than over nuclei of nonmigrating epidermis (Fig. 2). Three of the radioactive nuclei are indicated by arrows. DNA synthesis in these apical epithelial cells will decrease further the second day and essentially disappear by the third day post amputation. Magnification: × 400.

FIG. 2. Epidermis of the elbow region, treated as above with tritiated thymidine 3 hours before fixation. This epithelium is typical of epidermis elsewhere over the body and exhibits a high degree of thymidine incorporation in ∼ 15% of its nuclei. Three of the radioactive nuclei are indicated by arrows. This epithelium is 3–4 cells in thickness and has a basement membrane (*bm*) between it and the underlying connective tissue. A mucous gland in the dermis is seen in the center of the photomicrograph, with its duct opening to the surface. A region of skeletal muscle (*M*) appears in the lower half of the picture. Magnification: × 400.

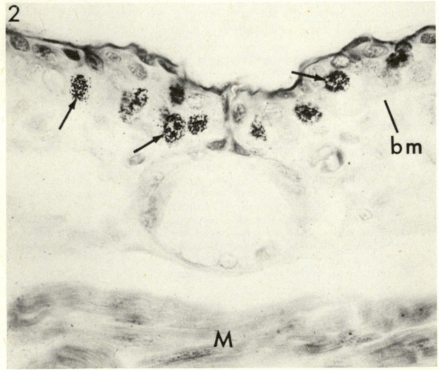

posed soft tissues and, except for the projecting bony surfaces, the wound is covered within a day after amputation by epithelium 3–8 cells in thickness. This distal migration results in a thinning of the side epithelium for a distance ~ 1.5 mm proximal to the tip. By 2 days, the thickened wound epithelium has reached the cut ends of the bones and covered the amputation surface completely. It forms an early apical epithelial cap 6–8 cells in thickness at this time. There is some hemorrhage and histolysis in the injured underlying tissues. Within 4 days, the white blood cells and macrophages which appeared in the area after amputation have begun to phagocytize the wound debris and the remaining muscle and connective tissue nuclei show no further sign of degeneration.

The intensity of thymidine incorporation by the epithelial cells migrating over the wound 1 day after amputation is considerably less (Fig. 1)[3] than that shown by nonmigrating limb epidermis (Fig. 2). Approximately 15% of the epidermal cells of the unamputated limb incorporate thymidine under the usual conditions of epidermal growth and replacement that occur continuously all over the body. By the second day after amputation, however, the percentage of nuclei incorporating thymidine in the apical limb epidermis covering the wound has dropped from 15% to less than 5% and these nuclei show even less intensity of labeling than that pictured in Fig. 1. By the third

---

[3] With the exception of the diagrams, all figures are unretouched photomicrographs of autoradiographs.

Fig. 3.   Low magnification photomicrograph of the distal portion of a regenerating limb 8 days post amputation, treated with tritiated thymidine 3 hours before fixation. White blood cells (w) have accumulated around the cut end of the amputated bone and apical epithelium has piled up lateral to this area, forming a cap over the distal end of the regenerate. The border between apical and proximal epithelium (at the large arrow) is relatively sharp with respect to incorporation of thymidine. This region (arrow) is magnified in Fig. 4. Labeled nuclei are indicated by small arrows and pigment cells appear at p. The junction between apical and proximal epithelium on the left is located more proximally and is not shown in the photograph. Magnification: × 90.

Fig. 4.   Higher magnification of the junction of the proximal (Pr) and apical (Ap) epithelium shown at the large arrow in Fig. 3. The heavily labeled nuclei in the proximal epithelium contrast strikingly with the unlabeled nuclei in the apical epithelium. Pigment cells in the dermis appear at p and a labeled fibroblast at f. The border between dermis and epidermis is at the double-ended arrow. Magnification: × 400.

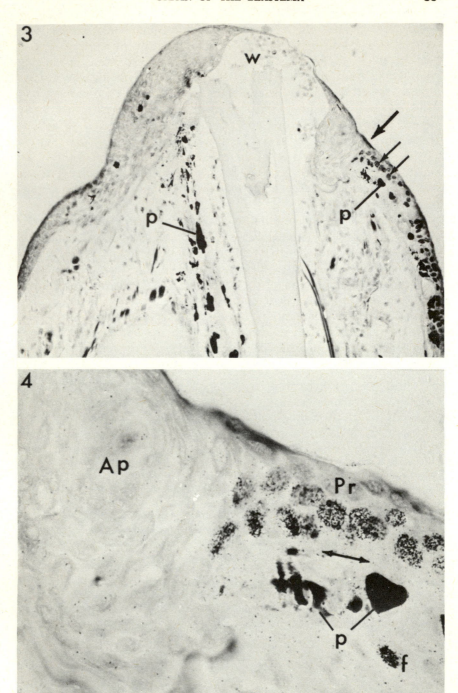

day after amputation, there is no longer any significant thymidine incorporation by the apical epithelial cap and this loss of the usual ability of epidermis to synthesize DNA characterizes the apical cap throughout the period of blastema formation to be described below.

Days 5–10 of regeneration. There is no thymidine incorporation in the internal tissues of the unamputated limb. As early as the fourth day after amputation, however, some of the cells in the muscle and connective tissue of the regenerating limb begin to synthesize DNA. These cells are diffusely located throughout the limb stump for a distance 1–2 mm proximal to the wound surface. As blastema formation proceeds, an increased number of cells in the muscle, endomysium, epimysium, nerve sheaths, periosteum, and loose connective tissue 1 mm proximal to the amputation surface begin to incorporate thymidine. Concomitant with the onset of DNA synthesis, the nuclei of these cells become larger and the cells become more mesenchyma-like in appearance, a transformation which leads to the production of a number of undifferentiated-appearing blastema cells by the end of this period. Evidence that the cells synthesizing DNA in these dedifferentiating tissues actually do give rise to the blastema cells will be presented in Series II. The volume of the limb stump increases rapidly after 7 days owing in part to division of these internal cells (Chalkley, 1954) and in part to transient edema (Singer and Craven, 1948).

During this period, the apical epidermal cap shows practically no DNA synthesis (Figs. 3 and 4). In the cap proper, which is 8–14 cells in thickness and lies distal to the cut ends of the bones, only occasional labeled cells are encountered in a section. However, the stump epithelium over an area 1.5 mm proximal to the apical cap shows a relatively high degree of thymidine incorporation at 5 days (25% of the cells

---

FIG. 5. Distal end of a regenerating limb 18 days post amputation, treated with tritiated thymidine 3 hours before fixation. The junction between proximal (*Pr*) and apical (*Ap*) epithelium is still clear cut (large arrow). Proximal epithelial cells show intense thymidine incorporation (top inset, × 400). A few apical epithelial cells show slight thymidine incorporation (lower inset, × 400) and have probably recently migrated into the apical cap (cf. Fig. 1). Numerous cells in the internal tissues of the stump now show intense thymidine incorporation (muscle, *M*; nerve sheaths, *N*; periosteum and endosteum, small arrows, lower part of picture). The newly formed blastema cells (*B*) are located near these dedifferentiating tissues and also show intense thymidine incorporation. An osteoclast (*Os*) is present in close contact with the bone, which has been partly reabsorbed. Magnification: × 125.

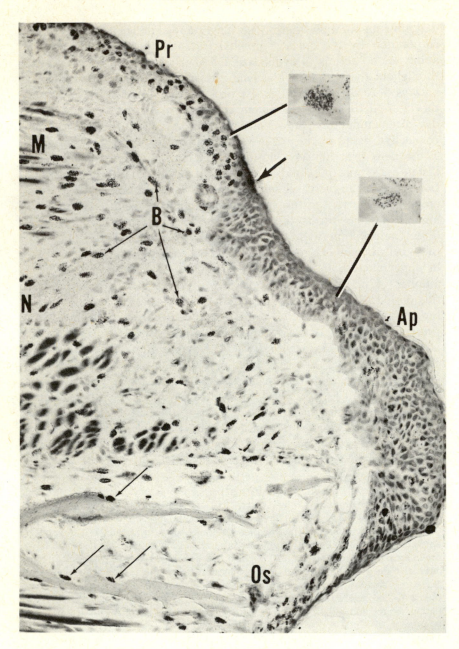

labeled) which increases further at 8 days (35% of the cells labeled). A fairly sharp border usually exists between the active proximal epithelium and the inactive apical epithelium (Fig. 4). By 10 days after amputation the proximal epithelium returns to a lower rate of DNA synthesis and remains at this level (~ 20% of the cells labeled). This distribution of thymidine incorporation correlates well with mitotic indexes in proximal and distal epidermis (Chalkley, 1954). The increased mitotic activity of the proximal epithelium is associated with a definite increase in the size of the apical cap 10–12 days post amputation; at the same time the inner mass of the regenerate is also increasing and this lateral expansion is accommodated by the proliferating proximal epithelium.

Days 10–20 of regeneration. During this period, the apical epidermal cap reaches its greatest thickness (12–15 cells). A few labeled cells may be found in the cap, but they generally have fewer grains exposed per nucleus than the cells of the proximal epithelium (Fig. 5). In the intensity of their radioactivity, they resemble the cells of the epidermis migrating over the wound surface 1–2 days post amputation (Fig. 1). Since the apical cap is increasing in size at this time, it seems likely that these occasional labeled nuclei belong to cells which have recently migrated into the cap from the mitotically active proximal epidermis. This interpretation is consistent with cell migration studies (Series II and III). A moderately sharp boundary between the apical cap epidermis and the proximal epithelium with respect to thymidine incorporation persists (Fig. 5). [It might be noted here that the area of actively proliferating proximal epithelium immediately lateral to the apical cap may be thicker than shown in Fig. 5 (*Pr*). Tangential sections through this proximal area could lead to the erroneous conclusion that the apical cap proper was incorporating thymidine.]

FIG. 6.   Higher magnification photomicrograph of an area of dedifferentiating muscle from a regenerate 12 days post amputation, fixed 3 hours after treatment with tritiated thymidine. Four nuclei in the intact muscle fiber (*M*) shown here have begun to synthesize DNA. It is more common, however, to find synthetic activity in the nuclei (small arrow) of dedifferentiating mononucleate muscle fragments. Magnification: × 400.

FIG. 7.   Higher magnification of a nerve (*N*) similar to that shown at *N* in Fig. 5. Schwann cells and the fibroblasts of the endoneurium and epineurium (three of which are indicated by small arrows) are synthesizing DNA. An area of intact muscle appears at *M*. From a regenerate fixed 20 days post amputation, 3 hours after thymidine injection. Magnification: × 400.

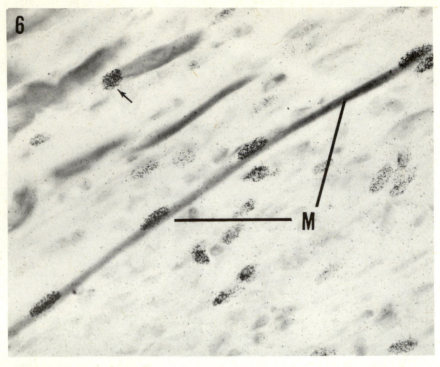

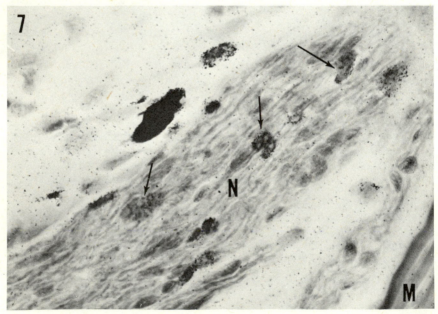

In contrast with the low degree of thymidine incorporation in the apical epidermis, a high degree of nucleoside utilization is exhibited by the dedifferentiating cells of the inner stump tissues and by the blastema cells. The process of dedifferentiation of muscle, endomysium, nerve sheaths, and other connective tissue (Figs. 5–7), including cartilage when present in the stump, is always accompanied by active DNA synthesis. Some of the nuclei in dedifferentiating muscle fibers may begin to synthesize DNA while the fiber is still multinucleated (Fig. 6), but it is more common to find synthetic activity in the rounded nuclei of the mononucleate muscle fragments derived from the syncytial muscle fibers. This is completely in keeping with Chalkley's observations of fewer mitoses in intact muscle fibers than in the mononucleated cells of the limb stump and, indeed, the pattern of DNA synthesis closely follows that of mitosis (Chalkley, 1954). The number of internal cells dividing increases rapidly between 7 and 19 days post amputation (Chalkley, 1954). It is interesting to note, by way of comparison, that the 5-day limb stump contains approximately 2500 blastema and connective tissue cells synthesizing DNA, the 10-day stump 5000, and the 15–20-day stump, 10,000–20,000.

During the 10–20-day period after amputation of the limb, the size of the regenerate increases and an early blastema comprised of un-differentiated-appearing cells accumulates distally. The increase in cellular mass due to cell proliferation is not expressed by a significant growth in length of the limb at first, but rather by a lateral accumula-

---

FIG. 8.   Higher magnification of a part of the apical epithelial cap from the regenerate shown in Fig. 5 (18 days post amputation). White blood cells (*w*) and macrophages containing pigment granules and other inclusions are present in this epidermis. The epidermal cells are acidophilic and their cytoplasm stains intensely with the Cason technique. The cytoplasm of the blastema cells (*B*) is basophilic and cannot be seen in this photomicrograph. The blastema cells show a high degree of DNA synthesis, whereas the apical epidermis does not. The amputated surface of the bone is at the arrow. Magnification: × 400.

FIG. 9.   Apical epithelium and part of the blastema of a regenerate 28 days post amputation treated with tritiated thymidine 3 hours before fixation. The nuclei of the epithelial cells have now begun to synthesize DNA and are rounder (cf. Fig. 8). Blastema cells (*B*) show intense thymidine incorporation and are actively dividing. Their cytoplasm is not stained in this preparation. Magnification: × 400.

Note added in proof: The areas in the epithelium which appear empty in this reproduction did contain lightly stained cytoplasm in the original figure. A similar loss of cytoplasmic detail has occurred in the reproduction of Figs. 14 and 15.

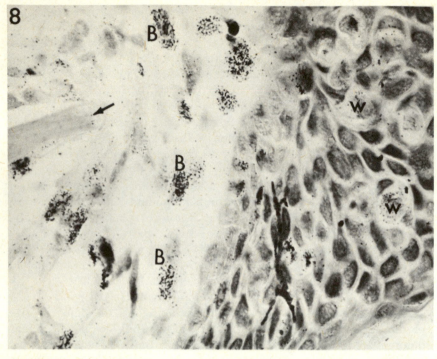

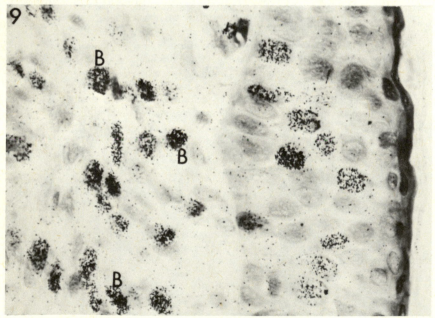

tion of blastema cells derived from the dedifferentiating inner stump
tissues. Longitudinal sections of limbs treated with tritiated thymidine
(Fig. 5, see also Fig. 10) emphasize the fact perhaps more clearly
than Chalkley's (1954) graphs of cross sections that the internal tissues
for a considerable distance (∼ 1 mm) proximal to the amputation
surface give rise to the blastema cells. The newly formed blastema
cells (B, Fig. 5) are found in close proximity to the dissociating muscle
and other formed inner tissues: they do not accumulate under the
apical epithelial cap until the end of this period.

Days 20–28 of regeneration. This period is characterized by the
distal elongation of the regenerate. The blastema is rapidly increasing
in size owing to cell proliferation, and the cells composing it are
actively incorporating thymidine (B, Fig. 9). The regenerate assumes
the shape of a cone and later will become paddlelike and redifferentiate
into a new limb.

In this period, the apical epidermal cap gradually loses the unusual
morphological features described in the previous section, although it
remains somewhat thickened (∼ 8 cells). The epithelial-blastemal
boundary becomes relatively smooth, a basement membrane is reestablished, and active thymidine incorporation begins in many of
the apical epithelial cells. Mitoses are more common and, except for
its thickening, the apical epidermis (Fig. 9) is now similar in its appearance to epidermis elsewhere. The greatest amount of thymidine
incorporation, however, still occurs proximal to the apical cap.
Chalkley (1954) has shown that the peak of epidermal mitotic activity
also continues to be located proximally until after the limb redifferentiates.

*Series II. Regenerating Limbs Treated with Thymidine at 5, 10, and 15
Days post Amputation and Fixed at Daily Intervals after Treatment*

The first experimental series demonstrated a high degree of DNA
synthesis in the dedifferentiating cells of the inner tissues of the
amputated stump and revealed that the apical epidermal cap loses the
usual ability of epithelium to synthesize DNA (Fig. 10). On the
premise that DNA synthesis is an important characteristic of cells
transforming into blastema cells, these data suggest that the blastema
originates from the internal tissues, and not from the inactive apical
epidermis. This important premise can be proved by tracing the cells
labeled at the time of exposure to the isotope. To this end, tritiated
thymidine was injected at 5, 10, or 15 days after amputation to label

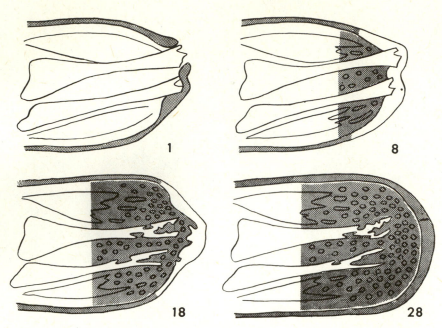

FIG. 10. Diagram summarizing the results of Series I. The regenerating limbs were treated with tritiated thymidine on the same day they were fixed. Sites of DNA synthesis are indicated by stippling. On the first day post amputation, the epidermis is the only limb tissue incorporating thymidine (1). The amount of thymidine incorporation decreases in migrating epidermal cells. Three days later, there is no significant thymidine incorporation by the apical epithelium. At 8 days post amputation (8), the internal cells proximal to the amputation surface are actively synthesizing DNA and a few blastema cells (small circles) have appeared. At 18 days, the number of dedifferentiating cells incorporating thymidine has increased and many blastema cells are present. These proliferating blastema cells are actively synthsizing DNA. By 28 days post amputation (28) the apical epithelium has begun to synthesize DNA again and the distal growth of the blastema has resulted in elongation of the regenerate.

the dedifferentiating cells of the internal tissues. Representative regenerating limbs were fixed at daily intervals after thymidine administration to follow the labeled cells as they transform into blastema cells. The description below will deal in detail with limbs exposed to tritiated thymidine on the tenth day of regeneration and fixed at 10, 11, 12, 13, 14, and 15 days post amputation. The nomenclature, 10, 10 + 1, 10 + 2, 10 + 3, 10 + 4, and 10 + 5 days, will be used to relate the time of fixation to the time of treatment of the regenerate. In two

additional experiments, limbs were treated at 5 days and fixed thereafter at daily intervals 5 through 10 days post amputation, and treated
at 15 days and fixed daily 15 through 20 days post amputation. The
results of these experiments were so similar to the 10-day treated series
that they will be referred to only briefly.

In the regenerating limb treated with thymidine on the tenth day
of regeneration, and fixed 3 hours thereafter, approximately 5000 nuclei
in the internal stump tissues are heavily labeled at the time of treatment. Three days later (13 days post amputation), the limbs in this
series contain twice as many labeled nuclei in the blastema and inner
stump tissues, and 4–5 days later (14–15 days post amputation) the
limbs treated on day 10 contain 12,000–18,000 labeled internal cells.
This increase in number is due to cell division and the label becomes
diluted. Major dilution of the isotope in the internal tissues, as indicated by decreased number of grains exposed over nuclei, occurs about
3 days after the initial injection. No labeled mitoses are seen 3 hours
after thymidine treatment, but after 1-day labeled mitoses appear
which represent divisions of cells that were synthesizing DNA at the
time of injection. As a rough estimate, it can be surmised that the
average fibroblast, muscle, or Schwann cell which incorporated thymidine at the time of injection has divided 3 days later. However, some
cells located in the proximal region of the regenerate may still be
heavily labeled 4–5 days after incorporating the thymidine and are
probably dividing less frequently than cells located more distally.

Almost all the mesenchyma-like cells that are formed in the period
10 + 1 to 10 + 5 days, and which have the morphological features of
blastema cells, are labeled (Fig. 11). It is thus apparent that they did

---

FIG. 11.   Blastema cells in a regenerate treated with tritiated thymidine 10
days post amputation and fixed 5 days later. Most of the blastema cells are labeled,
although less intensely than in Fig. 9. They have divided once since the injection
of thymidine and the label has become diluted. These blastema cells were derived
from the dedifferentiating inner tissues shown in Figs. 5–7. Note that osteoclasts
(arrows) are not labeled. Magnification: × 400.

FIG. 12.   Blastema cells in a 15-day regenerate from an animal injected with
tritiated thymidine before amputation. The internal limb tissues were not labeled
and the blastema cells show no radioactivity. Radioactive leucocytes (arrow)
which incorporated thymidine before amputation, at the time of injection, are
present in the regenerate. Osteoclasts (Os) contain labeled nuclei and were
probably derived by fusion of the mononuclear leucocytes. Contrast this figure
with Fig. 11. Magnification: × 400.

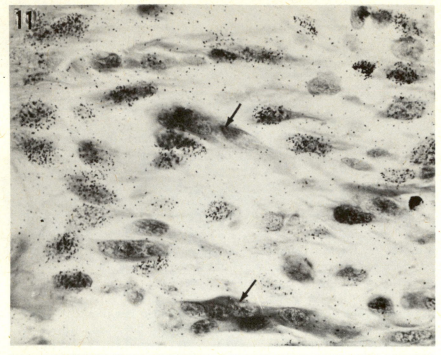

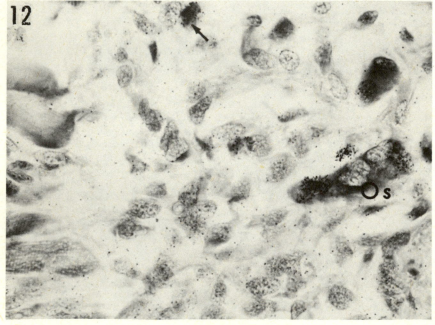

arise from cells actively synthesizing DNA at the time of injection
(Fig. 13). The results of the 5- and 15-day series are consistent with
this conclusion.

During the period under study, the apical epidermal cap increases
in thickness. The over-all number of labeled and unlabeled epithelial
cells which migrate into the cap is approximately equal to the number
of cells added to the cap during this period. At no time in any of the
experimental series did more cells appear to enter the apical cap than
would be required for its ensuing increase in size.

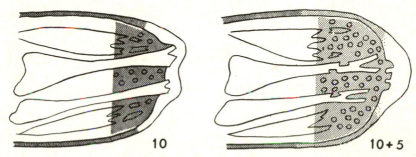

Fig. 13.   Diagram summarizing the results of Series II. The regenerating limb
on the left was exposed to tritiated thymidine 10 days after amputation and fixed
the same day. Thymidine was incorporated by the dedifferentiating internal tissues
and blastema cells (small circles), but not by the apical epithelial cap. The
regenerating limb on the right was treated with thymidine at the same time and
fixed 5 days later (10 + 5). The internal cells which incorporated thymidine at
10 days have now given rise to numerous blastema cells which are well labeled,
although the isotope has been diluted by cell division. Some of the proximal epithelial cells that incorporated thymidine on day 10 have moved into the apical
cap to increase its size. Not shown in the diagram are the internal cells synthesizing DNA at day 10 which are located nearer the elbow region. These move distally
by 10 + 5 days.

*Series III. Regenerating Limbs Treated with Thymidine 1 Day before
Amputation and Fixed at Daily Intervals after Amputation*

The first two series of experiments have indicated that blastema cells
arise from the dedifferentiating tissues of the inner stump which are
actively synthesizing DNA during the early phases of regeneration. It
is possible, by timing the injection of thymidine properly, to label the
epithelium but not the muscle and connective tissue of the regenerate.
The labeled epithelium can be followed, and in this manner the con-

clusion that the apical epidermis does not transform into blastema cells can be supported by a third line of evidence. Salamanders were injected before amputation, and after a sufficient interval for the removal of circulating thymidine, their forelimbs were amputated (see Materials and Methods). The epithelium is the only limb tissue that incorporates thymidine under these circumstances. Newly forming blood cells in their sites of origin elsewhere in the body are also labeled. Regenerating limbs of animals injected before amputation were fixed at 1, 3, and 6 hours and at 1, 2, 3, 4, 5, 6, 7, 8, 10, 12, 15, 20, and 28 days post amputation.

Days 0–4 of regeneration. One hour after amputation, the wound is still open and there is as yet no apparent migratory activity by the epithelium. By 3 hours, the epithelium immediately adjacent to the wound migrates as a mass several cell layers in thickness to the edges of the cut bones, and by 6 hours most of the tip is covered except the bone. Migrating epithelium continues to pile up lateral to the projecting radius and ulna until the cut ends are reached. At 1 day, a thin sheet of labeled epithelial cells may be seen migrating over the cut bony surfaces (Fig. 14). The labeled migrating epidermis forms an apical cap 6–8 cells in thickness 2 days after amputation. The first major dilution of the label in the epithelium occurs 3–4 days post amputation.

Days 5–10 of regeneration. The cells in the apical epithelial cap labeled by the preamputation injection of thymidine show no further change in the intensity of their radioactivity and presumably have divided only once. By 6 days post amputation, some dilution of label by a second cell division has begun proximally. This second dilution of isotope in the proximal epidermis does not become extensive until 10 days post amputation and is limited to the same proximal area ($\sim$ 1.5 mm from the wound surface) which reached a peak of new DNA synthesis at about 8 days. Thus, there is a close correlation between sites of DNA synthesis (Series I) and evidence of cell division and isotope dilution (Series II and III).

Beginning shortly after amputation there is an extravasation of labeled polymorphonuclear and mononuclear leucocytes throughout the limb which increases to $\sim$ 5,000 cells by 5 days after amputation. These labeled blood cells remain in the limb tissues until 12–15 days post amputation and then disappear. The osteoclasts present around fragments of bone are labeled (*Os*, Fig. 12) and probably originated from fusion of mononuclear leucocytes. A separate analysis of the origin and subsequent fates of these labeled blood cells will be published. Let it suffice to say here that they apparently do not divide after leaving the blood stream and may be distinguished from epithelium by their more intense radioactivity. In early stages they might be confused with regeneration cells, but since these labeled leucocytes are disappearing at the time most of the blastema cells are appearing (12–15 days post amputation) and are not found in later regenerates in significant number, it seems reasonable to conclude that they did not become blastema cells.

Days 10–28 of regeneration. The apical cap epithelium, which has not been actively proliferating, is still well labeled at 15 days (Fig. 15). A few more lightly labeled cells move into the periphery of the cap from the proximal epithelium between 10 and 15 days when the cap is increasing in size. The label in the epithelium ~ 1.5 mm proximal to the tip becomes more diluted owing to the active cell division there. At the elbow region epithelial cells not participating in the regenerative process are still well labeled. These cells begin to move toward the outer keratinized surface of the epithelium 12–15 days post amputation.

The blastema cells that appear in this period are not labeled (Fig. 12), whereas in Series II they were well labeled (Fig. 11).[4] The main difference in these two series is that the dedifferentiating internal tissues were labeled in Series II and were not labeled in the present case. On the other hand, the apical epithelium was not labeled in

[4] Figure 11 illustrates the 10–15 day series. However, essentially all the blastema cells present at 20 days are labeled if the internal tissues were labeled at 15 days, and the same was true in the 5- through 10-day series.

FIG. 14. Distal end of a regenerate from an animal injected with tritiated thymidine before amputation and fixed 1 day after amputation. The epithelium was labeled by the preamputation treatment. A sheet of labeled epithelium (arrows) can be seen migrating over the cut surface of the radius and a thick layer of apical epithelium appears lateral to the bone. Magnification: × 300.

FIG. 15. Distal end of a regenerate from an animal injected with tritiated thymidine before amputation and fixed 15 days after amputation. An apical epithelial cap has formed over the cut ends of the bone. The labeled cells have divided once and are not as intensely labeled as they were at day 1 (Fig. 14), and some of them have begun to move toward the outer surface of the epidermis. No transformation of these labeled epithelial cells into blastema cells can be demonstrated. Magnification: × 300.

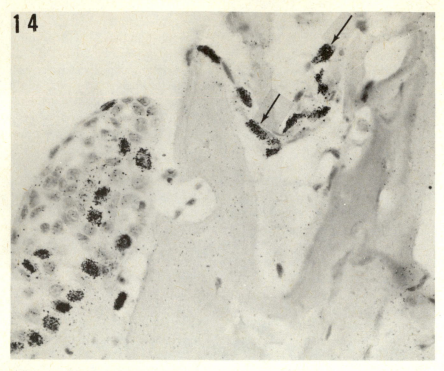

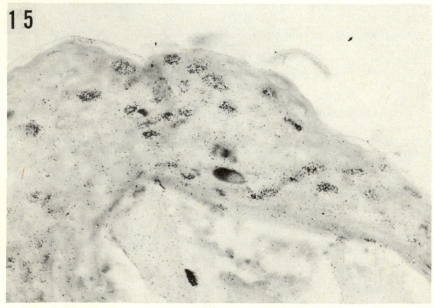

Series II, but is labeled in Series III (compare Figs. 13 and 16). Thus, these two series complement each other in supporting the conclusion that the blastema is derived from the inner limb tissues and not from the apical epithelial cap.

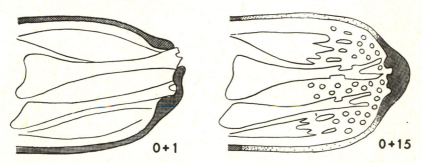

FIG. 16. Diagram summarizing the results of Series III. The animals were injected with tritiated thymidine before amputation to label the epidermis. The internal tissues of the limb are not labeled by such an injection. One day after amputation (0 + 1) the epidermis of the limb stump has migrated to cover the amputation surface. The apical epithelial cells do not continue to divide and therefore the radioactive label does not become very diluted in the apical cap (0 + 15). The proximal epithelium does proliferate extensively, and by 15 days post amputation (0 + 15) the radioactive thymidine incorporated before amputation has been considerably diluted in this region. Although the apical cap was well labeled during blastema formation, the blastema cells (small circles) are not labeled. They arose from the dedifferentiating inner tissues. Not shown in the diagram are the leucocytes which label in their tissues of origin and then migrate through the limb after amputation.

By 20–28 days after the initial injection of thymidine, much of the labeled epithelium all over the body has been lost by desquamation. The labeled cells in the apical cap can also be seen undergoing desquamation and by 28 days there are only a few labeled cells remaining there, all of these peripherally located.

DISCUSSION

The three series of experiments reported here, when considered together, provide a dynamic picture of cell proliferation and migration during the early phases of limb regeneration in *Triturus viridescens* (see Figs. 10, 13, and 16). They emphasize in a more dramatic manner than has heretofore been possible the important fact discovered by

Chalkley (1954) that dedifferentiation in blastema formation is accompanied by cell proliferation. The cells comprising the muscle, endomysium, periosteum, nerve sheaths, and loose connective tissue of the distal limb stump begin to synthesize DNA 4–5 days post amputation. Chalkley (1954) reported the first peak in cell division at 7 days post amputation, but he did not study earlier stages. This evidence of DNA synthesis in the inner tissues of the stump demonstrates that the dedifferentiating cells are not degenerating, a fact which has been recently emphasized by autoradiographic studies of protein synthesis (Bodemer and Everett, 1959) and electron microscopic studies (Hay, 1959). There are no significant cell deaths, as indicated by pycnotic nuclei, after the first few days post amputation. The leucocytes attracted into the area during the first 5 days do not increase in number after this interval and are probably concerned with removing wound debris rather than dying cells. It is misleading to describe this process as histolysis, because it directs attention away from the most important aspect of blastema formation: the release of living cells from the confines of their previous organization with accompanying active mitosis of these cells.

Prior to Chalkley's study in 1954, many investigators had concluded that blastema formation occurred without mitosis of the formed stump tissues, and this fundamental misunderstanding led some workers to postulate a partly hematogenic origin of the blastema (Colucci, 1884; Hellmich, 1930; Ide-Rozas, 1936; and others) and others to suggest an epidermal origin (Godlewski, 1928; Rose, 1948). Chalkley's careful analysis (Chalkley, 1954) clearly demonstrated that mitosis does occur in the muscle, epimysium, endomysium, periosteum, nerve sheaths, and other connective tissue of the limb stump as the cellular components of these tissues transform into blastema cells. It is apparent from Chalkley's study and from the present investigation that cell division in the proximal internal tissues begins early enough to account for the appearance of a blastema without postulating an epidermal or blood cell origin of cells or, for that matter, a reserve cell or purely fibroblast origin (Fritsch, 1911; Weiss, 1939; and others). The cell-tracing experiments reported here (Series II and III) leave no reasonable doubt that blastema cells do actually arise from the dedifferentiating inner tissues.

Dedifferentiation, as it occurs in the regenerating limb, i.e., the transformation of formed cell types into undifferentiated-appearing blastema cells, can now be said to consist of the following cytological and metabolic events: (1) The loss of specialized intracellular and extracellular products that were formed during cellular differentiation. Various aspects of this loss of structure have been described with the light microscope (see Butler, 1933; David, 1934; Thornton, 1938) and with the electron microscope (Hay, 1958, 1959). (2) Enlargement of nuclei and nucleoli with the commencement of active nucleic acid synthesis. The nucleocytoplasmic ratio increases and some cytoplasm is lost (see Hay, 1959, and others). (3) Increased cytoplasmic ribonucleoprotein granules and simplification of cytoplasmic organelles (Hay, 1958). The cytoplasm becomes diffusely basophilic as seen in stained preparations observed with the light microscope. At the same time there is an increase in protein synthesis by these cells (Bodemer and Everett, 1959). These morphological and metabolic features are consistent with the idea that dedifferentiating cells are beginning a phase of intracellular protein and nucleic acid synthesis preparatory to active cell proliferation. The division of the dedifferentiating cells that incorporated thymidine (Series II) substantiates this conclusion.

Neither the blood cells entering the limb nor the apical epidermis acquire the cytological features described above. Intracellular products such as tonofilaments do not disappear from apical epidermal cells (Salpeter and Singer, 1960; Hay, 1960). The cytoplasm of these epidermal cells continues to have an affinity for acid dyes (Fig. 8). Although protein synthesis persists (Bodemer and Everett, 1959), DNA synthesis decreases in the apical epithelium. Confidence in cytological criteria of dedifferentiation is upheld by cell-tracing studies, for if blood and epidermis are labeled before amputation (Series III), no transformation of these labeled cell types into blastema cells can be demonstrated. On the other hand, transformation of labeled inner tissues into blastema cells can readily be demonstrated (Series II)

No serious attention has been given the hematogenic theory of origin of the blastema since the irradiation experiments of Butler and others (Butler, 1935; Brunst and Chérémetiéva, 1936; Butler and O'Brien, 1942). However, Rose's work in 1948 continues to have considerable influence in the field of regeneration and a detailed review of his evidence for epidermal dedifferentiation in the light of the information gained in the present study seems worth while. The first line of evidence came from the fact that Nile blue dye placed on the periphery of the wound epithelium 2–7 days post amputation migrated into the apical epithelium. Our study of labeled epithelium supports Rose's conclusion that proximal epidermal cells are migrating into the cap. The correlated series of experiments (I, II, and III) permit the following over-all picture of epidermal migrations: proximal epidermis migrates to cover the wound surface and begins to form a thickened cap the day after amputation. This cap increases further in size 10–15 days post amputation, and the cells composing it cease to synthesize DNA. The majority of these apical cells probably divide only once after entering the cap, for the isotope does not become greatly diluted in apical epithelial cells which were permitted to incorporate thymidine before migration (Series III); such an apical cap remains well labeled and so can easily be traced. The cap increases enormously in size primarily because the limb epithelium proximal to the amputation surface is actively synthesizing DNA (reaching a peak at 8 days), proliferating, and supplying cells to the wound epithelium. In long-term experiments (Series III), labeling in this proliferating area of proximal epithelium becomes diluted ~ 10 days after injection of thymidine. If counts of epithelial cells moving into the cap are made (Series II), the number would seem to approximate the number of cells gained by the cap and not to exceed it, as would have to be the case if epidermis entered the blastema. Rose did not, however, rely on his dye studies for the conclusion that epidermis enters the blastema, but rather extended his investigation to quantitative considerations.

Rose calculated that in the 16–20 day period post amputation there was a loss of ~ 18,000 cells from the apical epidermis at the same time a blastema of about the same number of cells suddenly appeared. His data, as presented, are significant, and it is unlikely that selection of cases (Chalkley, 1959) seriously affected their validity. Rose believed, as indeed many others did also, that there was not enough cell division present to account for the appearance of the blastema. Chalkley's study (1954) showed that there was a considerable amount of mitosis in the internal tissues ~ 0.5 mm proximal to the wound surface, and the present study has shown that active DNA synthesis extends over a considerable proximal area (~ 1 mm). Dedifferentiation of internal tissues and accompanying cell division give rise to blastema cells continuously from the 5–10-day period post amputation through 20 days. There is, therefore, no sudden appearance of blastema cells as such, but there is a moderately abrupt distal accumulation of a compact blastema about 20 days post amputation. Since Rose counted as blastema only those cells distal to the cut ends of the bones, the phenomenon he probably measured was distal accumulation of blastema cells. The resulting elongation of the regenerate is accompanied by a thinning of the apical epidermis, which at this time also begins to synthesize DNA again. The regenerate then rapidly forms a conical growth which later assumes the shape of a paddle and redifferentiates. We feel certain that Rose's calculations for the epidermis represent this thinning out of the apical cap with the beginning of blastema elongation, but we are at a loss to explain why in his study the radius of the wound epithelium remained the same if the blastema was indeed growing. The answer to this may be that Rose chose the end of the basement membrane as the boundary between wound epithelium and "old" epithelium. We doubt if this is a reliable index of the old/new boundary, because there is so much growth of proximal epithelium during the lateral expansion of limb stump. The basement membrane might well be extended distally during the process.

Rose *et al.* (1955) obtained limb regenerates from irradiated newt forelimbs which were stripped, allowing viable shoulder epidermis to migrate over the amputated stump. Hay (1952) reported polyploid blastema cells in a diploid tadpole limb which had received a tetraploid skin graft. In the latter case, the transplant was placed proximally with the idea that only the epidermis of the transplant would migrate distally. Chalkley (1959) suggested that some viable subepidermal cells might have accompanied the replacement epithelium in Rose's experiment. In the present study, we have been impressed by the fact that some of the fibroblasts synthesizing DNA during blastema formation were located as far proximally as the elbow. These apparently migrate to the amputation surface at the distal ends of the radius and ulna. Polyploid fibroblasts were included in our proximally located skin grafts (Hay, 1952) and might also have migrated into the blastema. Both these experiments (Rose *et al.*, 1955; Hay, 1952), as well as the transplantations of labeled epidermis being undertaken currently in other laboratories (O'Steen, 1960; Riddiford and Singer, 1960), involve considerable experimental manipulation of the regenerating limb. The experiments reported here did not involve irradiation or transplants and have more validity with respect to the events that occur in normal *Triturus* limb regeneration than any experiment which disrupts the process of regeneration. The ends of the cut radius and ulna were not even trimmed after amputation (as is the practice of many workers), and in this respect as well as with regard to temperature and amputation level, our experimental conditions were identical with those under which Rose (1948) described epidermal dedifferentiation as a feature of normal regeneration.

The fact that the apical cap did not incorporate significant amounts of thymidine during the early stages of regeneration surprised us for, like many other workers, we had assumed it was actively growing. Yet Chalkley in 1954 clearly demonstrated the fact that the proximal epidermis exhibits a high rate of mitosis and that there is very little cell division in the apical epidermis distal to the amputation surface prior to 19 days of regeneration. Other workers (Litwiller, 1939; Schmidt, 1958) have noted a similar distribution of mitosis in the epidermis of the regenerate. Our data on DNA synthesis (Series I) are completely consistent with these calculations of cell division. The fact that some cells which incorporate thymidine in the proximal epidermis and then migrate into the apical cap do divide (Series II) would account for occasional mitoses in the cap. In Series III, it was demonstrated that division of those cells in the apical cap which synthesized DNA before migrating distally does not continue after one dilution of the label. The proximal epidermis, on the other hand, proliferates extensively and the label becomes very diluted. Both Chalkley (1954) and Rose (1948) noted distal migration of proximal epidermis. The fact that the apical cap forms from proximal epidermis, therefore, should not come as a surprise to anyone acquainted with the earlier literature. Indeed, mitosis is not generally associated with wound closure by epithelial migration. The cells around the margin of a wound proliferate (Arey, 1932). The phenomena that are unique in limb regeneration are that proximal epithelium continues to migrate into the wound area and that the cells in the apical cap do not regain their usual rate of cell division until after blastema formation and elongation of the regenerate.

It is tempting to speculate that the cells in the apical cap do not resume their normal mitotic rate because they become involved in some special function in the regenerative process. The histological appearance of the inner border of this epithelium is highly unusual during the 10–20-day period post amputation. Although Rose's interpretation is no longer tenable, his original description of this remarkable epidermis has been amply confirmed. He noted "small tongues of cells protruding into the region of regeneration cell formation" and large "fluid spaces . . . between the swollen basal cells" (Rose, 1948). With the electron microscope, it is apparent that the contact between apical epithelium and blastema is very intimate during the early stages of regeneration (Salpeter and Singer, 1960; Hay, 1960). Bodemer (1958) first demonstrated that the wound epidermis actively participates in removing debris such as charcoal placed inside the limb, and Singer and Salpeter (in press) also have shown that apical epidermal cells are capable of migrating inward to encircle foreign objects. It is a common observation that the apical cap contains many macrophages and leucocytes (Figs. 3 and 8), as well as debris such as bone fragments. When degeneration is extensive in the inner tissues, as in beryllium poisoning, these features are even more prominent (Scheuing and Singer, 1957). It seems reasonable to conclude that 'the apical epidermis can play some role in the removal of wound debris. These special wound conditions might influence the mitotic activity of this epithelium and they could account for the unusual appearance of the inner border of the apical cap of the early blastema. It seems likely, however, that the apical cap has some additional role in regeneration, a role of more fundamental importance in the development of the limb. The possibility that the epithelium attracts the blastema cells distally (Polezajew, 1936) has received considerable support recently by the work of Thornton (1956, 1957, 1958). The orientation of the developing regenerate can be changed by moving the apical epithelial cap of larval *Amblystoma* limbs (Thornton, 1960). An apical epithelial cap is a feature not only of all normally regenerating limbs, but also of limbs induced to grow in frogs (Rose, 1944, 1945), induced supernumerary limbs (Bodemer, 1958; Ruben and Frothingham, 1958; Butler and Blum, 1960), and normally developing embryonic limbs (Saunders, 1948; and others). Whether the specializations of the apical epidermis described here are of fundamental importance in development of the limb, or are merely related to the early wound activities in amputated limbs alone, might be determined by comparative studies of various types of developing limbs.

SUMMARY

The present study has utilized autoradiography to detect incorporation of tritiated thymidine by the cells of the regenerating *Triturus viridescens* limb during blastema formation and to follow the subsequent migration of cells labeled by the isotope. New information was obtained on sites of DNA synthesis during blastema formation and the role of the internal tissues and epithelium was re-evaluated by tracing, for the first time, the actual fate of labeled cells during normal regeneration.

Three series of experiments were performed. In the first, regenerating limbs 1–28 days post amputation were fixed the same day that the animals were injected with tritiated thymidine. This experiment revealed that DNA synthesis begins 4–5 days after amputation in the dedifferentiating muscle, endomysium, epimysium, periosteum, nerve sheaths, and loose connective tissue of the stump for a distance ∼ 1 mm proximal to the amputation surface. The number of cells in the inner tissues synthesizing DNA increases rapidly 10–20 days post amputation. The epidermis which migrates over the wound surface ceases to synthesize DNA within about 2 days. Epithelium proximal to the amputation surface, however, reaches a high level of DNA synthesis 8 days after amputation, and its cells migrate distally to increase the size of the apical cap 10–15 days post amputation. During blastema formation, no more than 2% of the cells in the apical cap incorporate thymidine, these rather feebly. After the blastema is established and the regenerate has begun to elongate, the apical cap thins out and its cells begin to synthesize DNA again.

In the second experimental series, limbs were treated with tritiated thymidine during regeneration (5, 10, and 15 days post amputation) and representative limbs were fixed at daily intervals after treatment. The dedifferentiating inner cells incorporated a large amount of thymidine on the day of injection, whereas the apical epithelium did not. Almost all the blastema cells that appeared subsequently were labeled, and it can be concluded that they were derived from the dedifferentiating internal tissues.

In the third experimental series, animals were injected with tritiated thymidine before their limbs were amputated. Epidermis is the only limb tissue that labels under these circumstances. Blood cells are labeled in their sites of origin, the liver and spleen. After amputation the labeled epidermis migrates over the wound surface and forms an apical cap which remains well labeled throughout blastema formation. The isotope becomes diluted in that area of proliferating proximal epithelium shown in Series I to be actively synthesizing new DNA. Blood cells labeled at the time of injection extravasate into the wounded limb tissues, but disappear from the limb 12–15 days post amputation.

The blastema cells which form in these limbs that contained labeled blood and apical epithelium are not labeled.

In the discussion, the concept that the morphological and physiological changes which take place in dedifferentiating tissues make possible a phase of active cellular proliferation, is emphasized. The apical epidermal cap does not exhibit the DNA synthesis or any of the other essential features of dedifferentiating cells and, when this epidermis is labeled by appropriate treatment, no transformation of epithelial cells into blastema cells can be demonstrated. Indeed, the cells of the apical cap give evidence of being more highly differentiated than epidermis elsewhere.

REFERENCES

AREY, L. B. (1932). Certain basic principles of wound healing. Anat. Record 51, 299–313.

BODEMER, C. W. (1958). The development of nerve-induced supernumerary limbs in the adult newt, Triturus viridescens. J. Exptl. Zool. 102, 555–582.

BODEMER, C. W., and EVERETT, N. B. (1959). Localization of newly synthesized proteins in regenerating newt limbs as determined by radioautographic localization of injected methionine-S^{35}. Develop. Biol. 1, 327–342.

BRUNST, V. V., and CHÉRÉMETIÉVA, E. A. (1936). Sur la perte locale du pouvoir régénérateur chez le Triton et l'axolotl causée par l'irradiation avec les rayons X. Arch. zool. exptl. génér. 78, 57–67.

BUTLER, E. G. (1933). The effects of x-radiation on the regeneration of the forelimbs of Amblystoma larvae. J. Exptl. Zool. 65, 271–315.

BUTLER, E. G., and BLUM, H. F. (1960). Personal communication.

BUTLER, E. G., and O'BRIEN, J. P. (1942). Effects of localized x-radiation on regeneration of the urodele limb. Anat. Record 84, 407–413.

CHALKLEY, D. T. (1954). A quantitative histological analysis of forelimb regeneration in Triturus viridescens. J. Morphol. 94, 21–70.

CHALKLEY, D. T. (1959). The celluar basis of limb regeneration. In "Regeneration in Vertebrates" (C. S. Thornton, ed.), pp. 34–58. Univ. of Chicago Press, Chicago, Illinois.

COLUCCI, V. (1884). Intorno alla rigenerazione degli arti e della coda nei Tritoni. Studio sperimentale. Mem. ric. accad. sci. inst. Bologna [Ser. 4] 6, 501–566.

CRONKITE, E. P., BOND, V. P., FLIEDNER, T. M., and RUBINI, J. R. (1959). The use of tritiated thymidine in the study of DNA synthesis and cell turnover in hemopoietic tissues. Lab. Invest. 8, 263–277.

DAVID, L. (1934). La contribution du matériel cartilagineux et osseux au blastème de régénération des membres ches les amphibiens urodèles. Arch. anat. microscop. 30, 217–234.

EVERETT, N. B., REINHARDT, W. O., and YOFFEY, J. M. (1960). The appearance of labeled cells in the thoracic duct lymph of the Guinea Pig after the administration of tritiated thymidine. Blood 15, 82–94.

FIRKET, H. (1958). Recherches sur la synthèse des acides désoxyribonucléiques et la préparation à la mitose dans des cellules cultivées *in vitro*. *Arch. biol.* (*Liège*) **69**, 1–166.

FIRKET, H., and VERLY, W. G. (1958). Autoradiographic visualization of DNA in tissue culture with tritium-labeled thymidine. *Nature* **181**, 274–275.

FRITSCH, C. (1911). Experimentelle Studien über Regenerationsvorgänge des Gliedmassenskelets der Amphibien. *Zool. Jahrb.* **30**, 377–472.

GODLEWSKI, E. (1928). Untersuchungen über Auslösung und Hemmung der Regeneration beim Axolotl. *Wilhelm Roux' Arch. Entwicklungsmech. Organ.* **114**, 108–143.

GREULICH, R. C. (1960). Evidence for direct and indirect radiation injury from *in vivo* administration of tritiated thymidine. *Anat. Record* **136**, 336–337.

HAY, E. D. (1952). The role of epithelium in amphibian limb regeneration, studied by haploid and triploid transplants. *Am. J. Anat.* **91**, 447–482.

HAY, E. D. (1958). The fine structure of blastema cells and differentiating cartilage cells in regenerating limbs of *Amblystoma* larvae. *J. Biophys. Biochem. Cytol.* **4**, 583–592.

HAY, E. D. (1959). Electron microscopic observations of muscle dedifferentiation in regenerating *Amblystoma* limbs. *Develop. Biol.* **1**, 555–585.

HAY, E. D. (1960). The fine structure of nerves in the epidermis of regenerating salamander limbs. *Exptl. Cell Research* **19**, 299–317.

HAY, E. D., and FISCHMAN, D. A. (1960). Origin of the regeneration blastema of amputated *Triturus viridescens* limbs, studied by autoradiography following injections of tritiated thymidine. *Anat. Record* **136**, 208.

HEATH, H. D. (1953). Regeneration and growth of chimaeric amphibian limbs. *J. Exptl. Zool.* **122**, 339–366.

HELLMICH, W. G. (1930). Untersuchung über Herkunft und Determination des regenerative Materials bei Amphibien. *Wilhelm Roux' Arch. Entwicklungsmech. Organ.* **121**, 135–203.

HUGHES, W. L., BOND, V. P., BRECHER, G., CRONKITE, E. P., PAINTER, R. B., QUASTLER, H., and SHERMAN, F. G. (1958). Cellular proliferation in the mouse as revealed by autoradiography with tritiated thymidine. *Proc. Natl. Acad. Sci. U. S.* **44**, 476.

IDE-ROZAS, A. (1936). Die cytologischen Verhältnisse bei der Regeneration von Kaulquappenextremitäten. *Wilhelm Roux' Arch. Entwicklungsmech. Organ.* **135**, 552–608.

KARCZMAR, A. G., and BERG, G. C. (1951). Alkaline phosphatase during limb development and regeneration of *Amblystoma opacum* and *Amblystoma punctatum*. *J. Exptl. Zool.* **117**, 139–164.

LEBLOND, C. P., MESSIER, B., and KOPRIWA, B. (1959). Thymidine -H3 as a tool for the investigation of the renewal of cell populations. *Lab. Invest.* **8**, 296–308.

LITWILLER, R. (1939). Mitotic index in regenerating amphibian limbs. *J. Exptl. Zool.* **82**, 273–286.

MESSIER, B., and LEBLOND, C. P. (1957). Preparation of coated radioautographs by dipping sections in fluid emulsion. *Proc. Soc. Exptl. Biol. Med.* **96**, 7–10.

O'STEEN, W. K. (1960). Radioautographic studies of cell proliferation in normal and regenerating tissues of the adult newt. *Anat. Record* **136**, 253–254.

PAINTER, R. B., and DREW, R. M. (1959). Studies on deoxyribonucleic acid metabolism in human cancer cell cultures (HeLa). *Lab. Invest.* **8**, 278–285.

PAINTER, R. B., DREW, R. M., and HUGHES, W. L. (1958a). Inhibition of HeLa growth by intranuclear tritium. *Science* **127**, 1244–1245.

PAINTER, R. B., FORRO, F., and HUGHES, W. L. (1958b). Distribution of tritium labeled thymidine in *Escherichia coli* during cell multiplication. *Nature* **181**, 328–329.

PLAUT, W. (1959). The effect of tritium on the interpretation of autoradiographic studies on chromosomes. *Lab. Invest.* **8**, 286–295.

POLEZAJEW, L. N. (1936). Die Rolle des Epithels bei der Regeneration und in der normalen Ontogenese der Extremitäten bei Amphibien. *Zool. Zhur.* **15**, 277–291.

RIDDIFORD, L. M., and SINGER, M. (1960). Autoradiographic studies of tritiated thymidine infused into the blastema of the early regenerate in the adult newt, *Triturus. Anat. Record* **137**, 388.

ROSE, S. M. (1944). Methods of initiating limb regeneration in adult anura. *J. Exptl. Zool.* **95**, 149–170.

ROSE, S. M. (1945). The effect of NaCl in stimulating regeneration of limbs of frogs. *J. Morphol.* **77**, 119–139.

ROSE, S. M. (1948). Epidermal dedifferentiation during blastema formation in regenerating limbs of *Triturus viridescens. J. Exptl. Zool.* **108**, 337–362.

ROSE, F. C., QUASTLER, H., and ROSE, S. M. (1955). Regeneration of x-rayed salamander limbs provided with normal epidermis. *Science* **122**, 1018–1019.

RUBEN, L. N., and FROTHINGHAM, M. L. (1958). The importance of innervation and superficial wounding in urodele accessory limb formation. *J. Morphol.* **102**, 91–118.

SALPETER, M., and SINGER, M. (1960). Differentiation of the submicroscopic adepidermal membrane during limb regeneration in adult *Triturus*, including a note on the use of the term basement membrane. *Anat. Record* **136**, 27–32.

SAUER, M. E., and WALKER, B. E. (1959). Radioautographic study of interkinetic nuclear migration in the neural tube. *Proc. Soc. Exptl. Biol. Med.* **101**, 557–560.

SAUNDERS, J. W. (1948). The proximo-distal sequence of origin of the parts of the chick wing and the role of the ectoderm. *J. Exptl. Zool.* **108**, 363–403.

SCHEUING, M. R., and SINGER, M. (1957). The effects of microquantities of beryllium ion on the regenerating forelimb of the adult newt, *Triturus. J. Exptl. Zool.* **136**, 301–328.

SCHOTTÉ, O. E. (1939). The origin and morphogenetic potencies of regenerates. *Growth Suppl.* **1939**, 59–76.

SCHMIDT, A. J. (1958). Forelimb regeneration of thyroidectomized adult newts. II. Histology. *J. Exptl. Zool.* **139**, 95–136.

SIDMAN, R. L., MIALE, I. L., and FEDER, N. (1959). Cell proliferation and migration in the primitive ependymal zone; an autoradiographic study of histogenesis in the nervous system. *Exptl. Neurol.* **1**, 322–333.

SINGER, M., and SALPETER, M. (in press). The role of the wound epithelium in vertebrate regeneration. *Intern. Symposium on Growth, Purdue Univ., Lafayette, Indiana, 1960.*

SINGER, M., and CRAVEN, L. (1948). The growth and morphogenesis of the regenerating forelimb of adult *Triturus* following denervation at various stages of development. *J. Exptl. Zool.* **108,** 279–308.

TAYLOR, J. H., WOODS, P. S., and HUGHES, W. L. (1957). The organization and duplication of chromosomes as revealed by autoradiographic studies ·using tritium labeled thymidine. *Proc. Natl. Acad. Sci. U. S.* **43,** 122–128.

THORNTON, C. S. (1938). The histogenesis of muscle in the regenerating forelimb of larval *Amblystoma punctatum. J. Morphol.* **62,** 17–47.

THORNTON, C. S. (1956). The relation of epidermal innervation to the regeneration of limb deplants in *Amblystoma* larvae. *J. Exptl. Zool.* **133,** 281–300.

THORNTON, C. S. (1957). The effect of apical cap removal on limb regeneration in *Amblystoma* larvae. *J. Exptl. Zool.* **134,** 357–382.

THORNTON, C. S. (1958). The inhibition of limb regeneration in urodele larvae by localized irradiation with ultraviolet light. *J. Exptl. Zool.* **137,** 153–180.

THORNTON, C. S. (1960). Influence of an eccentric epidermal cap on limb regeneration in *Amblystoma* larvae. *Develop. Biol.* **2,** 551–569.

WEISS, P. (1939). "Principles of Development." Holt, New York.

Reprinted from *Develop. Biol.*, **1**, 327–342 (1959)

6

Localization of Newly Synthesized Proteins in Regenerating Newt Limbs As Determined by Radioautographic Localization of Injected Methionine-S^{35} [1]

CHARLES W. BODEMER AND NEWTON B. EVERETT

Department of Anatomy, School of Medicine, University of Washington, Seattle, Washington

Accepted August 4, 1959

INTRODUCTION

Exhibiting as it does such fundamental processes as wound healing, dedifferentiation, differentiation and growth, the regenerating amphibian limb provides an admirable object for the study of relations between metabolism and cellular activities. A thorough understanding of the metabolic events associated with each of these stages of regeneration possesses value for a more profound insight into both the regenerative process itself and the physiology of the individual cell.

Protein synthesis constitutes one of the most basic features of cellular metabolism, yet relatively few studies have been devoted to the protein metabolism of the regenerating limb and none specifically to protein synthesis. Although determinations have been made of the glutathione content (Orekhovitch, 1934), the amino nitrogen (Vladimirova, 1935; Orekhovitch, 1936), and cathepsins (Jensen *et al.*, 1956) of the regenerating limb and/or tail, these studies provide no information regarding the sites of protein synthesis in the regenerating limb and allow no correlation between the metabolic changes and histological events characterizing successive stages of regeneration. Radioautographic visualization of injected methionine-S^{35} provides an excellent means for establishing the desired correla-

[1] Supported by research grants from the United States Public Health Service, Department of Health, Education, and Welfare (RG-5820 and H-1530).

327

tion between localization of newly synthesized protein and the morphological events of limb regeneration. In addition, this technique allows, through subjective appraisal of the density of the radioautographic image, some estimate of the amount of new protein present at the various sites.

This investigation is part of a general program devoted to analysis of the protein metabolism of the regenerating amphibian limb. It reveals the relative intensity of amino acid uptake and the various loci of accumulation of new protein during periods ranging from that immediately following amputation through the time of differentiation of new tissues in the regenerating limb.

MATERIALS AND METHODS

This investigation is based upon 25 adult newts, *Triturus viridescens*, obtained from the vicinity of Petersham, Massachusetts. Each newt, weighing 1.5–1.6 gm, was anesthetized in a 30 % solution of a saturated chloretone solution. Both forelimbs were amputated a short distance proximal to the elbow. When the regenerates had attained the desired stage of development, the newts were given intraperitoneal injections of 9 μc of DL-methionine-S^{35} containing 0.04 mg methionine. Preliminary chemical analyses revealed that for the entire limb regenerate at room temperature the interval for maximal uptake of injected methionine-S^{35} is at or near 6 hours after injection. Accordingly, all the newts used for this investigation were sacrificed 6 hours after injection of the methionine-S^{35}.

Both limbs were removed from the injected newts at the following intervals after amputation: day 1 (2 animals); day 3 (2 animals); day 5 (3 animals); day 7 (3 animals); day 10 (2 animals); day 13 (3 animals); day 17 (3 animals); day 21 (4 animals); day 27 (3 animals). The consistency of findings within each of the groups selected for study encourages the belief that the sample size provides a background adequate for generalization. The removed limbs were immersed in Bouin's fixative for 48 hours. The fixed tissues were decalcified, embedded in paraffin, and sectioned serially at 8 μ. After staining in hematoxylin and eosin, coated radioautographs (Gross *et al.*, 1951) were prepared using NTB-3 emulsion (Eastman Kodak), stored under refrigeration in a light-proof box for 96 hours, then developed and mounted for histological study.

On the basis of their critical position within the regeneration process, regenerates at certain postamputation intervals were selected for detailed description. The observations recorded in the text represent a composite of the findings for an entire sample at a particular interval. Similarly, the sections photographed were selected for illustrating best the characteristics of a given group.

OBSERVATIONS

Wound-healing Phase

Within 24 hours after amputation the transected surface of the limb is covered with a well-defined wound epithelium. This epithelium produces an intense radioautographic reaction in contrast with the slight reaction produced by the other components of the limb (Fig. 1). During the succeeding days comprising the wound-healing phase of regeneration the wound epithelium maintains a higher concentration of radioactivity than the other limb tissues, including the normal epidermis with which it is continuous.

Dedifferentiation Phase

By the fifth day after amputation wound healing is complete and there is a moderate accumulation of fibrocellular tissue beneath the wound epidermis. Both the wound epidermis and the subepidermal aggregation of cells are characterized by a strong radioautographic reaction (Fig. 2). Of particular interest on the fifth day is the now apparent involvement of the internal limb tissues in the regenerative process. The periosteum of the transected humerus has become strongly radioactive, although the osteocytes located within the enclosed lacunae of the humerus produce a negligible radioautographic reaction (Fig. 2). The reduction pattern produced by the brachial muscles is of considerable interest. There is apparent on the fifth day after amputation a definite proximodistal gradient of reactivity in the transected brachial muscles. The proximal portions of these muscles produce only a slight reduction of the overlying emulsion (Fig. 3). More distally, toward the transected ends of the muscles, the density of reduction increases, and in the immediate area of the free, distal ends the reactivity of the muscles is substantially greater than elsewhere along their length (Fig. 4). On the fifth day dissociation of the damaged muscles is apparent. Sarcoplasmatic fragments of varying size are released from the cut ends of the muscles. These

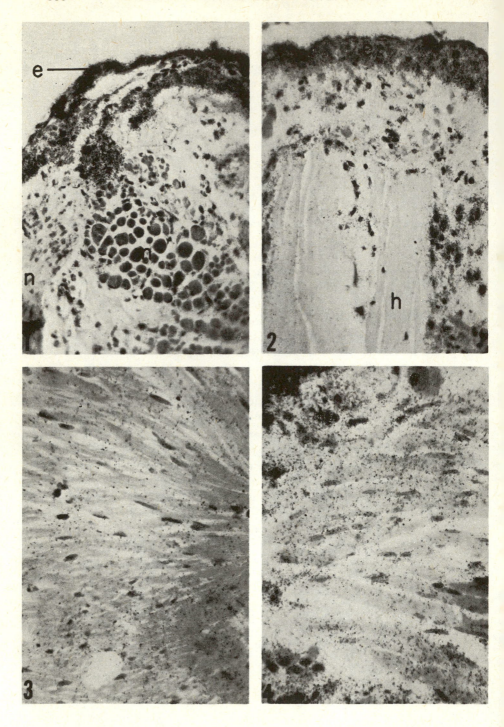

sarcoplasmatic fragments may be small straplike masses of sarcoplasm surrounding a single nucleus or large multinucleate fragments. The role of the sarcoplasmatic fragment, and, indeed, of the isolated cell, in regeneration of muscle has long been debated, and its importance in formation of the amphibian regeneration blastema is still undecided. It is therefore of some interest to note that the radioautographs reveal that the sarcoplasmatic fragments apparently released from the transected arm muscles produce a dense radioautographic image (Figs. 5, 6). Connective tissue cells located along the endomysial planes of the cut muscles, like the sarcoplasmatic fragments, display a pronounced concentration of radioactivity (Fig. 6).

A gradient in the density of reduced emulsion exists along the proximodistal axis of the limb nerves, and at their transected ends the nerves produce a reduction pattern comparable with that characteristic of the brachial muscles (Figs. 5, 6). The strong reactivity appears to be confined largely to the connective tissue cells of the various nerve sheaths, including probably the Schwann cells.

A blastema has begun to form by the tenth day after amputation. The undifferentiated blastemal cells display a pronounced concentration of radioactivity. The thickened epidermis overlying the blastema, however, remains the most reactive component of the regenerating limb (Fig. 7). By the tenth day dissolution of the humerus is well underway. The lacunae situated within the peripheral part of the degenerating bone become opened by disintegration of their bony walls and the osteocytes contained therein are liberated into the area of the forming blastema. Prior to the complete breakdown of the lacunal walls, and at the time of their release, the osteocytes are overlain by considerably more reduced silver grains than are

FIG. 1.   A longitudinal section through a regenerating limb 24 hours after amputation which illustrates the heavy concentration of radioactivity in the wound epithelium; e, wound epidermis; m, muscle; n, nerve.

FIG. 2.   A section through the regenerating limb 5 days after amputation illustrating the radioautographic reaction produced by the periosteum and the fibrocellular tissue underlying the wound epithelium; h, humerus.

FIG. 3.   A section illustrating the negligible radioautographic image produced by the proximal portions of the transected arm muscles on the fifth day after amputation.

FIG. 4.   A section through the distal end of the muscles illustrated in Fig. 3. Note the much greater concentration of radioactivity in the terminal portion of the muscles.

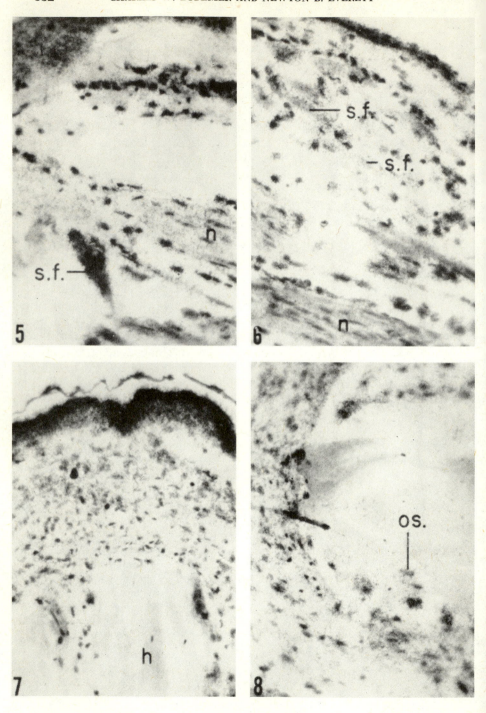

those osteocytes located within the more central portions of the humerus (cf. Figs. 7, 8). It is the dedifferentiating osteocyte, then, that is characterized by a pronounced incorporation of labeled amino acid.

Growth Phase

Following establishment of a definitive blastema, extensive dedifferentiation ceases and growth and mitosis become the dominant features of the regenerative process. The entire regenerate is characterized by a high level of methionine-S^{35} incorporation and the radioautographic image is generally of even intensity throughout the blastema. It is of considerable significance, however, that in the immediate vicinity of the end of the large brachial nerve the intensity of the radioautograph appears to be greater than in those areas of the blastema more distant from the end of the nerve (Fig. 9).

Differentiation Phase

Cartilage is the first tissue to differentiate within the regeneration blastema. By the twenty-first day distinct condensations of precartilage may be observed. The rounded cells, which at this time have already been surrounded by deposited matrix, are overlain by dense reduced emulsion, as are the more fusiform cells of the perichondrium (Fig. 10). The developing matrix displays some concentration of radiomaterial.

In the 21-day regenerate early muscle anlagen are present as discrete bundles of elongate cells. These condensed cells are intensely radioactive (Fig. 11). In addition to the independent muscle

---

FIG. 5. A section through the regenerate revealing the intense radioautographic reaction produced by a large, multinucleate, sarcoplasmatic fragment (*s.f.*) released from the transected muscles of the arm on the fifth day after amputation; *n*, nerve.

FIG. 6. A section through the regenerating limb 5 days after amputation illustrating the concentration of radioactivity within the sarcoplasmatic fragments (*s.f.*) and the end of the peripheral nerve (*n*).

FIG. 7. A section through a 10-day limb regenerate which illustrates the relatively uniform radioautographic reaction produced by the blastema; *h*, humerus.

FIG. 8. A section through the end of the transected humerus which reveals the concentration of radioactivity characteristic of dedifferentiating osteocytes (os.).

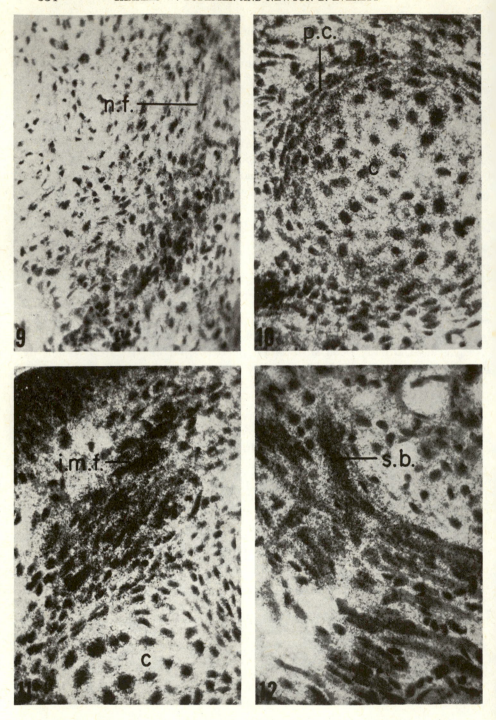

anlagen formed as condensations within the regenerate, new muscle forms by sarcoplasmatic budding from the transected ends of the arm muscles. These sarcoplasmatic buds, like the isolated muscle anlagen produce an intense reduction of the emulsion (Fig. 12).

The observations recorded above are summarized in Fig. 13. As assessed by subjective appraisal of the density of reduced emulsion overlying the tissues, there is a definite pattern to the localization of labeled proteins during the different stages of limb regeneration. During the first phase the epidermis is the most active tissue of the limb. Emerging from an initial state of relative quiescence, the internal limb tissues reveal a strong radioautographic reaction during the dedifferentiative phase. The reactivity of tissues of the blastemal regenerate appear about equivalent. At the time of histodifferentiation in the regenerate the density of the image increases significantly over those cells undergoing differentiation.

DISCUSSION

Although the technique of radioautography is well-established, the use of methionine-S^{35} for radioautographic localization of sites of protein synthesis is relatively new. Methionine is incorporated into certain proteins (Tarver, 1954) and, unlike bicarbonate-C^{14} (Greulich and Leblond, 1953) and glycine-C^{14} (Kemp, 1956), apparently does not also serve as a direct precursor of carbohydrates. Methionine may become quickly converted into cystine (Tarver and Schmidt, 1939; Forker *et al.*, 1951; Leblond *et al.*, 1957), but this should not substantially alter observations regarding the sites of its incorporation. A complicating factor is contributed by the fact that methionine may be adsorbed onto the surfaces of protein (Levin *et al.*, 1956); this adsorbed amino acid may be removed, however, by TCA-fixation

---

Fig. 9.   A section through the 13-day regenerate which illustrates the greater radioautographic reaction produced by those cells near the end of the regenerating nerves as compared with the more distant blastemal cells; *n.f.*, nerve fiber.

Fig. 10.   A section illustrating the intense radioautographic reaction produced by differentiating cartilage (*c*) in the 21-day regenerate; *p.c.*, perichondrium.

Fig. 11.   A section which illustrates the high reactivity of the independent muscle fibers (*i.m.f.*) differentiating within the 21-day regenerate; *c*, cartilage.

Fig. 12.   A section illustrating the strong radioautographic reaction produced by sarcoplasmatic buds (*s.b.*) emerging from the transected muscles of the arm.

and thorough lipid extraction (Tarver, 1954; Zamecnik *et al.*, 1956). Of particular relevance to the procedure employed in this investigation are the observations of Leblond *et al.* (1957) regarding the fixative employed for radioautography of methionine-S^{35}. Comparing radioautographs produced by TCA-fixed tissues and tissues fixed in Bouin's fluid, these authors detected no appreciable differences in the pattern or density of the radioautographic image. They noted also that the regular paraffin-embedding procedure provides for extraction of lipids as stipulated by Tarver (1954). There is thus justification for confidence in the techniques used for this study of protein synthesis, both in terms of the isotope employed and the mode of preparation of tissues. Some qualification may be necessary, however, regarding the possibility of incorporation of S^{35} in the form of free sulfate liberated by oxidation of methionine in excess of the physiological limit. In a study of protein synthesis in the growing rat, Bélanger (1956) noted the incorporation of S^{35} as sulfate characteristically into the walls of blood vessels, goblet cells, and cartilage. It is difficult, dealing with developing cartilage, as in this investigation, to distinguish that radioactivity which is attributable to synthesized protein and that which might be ascribed to incorporated

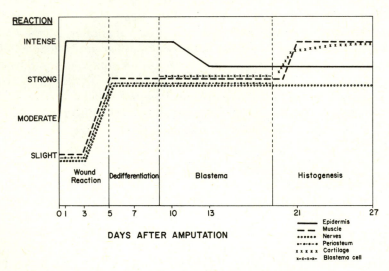

FIG. 13. Radioautographic reaction produced by S^{35}-methionine incorporated into regenerating newt limbs 6 hours after injection.

free sulfate. No determination was made of the plasma methionine in the newts studied, and it is not known whether the dosage employed exceeded their physiological limit. We did not observe a significant concentration of radioactivity in the walls of the blood vessels of the amputation stump. Thus, although the injected dose of the radioactive amino acid may have exceeded somewhat the physiological limit of the newts, and some of the radioactivity in the cartilage matrix of the regenerate may have been produced by free S^{35} liberated from oxidized methionine, it is believed that most of the radioactivity may be attributed to synthesis of new protein.

The radioautographic reaction produced by tissues following injection of methionine-S^{35} reveals sites of localization of newly generated proteins. It is probable that most of these proteins are actually synthesized at the sites of positive radioautographic images. Methionine-S^{35} is not, however, incorporated into all proteins (Fisher, 1954), and does not provide an accurate indication of *all* the proteins synthesized during the period of its availability. It may also be questioned whether the radioautographic reaction at each site is produced entirely by new proteins generated at that site. It is known, for example, that there is a transfer of labeled proteins from the liver to the carcass in rats following injection of glycine-N^{15} (Shemin and Rittenberg, 1944). In addition, it is possible that serum albumin, labeled elsewhere by incorporation of the methionine-S^{35}, may pass through the extravascular space and, either directly or by phagocytosis, enter a cell and contribute to a positive radioautographic reaction. Protein turnover may also account for some of the observed radioautographic reactions. Thus, it is perhaps wise at this time to consider the concentrations of radioactivity revealed by the radioautography following administration of methionine-S^{35} as localizations of recently labeled protein without stipulating that all this protein is synthesized at the site of the radioautographic reaction.

As judged by subjective appraisal of the density of reduced silver overlying the tissues of the regenerating limb, the sites of most active methionine-S^{35} incorporation appear to follow a definite sequential pattern (cf. Fig. 13). Extrapolating from the observations, it may be considered that after amputation there is a gradual increase in the magnitude of incorporation of this amino acid in the regenerate as a whole. The magnitude of uptake appears to attain a high level at the time of blastema formation and to reach its maximal intensity coincident with histodifferentiation in the regenerate. This subjective interpretation receives support from the quantitative study of the amino acid content in the regenerating axolotl limb. Vladimirova (1935) demonstrated that the amino nitrogen increases gradually after amputation, reaches its maximum at the three-digit stage, and thereafter recedes to the normal concentration. A comparable pattern has been established for the concentration of glutathione (Orekhovitch, 1934) and the alkaline phosphatase concentration of the regenerating limb (Ghiretti, 1950; Karczmar and Berg, 1951). The histological data presented by Karczmar and Berg reveal an increased alkaline phosphatase content in dedifferentiating and differentiating tissues. Thus the pattern of amino acid uptake described herein appears to conform to a general metabolic pattern within the regenerating limb, a sequential pattern which proceeds apace with the sequence of morphological events in regeneration.

A feature of interest emerging from this study is the demonstration of an increased concentration of radioactivity in dedifferentiating tissues and cells. Dedifferentiating osteocytes enlarge and simultaneously produce a greater radioautographic reaction. The reactivity of the endomysial connective tissue cells and the connective tissue cells of the nerve sheaths similarly increases during the dedifferentiation phase. Nucleated sarcoplasmatic fragments released from the ends of the transected muscles are intensely radioactive, as are the sarcoplasmatic buds emerging in continuity with the ends of the transected muscle. Accented amino acid incorporation then, appears to be an intrinsic and constant feature of the dedifferentiation process.

Despite much research devoted to its analysis, the exact origin of the regeneration blastema remains unknown. Much of the uncertainty stems from an inadequate knowledge of the behavior of muscle cells in blastema formation and their relation to the new muscle of the regenerate. Indeed, the role of the independent muscle cell or sarcoplasmatic fragment in normal regeneration of muscle remains debated.[2]

---

[2] This question was discussed in an earlier paper on this subject (Bodemer, 1958) and will not be dwelt upon here. References to more extensive analyses of this problem may also be obtained in this work.

Information necessary to appraise properly the possible participation of sarcoplasmatic fragments in blastema formation is whether these are viable fragments or merely dissociated moribund masses. While not answering this fundamental question, our observations reveal that when they are identifiable as independent structures, the sarcoplasmatic fragments are always overlain by a dense area of reduced silver. It is unknown whether the reactivity can be ascribed entirely to protein synthesis within the fragments or whether the concentration of radioactivity occurred before or after their liberation as independent entities. The evidence suggests, however, that these are metabolically active fragments, a suggestion supported by the recent electron microscope study by Hay (1959). The possibility of active protein synthesis within these nucleated fragments argues in favor of their continued existence subsequent to the assumption of an independent habitus. It is not possible on the basis of the evidence presented here to ascertain the fate or potentialities of the sarcoplasmatic fragments; the observations do, however, tend to support the belief that the sarcoplasmatic fragments are viable and are not, therefore, incapable of contributing to the blastema.

It is reasonable to assume that much of the amino acid uptake observed in this study is related to cell growth and/or preparation for cell division. The intense incorporation of methionine-S^{35} into the newly generated wound epithelium could be accounted for in this way. The wound epithelium increases from a very thin layer of cells to a definite epidermal cap of considerable thickness during the first 2 weeks after amputation. Extensive mitosis and cell growth are necessary concomitants of this phenomenon, and protein synthesis would be expected to be quite pronounced during this period. A preponderance of the protein synthesis in the blastemal cells is probably for growth and division, and growth of newly differentiated muscle and cartilage would also require intensive protein synthesis. There is good reason to believe, then, that synthetic processes associated with cell growth and mitosis are the dominant factors operative in establishing the intensity of amino acid incorporation in the various components of the regenerating limb. According to this interpretation the sequential pattern of loci of labeled proteins might derive from the sequential involvement of the various tissues in growth and division.

Other factors may contribute to the magnitude of S^{35} incorporation at certain stages in limb regeneration. Many investigations have established that differentiating embryonic cells synthesize proteins specific for the cell type (Ebert, 1952), and there is reason to suspect that histodifferentiation in the limb regenerate is accompanied by synthesis of specific proteins (DeHaan, 1956). Since the blastemal cell appears to be relatively undifferentiated immunologically it is tempting to speculate that there may be a basic alteration in the proteins of the cell in association with dedifferentiation. This raises the unsettled question of the true meaning of dedifferentiation. True (functional or physiological) dedifferentiation of a cell, resulting in its reassumption of wide tissue-forming potencies, would seem to require a change in its complement of specific proteins. Should this process occur, it might conceivably account for some of the methionine-S^{35} incorporation revealed by radioautographs of dedifferentiating cells.

The increased incorporation of S^{35} into the transected brachial musculature and its variation along a proximodistal gradient suggests also that some of the radioautographic results may reflect a wound reaction, either in the form of increased protein synthesis or augmented protein turnover. Since not all of the damaged musculature undergoes dissociation, it seems likely that some of the observed labeling of protein may be attributed to a wound reaction and intrinsic restorative processes within the muscle.

Of considerable interest is the observation that there appears to be a gradient in the intensity of methionine-S^{35} incorporation within the blastema which is based upon the location of the end of the larger arm nerves. Quantitative experiments now in progress tend to confirm the radioautographic indications. The influence of the nerve in amphibian limb regeneration is profound (Singer, 1952). Not only is it essential for the inauguration of regeneration, but in later stages the nerve affects mitotic activity and the volume attained by the regenerated limb. Our observations suggest that the magnitude of amino acid uptake might be included within the array of phenomena embraced by the nervous influence, and it may well be that the nervous influence on mitosis and blastemal growth derives from a more basic influence on the magnitude of protein synthesis within the blastemal cell. At this time this is scarcely more than speculation; more definite evidence comprises the goal of investigations now underway in this laboratory.

SUMMARY

Sites of amino acid incorporation within the regenerating forelimb of adult *Triturus viridescens* were determined by the radioautographic localization of injected methionine-S^{35}. The limb regenerates studied ranged in age from 24 hours to 27 days.

During the first few days after amputation the epidermis only produces an intense radioautographic reaction. The intensity of the reaction of the mesodermal tissues increases by the fifth day, when the periosteum of the transected humerus and the terminal portions of the cut muscles and nerves display a concentration of radioactivity. Fragments of sarcoplasm detached from the damaged muscles produce a strong reaction, suggesting that they are not moribund fragments. Those osteocytes apparently being released from the degenerating bone are highly radioactive. The intensity of the radioautographic reaction is generally even throughout the blastema during the growth phase. The radioautographs suggest that during the growth phase the peripheral nerve may exert an influence on protein synthesis within the undifferentiated blastemal cell. In later stages the more intense reduction of emulsion is produced by those cells undergoing differentiation.

The observations thus suggest that increased incorporation of amino acid is characteristic of the regenerate generally, and comprises a significant feature of dedifferentiating as well as differentiating cells.

REFERENCES

BÉLANGER, L. F. (1956). Autoradiographic visualization of the entry and transit of S^{35}-methionine and cystine in the soft and hard tissues of the growing rat. *Anat. Record* **124**, 555–580.

BODEMER, C. W. (1958). The development of nerve-induced supernumerary limbs in the adult newt, *Triturus viridescens*. *J. Morphol.* **102**, 555–582.

DeHAAN, R. L. (1956). The serological determination of developing muscle protein in the regenerating limb of *Amblystoma mexicanum*. *J. Exptl. Zool.* **133**, 73–86.

EBERT, J. D. (1952). Appearance of tissue-specific proteins during development. *Ann. N. Y. Acad. Sci.* **55**, 67–84.

FISHER, R. B. (1954). "Protein Metabolism." Methuen, London.

FORKER, L. L., CHAIKOFF, I. L., ENTENMAN, C., and TARVER, H. (1951). Formation of muscle protein in diabetic dogs studied with S^{35}-methionine. *J. Biol. Chem.* **188**, 37–48.

GHIRETTI, F. (1950). On the activity of acid-and alkaline phosphatase during tail regeneration in *Triturus cristatus* (Laur.). *Experientia* **6**, 98–100.

GREULICH, R. C., and LEBLOND, C. P. (1953). Radioautographic visualization of radiocarbon in the organs and tissues of newborn rats following administration of C[14]-labeled bicarbonate. *Anat. Record* **115**, 559–586.

GROSS, J., BOGOROCH, R., NADLER, N. J., and LEBLOND, C. P. (1951). The theory and methods of the autographic localization of radio-elements in tissues. *Am. J. Roentgenol.* **65**, 420–458.

HAY, E. D. (1959). Fine structure of dedifferentiating muscle in regenerating salamander limbs. *Anat. Record* **133**, 287 (Abstract).

JENSEN, P. K., LEHMANN, F. E., and WEBER, R. (1956). Catheptic activity in the regenerating tail of *Xenopus* larvae and its reaction to histostatic substances. *Helv. Physiol. Acta* **14**, 188–201.

KARCZMAR, A. G., and BERG, G. G. (1951). Alkaline phosphatase during limb development and regeneration of *Amblystoma opacum* and *Amblystoma punctatum*. *J. Exptl. Zool.* **117**, 139–163.

KEMP, N. E. (1956). Localization of glycine-2-C[14] injected into adult female frogs. *J. Exptl. Zool.* **133**, 227–239.

LEBLOND, C. P., EVERETT, N. B., and SIMMONS, B. (1957). Sites of protein synthesis as shown by radioautography after administration of S[35]-labeled methionine. *Am. J. Anat.* **101**, 225–272.

LEVIN, W. G., PERRY, J. E., and BLOCKER, T. G., JR. (1956). Adsorption of sulfur-35 labeled L-methionine by serum proteins *in vitro*. *Texas Repts. Biol. and Med.* **14**, 372–375.

OREKHOVITCH, W. N. (1934). Zur Frage über die Aktivierung der Proteolyse in den regenerierenden Geweben. *Z. physiol. Chem. Hoppe-Seyler's* **224**, 61–66.

OREKHOVITCH, W. N. (1936). Über die Wirkungsbedingungen des Kathepsins in den Geweben regenerierender Organe von Amphibien. *Biochem. Z.* **285**, 285–289.

SHEMIN, D., and RITTENBERG, D. (1944). Some interrelationships in general nitrogen metabolism. *J. Biol. Chem.* **153**, 401–421.

SINGER, M. (1952). The influence of the nerve in regeneration of the amphibian extremity. *Quart. Rev. Biol.* **27**, 169–200.

TARVER, H. (1954). Peptide and protein synthesis. Protein turnover. *In* "The Proteins" (H. Neurath and K. Bailey, eds.), Vol. 2, Part B, pp. 1199–1292. Academic Press, New York.

TARVER, H., and SCHMIDT, C. L. A. (1939). The conversion of methionine to cystine: experiments with radioactive sulfur. *J. Biol. Chem.* **130**, 67–80.

VLADIMIROVA, E. A. (1935). Soderzhanie aminokislot v regeneriruyushchikh konechnostyakh aksolotley na raznykh stadiyakh regeneratsii. *Trudy Lab. Eksptl. Zool. Morfol. Zhivotnykh Akad. Nauk. S.S.S.R.* **4**, 163–167.

ZAMECNIK, P. C., KELLER, E. B., LITTLEFIELD, J. W., HOAGLAND, M. B., and LOTFIELD, R. B. (1956). The mechanism of incorporation of labeled amino acids into protein. *J. Cellular Comp. Physiol.* **47**, (Suppl.), 81–102.

Reprinted from *J. Exptl. Zool.*, **145**, 43–47 (1960)

7

Preblastemic Changes of Intramuscular Glycogen in Forelimb Regeneration of the Adult Newt, *Triturus viridescens*[1]

ANTHONY J. SCHMIDT

University of Illinois, College of Medicine, Department of Anatomy, Chicago, Illinois

The biodynamics of regenerating systems long have challenged students of biology. Fundamental to this challenge is the understanding of the molecular events and controlling mechanisms that direct the regeneration of tissues, organs, and complex organized structures such as a tail or a limb. Indeed, contemporary investigators are exploring nerve mediators (Singer, '59; Taban, '55), muscle protein (DeHaan, '56; Laufer, '59), and protein synthesis (Bodemer and Everett, '59) in amphibian regenerating systems. Of special interest is the growing recognition given to the role of polysaccharides in such regeneration-affiliated fields as wound-healing (Jackson, '58), induction (Grobstein, '55), and growth (Harris, '58), among others.

The subject of our study is the distribution of glycogen in the tissues of a regenerating forelimb of the adult newt. The basis of this investigation is the early study of Okuneff ('33) who reported a considerable increase in lactic acid during the early phases of the formation of a blastema in the regenerating axolotl limb. It is known that in muscle, anaerobic glycolysis results in an equivalent increase in lactic acid (Dickens, '51). Therefore, we sought to determine if the preblastemic rise in lactic acid could be visualized histochemically by a depletion of tissue glycogen.

METHODS AND MATERIALS

We studied the regenerating forelimbs of the adult newt, *Triturus viridescens*. The newts were collected from ponds in the vicinity of Petersham, Massachusetts. One series of animals was thyroidectomized 15 days prior to amputation through both forelimbs. A normal series was composed of euthyroid newts whose forelimbs were amputated as above. The operative procedures have been described in detail earlier (Schmidt, '58a).

The experimental animals were maintained at a constant temperature of $20 \pm 1°C$ within individual fingerbowls containing pre-aerated distilled water. The animals were exposed daily to 8 hours of diffuse light from a 6-watt fluorescent lamp. Once a week the newts were fed Tubifex, following which the water in the fingerbowls was changed.

The method of maintenance described is an improvement over that reported earlier (Schmidt, '58a). Subjects of this modification were two groups of adult newts — thyroidless and normal — maintained for over 300 days with less than a 5% mortality.

Observations were made, and concomitant samples were taken at 24 hours post-amputation, and every three-day period thereafter until the 37th day of regeneration. By the 37th day of regeneration, the majority of the limbs bore digits. Of the experimental population, 39 limbs from the thyroidless series, and 20 limbs from the normal series were studied for this report.

The microtechnical procedure was that of freeze-substitution (for details and other approaches, see Neder and Sidman, '58, and Pearse, '60). The regenerates sampled were quenched in isopentane chilled

[1] This investigation was supported in part by an institutional grant from the American Cancer Society, Inc., and in part by a Public Health Service research grant, RG-6208(R1), from the National Institutes of Health.

159

with liquid nitrogen to approximately −160°C. The quenched tissue was freeze-substituted and fixed in a solution of absolute alcohol saturated with picric acid kept at about −70°C in a dry-ice chamber for from 5 to 8 days. All limbs were subsequently decalcified with Jenkin's fluid, dehydrated in alcohols, cleared, and double-embedded in celloidin and paraffin.

The embedded material was sectioned serially at 10 μ, and stained with the McManus periodic acid-Schiff (PAS) procedure for polysaccharides, as described by Lillie ('54). Every 10th through 14th section was applied to a separate slide for other tests, one of which was a salivary amylase digest for glycogen.

RESULTS

The microscopic examination of the sectioned regenerates is concerned here only with the PAS-staining material which, by location and ready removal with salivary amylase (fig. 2) we believe to be glycogen. Since no differences in the PAS staining could be found between the regenerates of the thyroidless and euthyroid newts, the following examples, selected as being the most representative of the observations made, are from the series of thyroidless newts.

A detailed histological report on the progress of regeneration in the adult newt has been presented earlier (Schmidt, '58b). Here, our sole intent is to describe the distribution of glycogen in the preblastemic regenerates.

At 24 hours, and through the 4th day post-amputation (figs. 1, 2, 3, and 4), the muscle fibers adjacent to the amputation surface are PAS-negative, and thus depleted of glycogen. The muscle fibers appear structurally normal, with internal myofibrillae and surrounding mysial sheaths clearly observable (fig. 4). The fibrous components of muscle tissue are frayed and disorganized at the cut ends of the amputation surface. Globules of glycogen are visible between some of the muscle fibers.

At 7 and 10 days post-amputation (figs. 5 and 6), intact muscle is found at a variable distance from the amputation surface. No intact glycogen-depleted myo-

fibers are readily discerned. Rather, frayed mysial sheaths, and perhaps myofibrillae can be found distal to the intact muscle. In and around this latter region small vesicles or bubbles can be observed, this activity continuing (figs. 7 and 8) through the formation of the blastema.

At 16 days post-amputation (fig. 8), an accumulation of cells beneath the apical epithelium heralds the formation of a blastema. Figure 8 is particularly unique in illustrating numerous bubbles trailing off the frayed end of some muscle fibers. This bubbling may be an extreme manifestation of the bubbling noted earlier.

Continued observations are concerned with the formation of the blastema, and warrant separate consideration.

DISCUSSION

There is little doubt that the intramuscular PAS-positive material is glycogen. As well as being PAS-positive, the intramuscular substance is readily digested with salivary amylase, a recognized test for glycogen (Pearse, '60).

The results obtained in our study demonstrate a post-operative depletion of glycogen from the muscle fibers adjacent to the amputation wound. The glycogen depletion is present for at least 4 days following the amputation, but is no longer clearly discernible by the 7th day of regeneration. Observations made on the 7th and consecutive days show a delineation between the intact stump muscle and the former glycogen-depleted distal muscle, the only evidence of the latter being frayed and disorganized strands of mysial connective tissue, possibly some myofibrillae, and a vesicular or bubbly picture.

As was pointed out in the previous section, the bubbling is particularly extreme in the 16-day regenerate illustrated in figure 8. This apparent boiling activity may well be a manifestation of muscle autocatalysis, or particularly, of a proteolytic activity that would be favored in the prevailing acid environment (Okuneff, '28). We are pursuing this observation in some detail since we feel that this activity bears significantly on the formation of the blastema.

There is no doubt that our results complement the investigations of Okuneff ('33) who demonstrated a marked increase in lactic acid during the preblastemic phase of urodele limb regeneration. A low respiratory quotient for this phase of regeneration has been reported (Ryvkina, '45): it is during anaerobic glycolysis that a large quantity of lactic acid is formed, and there is an intracellular depletion of glycogen equivalent to the increase in lactic acid (Dickens, '51). In the early phases of regeneration of the newt limb, anaerobic glycolysis prevails, and the source of glycogen is within the muscle fibers approximating the amputation surface.

From the evidence at hand, we can hypothesize an environmental prerequisite extant during the wound-healing and dedifferentiative (regressive) phases leading to the formation of regeneration blastema in the adult newt. Upon severing the limb, local trauma, blood clot formation, and epithelial wound closure follow in succession. Since the continuity of blood vessels no longer exists, there is vascular stasis, with a resulting hypoxia of the tissues adjacent to the amputation wound. The vast stores of glycogen within the muscle fibers continue to be metabolized, and the glycolysis, taking place in an anaerobic environment, leads to the formation of large quantities of lactic acid. The lactic acid can readily accumulate in the distal stump due to the lack of a venous return. As a further complication of the vascular stasis, the muscle glycogen cannot be replenished as the stores are utilized, contributing to myofiber autolysis.

We may conclude that the anaerobic glycolysis in the distal limb stump is, at least in part, responsible for the local acid environment that is compatible with the wound- and regressive phase metabolism directed toward the formation of the regeneration blastema.

SUMMARY

1. The course of regeneration of the forelimb of the adult newt was pursued histochemically with the PAS staining procedure for polysaccharides.

2. The distribution of intracellular glycogen was verified by comparing sections digested with salivary amylase to none-treated sections.

3. We observed a depletion of intramuscular glycogen adjacent to the amputation wound at 24 hours and lasting through at least 4 days post-amputation. It is proposed that the glycogen of these myofibers is metabolized anaerobically leading to lactic acid, and in this way contributing to the local acidity of the preblastemic phases in regeneration.

4. By the 7th day post-amputation and beyond, the glycogen-depleted muscle fibers are no longer discernible; instead, frayed mysial sheaths, possibly myofibrillae, and a bubbly appearance prevails.

5. A case of severe bubbling at the distal end of some muscle fibers is illustrated, suggesting autocatalysis of these fibers.

LITERATURE CITED

Bodemer, C. W., and N. B. Everett 1959 Localization of newly synthesized proteins in regenerating newt limbs as determined by radioautographic localization of injected methionine-S[35]. Develop. Biol., 1: 327–342.

DeHaan, R. L. 1956 The serological determination of developing muscle protein in the regenerating limb of Amblystoma mexicanum. J. Exp. Zool., 133: 73–86.

Dickens, F. 1951 Anaerobic glycolysis, respiration, and the Pasteur effect. In: The Enzymes, J. B. Sumner and K. Myrbäck, eds. Academic Press, N. Y., vol. II, pt. I, pp. 624–683.

Feder, N., and R. L. Sidman 1958 Methods and principles of fixation by freeze-substitution. J. Biophys. Biochem. Cytol., 4: 593–602.

Grobstein, C. 1955 Tissue interaction in the morphogenesis of mouse embryonic rudiments in vitro. In: Aspects of Synthesis and Order in Growth, D. Rudnick, ed. Princeton University Press, Princeton, N. J., pp. 233–256.

Harris, M. 1958 Selective uptake and release of substances by cells. In: The Chemical Basis of Development, W. D. McElroy and B. Glass, eds. The Johns Hopkins Press, Baltimore, Md., pp. 596–626.

Jackson, D. S. 1958 Some biochemical aspects of fibrogenesis and wound healing. New England J. Med., 259: 814–820.

Laufer, H. 1959 Immunochemical studies of muscle proteins in mature and regenerating limbs of the adult newt, Triturus viridescens. J. Embryol. Exp. Morph., 7: 431–458.

Lillie, R. D. 1954 Histopathologic Technic and Practical Histochemistry. McGraw-Hill Book Co., The Blakiston Division, New York.

Okuneff, N. 1928 Über einige physiko-chemische Erscheinungen während der Regeneration. I. Mitteilung: Messung der Wasserstoffionenkonzentration in regenerierenden Extremitäten des Axolotl. Biochem. Zeitschr., 195: 421–427.

——— 1933 Über einige physiko-chemische Erscheinungen während der Regeneration. V. Mitteilung: Über den Milchsäuregehalt regen-

46 ANTHONY J. SCHMIDT

erierender Axolotlextremitäten. Ibid., 257: 242–
244.
Pearse, A. G. E. 1960 Histochemistry, Theoretical and Applied, 2nd ed. Little, Brown and
Co., Boston, Mass.
Ryvkina, D. E. 1945 Respiratory quotient in regenerating tissues. C. R. Acad. Sci., U.R.S.S.,
49: 457–459.
Schmidt, A. J. 1958a Forelimb regeneration of
thyroidectomized adult newts. I. Morphology.
J. Exp. Zool., 137: 197–226.

——— 1958b Forelimb regeneration of thyroidectomized adult newts. II. Histology. Ibid.,
139: 95–136.
Singer, M. 1959 The acetylcholine content of
the normal forelimb regenerate of the adult
newt, Triturus. Develop. Biol., 1: 603–620.
Taban, C. 1955 Quelques problèmes de régénération chez les Urodèles. Rev. Suisse Zool.,
62: 387–468.

PLATE 1

EXPLANATION OF FIGURES

1 A limb sampled at 24 hours post-amputation (a512-9). The arrow indicates intact muscle fibers containing glycogen (PAS-positive material), while the distal fibers (M) are depleted of their glycogen (PAS-negative).

2 Illustrated is a section from the same limb as in figure 1 (a512-9). The section was treated with salivary amylase before staining with PAS, therefore, the muscle fibers (M) are devoid of glycogen.

3 A limb sampled at 4 days post-amputation (a512-7). The arrow indicates intact muscle fibers containing glycogen, while the distal fibers (M) are depleted of their glycogen.

4 This is a high magnification of the section in figure 1 (a512-9). It illustrates the intactness of the distal, glycogen-depleted muscle fibers. Myofibrillae (arrow) are readily visible. Note the dark globules of glycogen between the muscle fibers.

5 A limb sampled at 7 days post-amputation (a512-6). The arrow indicates intact muscle fibers containing glycogen. Distally, no intact glycogen-depleted muscle fibers are discernible, rather, frayed fibrous material and small bubbles are seen.

6 A limb sampled at 10 days post-amputation (a516-5). The arrow indicates intact muscle fibers containing glycogen. Distally, a short arrow points to fraying mysial sheaths and concomitant bubbling. Compare with figure 5 to note the variable distance between the intact muscle fibers and the apical epithelium that one finds at these times post-amputation.

7 A limb sampled at 13 days post-amputation (a513-10). This high magnification illustrates (arrow) fraying mysial sheaths and bubbles distal to intact, glycogen-containing muscle fibers, that is typical during this regressive phase in regeneration.

8 A limb sampled at 16 days post-amputation (a513-7). A short arrow directs attention to the extreme amount of bubbling taking place at the distal end of some muscle fibers. A subapical accumulation of cells (B) initiating a blastema may be noted.

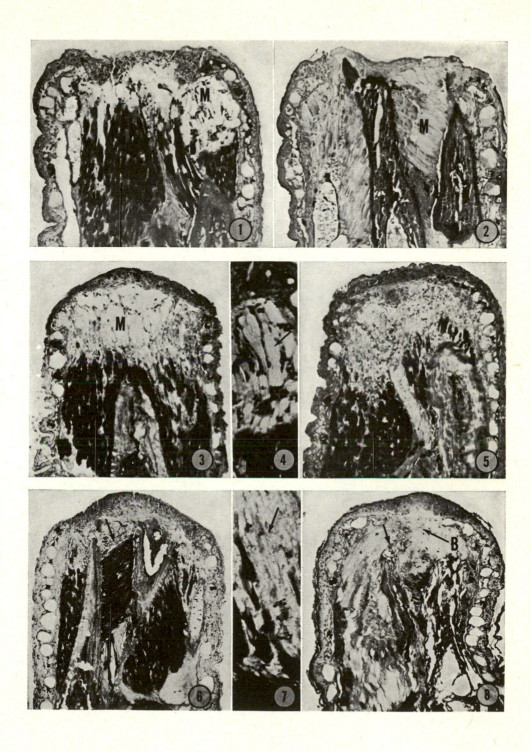

Reprinted from *Develop. Biol.*, **17**, 571–583 (1968)

8

Collagenolytic Activity in Regenerating Forelimbs of the Adult Newt (*Triturus viridescens*)[1-3]

HERMES C. GRILLO, CHARLES M. LAPIÈRE,[4] MARC H. DRESDEN,[5]
AND JEROME GROSS

*Departments of Surgery and Medicine, Harvard Medical School at the
Massachusetts General Hospital, Boston, Massachusetts 02114*

Accepted November 27, 1967

INTRODUCTION

The complete regeneration of functional limbs after amputation in the amphibian is dependent upon two major factors: a phase of "dedifferentiation" of the stump tissues preceding blastema formation (Chalkley, 1959; Hay and Fischman, 1961) and an adequate nerve supply (Singer, 1960). The first phenomenon is characterized by breakdown of intercellular substances and a loss of differentiated cellular characteristics. The cells appear to revert to a more primitive state, whereupon many of them migrate distally and aggregate to form the early blastema. It has been suggested that breakdown of the extracellular matrix is crucial to formation of the blastema from the freed cells (Butler and Puckett, 1940).

An increase in hydrolytic enzymes such as cathepsin, acid phosphatase, and peptidases in the amputation stump during the early phases of regeneration have been described by numerous investigators (see reviews by Flickinger, 1967; Hay, 1966; Lehmann, 1961; Weber,

[1] This investigation was supported by Public Health Service Research Grant CA-03638 from the National Cancer Institute, Grant AM3564 from the United States Public Health Service and Grant AM5142 from the National Institute of Arthritis and Metabolic Diseases.

[2] This is publication No. 451 of the Robert W. Lovett Memorial Group for the Study of Disease Causing Deformities, Massachusetts General Hospital, Boston, Massachusetts.

[3] This work in preliminary form has been reported earlier in a symposium (Grillo, 1964).

[4] Present address: Service de Dermatologie, Hôpital de Bavière, Liège, Belgium.

[5] Research Fellow of the Atomic Energy Commission.

571

1967). These enzymes, however, are functional at acid pH and have not been associated with any specific tissue substrate. Because digestion of the extracellular protein, collagen, requires specific collagenases such as those detected in certain tadpole tissues (Gross and Lapière, 1962), mammalian uterus (Gross *et al.*, 1963; Jeffrey and Gross, 1967), skin wounds (Grillo and Gross, 1967), normal skin (Eisen *et al.*, 1968) and bone (Walker *et al.*, 1964), the tissues of the amputation stump in the newt *Triturus viridescens* have been assayed by a tissue culture technique used in previous studies (Gross and Lapière, 1962). This assay involves the maintenance of living, sterile tissue on reconstituted collagen gels in a medium of physiologic salt solution. Collagenolytic activity is detected by breakdown of the collagen substrate, observed visually or by the release of soluble collagen peptide fragments into the medium. The method has been shown to be essentially specific for collagenase.

METHODS

Mature *Triturus viridescens* were obtained from Connecticut Valley Biological Supply Company and maintained in groups of 3–6 at room temperature in large bowls containing distilled water. After a period of observation to detect fungal contamination, both forelimbs were amputated evenly with sharp scissors just above the elbow joint, and the animals were returned to their bowls for periods ranging from 1 to 30 days. When fungus was observed, animals were discarded.

Twenty-four hours before the regenerating limbs were collected, the newts were placed in water containing 1,000,000 units of penicillin and 0.4 gm of streptomycin per liter. The entire forelimb was rapidly removed and placed in amphibian Tyrode's solution (NaCl 8.0 gm, KCl 0.20 gm, $CaCl_2$ 0.20 gm, $MgCl_2$ 0.10 gm, NaH_2PO_4 0.05 gm, $NaHCO_3$ 1.0 gm, glucose 2.0 gm, water 1500 ml) containing 2000 units of penicillin per milliliter. The whole forelimbs were dissected under a microscope using a watchmaker's forceps and a fine scalpel, keeping the tissues moist at all times in Tyrode's solution.

Whole regenerating limbs were sampled just proximal to the original amputation site (Fig. 1a). Where several samples of the regenerate and proximal limb tissues were taken serially, sample A consisted of the blastema tip, B the base of the regenerating blastema (included in a proximal slice of tissue approximately 1 mm wide), and C and D serial sections of proximal forearm tissues each 1 mm thick (Fig. 1b).

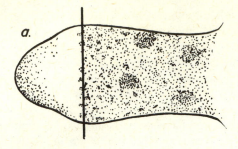

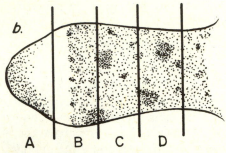

FIG. 1. Diagrams indicating plane of sampling of regenerating forelimb of adult newt. (a) Whole tip is sampled at proximal border of blastema. (b) Serial samples are approximately 1 mm wide and include: A, blastema tip; B, regenerating plate; C and D, proximal unwounded tissues.

Tissues were planted promptly on collagen gel substrates. Samples of tip of regenerate (A) or of whole regenerate were placed with raw surfaces on the gel surface. Larger regenerates were slit longitudinally and similarly placed with cut surface on the gel.

Cultures. Acid-extracted guinea pig dermal collagen dissolved at neutral pH was used to prepare sterile, thermally reconstituted gels for the culture substrate, as previously described (Gross and Lapière, 1962). Concentration of collagen in the gels was 0.025%, 0.05%, or 0.10%. Dilutions from 0.4% collagen solution in phosphate-buffered (pH 7.6) 0.4 M NaCl were made with Tyrode's solution altered from the given composition to reduce sodium appropriately (composition: KCl 0.26 gm CaCl$_2$ 0.26 gm, MgCl$_2$ 0.135 gm, NaHCO$_3$ 1.35 gm, glucose 2.7

gm, water 1500 ml). Amino acids were added as nutrients. Salt-extractable guinea pig dermal collagen labeled *in vivo* with glycine-^{14}C was used for quantitative experiments. Collagen solution (125 μl) was gelled in each Lucite microslide chamber. Cultures were incubated at 27°C or 37°C for 3 or 4 days in a humidified atmosphere of 90% O_2 and 10% CO_2.

A few samples were incubated at 31°C, which is considered to be optimal for the culture of amphibian tissues (Stephenson, 1966). Visible lysis was recorded on a scale of 1+ to 4+: 1+ indicated definite but minimal lysis, 4+ total lysis of the gel and 2+ and 3+ intermediate stages (Fig. 2). Lysis appeared initially after 1 day of incubation and

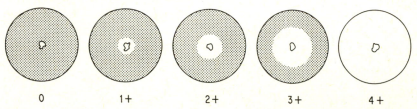

0   1+   2+   3+   4+

FIG. 2. Diagram of scale of visible lysis. The gel on the left is not affected by the explant at its center. 1+ indicates definite but minimal lysis; 4+ indicates full lysis of the gel; 2+ and 3+ show intermediate amounts of lysis.

increased up to 3 to 4 days. Radioactivity in solution from degraded collagen was determined after centrifugation of the total culture at 34,000 g for 15 minutes at 27°C. The radioactivity in the supernatant of each culture was measured in Bray's mixture (Bray, 1960) using an automated scintillation counter and compared with collagen gel blanks exposed to 0.01% trypsin solution.

The necessity for cellular viability in production of collagenase was evaluated by freezing and thawing tissues three times. The regenerating tissue explants on filter paper moistened with Tyrode's solution were placed in a beaker which was frozen in a mixture of CO_2, ice, and ethanol.

Cultures which showed bacterial or fungal contamination were discarded.

RESULTS

Course of Collagenolytic Activity during Regeneration

Collagenolytic activity was not found in intact forelimb tissues removed at amputation. By 5 days after amputation, variable amounts

TABLE 1

COLLAGENOLYTIC ACTIVITY OF REGENERATING NEWT FORELIMBS
MEASURED AS VISIBLE LYSIS AT DIFFERENT TIMES[a]

Day after amputation	Number of cultures	Frequency of lysis		Extent of lysis		
		Number of cultures showing lysis	Percent of cultures showing lysis	Sum of +'s in all cultures	Total +'s possible	Percent of total possible lysis occurring
0	8	0	0	0	32	0
	8	0	0	0	32	0
5	12	1	8	1	48	2
	7	2	28	2	28	7
10	12	8	67	16	48	33
	8	4	50	10	32	31
15	12	9	75	11	48	22
	8	4	50	12	32	37
20	12	5	41	5	48	10
	8	6	75	8	32	25
25	12	2	16	2	48	4
	8	2	25	7	32	21
30	12	0	0	0	48	0
	12	3	25	3	48	6

[a] Cultures were incubated for 4 days, the first group at 27°C and the second at 37°C.

of activity were demonstrable as visible lysis from explants of whole regenerates (Tables 1 and 2). The frequency and amount of collagenolytic activity rose at the 10- and 15-day interval after amputation to a peak at 15 days. After that, activity tended to fall slowly and irregularly until 30–35 days.

Series of cultures incubated at 27°C and 37°C were compared (Table 1) and failed to demonstrate any regular acceleration or accentuation of activity at higher temperature. Quantitative measurements (Table 2) of enzyme activity by release of radioactivity from collagen gels labeled with glycine-^{14}C conformed with the early and peak pattern of activity observed as visible lysis. Later in the course of regeneration, quantitatively measured activity appeared to be greater in amount. The total amount of tissue in the explants increased, however, in the later phases of regeneration as a full limb developed.

Localization of Lytic Activity

Cultures of serially sectioned regenerating limbs and proximal tissues at 15 days after amputation—the period of peak collagenolytic

TABLE 2

COLLAGENOLYTIC ACTIVITY OF REGENERATING NEWT FORELIMBS AT
DIFFERENT TIMES AFTER AMPUTATION MEASURED AS
RELEASE OF LABEL FROM ^{14}C-COLLAGEN GELS[a]

Day	Number of cultures	Visible lysis (av +'s)	Average cpm
0	7	0	239.9
	10	0	786.5
5	9	1.5	454.4
	10	1.8	1471.4
10	6	1.6	483.1
	6	3.5	3166.5
15	10	2.8	1082.2
	10	4.0	3317.6
20	6	1.3	556.2
	9	1.7	1915.0
25	10	1.8	704.9
	8	0.9	1304.6
30	6	2.8	976.8
	9	1.0	1072.1
35	10	2.1	827.1
	7	1.1	1322.8
60	12	0.75	662.3

[a] Cultures were incubated at 31°C for 96 hours and 72 hours, respectively, in these two experiments. 1485 and 5554 cpm/substrate gel, respectively, both 0.025% collagen. Average number of +'s (scale: 0 to 4+) is recorded. Counts per minute are values above blanks. Incubation of gels with 0.01% trypsin released only an average of 105.5 cpm. These experiments are additional to those tabulated in Table 1.

activity both in frequency and extent—clearly placed the maximal enzymatic production in the regenerating plate of the proximal blastema (zone B) (Table 3, Fig. 3). The tip of the blastema was also highly active. Explants of tips (zone A) were of markedly lesser volume than the wider proximal regenerating tissue mass in zone B, which could readily account for the smaller amount of collagen degradation. The section of tissue (zone C) proximal to the growing blastema displayed moderate activity which fell sharply in most cases to intact tissue levels (zone D).

Quantitative measurements of enzyme activity on ^{14}C-labeled collagen gels demonstrated good correlations between observations of visible lysis and release of radioactivity in each of the zones (Table 4). As more concentrated collagen gels were used as substrates, direct visualization of lytic activity became more difficult because of in-

169

TABLE 3

COLLAGENOLYTIC ACTIVITY OF SERIAL ZONES OF 15-DAY REGENERATING
NEWT FORELIMBS MEASURED AS VISIBLE LYSIS[a]

| | | Frequency of lysis | | Extent of lysis | | |
Zone	Number of cultures	Number of cultures showing lysis	Percent of cultures showing lysis	Sum of all +'s in all cultures	Total +'s possible	Percent of total possible lysis occurring
A	12	6	50	12	48	25
B	12	10	83	16	48	33
C	12	6	50	6	48	12
D	12	0	0	0	48	0
A	12	2	17	2	48	4
B	12	7	58	10	48	20
C	12	2	17	3	48	6
D	12	0	0	0	48	0
A	20	17	85	30	80	37
B	20	18	90	33	80	41
C	20	10	50	13	80	16
D	20	7	35	18	80	10
A	16	1	6	1	64	2
B	16	10	60	10	64	16
C	12	6	50	6	48	13
D	16	1	6	2	64	3

[a] Cultures were incubated for 4 days at 27°C.

creasing density, but measurements of release of radioactivity continued at comparable levels (Table 4). When regenerating limbs were split longitudinally and cultured on the gels, lytic activity appeared initially around the blastema and spread gradually from this region (Fig. 4).

Epithelial and mesenchymal tissues were not separated in these experiments.

Cellular Viability

Living cells were necessary for production of collagenolytic activity. Freezing and thawing whole tips of 15-day regenerating limbs prevented visible collagenolytic activity on subsequent culture in nearly all cases (Table 5) and little radioactivity was released from labeled collagen. The amount of degradation was of the same degree as that

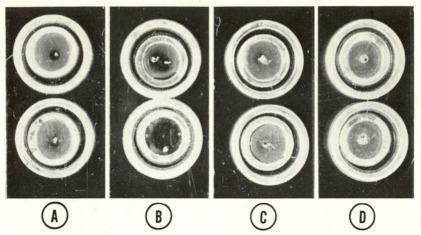

FIG. 3. Cultures of 15-day forelimb regenerates from adult newts serially sampled as indicated in Fig. 1b, on reconstituted collagen gels after 48 hours. A dark zone surrounding an explant indicates lysis. The lighter areas are residual intact collagen gels. Zone A is from the tips of the regenerates and shows 1+ lytic activity. The samples containing the regenerating plate (zone B) of proximal blastema have caused 2+ and 3+ lysis of collagen. The upper sample has separated into two fragments, and the lower has drifted eccentrically in the liquefied gel. Zone C is proximal to the line of amputation. There is a trace of visible lysis above ($\pm$) and none below. Zone D samples show no visible lysis. See Fig. 1b for diagrammatic definition of zones A–D.

produced by 0.01% trypsin. This contrasted with consistent and extensive lytic activity of living control cultures of 15-day regenerates.

DISCUSSION

The time course of production of a collagenolytic enzyme which appears in the regenerating tissues of the amputated forelimb of the adult newt and which degrades native collagen at physiologic pH, parallels the period in which the bony matrix regresses, proximal tissues are broken down, myofibrils disappear, syncytial fibers break up, and in which Schwann cells and connective tissue cells are released into the early blastema, with subsequent rapid proliferation of these embryonic appearing cells (Hay, 1966). As the regenerating mass increases, a new limb begins to differentiate, with rapid remodeling of cellular and structural elements. In the adult newt these phases are accomplished in the first 3–4 weeks after amputation. It is tempting to assign a central role to enzymes which degrade the intercellular

TABLE 4
COLLAGENOLYTIC ACTIVITY OF SERIAL ZONES OF 15-DAY REGENERATING
NEWT FORELIMBS MEASURED AS VISIBLE LYSIS AND BY
RELEASE OF LABEL FROM ^{14}C-COLLAGEN GELS[a]

Collagen conc. of gel (%)	Zone	Number of cultures	Visible lysis (av +'s)	Average cpm
0.025	A	7	0.14	67
	B	7	1.71	425
	C	7	1.01	216
	D	7	0.14	146
	A	7	0.86	313
	B	8	1.13	331
	C	6	0.42	227
	D	6	0	101
	A	15	1.50	259
	B	15	2.46	379
	C	16	1.03	342
	D	16	0.13	69
0.05	A	6	0.33	387
	B	8	0.56	292
	C	6	0	161
	D	6	0	70
0.1	A	7	0	64
	B	7	0	245
	C	6	0	166
	D	7	0	147

[a] Cultures were incubated for 96 hours at 37°C; 1485 cpm/substrate gel, 0.025% collagen, and other values proportional. Visible lysis was recorded from 0 to 4+. Average number of +'s is recorded. Counts per minute are values above blanks. These experiments are additional to those tabulated in Table 3.

matrix in facilitation of these structural rearrangements. Collagenolytic activity, however, is not limited to amphibian regeneration. Intensification of collagenase production in epithelium and its appearance in new connective tissue have been documented during repair of mammalian wounds (Grillo and Gross, 1967). with a very different final organization of healed tissues—namely, cicatrization.

In the time-course experiments reported, the amount of tissue in the explants was not quantitated, because of its complex nature: the entire

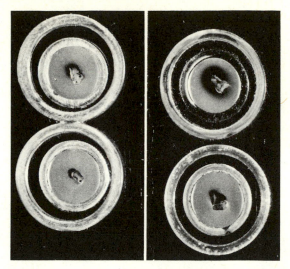

FIG. 4. Cultures of 15-day regenerating forelimbs slit longitudinally and including both blastema and proximal unwounded tissues, after 24 hours of culture, cut side down, on reconstituted collagen gels. The regenerating tip points toward the upper left in each case. Lysis has begun uniformly around the tips, rather than in relation to proximal, unwounded tissues of the forelimb.

TABLE 5

COLLAGENOLYTIC ACTIVITY IN 15-DAY REGENERATES OF NEWT
FORELIMBS AFTER FREEZING AND THAWING[a]

Sample	Number of cultures	Visible lysis			Average cpm
		Sum of +'s in all cultures	Total +'s possible	Percent of total possible lysis occurring	
Untreated tissues	3	9	12	75	171
	7	26	28	93	
Frozen and	13	2	52	4	29
thawed tissues	7	0	28	0	
Trypsin, 0.01%	2	0	8	0	34
	2	0	8	0	
Blank gels	2	0	8	0	0
	2	0	8	0	

[a] Cultures were incubated for 96 hours at 31°C; 1485 cpm/substrate gel, 0.025% collagen. Visible lysis was recorded on scale 0 to 4+. Counts per minute are values above blanks.

regenerate was examined. In late phases of regeneration the total amount of tissue was increasingly large—and probably gave an apparently high index of activity at these times. Gross and Lapière (1962) showed a direct relationship between amount of tissue and amount of collagenolytic activity.

The apical epithelial cap is essential to the regenerative process, although it does not seem to contribute cells to the blastema (Hay, 1966). Proteolytic enzymes active at acid pH appear to be present in the epithelium during dedifferentiation and blastema formation (Schmidt, 1966; Singer and Salpeter, 1961). Epithelium and mesenchyme were not separately studies in these experiments; epithelium alone has been shown to produce collagenase actively in other circumstances, such as in the tadpole tail fin (Eisen and Gross, 1965) and in mammalian skin (Grillo and Gross, 1967). While unwounded connective tissue has not been found to produce collagenase activity either in amphibians (Eisen and Gross, 1965) or in mammals (Grillo and Gross, 1967), repairing connective tissue does possess this capability (Grillo, unpublished; Grillo and Gross, 1967). It is possible that the blastema itself possess such potential in the newt.

Living cells are necessary for production of the enzyme. This was demonstrated by nearly total absence of activity following freeze-thawing of regenerates prior to culture. This is consistent with the behavior of collagenase-producing tissues in all other systems studied by these methods.

The precise role of collagenase in the process of limb regeneration remains to be defined. Studies now in progress will attempt to define further the mode of action of this enzyme. It may serve to facilitate loosening of cell attachments or disposition of structural elements in the proximal breakdown process and in tissue remodeling as structural regeneration proceeds. A less likely function might be prevention of cicatrization which in itself might constrain regeneration.

SUMMARY

A collagenolytic enzyme capable of degrading reconstituted collagen fibrils at neutral pH and physiological temperature, not demonstrable in the intact forelimb of the adult newt (*Triturus viridescens*) appears in cultures of healing and regenerating tissues of the amputation stump. Tissues proximal to the regenerating region are inactive. The amount of enzyme produced, measured by visible breakdown of thermally reconstituted mammalian collagen gels, and confirmed by release of radioactivty from ^{14}C-glycine-labeled collagen gels, rises to a peak at 15 days and remains at a high level during the phases of dedifferentiation of the stump and blastema formation, but gradually falls after 20 days as early digital differentiation occurs. Living cells are required for collagenase production since freezing and thawing inactivates the explants.

The authors wish to acknowledge the excellent assistance of Mrs. Hilliard Macomber and Miss Judith Hothan.

REFERENCES

BRAY, G. A. (1960). A simple efficient liquid scintillator for counting aqueous solutions in a liquid scintillation counter. *Anal. Biochem.* 1, 279–284.

BUTLER, E. G., and PUCKETT, W. O. (1940). Studies on cellular interaction during limb regeneration in Amblystoma. *J. Exptl. Zool.* 84, 223–237.

CHALKLEY, D. T. (1959). The cellular basis of regeneration. *In* "Regeneration of Vertebrates" (C. S. Thornton, ed.), pp. 34–58. Univ. of Chicago Press, Chicago, Illinois.

EISEN, A. Z., and GROSS, J. (1965). The role of epithelium and mesenchyme in the production of a collagenolytic enzyme and a hyaluronidase in the anuran tadpole. *Develop. Biol.* 12, 408–418.

EISEN, A. Z., JEFFREY, J. J., AND GROSS, J. (1968). Human skin collagenase: Isolation and mechanism of attack on the collagen molecule. *Biochem. Biophys. Acta*

FLICKINGER, R A. (1967). Biochemical aspects of regeneration. *In* "The Biochemistry of Animal Development" (R. Weber, ed.), Vol. II, pp. 303–337. Academic Press, New York, in press.

GRILLO, H. C. (1964). Aspects of the origin, synthesis and evolution of fibrous tissue in repair. *In* "Advances in Biology of Skin" (R. Billingham and W. Montagna, eds.), Vol. V, pp. 128–143. Pergamon Press, London.

GRILLO, H. C. (1967). Collagenolytic systems in healing mammalian wounds. Workshop in the healing of osseous tissue (1965), pp. 143–147. Natl. Res. Council—Natl. Acad. Sci., Washington, D.C.

GRILLO, H. C., and GROSS, J. (1967). Collagenolytic activity during mammalian wound repair. *Develop. Biol.* 15, 300–317.

GROSS, J., and LAPIÈRE, C. M. (1962). Collagenolytic activity in amphibian tissues; a tissue culture assay. *Proc. Natl. Acad. Sci. U.S.* 48, 1014–1022.

GROSS, J., LAPIÈRE, C. M., and TANZER, M. L. (1963). Organization and disorganization of extracellular substances; the collagen system. *Symp. Soc. Study Develop. Growth* 21, 175–202.

HAY, E. D. (1966). "Regeneration." Holt, New York.

HAY, E. D., and FISCHMAN, D. A. (1961). Origin of the blastema in regenerating limbs of the newt *Triturus viridescens. Develop. Biol.* 3, 26–59.

175

JEFFREY, J. J., and GROSS, J. (1967). Isolation and characterization of a mammalian collagenolytic enzyme. *Federation Proc.* **26**, 670. (Abstract.)

LEHMANN, F. E. (1961). Action of morphostatic substances and the role of proteases in regenerating tissues and in tumour cells. *Advan. Morphogenesis* **1**, 153–187.

SCHMIDT, A. J. (1966). "Molecular Basis of Regeneration:Enzymes." Univ. of Illinois Press, Urbana, Illinois.

SINGER, M. (1960). Nervous mechanisms in the regeneration of body parts in vertebrates. *In* "Developing Cell Systems and Their Control" (D. Rudnick, ed.), pp. 115–133. Ronald Press, New York.

SINGER, M., and SALPETER, M. M. (1961). Regeneration in vertebrates: the role of wound epithelium. *In* "Growth in Living Systems" (M. X. Zarrow, ed.), pp. 277–311. Basic Books, New York.

STEPHENSON, N. G. (1966). Effects of temperature on reptilian and other cells. *J. Embryol. Exptl. Morphol.* **16**, 455–467.

WALKER, D. B., LAPIÈRE, C. M., and GROSS, J. (1964). A collagenolytic factor in rat bone promoted by parathyroid extract. *Biochem. Biophys. Res. Commun.* **15**, 397–402.

WEBER, R. (1967). Biochemistry of amphibian metamorphosis. *In* "The Biochemistry of Animal Development" (R. Weber, ed.), Vol. II. pp. 227–301. Academic Press, New York.

Outgrowth and Differentiation of the Blastema

II

Editors' Comments on Papers 9 Through 16

What kinds of information are regarded by cells in a freshly amputated stump as a signal to regenerate? This question can be approached in two basic ways. One either deletes factors from limbs that can regenerate, or adds them to limbs that cannot.

Both types of experiment have been instructive. Eliciting regeneration by specific acts to limbs that would not otherwise do so is a particularly dramatic demonstration; developing frogs, which lose their powers of limb regeneration as they mature, have long been a favorite organism for this type of experiment.

S. Meryl Rose reports experiments in which regenerative responses were elicited in otherwise incompetent frog limbs by traumatizing the distal epidermal–dermal covering. It has been known for some time that animals that do not regenerate limbs seal the amputation plane with epidermis, which is quickly followed by dermis. Rose demonstrates that if the dermatization of the cut surface is discouraged, either by destroying surface layers osmotically, or by regularly peeling away the distal cover, a regenerate of sorts is formed.

L. W. Polezhayev in his laboratory in Moscow has for years investigated similar phenomena. The review article included here (which is itself a review of a larger work) is a comprehensive and systematic survey of such efforts, made primarily in various European laboratories. He points out that tissues simply age and thereby become less competent to regenerate; most of the techniques to prolong regenerative ability involve trauma of various sorts.

However, while retention of larval cellular characteristics appears important in determining the fate of an amputated limb, other factors are clearly at work. Emile Guyénot and Oscar Schotté demonstrate that limbs may be caused to appear on unexpected locations on salamanders by deviating sciatic or brachial nerves to a position under the skin. This is an intriguing demonstration of the capabilities inherent in nerve tissue with respect to the production (in this case, ontogenetic) of new limbs. But since nerve deviations to some parts of the subsurface of the organism do not give rise to good limbs, one must return to the notion that there must be a basic competence to respond within the cells themselves. Possibly the abortive growths that arise from less favored areas could be caused to regenerate by the same methods that work with frogs. Also, deviation of nerves to the dorsal crest or tail gives rise to growths which resemble the area to which the nerve was deviated more than the limb from which the nerve originally came. Thus the capacities of the cells themselves, rather than that of the permissive nerve, determines the nature of the final outgrowth, or lack of it.

If it is indeed the cells themselves that control their own fate, can this concept be extended to cells within the blastema itself, even though they appear identical? The report by J. Faber appears to answer this question in the affirmative. He transplanted whole blastemas and portions of them to one of the areas identified by Guyénot as not possessing limb-forming capacities of its own (the center of the back). The nature of the subsequent outgrowth varied according to the part of the blastema that was transplanted. These results, coupled with carbon-marking experiments led him to postulate an "apical proliferation centre" which both gives rise to new cells and specifies that they will form distal parts, regardless of organismic influence. Stump tissues appeared to him much less important morphogenetically. In a later paper, David L. Stocum reexamines Faber's notion of a distal organization center that specifies distal structures. Using the same methodology as Faber, but more precise in his assay of subsequent morphogenesis of transplanted blastemata, he suggests that resorption may have been responsible for the distalization of Faber's transplants of early blastemata. He observed a remarkable degree of self-organizing ability in transplants of whole blastemata and was able to correlate subsequent development of parts of transplanted blastemata with the location from which they were taken. In the first paper of this series, Stocum (1968) also showed that whole blastemata, provided they had an epithelial covering, were able to differentiate *in vitro*. This demonstration is of special interest since, of course, there is no contact whatever with any hypothetical organismic influence in this case. Such an illustration of understanding of factors necessary to control the formation of a blastema is conspicuously lacking at present.

It has been noticed by those familiar with the phylogenetic distribution of competence to regenerate limbs that the ability to do so seems generally associated with organisms that do not make much use of their limbs. Most urodeles spend much of their lives in water; even on land they locomote with a flexure of the body that is basically a swimming motion. And while adult anurans are efficient in their terrestrial movements, frogs also become progressively unable to regenerate limbs as they mature. Speculations along these lines, concerned with responses to evolutionary pressures, become less entertaining when attempts are made to include other structures that urodeles are uniquely able to regenerate. These include elements of the jaw, the lens, and an accurate tail regenerate. The report by Randall Reyer is an attempt to identify forces that prevent the dorsal iris of a newt's eye from regenerating a lens when it would be inappropriate to do so. Tuneo Yamada has been especially active in identifying important aspects of this system and, with Tanaka and Albright, reports on the time and places where specific lens proteins reappear.

179

Urodele tails also regenerate, and as opposed to reptiles, produce an accurate copy of the missing part. Sybil Holtzer elegantly examines a most intriguing observation with this system. (Unfortunately, space limitations in this volume have dictated that substantial omissions be made from this paper.) When the tail of urodele larvae is amputated, the regenerated tissue is not precisely characteristic of the missing part. Instead of regenerating notochord, which would be a chronologically accurate restitution of the missing part, the new tail contains vertebral centra, which do not appear in the intact animal until a later time. While there may be little teleological point in regenerating structures which must shortly undergo substantial changes anyway (thus this is a special case), it is possible to speculate on the nature of genetic pools involved generally in regeneration. Clearly, this restitution does not involve a simple replay of the gene complex that gave rise to the tail, but an arousal of new portions. If, then, the ability to regenerate appendages generally involves new genetic segments, which may have arisen only in organisms which display the ability to regenerate, possibilities for achieving regeneration of limbs in higher vertebrates may be slim indeed.

References

Stocum, D. L., 1968. The urodele limb regeneration blastema: a self-organizing system. Develop. Biol., *18:* 441–456.

Reprinted from *J. Exptl. Zool.*, **95**, 149–170 (1944)

METHODS OF INITIATING LIMB REGENERATION IN ADULT ANURA [1]

9

S. MERYL ROSE

Department of Zoology, Smith College, Northampton, Massachusetts

THREE PLATES (TWELVE FIGURES)

INTRODUCTION

It is a general principle that a regenerant arises from or close to an open wound. It has been shown experimentally in widely separated groups that when an amputation wound is covered, regeneration is inhibited. The coelenterate, Tubularia, will not regenerate when the cut end is covered with sand. (Loeb, 1892). A glass capillary over the end of a Tubularia stem also prevents regeneration (Barth, '38). The flatworm, Planaria lugubris, does not regenerate if the freshly cut surface is obliterated by peripheral tissue pinching over the cut surface (Brønsted, '39). An insect, one of the phasmids, does not regenerate appendages if a chitinous covering is allowed to form over the stump (Schaxel, u. Adensamer, '23). And, lastly, among the Amphibia, regeneration is prevented when skin is placed over a fresh amputation surface (Tornier, '06; Schaxel, '21; Godlewski, '28).

Among the Amphibia, newts and salamanders and frog tadpoles can regenerate limbs. They do not do so if a piece of skin is sewed over the amputation surface. The adult frog does not replace limbs and it has a thicker skin than the other forms. Is this thicker, more water resistant skin in some way responsible for the failure of limb regeneration? An interesting correlation has been observed between increase in skin thickness and decrease of regenerative ability in the frog tadpole (Poležajew, '39b; confirmed for Rana clamitans in our laboratory by Hickok — unpublished).

Wound closure in an adult frog is rapid. After limb amputation an epithelium, presumably from the epidermis at the periphery of the wound, spreads over the wound. This is not peculiar to the non-regenerating animals. It is also usual for regenerating animals to have their wounds rapidly covered by an epithelium. The striking difference in the wound healing of an adult frog is that the entire

[1] Contribution from the Department of Zoology, Smith College, No. 204.

149

skin, epidermis and dermis, follows the epithelium and closes over the stump within a week or two. The question is whether the rapid wound closure by the complete old skin prevents regeneration. The results of an initial attempt to prevent rapid healing of adult frog amputation wounds have been reported (Rose, '42).

Tadpole limbs too old to regenerate without special treatment have been found to regenerate sometimes after two quite different treatments. Some limbs too old to regenerate have regained the power after transplantation from their normal position to another part of the body of the same animal (Liosner, '31). Another effective method of stimulating regeneration in old tadpole limbs is by injuring the limb stump by cutting with a scalpel (Polejaiev, '36; Poležajew, '39a [2]). How these methods may stimulate regeneration will be discussed below.

MATERIALS AND METHODS

The frogs used in these experiments were three species of Rana: clamitans, palustris and pipiens.

In the experiments to be described amputations were usually performed through the mid forearm. The wounds of some experimental animals were treated with a saturated NaCl solution. These animals were made to swim in a salt bath 1 to 3 minutes for each treatment. They were subjected to between three and five treatments in the first 24 hours after amputation. On the following days they received two treatments per day.

Other experimental animals had their wounds kept open by repeated removal of skin. At the time of amputation the skin was removed back from the periphery of the wound surface for a few millimeters and was removed from this region each time it reappeared. At the same times the epithelium over the amputation surface was peeled away.

During some experiments frogs were kept in groups of ten in aerated aquaria. The most satisfactory aquarium was one 15 inches square in which water was left standing to a height of 3 or 4 inches. Stones were placed in the aquaria in such a way that when the frogs sat on them the amputated limb was just under the surface of the water. The water was changed every 2 or 3 days. For other experiments frogs were kept in quart glass jars, the metal covers of which were punctured with holes for the passage of air. Two to three hundred cubic centi-

---

[2] There are several ways in which this author's name is spelled when transposed from the Russian: Polezaiev, Polejaiev, Poleźajew and Poleszajew.

meters of water was kept in the jars and changed twice daily. Both types of containers were satisfactory.

In the first year of this work great difficulty was experienced in maintaining the health of both stock and experimental animals. Originally they were force-fed ground beef two or three times a week. After a few months the animals were obviously unhealthy. The hind legs of many twitched after handling. As time went on extreme muscle contractures appeared all over the body. Animals responded to slight stimuli by stiffening and quivering as do animals treated with strychnine. After a few weeks the hind limbs bones were so weak that the contractures often caused the bones to break. The same course was run by animals maintained on a pig and beef liver diet. However, none of the symptoms appeared in some frogs which were given flies in addition to liver and which were sunned for a few hours every second or third day. The condition was apparently caused by abnormal calcium metabolism and was probably at least partially due to a vitamin D deficiency.

Since the diet was apparently deficient in vitamin D and possibly in calcium, it was thought that it might be well to supplement with other vitamins and with milk and bone meal. The experiments reported in this paper were performed on frogs which received in addition to liver twice a week, 1 or 2 cc. each week of the following mixture: 25 cc. of milk, about 5 gm. of SAS, which contains among other things dried beef scraps and bone meal, a few cubic centimeters of cod liver oil, a powdered brewers' yeast tablet and the contents of two vitamin capsules, Sears' Super Kaps. The vitamin mixture was force-fed in a pipette through the mouth and oesophagus directly to the stomach. With this diet frogs grow rapidly and remain healthy.

Epidemics also took their toll during the early days of experimentation. At one time a fungal growth appeared regularly on amputated limbs. It was found that this could be completely prevented by keeping frogs after amputation in a solution of 0.15% NaCl in tap water. This solution in itself does not cause frog limb regeneration.

Another time frogs were dying throughout the laboratory from what was apparently a bacterial infection of the limb stump. This was prevented by keeping the frogs after limb amputation in a 0.015% solution of sulfa-diazine in tap water. This too does not cause limb regeneration.

The skin removal and salting experiments reported here were performed on animals which grew and were maintained in a healthy condition.

Regenerants and limb stumps were studied histologically after de-calcification in Bouin's fixative for approximately 3 weeks, imbedding in paraffin, sectioning at 10 μ and staining with Mallory's triple stain.

EXPERIMENTAL SECTION

Normal healing

Healing after limb amputation has been observed in 150 frogs. The species used were catesbiana, clamitans, palustris, pipiens and sylvatica of the genus Rana. The individuals ranged in developmental stage from just metamorphosed to fully grown.

Amputations were performed through either the left or right fore-arm. A few days after the operation a smooth, clear epithelium was seen extending from the skin border over the end of the limb. Slowly the extent of the area covered by epithelium only decreased as a complete skin migrated in and replaced the epithelium. This replacing skin was not regenerated but was old skin which had originally been at the periphery of the wound. Within 1 or 2 weeks the stump was completely sealed, or almost so, by a skin cover. If a small central region remained free of old skin, scar tissue filled in later or skin regenerated there. Sometimes a scar projected out beyond the level of amputation, but usually the stump retained a smoothly rounded surface.

The character of the tissues which seal the stump may be seen in figures 1 and 2. These are sections through the distal ends of healed forelimb stumps of a first year Rana palustris adult. The limb shown in figure 1 was fixed 10 days after amputation, and the limb in figure 2 at 39 days. Both limbs are sealed by old skin. The new tissue which formed below the skin is almost entirely connective tissue. Collagenous fibers which stain blue with the aniline blue of Mallory's triple stain make up the bulk of this scar tissue. In both cases new cartilage has formed around the cut end of the bone. The fibrous scar in figure 1 consists of loose fibers. It is continuous with muscle and connective tissue laterally and completely covers the cut end of the radio-ulna. The fibrous scar in figure 2 is composed of more densely packed fibers. It is not unusual to find islands of cartilage in such a scar.

Well-developed lymph spaces and loose skin are clearly seen in figure 2 and in figure 11, a section of a normal, unamputated wrist and hand region of a small adult Rana sylvatica. The looseness of this skin may facilitate rapid wound closure because it is free to slide over the underlying tissues.

184

If blastemas were present on these amputated limbs, they would occupy the space now taken by the scar. Furthermore the complete skin covering would be lacking.

Regeneration after salt treatments

Since skin over the frog limb stump might be responsible for the failure of regeneration and the formation of scar, methods for preventing skin migration were sought. The first method involved placing frogs with amputated limbs in a saturated solution of NaCl for short periods. The treatments were given twice a day for periods of 1 to 3 minutes. The salt treatment was chosen because the outer cells at the amputation surface could be injured or destroyed without injury to more internal cells. The treatments also prolonged the period of irritation subsequent to amputation.

Healing was greatly delayed. The wound usually remained free of epithelium for more than 2 days. Often during the treatments of the first and second day after amputation there was bleeding from the amputation surface. The old skin of these animals did not close off the wound in 1 or 2 weeks as had happened after amputation alone. In fact, the amputation surface was never covered by old skin. Instead, within 20–30 days a regeneration blastema covered only by an epithelium appeared and grew out from the skin-free region.

Twenty-five salt-induced regenerants have been studied histologically. Figures 5 and 8 represent the extremes in types of regenerants observed. Figure 5 is a section showing a very poorly-formed regenerant. The frog from which it was taken was a recently metamorphosed Rana clamitans. It received four salt treatments the first day and four the second day. Thereafter for 30 days it was subjected to two treatments a day. The limb was fixed 10 months later. Approximately the upper three-fourths of figure 5 is regenerant. The new skeleton is a single piece of cartilage attached to the old bone. The tissue between the cartilage and the skin is quite abnormal. Instead of being mostly muscle it is almost entirely dense fibrous tissue. Just a few muscle fibers have developed. The most nearly normal regenerated tissue is the skin. Epidermis, glands and dermis differentiated.

Figure 8 is a section of one of the better regenerants obtained after salt treatments. It grew from the forearm region of a first year adult Rana clamitans. The salt treatments in this case were discontinued after only 5 days. There were five treatments the first day, four the second and two each on the following 3 days. The regenerant was fixed 5 months after amputation. All of figure 8 is regenerant. The regen-

erated distal end of the radio-ulna appears normal. Beyond it two cartilaginous carpals may be seen. At the tip of the regenerant are several
smaller and abnormal cartilaginous elements. Good muscles, tendons,
ligaments and skin are present.

The best of the salt-induced regenerants show a striking disparity
between proximal and distal portions. The proximal regions are normal or nearly so while the more distal regions are always abnormal.
Figure 3 is a regenerated wrist region of another first year Rana clamitans. The arm was amputated in the mid-forearm region and the animal treated in a 0.3% NaCl solution each day for 11 days beginning
with the second day after amputation. The arm was allowed to grow
for 5 months. The regenerated forearm and wrist region are quite
normal and contain all of the tissues and structures normally found
in these regions. Skin may be seen extending across the base of figure
3. Directly above it is muscle and above the muscle are the skeletal
elements. The bony radio-ulna with its narrow cavity and cartilaginous
distal end articulates with two carpals, parts of which may be seen on
the right side of the figure. Ligaments connect the skeletal elements.
Immediately above the radio-ulna is more muscle and still farther up
a blood vessel and nerve cut longitudinally. In short the regenerated
wrist region contains all of the tissues normally found in that region
and their arrangement is normal. It was a functional wrist. Four other
regenerated wrists as good or almost as good as those shown in figures
3 and 8 have been obtained. Distal to the two carpals shown, regeneration was restricted to several blocks of cartilage, some fibrous tissue
and skin.

The imperfectly regenerated distal tip of a limb may be seen in figure
4, which is a more magnified view of the upper left portion of figure 8.
This regenerant was fixed after 5 months and no growth or change
in form had been observed for a month. Differentiation was nearly complete at the time of fixation. At the tip of the limb only a few small
skeletal elements, probably representing metacarpals and phalanges,
are present. Two of these are shown in figure 4. The lower one articulates with a carpal, the distal end of which is at the base of the
figure. These distal structures are small relative to the larger more
proximal ones (fig. 8). Furthermore, the tissues in the distal region
are not completely differentiated and those which are, appear to be
young. For example, the chondrocytes are close together and little
matrix has been deposited between them (fig. 4). Chondrocytes are
apparently still differentiating at the tip of the distal cartilage. A
young tendon with its associated partially differentiated muscle fibers is

attached to the left distal region of the outermost cartilage. Surrounding the muscle and tendon are undifferentiated cells. More of these mesenchyme cells are to the right of the cartilages. A number of pigment cells are visible throughout the figure. The region is abnormal in that differentiation is incomplete and the structures are small. The distal tip of such a regenerant never develops much further. The few remaining mesenchyme cells become fibroblasts and collagenous fibers form between them. It should be noted that the epidermis and skin glands are well formed and also that a large amount of irregularly arranged fiber has developed in the dermal region. Characteristically the skin is heavy over these regenerants while regeneration is slowing down and stopping. The question arises: — Is this relatively early differentiation of dermal fiber connected with the retardation and suppression of further regeneration? If regenerated skin in a frog is the agent inhibiting further regeneration, it is probably the dermal layer which is responsible because in normally regenerating limbs, e. g., those of urodeles, the dermis forms very late, after other tissues have differentiated. The epidermis, on the other hand, is probably not acting as an inhibitor because it becomes thick very early in both Urodela and the salt-treated Anura. This will be discussed below.

Five first year adult Rana clamitans had four toes of a hind foot cut off level with the shortest toe, which was not amputated. The frogs received daily baths in 0.3% NaCl solution for 12 days. The baths lasted for an hour to an hour and a half each day. The cut toes of all animals regenerated. They did not attain full size, but in every case the normal relative size relationships were reestablished. The fourth toe grew more than the others. The fifth, third and second toes graded off in that order. A section of one of these feet may be seen in figure 10.

Regeneration after skin removal

Salt treatments were obviously doing two things. They were irritating the stump, causing destruction of tissue and producing hyperaemia. They were also preventing skin closure. The injury and irritation effects have not been investigated. So far only an attempt has been made to determine whether at least a part of the effect of treating with salt is by way of prevention of skin closure. A test was made by cutting the skin back and preventing it from covering the end of the stump. In this way skin was kept from sealing off the end of the stump for a number of days, as long or longer than was necessary for blastemas to form on salt-treated animals. If the salt produced its effect by keeping the end of the stump free of skin, then removal of skin with a knife should also induce regeneration.

Thirty-five animals had a limb amputated between elbow and wrist. At the same time a few millimeters of skin were peeled back from the amputation surface and cut away. Periodically the skin which had grown down around the stump, and the epithelium over the amputation surface were removed. Some regeneration occurred in all cases. However, it was never as complete or normal as the regeneration following salt treatments.

Figure 7 is a photograph of a section through a regenerant which grew on a Rana clamitans after such treatment. This frog was 5 cm. long from tip of snout to posterior end of urostyle. It was collected in the spring. Judging from its size it had metamorphosed early in the previous year. Older frogs have also been used and they too regenerate. However, their larger limbs do not as readily lend themselves to histological preparation. For that reason most of the study has been made on small frogs. The regenerant shown in figure 7 was obtained by peeling the skin back from the wound at amputation time and again at the end of the second, third and fourth weeks after amputation. The regenerant was fixed 1 year after amputation. Prior to regeneration there was much resorption. One month after amputation the regression was to within a few millimeters of the elbow. Regeneration started from that point, which is about ¼-inch from the base of the figure. Skin, bone, cartilage and a large amount of fibrous tissue formed. It is to be noted that muscle is not well represented in the regenerant. The only muscle which differentiated may be seen as a dark island surrounded by scar tissue on the right side of the figure. The skin is unusual in that the dermis blends with the great mass of fiber between skeleton and skin. An abundance of scar tissue and cartilage was usual following skin removal.

A similar picture is seen in figure 6. This animal, also a Rana clamitans, was younger, only 4 months past metamorphosis. The arm was amputated in the mid radio-ulna region and the skin was cut back to the original edge of the amputation surface every 4 days. This was done three times, thus allowing the regeneration process 12 days to get started before skin could encroach on the amputation surface. Actually the amputation surface remained free of old skin, since no further migration occurred after the third cutting. The limb was fixed 8 months later (fig. 6). All but the lower ¼ inch of the figure is regenerant. An abnormal distal end of the radio-ulna had formed. Beyond this are two sets of cartilages. A large amount of fibrous tissue may be seen between the cartilages and between the cartilages and skin with which it blends.

Regeneration also occurred when only the tip of the stump was repeatedly opened. In these cases skin was allowed to cover all but the center of the amputation surface and the resulting outgrowth. The upper half of figure 9 is regenerated tissue which arose after such treatment. The animal was a first year Rana clamitans. The tip of the stump and of the regenerant was removed on the 9th, 18th, 27th, 38th, 45th, 57th, 63rd and 72nd days after amputation. The limb was fixed at 5 months. Regeneration in all of the animals so treated was poor. Skin, scar and cartilage developed.

Rana pipiens with open wounds, whether produced by salt treatment or cutting, regenerate very poorly. Almost all the new tissue is scar and cartilage.

Regeneration from untreated open wounds

A fortunate accident provided further evidence that an amputation surface must remain free of skin if regeneration is to commence. Two groups of five second-year Rana palustris had their arms amputated at the same time through the forearm. Both groups were left untreated in aerated water. But one group had received the usual vitamin ration while the other group had been neglected for several weeks. The group without added vitamins did not heal normally. A complete skin did not migrate and seal off the stumps. Only the migrated epithelium served as a covering. The hind legs of the vitamin-starved animals began to twitch 15 days after amputation. This is a symptom of vitamin D deficiency as explained above. The deficiency was alleviated by addition of the vitamin mixture to the diet. Surprisingly enough, the deficient animals with stumps covered only by epithelium began to regenerate. Figure 12 shows one of the regenerants so obtained. This one was fixed 8 months after amputation. The new distal end of the radio-ulna, hand cartilages, muscle and skin may be seen. The picture is very different from that of a healed palustris stump (cf. figs. 1 and 2). Here again, in figure 12, much too much fibrous tissue formed in the distal region of the regenerant. Apparently in the later stages of regeneration the distal undifferentiated blastema cells develop into fibrous tissue instead of other tissues as well. This type of tissue, as noted above, forms early in untreated limbs when old skin migrates over the end of the stump. The regenerants obtained after preventing skin closure in one way or another, do not form a fibrous scar immediately, but an unusually large amount of fibrous tissue forms later during regeneration at the time when the process is being suppressed.

DISCUSSION

It now seems important to know how skin closure may prevent regeneration. Much has been learned from the tadpole during the period when it loses the ability to regenerate limbs. At that time two changes are occurring in the skin. First it is slowly increasing in thickness. This is true of both epidermal and dermal layers. Another change is the freeing of the skin from the underlying tissues concurrent with production of lymph spaces (Poležajew, '39b). The separation of the skin may be extremely important in suppressing regeneration. Tadpole limbs do not become refractory to the regeneration stimulus over their entire length at the same time. More proximal limb segments lose regenerative ability earlier and at the very time when lymph spaces are developing (Schotté and Harland, '43). The freeness of the adult skin enables it to pinch down over most of the stump within a few days. The result of these two skin changes, a thickening and a loosening, is that the amputated limb is sealed by a fairly heavy skin before regeneration can occur. Prior to the formation of lymph spaces the skin is bound to the underlying tissues and is not free to migrate over the stump. Then regeneration can start unhampered by the skin. The author has noticed that some tadpole limbs which do not regenerate completely develop small pointed outgrowths from a part of the amputation surface. This is invariably from that part of the surface not covered by a complete skin.

It would be wrong to think of the skin as the only tissue which has changed in such a way that it prevents regeneration. Transplantation studies made on tadpoles at the time of loss of regenerative ability have shown that regeneration-suppressing changes are occurring in the underlying tissue as well. It is possible to transplant the mesodermal core of a hind limb blastema from one tadpole to a leg amputation surface of another. The host provides the epithelium for the transplanted mesoderm. By making different combinations it was learned that both epithelium and mesoderm participate in the aging process which causes regenerative ability to diminish and disappear. (Poležajew, '39b) When both epithelium and core were from tadpoles young enough to regenerate, new limbs formed. However, a mesodermal core from an older tadpole even in combination with young epithelium did not regenerate. Likewise young mesoderm covered by an epithelium from a tadpole too old to regenerate did not regenerate. Evidently epithelium and mesoderm both age in such a way that they will not allow limb regeneration. Whatever the change in epithelium and mesoderm may be, the present work shows it can be overruled by a wound which remains open for a long time.

The first investigator to prevent regeneration with a skin seal believed the skin restricted regeneration by its inelasticity (Tornier, '06). Recently the same conclusion has been reached with regard to the fibrous scars which form over x-rayed axolotl limbs (Litschko, '34) and over amputated limbs of tadpoles too old to regenerate. (Schotté and Harland, '43). The fact that structures such as nerves are bent in a curve under the scar indicates that they are mechanically restricted and can not grow distally. The idea that these scars and the accompanying tissue distortions are a symptom not an initial cause of regeneration failure will be advanced below.

Other observations belie the skin's mechanical restrictive action. First of all, if skin is to prevent axolotl tail regeneration, it must be sewed over the wound surface soon after the amputation, usually within 1 to 3 days (Godlewski, '28). During this time skin could not act as a mechanical block because there is no outward growth to be blocked. Instead, this is a time of dedifferentiation and resorption. On the other hand, if the amputation wound is left open for a few days and then covered with skin, regeneration is not prevented. The skin in the second case is present when outgrowth begins. Instead of restricting the growth, it allows the regenerant to pass through. Skin certainly does not act mechanically in preventing regeneration.

The nature of skin rupture by a regenerant in a newt has been observed carefully (Taube, '23). When feet were removed from limbs and the limbs sewed into a pocket of abdominal skin, it was possible for most limbs to reach the outer surface and continue to regenerate the feet. The passage through the skin was accomplished by the regenerant in the following way. Four to 7 days after the operation the epidermis of the abdominal skin over the end of the limb stump had partially disappeared. At the same time the deeper dermal layer closer to the regenerant was still intact. In 13 days the dermis also thinned out and subsequently disappeared. That the regenerant did not tear through the skin is indicated by the fact that the skin layer closer to the regenerant was not the first to start to disappear.

Another lead as to how active regenerants might pass through skin comes from detailed observations of tadpole tail regeneration (Naville, '24). A layer or two of connective tissue was sometimes seen applied to the newly-migrated epithelium over the amputation surface. But when the regenerant began to grow, this basal membrane disappeared. A similar membrane did not reappear until regeneration was completed. Apparently the fiber is dissolved during active regeneration. Even when axolotl tail stumps are covered with skin and regeneration is prevented thereby, some dermis disappears from the transplant. However, it is soon replaced by regenerated dermis and further regeneration does not occur (Godlewski, '28).

If skin does not act as an inelastic obstruction in the path of the developing blastema, how does it inhibit regeneration? Whenever regeneration occurs, e. g., in Urodeles, an epithelium from the epidermis at the border of the wound migrates over the cut ends of muscle, bone and other tissues. No dermis lies over the ends of the cut tissues. Only the epithelium separates them from the external environment. When skin is transplanted over an amputation wound, not only an epithelium stands between the cut ends and the environment but a dermal layer as well. This dermal layer was thought to be important as the regeneration-suppressing layer, but the idea was discarded because the belief as to its mode of action was shown to be wrong. It was claimed that reserve cells passed from the epidermis into the blastema (Godlewski, '28). When a dermis was present the cell passage was blocked. In the light of more recent studies this idea of the method of dermis inhibition is seen to be illusory. By combining an x-rayed axolotl limb, incapable of regeneration, with various unirradiated limb tissues, regenerated limbs have been obtained from several tissue sources. Not only skin, particularly the dermal layer, can serve as the source, but also muscle and bone will provide the cells for regeneration (Umanski, '37; Thornton, for muscle, '42). The epidermis can not be considered the only source of blastema cells. If the dermis does suppress regeneration, it must do so in some way other than by blocking the passage of epidermal reserve cells.

In more recent times the emphasis has been placed not upon the fact that dermis is lacking during regeneration but upon the fact that the amputation surface is covered by a new epithelium. If axolotl tails are completely covered with a skin flap immediately after amputation, as in Godlewski's experiments, the tail does not regenerate. However, if a small piece of the skin remains unattached at an edge, regeneration occurs (Efimov,[3] '31). In the former case, after complete coverage, a new epitheluim does not migrate over the end of the stump. With incomplete coverage new epithelium mounts the stump under the skin transplant and regeneration proceeds. The conclusion has been drawn that a new epithelium is necessary for regeneration. The ex-

[3] Also spelled Jeffimoff.

periment hardly justifies the conclusion. When regeneration occurred not only was there a new epithelium but also an open wound without a complete skin between the opening and the inner tissues. Either the new epithelium or the absence of a complete skin or the absence of dermis might be important in initiating regeneration. It would be unwise to conclude from such an experiment that one of the three possibilities is most important.

Another experiment was designed to show whether an epithelium is necessary for the continuation of regeneration already started. Axolotl limbs were amputated near the ankle. After regeneration had progressed for a time, skin was removed from the entire leg and the epithelium from the regenerant. The limb was then sewed into a pocket on the tail. The operation prevented migration of a new epithelium over the surface of the regenerant. It also inhibited regeneration (Poležajew and Faworina, '35). The conclusion was that that an epithelium is necessary during regeneration. Possibly the result — no regeneration — was as greatly influenced by the conditions inside the tail as it was by the lack of an epithelium. Here again the experiment is not conclusive because more than one experimental factor may have influenced the result. In fact, later work indicates that a skin pocket may sometimes cause abnormal regeneration. Amputated limbs, even with skin, sometimes regenerate imperfectly if they are kept for only a part of the regeneration period in a skin pocket (Morosow, '38). Even though the above experiments do not fully warrant the conclusions concerning the necessity of an epithelium, the epithelium may still be very important. Further experiments are necessary.

Other experiments seem to indicate that skin epithelia from different body regions differ in their ability to support limb regeneration. If skin is transplanted to the base of a skinned and amputated axolotl limb, an epithelium migrates from the transplant over the regenerant. When skin from head or back covers the base of the limb, regeneration is inhibited. On the other hand, tail, abdominal and limb skin do not inhibit limb regeneration (Efimov, '33; Poležajew and Faworina, '35). Apparently the epithelia furnished by head and back skin differ from the epithelia furnished by other regions. An inkling as to how an epithelium may inhibit regeneration comes from two sources. First, in axolotls an epithelium from a dark animal can induce a pigmented dermis even though that dermis develops from tissues of the white host (Kolodziejsky, '28). If an epithelium can induce its own type of dermis, epithelium from transplanted head skin might inhibit limb regeneration by inducing head dermis on a limb. Head dermis is very thick (Umanski, '37). Umanski goes so far as to say that head skin transplanted directly to the border of an amputation wound suppresses regeneration on account of its thick corium. He further states, "Verkleinerung des Coriums ermöglicht die Regeneration." The point being introduced here is that head skin at some distance from an amputation surface may suppress regeneration because an epithelium from it migrates and may induce a dermis too thick to support regeneration. This hypothesis is new and not strongly supported. It may well be that a foreign epithelium inhibits limb regeneration in some more direct way.

In recent years the ideas about the skin as an inhibitor of regeneration have centered around the role of the epithelium. It is quite generally believed that regeneration starts and proceeds only when there is an epithelium covering the wound. A further belief is that epithelia derived from skin of only some regions of the body will permit regeneration. It is pointed out in the discussion above that these views may be true but that they are not proven. Since, as shown in the present work, frogs will regenerate limbs when their wounds are covered with a migrated epidermis but not when covered with a whole skin, i. e., epidermis plus dermis, the dermis must be suspected as the part of the skin which inhibits regeneration. Evidence as to how the dermis may function in suppressing regeneration is available.

It is known that an open surface leads to dedifferentiation of existing structures in the vicinity of the surface. Dedifferentiation is a necessary step in limb regeneration. It provides the cells for redifferentiation and growth. Dedifferentiation after wounding has been shown in exaggerated form in denervated limbs of Amblystoma larvae. An intact denervated limb maintains itself for a long period. If, however, a denervated limb is cut, even in the finger region, the limb will dedifferentiate near the open end and continue to do so until the limb has disappeared. It is of interest that these denervated and wounded limbs show a repaired epidermis but the dedifferentiating region remains dermis-free (Butler and Schotté, '41).

Further evidence that dermis may suppress dedifferentiation comes from x-ray work. Amblystoma larval limbs x-rayed and amputated at the same time dedifferentiated completely and the limbs were resorbed (Butler and Puckett, '40). The photographs of sections of such limbs show a very thin dermis or none at all distally. Limbs which have regenerated for a time and are then x-rayed respond differently. They undergo neither regeneration nor resorption. The striking thing is that heavy dermal fiber forms around the end of the limb soon after x-radiation. Possibly the dermal layer maintains the constancy of the internal environment, allowing neither excess dedifferentiation of tissue nor excess growth.

When large axolotl limbs are irradiated and amputated at the same time the tissue response is quite different. A collagenous fiber scar forms distally and regeneration is completely inhibited (Litschko, '34). The appearance of the x-ray induced scars in the axolotl is similar to those normally arising after frog limb amputation. Evidence is presented that x-radiation stimulates fiber growth in these older axolotls and thereby prevents regeneration. In fact, Litschko reports that scar removal allows x-radiated limbs to regenerate.

The trauma coincident with amputation has been considered to be the regeneration stimulus. But the evidence available indicates that trauma is only one factor in stimulation. Schaxel has described a case which illustrates the point. A hind limb of a young axolotl was completely removed and skin sewed over the wound. No regeneration had occurred in 6 months. Then a skin piece, 4 mm. in diameter, was removed from the hind-limb region of the same animal. At the same time the tissues down to the pelvic border were loosened. The surrounding skin immediately began to cover the wound, but closure was not complete before a blastema arose. This blastema, however, was inhibited and regeneration ceased before anlage formation. Again the same region on the same animal was opened but a larger skin piece, 6 mm. in diameter, was removed and much greater internal injury was caused. This time a normal blastema appeared and regeneration was completed (Schaxel, '21). Similarly, when a femur is carefully removed from an axolotl limb with very little injury to the surrounding tissues, the wound closes over without femur regeneration. Only connective tissue and some cartilage replace the femur (Schaxel and Bohmel, '28). The above cases demonstrate that regeneration occurs when skin is removed and when internal tissues are injured sufficiently. It is well established that an opening in the skin is necessary for regeneration. The above examples seem to indicate that internal trauma is also necessary. A further experiment proves it. Poleszajew deviated a hind limb nerve of an axolotl to the body surface near the base of the limb. As previously shown, a limb regenerated over the nerve (Locatelli, '23). In Poleszajew's experiment skin was sewed over the region. In 2 months no regeneration had occurred. Then the skin was carefully removed. Still there was no regeneration of a limb. Skin regenerated and only a small protuberance developed

over the nerve. If, instead of careful skin removal, the skin was removed and the tissue underneath wounded, a limb regenerated (Poles-
žajew, '33). This proves that injury of tissues below the open skin
is also a factor in the stimulation of regeneration.

It has been thought that regeneration may be stimulated by the
release of growth-promoting substances from cut cells. Such may
be the case. But if it is, these wound hormones only act on tissues
not suppressed by skin. An additional hypothesis should be considered. Since regeneration will start and continue only with an epithelial layer minus dermis, this new skin covering must be playing
an active role in regeneration. It may be that this new epithelium does
not present as great a barrier to the diffusion of some substances between tissue and environment as did the old skin. In support of this
it is known that tadpoles and frogs differ greatly in their ability
to resist the passage of water through their skins (Adolph, '30). Possibly water itself is a substance which by changes in concentration
internal to an amputation surface favors regeneration. It is known
that increase in the water content of sea water will hasten the regeneration of three marine organisms (Loeb, 1892; Goldfarb, '07, '14).
Such a change might reverse the normal chronological progression
of tissues from a more liquid to a dryer state and cause them to revert
to an embryonic condition in which they grow more rapidly. The tadpole, at the time when regenerative ability is being lost, is rapidly becoming dryer (Glaser, '38). Whatever may be the physico-chemical
chain of events following cutting of tissues, skin removal and the formation of a new epithelium, a knowledge of them is necessary for
further advances in our understanding of the early phase of regeneration. A possible clue as to the direction in which knowledge may lie
comes from studies of the regeneration stimuli operative at a cut
surface in the coelenterate, Tubularia. There, oxygen, a regeneration
stimulant, enters through the amputation surface (Miller, '37; Barth,
'38; Zwilling, '39; Rose and Rose, '41) and regeneration inhibiting substances leave (Rose and Rose, 41; Goldin, '42). Something of the sort
may occur at the amputation surface in an Amphibian.

Certain similarities appear in the immediate effects of the different
methods of stimulating regeneration in frogs. When a tadpole limb
too old to regenerate was transplanted to another tadpole or even
to the back of the same tadpole, the power of regeneration was sometimes regained. Some dedifferentiation and intense hyperaemia were
observed after transplantation (Liosner, '31). Likewise traumatized

limbs of old tadpoles showed evidence of hyperaemia, oedema and dedifferentiation before they regenerated (Polejaiev, '36; Poleźajew, '39a). The same was true of the adult limbs in the present work after salting and skin removal. All of these treatments which evoke latent regenerative ability are drastic. They may have their effect not only by causing removal or dedifferentiation of inhibiting skin but also by producing internal trauma. This may lead to dedifferentiation of more internal tissues, which in a frog do not as readily respond to the amputation stimulus as do the young tadpole and urodele internal tissues.

The present work has demonstrated that regeneration of the adult frog limb may be initiated. Dedifferentiation, blastema formation, rapid growth and differentiation follow the treatments. But the regenerants are always abnormal distally. Abnormal distal regeneration has been reported before. A knowledge of how it was produced in the two cases to be cited may aid in the interpretation of its cause in the present case. Urodele limb blastemas developing over deviated nerves were small when skin encroached upon them from the side. Such blastemas developed into limbs with fewer than the normal number of digits (Guyénot and Schotté, '23). Strikingly like the frog limb regenerants are some in axolotls which were inhibited by head or body skin (Poleźajew and Faworina, '35). These were produced by removing the skin from an already regenerating limb and replacing it with head or body skin. The cases of abnormal distal regenerants so like the frog regenerants were produced by skin inhibition which did not begin until regeneration had proceeded uninhibited for some days. Perhaps the frog regenerants also suffered skin inhibition during the later part of the regeneraton period. They, of course, were regenerating a skin which itself would be expected to inhibit regeneration. It was observed that skin had differentiated by the time rapid growth had ceased. Work is now in progress which demonstrates that better regenerants may be obtained if the skin is not allowed to develop a heavy dermis. Besides skin there was an abnormally large amount of collagenous fibers at the distal ends of the regenerants by the time they stopped growing. This may have been the result of poor growing conditions. In tissue cultures when conditions unfavorable for growth occur fibers are produced. This fibrous tissue may itself further inhibit regeneration. The barrier blocking better frog limb regeneration seems to be the early development of the dermis and fibrous scar tissue.

SUMMARY

1. Limb stumps of adult frogs heal rapidly with a covering of old skin after limb amputation. Regeneration is limited to the formation of scar tissue, cartilage and skin.

2. When amputation surfaces remain open because of treatment with NaCl or because skin is cut away or for other reasons, regeneration begins.

3. After amputation through the mid forearm region, regeneration of a normal distal forearm and wrist region may be obtained from NaCl treated frogs.

4. The regenerants are abnormal distally in the hand region.

5. The manner in which skin may inhibit regeneration is discussed.

LITERATURE CITED

ADOLPH E. F. 1930 Living Water. Quart. Rev. Biol., vol. 5, pp. 51–67.

BARTH, L. G. 1938 Oxygen as a controlling factor in the regeneration of Tubularia. Physiol. Zool., vol. 11, pp. 179–186.

BRØNSTED, H. V. 1939 Regeneration in Planarians investigated with a new transplantation technique. Biol. Med., vol. 15, pp. 1–39.

BUTLER, E. G., AND W. O. PUCKETT 1940 Studies on cellular interaction during limb regeneration in Amblystoma. J. Exp. Zool., vol. 84, pp. 223–239.

BUTLER, E. G., AND O. E. SCHOTTÉ 1941 Histological alterations in denervated non-regenerating limbs of urodele larvae. J. Exp. Zool., vol. 88, pp. 307–341.

EFIMOV, M. I. 1931 Die Materialien zur Erlernung der Gesetzmässigkeit in den Erscheinungen der Regeneration. Z. Exp. Biol. (russ.) vol. 7, pp. 352–367 (reviewed by Poleźajew und Faworina, '35).

———— 1933 Die Rolle der Haut in Prozess der Regeneration eines Organs beim Axolotl. Z. Biol. (russ.) vol. 2 (reviewed by Poleźajew und Faworina, '35).

GLASER, O. 1938 Growth, time and form. Biol. Rev., vol. 13, pp. 20–58.

GODLEWSKI, EMIL JUN. 1928 Untersuchungen über Auslösung und Hemmung der Regeneration beim Axolotl. Arch. für Entw-mech., vol. 114, pp. 108–143.

GOLDFARB, A. J. 1907 Factors in the regeneration of a compound hydroid, Eudendrium ramosum. J. Exp. Zool.. vol. 4, pp. 317–356.

———— 1914 Changes in salinity and their effects upon the regeneration of Cassiopea xamancha. Papers from the Tortugas Lab. of Carn. Inst. of Wash. vol. 6, pp. 85–94.

GOLDIN, A. 1942 Factors influencing regeneration and polarity determination in Tubularia crocea. Biol. Bull., vol. 82, pp. 243–254.

GUYÉNOT, E., AND O. SCHOTTÉ 1923 Relation entre la Masse du bourgeon de régénération et la Morphologie du Régénérat. C. R. Soc. Biol., vol. 89, pp. 491–493.

KOLODZIESKY, M. L. 1928 Untersuchungen über die Beteiligung der transplantierten Haut an der Regeneration. Bull. Inter. de l'Académie Polonaise des Sciences et des Lettres. Serie B. Sciences Naturelles, Zoologie, pp. 1–62.

LIOSNER, L. D. 1931 Über den Mechanismus des Verlusts der Regenerationsfähigkeit während der Entwicklung der Kaulquappen von Rana temporaria. Arch. für Entw-mech., vol. 124, pp. 571–583.

LITSCHKO, E. J. 1934 Einwirkung der Röntgenstrahlen auf die Regeneration der Extremitäten, des Schwanzes und der Dorsalflosse beim Axolotl. Trudy Lab. Eks. Zool. I Morf. Jiv., vol. 3, p. 136–139 (German Summary).

LOCATELLI, P. 1923 L'influenza del sistema nervoso sui processi rigenerativi. Giorn. di biol. e med. sperim, vol. 4.

LOEB, J. 1892 Untersuchungen zur physiologischen Morphologie. Würzburg.

MILLER, J. A. 1937 Some effects of oxygen on the polarity of Tubularia crocea. Biol. Bull., vol. 73, p. 369.

MOROSOW, l. I. 1938 Die Hemmung und Wiederherstellung des Regenerationsprozesses der Extremität beim Axolotl. C. R. de l'Acad. Sci. URSS. vol. 20, p. 207 (German summary).

NAVILLE, A. 1924 Recherches sur l'histogenèse et la régénération chez les Batraciens Anoures. (Corde dorsale et téguments). Arch. de Biol., vol. 34, pp. 235–343.

POLESZAJEW, L. 1933 Über Resorption und Proliferation sowie über die Verhältnisse der Gewebe zu einander bei der Regeneration der Extremitäten des Axolotls. Biol. Zhurn. URSS. vol. 2, p. 385 (German summary).

POLEŽAJEW, L. W. UNTER BETEILIGUNG VON W. N. FAWORINA 1935 Über die Rolle des Epithels in den anfänglichen Entwicklungsstadien einer Regenerationsanlage der Extremität beim Axolotl. Arch. für Entw-mech. vol. 133, pp. 701–727.

POLEJAIEV, L. W. 1936 Sur la restauration de la capacité régénèrative chez les Anoures. Arch. d'Anat. Micr., vol. 32, pp. 437–463.

POLEŽAJEW, L. W. 1939a On the mode of restoration of regenerative power in the limbs of tailless Amphibians. C. R. (Dok.) de l'Acad. des Sci. de l'URSS, vol. 22, pp. 648–652,

————— 1939b Über die Bedeutung des Epithels und Mesoderms beim Verlust der Regenerationsfähigkeit der Extremitäten bei den Anuren. C. R. (Dok.) Ac. Sc. URSS., vol 25, pp. 538–542.

ROSE, S. M. 1942 A method for inducing limb regeneration in adult Anura. Soc. Exp. Biol. and Med., vol. 49, pp. 408–410.

ROSE, S. M., AND F. C. ROSE 1941 The role of a cut surface in Tubularia regeneration. Physiol. Zool., vol. 14, 328–343.

SCHAXEL. J. 1921 Aufassungen und Erscheinungen der Regeneration. Arb. a. d. Geb. der Exp. Biol., vol 1, pp. 1–99.

SCHAXEL, J., AND W. ADENSAMER 1923 Über experimentelle Verhinderung der Regeneration bei Phasmiden. Zool. Anz., vol. 56, pp. 128–133.

SCHAXEL, J., AND WILFRED BÖHMEL 1928 Regenerations- und transplantationsstudien. II. Ersatzbildung nach Entnahme von Organteilen. Zool. Anz., vol. 78, pp. 157–163.

SCHOTTÉ, O. E., AND MARGARET HARLAND 1943 Amputation level and regeneration in limbs of late Rana clamitans tadpoles, Jour. Morph., vol. 73, pp. 1–35.

TAUBE, ERWIN 1921 Regeneration mit Beteiligung ortsfremder Haut bei Tritonen. Arch. für Entw-mech., vol. 49, pp. 269–315.

————— 1923 Über die histologischen Vorgänge bei der Regeneration von Tritonen mit Beteiligung ortsfremder Haut. Arch. für mikr. Anat., vol. 98, pp. 98–120.

THORNTON, C. S. 1942 Studies on the origin of the regeneration blastema in Triturus viridescens. J. Exp. Zool., vol. 89, pp. 375–390.

TORNIER, G. 1906 Kampf der Gewebe im Regenerat bei Begünstigung der Hautregeneration. Arch. für Entw-mech., vol. 22, pp. 348–369.

UMANSKI, E. 1937 Untersuchung des Regenerationsvorganges bei Amphibian mittels Ausschaltung der einzelnen Gewebe durch Röntgenbestrahlung Biol. Zhurn. URSS., vol. 6, pp. 757–758 (German summary).

ZWILLING, E. 1939 The effect of the removal of perisarc on regeneration in Tubularia crocea. Biol. Bull., vol. 76, pp. 90–103.

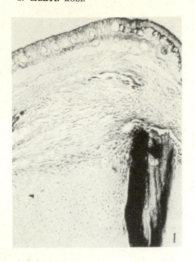
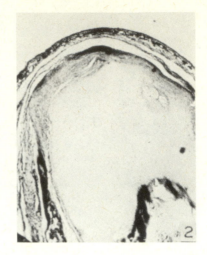
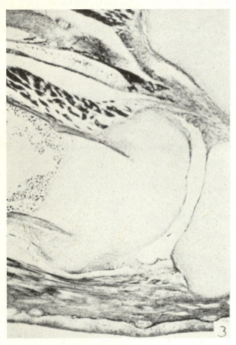
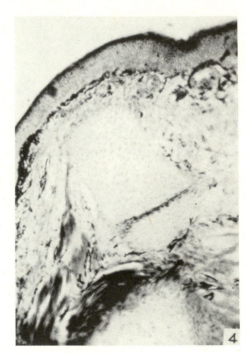

1 R. palustris (6/3/42)–7/2/42–7/12/42–L. Distal end of untreated arm stump 10 days after amputation. 35 ×.

2 R. palustris 6/3/42–7/12/42–R. Distal end of untreated arm stump 39 days after amputation. 35 ×.

3 R. clamitans–1–6/22/42–11/25/42. Wrist region of regenerant of salt treated animal. 100 ×.

4 R. clamitans–2–6/3/42–11/16/42. Distal end of arm regenerant of salt treated animal. 100 ×.

168

200

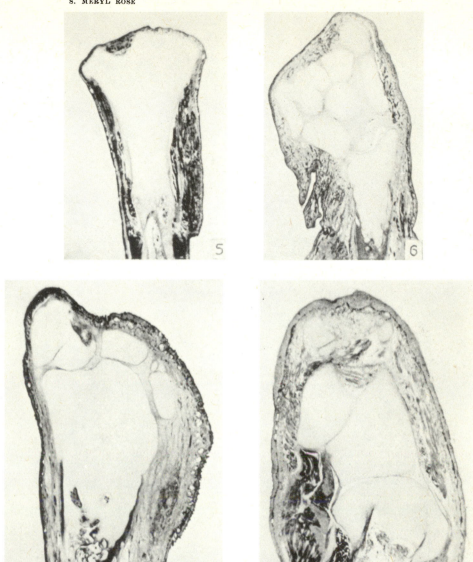

5 R. clamitans–4–6/6/42–4/13/43. Approximately the upper three fourths of the figure is arm regenerant and the lower fourth old tissue. Salt treated animal. 20 ×.

6 R. clamitans–1A–8/28/42–4/24/43. Approximately the upper four fifths of the figure is arm regenerant and the lower fifth is old tissue. Animal had had skin peeled away from amputation surface. 20 ×.

7 R. clamitans–C–5/5/42–4/23/43. Approximately the upper four fifths of the figure is arm regenerant and the lower fifth is old tissue. Animal had had skin peeled away from amputation surface. 20 ×.

8 R. clamitans–2–6/3/42–11/16/42. Arm regenerant of salt treated animal. 35 ×.

169

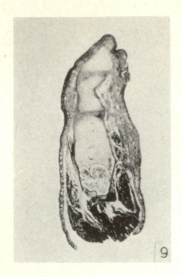

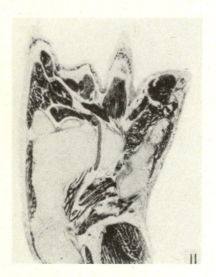

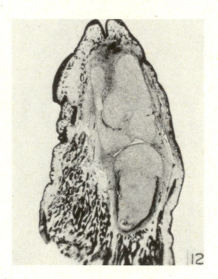

9   R. clamitans–5–6/25/42–11/16/42. Approximately the upper half of the figure is arm regenerant. Healing was prevented by repeated removal of the distal tip. 20 ×.

10   R. clamitans–L–6/22/42–4/23/43. Toe regenerants of four toes originally cut off level with the shortest toe which was not amputated (below in figure). Everything distal to level of tip of shortest toe is regenerated tissue. Salt-treated animal. 20 ×.

11   Hand region of an amputated arm of a first year Rana sylvatica. 20 ×.

12   R. palustris–1B–8/28/42–4/24/43. Arm regenerant of animal whose amputation wound did not close during a vitamin deficiency. 20 ×.

170

Copyright © 1946 by Cambridge University Press

Reprinted from *Biol. Rev.*, 21, 141–147 (1946)

10

THE LOSS AND RESTORATION OF REGENERATIVE CAPACITY IN THE LIMBS OF TAILLESS AMPHIBIA[1]

By L. W. POLEZHAYEV, Institute of Cytology, Histology and Embryology
of the Academy of Sciences of the U.S.S.R.

(*Received 16 November 1944*)

I. INTRODUCTION

Questions of the loss and restoration of regenerative capacity of complex organs which have been dealt with prior to this concern a very narrow group of problems in a single group of animals, namely, the Anura, and deal only with their limbs. The Anura are very suitable for experiments owing to the fact that during metamorphosis they lose their regenerative capacity in a way which is easy to follow experimentally. Another reason which makes Anura convenient is that to-day the mechanics of the development of limbs during normal ontogenesis have been well studied, as has also the process of regeneration in related animals which retain their regenerative capacity throughout their lives (Mangold, 1929; Polezhayev, 1935).

The term regenerative capacity is used here to mean the sum total of properties which make it possible for an organ to regenerate. Change in these properties may sometimes lead to the loss or restoration of the regenerative capacity. At least three forms of differentiation may be distinguished: organological (the general structure of the organ), histological (tissues) and cytological (cells). In normal ontogenesis all these processes of differentiation are interconnected but may be separated experimentally (Needham, 1942).

Dedifferentiation is the loss of some of the features which appeared earlier during differentiation. Dedifferentiation leads to simplified organization but not necessarily to the return of all the former morphogenetic potentialities to the cells. An organ may be regarded as dedifferentiated if some of the characteristics of its earlier structure have disappeared, although this must not be regarded as the dedifferentiation of the cells of which it is composed.

The word mesoderm will be used to indicate those parts of limbs which are of mesodermal origin, such as skeleton, musculature and connective tissue. My own experiments have, in the main, been made on one animal, *Rana temporaria*.

II. THE ONTOGENETIC LIMITS OF REGENERATIVE CAPACITY

The question of ontogenetic limits, or that stage of ontogenesis at which the loss of the regenerative capacity of the organs of Anura begins, has long been

the subject of study (Barfurth, 1895; Byrnes, 1904; Kammerer, 1905). It has been shown that forelimbs lose their regenerative capacity earlier than hindlimbs and the tail later still (Guyénot, 1927; Liosner, 1931). A comparative investigation of regenerative capacity in the order Anura is only just beginning (Kammerer, 1905; Ginzburg, 1941).

As a basis for the division of tadpoles into stages, Blacher's table (1928) will be used, to which I have made additions in stage I, and Liosner (1931) in stage II (Fig. 1). *Stage I*. The hindlimbs of the tadpole still have no articulated joint between the thigh and the shank. The foot is spatulate (stage Ia) or is spatulate with the small protuberances of fingers (stage Ib) or spatulate with larger finger protuberances (stage Ic). *Stage II*. Thigh and shank form an obtuse angle, and the rudimentary fingers are well defined. This stage is again divided into stages IIa and IIb in accordance with size and degree of development. *Stage III*. Thigh and shank form an acute angle. The belly is rounded, the forelimbs are not visible externally (stage IIIa), or the belly is flat, the elbows of the forelimbs are forcing their way through the opercular membrane and the hindlimbs are strongly developed (stage IIIb). *Stage IV*. One of the forelimbs has appeared, the hindlimbs have increased in size, the body has taken shape and the horny jaws have gone. *Stage V*. The tail is being resorbed. The tadpole is becoming a frog.

Microscopic examination (Polezhayev, 1940a) shows that at stage I the diaphysis of the skeletal elements of thigh and shank is developing cartilage, but the epiphysis has not yet chondrified. The musculature is represented only by the muscle rudiments. At stage IIa the epiphysis has chondrified, the diaphysis is ossifying, the thigh joint is articulate (but the knees are not articulate), muscle rudiments begin to change into fibres. At later stages the process of histological differentiation of the limbs increases progressively.

According to some authors (Barfurth, 1895; Liosner, 1931; Polezhayev, 1933) the hindlimbs of *Rana temporaria* lose their regenerative capacity at stage IIa, and according to others (Kammerer, 1905; Borssuk, 1935; Jakowlewa, 1938) at stage IIb. Special investigations have shown that the regenerative capacity depends on the level of amputation; the more proximal the amputation the poorer the regeneration (Jacomini, 1922; Polezhayev, 1939). This dependence of regeneration on the level of

[1] This article is a summary of a comprehensive monograph by the same author.

amputation may be connected with the fact that histological differentiation of the limbs of Anura takes place in a proximo-distal direction (Polezhayev, 1939, 1940a). When the amputation is made in the proximal part of the shank regenerative capacity is lost at stage IIa.

generate and the mesodermal tissues of the stump of the organ (muscle rudiments, skeleton) dedifferentiate. When the amputation is made at a later stage the limbs do not regenerate and the mesoderm tissues of the organ do not dedifferentiate. The question then arises as to whether the regenerative

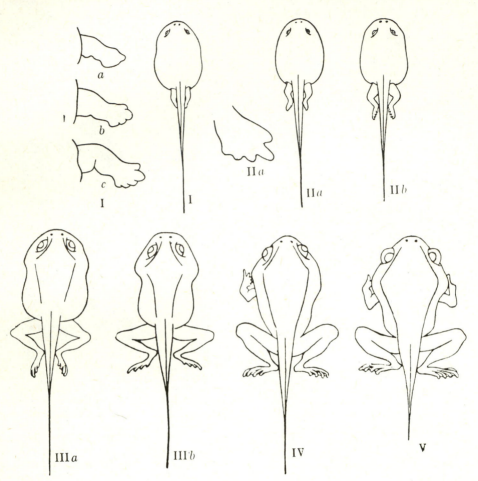

Fig. 1. The stages of metamorphosis of the tadpole

III. HISTOLOGICAL INVESTIGATION OF THE REGENERATION OF THE LIMBS OF TADPOLES AT VARIOUS STAGES OF METAMORPHOSIS

Although the problem is an old one, microscopical investigations of the process of regeneration in tadpoles' limbs is quite recent (Polezhayev, 1935, 1936, 1940a; Ide-Rozas, 1936). Leaving aside the data provided by cytological research which was done by Ide-Rozas, we have the following peculiar features of regeneration. Limbs amputated at stage I re-

blastema is formed from the dedifferentiated mesodermal tissue of the stump of the organ.

IV. THE RESTORATION OF REGENERATIVE CAPACITY IN THE LIMBS OF TADPOLES BY TRAUMATIZATION

Many experimenters have attempted to restore the regenerative capacity of the organs of the frog (Morgan, 1908), of man (Bier, 1923) and other animals, but all these attempts have ended in failure. The first positive results were obtained by Liosner (1931). When transplanting the limbs of tadpoles he raised the question as to whether the operation acted as a stimulating or an inhibiting factor. It proved to stimulate regeneration in four cases out of thirty-seven of limbs auto-transplanted on to the back at stage II*a*; the limbs regenerated while the other limb of each pair which was left in place but amputated at the same time did not regenerate. At later stages Liosner was unable to obtain similar results.

I turned my attention to the fact that the regenerative capacity of tadpoles' limbs is, as a rule, lost suddenly: after amputation (up to stage II*a*) the limbs either regenerate in full, or (after stage II*a*) do not regenerate at all, no blastema being formed on the amputated surface. If the loss of regenerative capacity is due to an insufficiency of regenerative material we should expect the formation of a regeneration blastema and full regeneration of the limb when we artificially provoke an accumulation on the amputated surface. In order to stimulate the accumulation of cell-forming material on the stumps (amputated at stage II*a*), the stumps were severely traumatized with a needle; other amputations performed at the same stage and same place were left untraumatized as controls. The result was that the traumatized limbs regenerated while the untraumatized ones did not do so (Polezhayev, 1933, 1935, 1936, 1939*a*, 1940*a*). A microscopical examination showed that the mesoderm of the control limbs had not dedifferentiated, whereas that of the traumatized limbs had done so: the muscular and skeletal elements had disassociated and divided into individual cells which came to the amputated surface and together with the epithelium formed the regeneration blastema.

V. THE MECHANICS OF THE RESTORATION OF REGENERATIVE CAPACITY IN THE LIMBS OF TADPOLES BY TRAUMATIZATION

To what is the formation of the regeneration bud in the traumatization experiment due? Is it due to increased proliferation under the influence of the products of cell destruction (wound hormones) or to the liberation of cells from the disassociating mesodermal tissue? Both the hindlimbs of a tadpole at stage II*a* were amputated. The left was used as a control, the right was traumatized either distally or proximally (pierced transversally). In both cases strong hyperaemia and inflammation were present, but regeneration took place only in that leg which was distally traumatized. In this case there was a strong dedifferentiation of the muscular rudiments and skeletal elements, which rose to the amputated surface, and the microscopical picture of regeneration was that described in the previous section. With proximal trauma, despite the abundance of the products of destruction in the limb tissues, the distal mesoderm did not dedifferentiate, a blastema was

not formed, and the limb did not regenerate. A count of the mitoses showed that they were approximately equal in the control and in the distally and proximally traumatized limbs (Polezhayev, 1939*a*). The result shows, therefore, that the restoration of regenerative capacity is due to the dedifferentiation of mesodermal tissue which comes to the amputated surface and not to proliferation.

VI. THE RESTORATION OF THE REGENERATIVE CAPACITY OF TADPOLES' LIMBS BY TRANSPLANTATION

Liosner's experiment mentioned above (1931) gave rise to some doubt (Jakowlewa, 1938), but it was repeated and confirmed by Polezhayev & Ginzburg (1939). We amputated both hindlimbs of tadpoles at stage II*a*; one was left as a control and the other (thigh and part of the shank) was transplanted autoplastically to the back or the tail. The result was that the control limb did not regenerate but the transplanted limb did. A microscopical examination showed that the muscular rudiments and skeletal elements in the transplanted limbs dedifferentiated strongly, but that there was no dedifferentiation in the control limb. Transplantation and traumatization, therefore, lead to the same results. This is another confirmation of the principle of equifinality, that is, that different processes leading to the same result are in the main determined by the same factors (Polezhayev, 1938*a*).

VII. PROLONGATION OF THE REGENERATIVE CAPACITY OF TADPOLES' LIMBS BY RE-REGENERATION

By the methods of traumatization and transplantation the regenerative capacity was restored to the limbs of tadpoles at stage II*a*. The question then arises as to whether it is possible to obtain regeneration at later stages. Experiments were made with the object of prolonging the regenerative capacity by the method of re-regeneration (Polezhayev, 1933, 1935, 1936). A right limb of tadpoles at stage I was amputated and allowed to regenerate. When differentiation was in organological and histological conformity with limbs at stage I the tadpoles had already reached stages II*b*–IV, with an individual difference in stages. An amputation was then made in the middle of the shank of the regenerated limb and in another limb for control. The latter did not regenerate, but the former did so even when the amputation was made at stage IV. In the last-mentioned case the regeneration took place when the animal was already a frog.

VIII. PROLONGATION OF THE REGENERATIVE CAPACITY OF TADPOLES' LIMBS BY REPEATED REGENERATION

At late stages in metamorphosis we were thus able to excite regeneration, but its source was the *young* tissue of a regenerated limb. Is it possible to find methods of prolonging the regeneration of the *old* tissue of the stump during the later stages? We made an attempt (Polezhayev & Morozov, 1941) to find a method for the biological 'rejuvenation' of the tissue of the stump in order to prolong the regenerative capacity. The formation of a regeneration blastema, as has been shown, depends on the dedifferentiation of the old tissue of the stump, but, on the other hand, we have data on the histolytic effect of a blastema (Orechowitsch & Bromley, 1934). It is to be assumed that the regeneration blastema which is formed will itself cause the dedifferentiation of the mesodermal tissue of the stump and consequently will 'rejuvenate' it.

The right hindlimbs of tadpoles in stage I were amputated, and several days later, when the animals had reached stages II*a*, II*b* and III*a*, the regeneration rudiments, which already had the form of large 'buds' or cones, were completely removed from the limbs. At the same time the left limbs were amputated. The result was that the left limbs did not regenerate, but the right, experimental, limbs regenerated even at stage III*a*. A microscopical examination of the latter showed that the mesodermal tissue had dedifferentiated. The source of regeneration was the old, 'rejuvenated' tissue of the stump.

IX. COMPARISON OF REGENERATION IN THE LIMBS OF ANURA AND URODELA

In restoring the regenerative capacity of the limbs of tadpoles the dedifferentiation of the mesodermal tissue plays a more important role than the process of proliferation. In connexion with this and with experiments on the determination of the regenerate (Polezhayev, 1934, 1936*a*, 1937) the question arose as to what is the role played by dedifferentiation and proliferation in the process of normal regeneration, since an earlier theory (Weiss, 1930; Bromley, 1930) held that in the formation of a blastema the chief role is played by the process of proliferation stimulated by the products of destruction, wound hormones, 'mitogenetic rays' and other factors. It was noted that 'the solution of the question concerning the formation of a regeneration bud should be obtained by experiment, calculating the mitotic coefficient of the subsequent stages of the primary phases in the development of the regeneration rudiments' (Polezhayev, 1935*a*).

On this basis a histological investigation of regenerating limbs of *Rana temporaria* at stage I, of tiny axolotls corresponding to the tadpole in size, and of fully grown axolotls was undertaken. At the same time the mitotic coefficient of the regenerating tissue was calculated (Polezhayev, 1939*a*, 1940, 1941; Polezhayev & Ginzburg, 1944). It was thus discovered that the regeneration bud in Anura and Urodela is formed when the mitotic coefficient is

very low, by means of the dedifferentiation of the mesodermal tissue of the stump. The further growth of the regenerated rudiments takes place when there is a high mitotic coefficient, that is, mainly by proliferation. These data were fully confirmed by Bassina (1940), who, however, was wrong in concluding that the flow of regeneration material to the blastema comes from the proximal part of the stump.

Data indicating that the regeneration bud is formed mainly by dedifferentiation and not by proliferation makes it necessary to re-examine the old theory of wound hormones. The significance of dedifferentiation in the processes of regeneration in the Urodela has been shown by a number of American zoologists working independently and using methods different from ours (Butler, 1933; Butler & Puckett, 1940; Thornton, 1938, 1942, and others).

Our experiments also showed points of difference between regeneration in Anura and in Urodela. The limbs of tadpoles regenerate in 8–10 days and those of axolotls in 25–35 days; the dedifferentiation in tadpoles is much more profound and the mitotic coefficient lower than in axolotls. Comparing all these data we may summarize by saying that regeneration of the limbs of Anura approaches morphallaxis in type, whilst in Urodela it is closer to epimorphosis. Dedifferentiation plays a more important role in the regeneration of Anura than of Urodela. The difficulty of dedifferentiation, therefore, more strongly influences the regeneration of Anura than of Urodela.

X. THE SIGNIFICANCE OF INTERNAL HUMORAL FACTORS IN THE LOSS OF REGENERATIVE CAPACITY IN TADPOLES' LIMBS DURING METAMORPHOSIS

Earlier investigators have shown that organs which have lost their regenerative capacity do not regain it when transplanted to an animal capable of regeneration, and, vice versa, that organs which can regenerate do not lose that capacity when transplanted to animals unable to regenerate (Guyénot, 1927; Liosner, 1931). This, however, does not decide whether the internal humoral media of an animal change during metamorphosis thus causing organs to lose their regenerative capacity. Borssuk (1935, 1936) tried to explain this but his data were not very accurate. We have carried out a further investigation of the problem (Polezhayev & Ginzburg, 1939) and in general can confirm Borssuk's data. The hindlimbs of *Rana temporaria* were amputated at stage II*a* in the proximal part of the shank and at the base, and the stumps (thigh and proximal part of the shank) were transplanted, in some cases by autoplasty, in others by homoplasty, to tadpoles at stage III*a*. In the first case the limbs regenerated under the influence of transplantation. In the second case, despite the stimulating effect of transplantation, the limbs did not regenerate and the mesodermal tissue of the transplant did not dedifferentiate. Under the influence of the internal humoral medium the histological differentiation of limbs at stage III*a* takes place more speedily than at stage II*a* and is a great hindrance to the dedifferentiation of the tissue. It therefore follows that during metamorphosis the internal humoral medium of the animal acquires the ability to accelerate histological differentiation and in this way hinder the dedifferentiation of the tissue, the formation of blastemata and the regeneration of the limbs.

XI. THE SIGNIFICANCE OF EPITHELIUM AND MESODERM IN THE LOSS OF THE REGENERATIVE CAPACITY OF TADPOLES' LIMBS

With the object of explaining the significance of the epithelium and mesoderm of limbs in the loss of regenerative capacity, experiments were made by transplanting mesoderm taken in stages I and III*a* under the epithelium at stages I, II*b*, III*a* and IV. In other words, epithelium and mesoderm of stages with and without regenerative capacity were combined (Polezhayev, 1939*b*). By combining stage I epithelium with stage I mesoderm (control experiment) the limbs regenerated well. When old mesoderm (stage III*a*) was combined with young epithelium (stage I) the limbs did not regenerate. When young mesoderm (stage I) was combined with old epithelium the result depended on the stage of the epithelium; the older the epithelium the poorer the result. In general, the epithelium like the mesoderm loses its ability to take part in the process of regeneration with age and the limb does not regenerate. A microscopical examination showed that the loss of the regenerative capacity of mesoderm depends on its histological differentiation. The loss of regenerative capacity in the epithelium consists in the gradual loss of its ability to bring out the dedifferentiation of the mesoderm and to join with the latter's cells to form a blastema. The changes in the mesoderm which take place with the loss of the regenerative capacity are more profound than those of the epithelium.

XII. THE SIGNIFICANCE OF THE NERVOUS SYSTEM IN THE LOSS OF REGENERATIVE CAPACITY OF TADPOLES' LIMBS

The limbs of *Rana temporaria* regenerate very rapidly (8–10 days); the same is true when they are transplanted. Special histological research with silver impregnation of the neural tissue by Ramon y Cajal's method shows that the nerves are clearly seen in the regenerated limbs, but are absent in regenerated stage I limbs which had been transplanted to the bellies of tadpoles (Polezhayev, 1939*b*). The transplanted limb regenerates more rapidly than the nerves can grow. It will be remembered

that for the regeneration of the limbs of Urodela the nervous system is necessary (Weiss, 1925; Schotté, 1926). The fact that the regeneration of tadpoles' limbs does not depend on the nervous system, or at least that of *Rana temporaria* does not, is probably to be explained by the morphogenetic peculiarities of their regenerative material and the early stage of the ontogenetic development of the animal. It has also been shown that the limbs even of newts can develop from their explanted presumptive material independently of the nervous system (Polezhayev, 1938). Experiments to prolong the regenerative capacity with regeneration excited at stage III and even stage IV (metamorphosis) show that the regeneration takes place irrespective of the ontonogenetic changes which originate in the nervous system.

XIII. THE CAUSES OF THE LOSS OF REGENERATIVE CAPACITY IN THE TISSUE OF TADPOLES' LIMBS

We saw that the loss of regenerative capacity of tadpoles' limbs depends on the changes in their internal humoral media. Can, however, the tissue of the limbs change and lose its regenerative capacity irrespective of changes in the internal humoral media? Both hindlimbs of a tadpole at stage I were amputated at the base. One was transplanted autoplastically (control), the other was transplanted to a tadpole at stage o when the limb is nothing more than a small conical outgrowth (experiment). The animals went through metamorphosis. In a fortnight the control animals reached a late stage of development, and the transplanted limbs differentiated strongly both organologically and histologically. The experimental animals only reached stage I of their development a month after the operation. The transplanted limbs were organologically as developed as they were at the time of the operation, but their histological development was greatly advanced although not to the same extent as in the control animals. The transplants of the control and experimental animals were amputated, but in neither case was there any regeneration. The result is to be explained by the progress of histological differentiation. The experiment showed that the process of histological differentiation of tadpoles' limbs which leads to their losing regenerative capacity may take place independently of the changes in the internal humoral media of the animals, but that the latter, in cases of normal development, greatly accelerate the histological differentiation and the process of the loss of regenerative capacity (Polezhayev & Ginzburg, 1941).

XIV. CONCLUSION

(1) *Causes of the loss of regenerative capacity in the limbs of Anura.* The loss of regenerative capacity in the limbs of Anura is caused by changes in their

mesoderm and epithelium, stimulating changes, in the course of the metamorphosis, in the internal humoral medium of the animal. Changes in the mesoderm are more profound than in the epithelium. The regenerative capacity in the mesoderm is lost in two stages: in the first stage histological differentiation occurs, and in the second stage cytological differentiation. Histological differentiation retains the cells within the tissues (this is the most important) and suppresses proliferation in the tissues (of secondary importance) which leads to the suppression of the formation of the regeneration blastema and of regeneration. At the beginning of the process of histological differentiation the cells retain morphogenetic potentialities. Later, cytological differentiation takes place and the cells lose at least part of their morphogenetic potentiality; the limb does not regenerate even when there is dedifferentiation of the mesodermal tissue. As it grows older the epithelium becomes less capable of causing dedifferentiation of the mesodermal tissue of the stump and of forming with it those connexions which are necessary for the formation of the regeneration blastema.

(2) *The possibility of restoration and prolongation of regenerative capacity in the limbs of Anura.* It is possible to prolong the regenerative capacity of tadpoles' limbs to any stage of metamorphosis and even after its completion. The main condition is that mesoderm capable of dedifferentiation be retained until the later stages of metamorphosis. At the present time two methods are recommended for prolonging the regenerative capacity of tadpoles' limbs: re-regeneration (amputation of part of a regenerating limb) and repeated regeneration (amputation of the whole regenerated part of a limb).

It is possible to restore lost regenerative capacity in tadpoles' limbs if conditions are provided which ensure the formation of a regeneration blastema. An accumulation of regeneration cell material on the amputation wound surface is essential for forming a regeneration blastema; the material must possess morphogenetic potentiality. The methods of traumatization and transplantation may restore the regenerative capacity by overcoming the process of histological differentiation of the tissues and effecting a dedifferentiation of the mesoderm. These methods were found to be applicable only at certain stages of growth, namely, II*a* and to some extent II*b*. To restore the regenerative capacity at later stages some new methods or changes in older methods must be employed. That this is possible has been shown by Rose (1942), who excited regeneration in adult frogs by salt treatment of the stumps of the amputated limbs. I do not know the details of this experiment, but it apparently amounts to accumulating regeneration material on the amputated surface in the form of slowly dedifferentiating mesodermal tissue of the limbs.

(3) *Comparison of the regenerative capacity of various classes of vertebrates.* Vertebrates of various classes may be ranged in a single series in accordance with their regenerative capacities, a series which almost coincides with the phylogenetic one. An analysis shows that their regenerative capacity depends on the character of the interrelations of histological and cytological differentiation and on the rate at which these occur in ontogenesis. In urodeles, fishes, and to some extent in reptiles (the tails of lizards), the regenerative capacity is wholly or partially retained irrespective of the histological and cytological differentiation of the organs. In anuras, and possibly in reptiles and mammals, the regenerative capacity is lost, at first in relation to the histological and cytological differentiation of the organs and later to the cytological differentiation of the organs. In birds the regenerative capacity is lost in relation to the cytological differentiation of the organs before histological differentiation begins. Apparently it is more difficult to restore the regenerative capacity of birds. It is probably possible to achieve this in mammals, at least in the early stages of ontogenesis.

(4) *Present and future objectives.* The ultimate object of the work discussed in this article is to develop methods which will enable the regenerative capacity in any group of vertebrates to be renewed. There are, however, two less distant aims, namely, (1) to work out methods of restoring the regenerative capacity in a single group, the Anura, and (2) to make a comparative investigation of the loss and restoration of regenerative capacity of organs in various groups of vertebrates.

XV. SUMMARY

(1) When the hindlimbs of *Rana temporaria* are amputated in the proximal part of the shank during stage I of metamorphosis they regenerate, but they lose their regenerative capacity at stage II. The regenerative capacity is retained longer in the distal than in the proximal parts of the limbs, which is due to the process of histological differentiation taking place in a proximal-distal direction. (2) When the amputation is made during stage I the mesodermal tissue of the limbs dedifferentiates, regenerative blastemata are formed and the limbs regenerate. When the amputation is made during a later stage the mesodermal tissue does not dedifferentiate, blastemata are not formed and the limbs do not regenerate. (3) By a traumatization or a transplantation it is possible to excite complete regeneration at those stages in which the opposite (control) limbs of the same animals do not regenerate. (4) The restoration of the regenerative capacity of the limbs of tadpoles is brought about by the dedifferentiation of the mesodermal tissue in the distal part of the stump and not by an acceleration of the process of cell proliferation. (5) The regenerative capacity of tadpoles' limbs may be extended by means of repeated regeneration to a late stage of metamorphosis, or indeed, up to the completion of this process. The renewal of the regenerative capacity is dependent on the retention of the mesodermal tissue of the limbs at a level of histological differentiation which permits of dedifferentiation. (6) The regeneration of the limbs of Anura and Urodela occurs in two phases: during the

first phase the blastemata are formed almost entirely without mitosis, by the direct liberation of cells from the dedifferentiating mesodermal tissue; during the second phase the blastemata are formed mainly through intensified cell proliferation. (7) Tadpoles' limbs regenerate much more quickly, their mesodermal tissue dedifferentiates much more profoundly, and the mitotic coefficient is much lower than in the case of Urodela (axolotls). The difficulty of dedifferentiation, therefore must have a stronger inhibiting influence on the regeneration of tadpoles' limbs than on those of axolotls. (8) During the process of metamorphosis the internal humoral medium of the animal undergoes changes due to which there is an acceleration of the histological differentiation of the limbs and a suppression of the possibility of dedifferentiation and the formation of blastemata. (9) During metamorphosis, together with the loss of regenerative capacity, changes take place in the properties of the epithelium and of the mesodermal tissue of tadpoles' limbs. Changes in the mesoderm have a stronger and more profound influence on the regenerative capacity than changes in the epithelium. (10) Changes in the nervous system have no important effect on the regenerative capacity of tadpoles' limbs. (11) The loss of regenerative capacity in tadpoles' limbs is connected primarily with histological differentiation. This may to a certain extent occur independently of the changes that take place in the internal humoral medium, but in a typical case it is accelerated by these changes. (12) The main factor causing a loss of regenerative capacity in tadpoles' limbs is the histological differentiation of the mesodermal tissue and its inability to dedifferentiate. (13) To restore the regenerative capacity of the limbs of Anura the formation of blastemata must be effected, for which it is primarily necessary to cause dedifferentiation in the mesodermal tissue of the stump.

XVI. REFERENCES

BARFURTH, D. (1895): *Roux Arch. Entw. Mech. Organ.* **1**.

BASSINA, J. A. (1940): Бюлл. эксп. биол. и мег. **10**, no. 5.

BIER, A. (1923): *Arch. klin. Chir.* **127**.

BLACHER, L. J. (1928): Труды Лаборатории экспериментальной биологии Московского Зоопарка, **4**.

BORSSUK, R. A. (1935): *Roux. Arch. Entw. Mech. Organ.* **133**. — (1936): *Arb. Inst. exp. Morphogenese, Moskau,* **5**.

BROMLEY, N. W. (1930): *Roux Arch. Entw. Mech. Organ.* **129**.

BUTLER, E. G. (1933): *J. Exp. Zool.* **65**.

BUTLER, E. G. & PUCKETT, W. O. (1940): *J. Exp. Zool.* **84**.

BYRNES, E. (1904): *Roux Arch. Entw. Mech. Organ.* **18**.

GINZBURG, G. I. (1941): *C.R. Acad. Sci., U.R.S.S.* **31**, no. 9.

GUYÉNOT, E. (1927): *Rev. suis. Zool.* **34**.

IDE-ROZAS, A. (1936): *Roux Arch. Entw. Mech. Organ.* **136**.

JACOMINI, E. (1922): *R.C. Accad. Bologna.*

JAKOWLEWA, T. M. (1938): Биологич. журнал, **7**.

KAMMERER, P. (1905): *Roux Arch. Entw. Mech. Organ.* **19**.

LIOSNER, L. D. (1931): *Roux Arch. Entw. Mech. Organ.* **124**.

MANGOLD, O. (1929): *Ergebn. Biol.*

MORGAN, T. H. (1908): *Amer. Nat.* **42**.

NEEDHAM, J. (1942): *Biochemistry and Morphogenesis.* Cambridge University Press.

ORECHOWITSCH, W. N. & BROMLEY, N. W. (1934): *Biol. Zbl.* **54**.

POLEZHAYEV, L. W.[1] (1933): Биологич. журнал, **2**. — (1934): *C.R. Acad. Sci. U.R.S.S.* **4**, nos. 8–9, p. 468. — (1935): Русский Архив Анатомии, Гистологии и Эмбриологии, **14**. — (1935*a*): Биологич. журнал, **4**. — (1936): *Arch. Anat. micr.* **32**. — (1936*a*): *Bull. Biol.* **70**. — (1937): *C.R. Acad. Sci. U.R.S.S.* **15**, nos. 6–7, p. 387. — (1938): **21**, no. 7, p. 357. — (1938*a*): Успехи соврем. биологии, **8**, no. 3. — (1939): *C.R. Acad. Sci. U.R.S.S.* **22**, no. 9, p. 644. — (1939*a*): **22**, no. 9, p. 648. — (1939*b*): **25**, no. 6, p. 538. — (1939*c*): **25**, no. 6, p. 543. — (1940): **27**, no. 5, p. 520. — (1940*a*): Experimental studies in the problem of the loss and restoration of regeneration capacity limbs to the Anurans. Monograph, Moscow: unpublished. — (1941): *C.R. Acad. Sci. U.R.S.S.* **30**, no. 4, p. 367.

POLEZHAYEV, L. W. & GINZBURG, G. I. (1939): *C.R. Acad. Sci. U.R.S.S.* **23**, no. 7, p. 733. — (1941): **30**, no. 6, p. 550. — (1944): **42**, no. 7, p. 315.

POLEZHAYEV, L. W. & MOROSOV, I. I. (1941): *C.R. Acad. Sci. U.R.S.S.* **30**, no. 7, p. 675.

ROSE, S. M. (1942): *Proc. Soc. Exp. Biol., N.Y.*, **49**, no. 3.

SCOTTÉ, O. (1926): *Rev. suis. zool.* **33**.

THORNTON, CH. S. (1938): *J. Morph.* **62**. — (1942): *J. Exp. Zool.* **89**.

WEISS, P. (1925): *Roux Arch. Entw. Mech. Organ.* **104**. — (1930): *Entwicklungsphysiologie der Tiere.* Dresden und Leipzig.

[1] The author's name, Полежаев, is variously transliterated as Polezhayev, Poléjaiev, Poleszajev, Poležajew and Poležaiev.

ADDENDUM

Since this article was sent to press the following papers dealing with the problem under discussion have come to my notice:

BARBER, L. W. (1944): *Anat. Rec.* **89**, 441. — (1944): *Anat. Rec.* **89**, 560.

GIDGE, N. M. & ROSE, S. M. (1944): *J. Exp. Zool.* **97**, 71.

MARCUCCI, D. E. (1916): *Arch. Zool. ital.* **8**, 89.

POLEZHAYEV, L. W. (1945): *C.R. Acad. Sci. U.R.S.S.* **48**, no. 3. — (1945): *C.R. Acad. Sci. U.R.S.S.* **49**, no. 6.

ROSE, S. M. (1944): *J. Exp. Zool.* **95**, 149. — (1945): *J. Morph.* **77**, 119.

SCHOTTÉ, O. E. & HAKLAND, M. (1943): *J. Exp. Zool.* **93**, 453. — (1943): *J. Morph.* **73**, 329.

THORNTON, C. S. (1944): *Anat. Rec.* **89**, 559.

Reprinted from *Compt. Rend. Soc. Biol.*, **94**, 1050–1052 (1926)

11

Présidence de M. L.-F. Henneguy.

M. le Président lit un décret du Président de la République approuvant la modification des statuts portant sur les articles 5 et 6 augmentant le nombre des membres associés et correspondants.

DÉMONSTRATION DE L'EXISTENCE DE TERRITOIRES SPÉCIFIQUES DE RÉGÉNÉRATION PAR LA MÉTHODE DE LA DÉVIATION DES TRONCS NERVEUX,

par EMILE GUYÉNOT et O. SCHOTTÉ.

L'année dernière, P. Locatelli a réalisé une très belle expérience qui a soulevé un intérêt légitime. Cet auteur coupait le nerf sciatique de Tritons et en déviait le bout proximal, de façon à le faire aboutir à la surface, dans une région plus dorsale. En ce point, pouvait se développer une patte hétérotopique plus ou moins complète. Nous avons répété l'expérience de Locatelli et sommes en mesure de la confirmer, tout en apportant des précisions sur le mécanisme de ce phénomène.

En opérant soit sur les nerfs de la patte antérieure, soit sur ceux de la patte postérieure (sur *Triton cristatus*), et en faisant aboutir le bout des nerfs sectionnés au niveau de la peau, *dans le territoire basilaire* du membre (c'est-à-dire toute la région de l'épaule ou du bassin), nous avons obtenu 13 résultats positifs

210

(9 dans l'épaule, 4 dans le bassin) dont 10 ont été déjà décrits (1). Au niveau de l'aboutissement du nerf se forme un petit bourgeon cellulaire qui s'accroît lentement, s'arrête souvent dans son développement ou, au contraire, arrive à former des doigts ou même une patte complète (figure). Dans ce dernier cas, l'étude histologique a montré la présence d'un squelette normal, à l'état cartilagineux, de nerfs, de vaisseaux et de muscles, bien que la musculature paraisse plus réduite que dans une patte en situation normale.

La continuité des nerfs avec leurs centres paraît indispensable. Dans 15 cas, nous avons retourné de la même manière les

A gauche patte hétérotopique ; au milieu développement de la crête ;
à droite, queue supplémentaire.

nerfs après avoir sectionné proximalement les branches du plexus, sans obtenir aucun résultat. De même, la transplantation des nerfs sans connexion avec les centres ou de nerfs attachés aux seuls ganglions rachidiens n'a été suivie d'aucun effet. Nous ne voulons pas tirer de conclusions de ces dernières expériences, touchant le rôle négatif que nous attribuons aux ganglions rachidiens dans la régénération, avant d'avoir fait l'examen histologique des parties transplantées. Ces expériences tendent néanmoins à montrer que les résultats positifs ne sont pas à mettre sur le compte d'un simple transport d'éléments cellulaires apportés par le nerf et ses gaines.

Les insuccès assez nombreux que nous avons observés paraissent tenir à plusieurs causes : étranglement cicatriciel du bourgeon par la peau, déviation des axones régénérés, mais surtout aboutissement des nerfs transplantés en dehors du territoire basilaire de la patte. L'un de nous (Guyénot, 1926) a récemment

(1) *C. R. Soc. Phys. Hist. nat. Genève*, 18 février 1926.

attiré l'attention sur cette notion que l'animal est, en réalité, une mosaïque de territoires ayant des capacités régénératives propres. Le museau du Triton ou de la Salamandre ne régénère que dans la mesure où on a laissé une partie du territoire correspondant (Guyénot et Vallette, 1925 ; Vallette, 1926). La patte n'est plus régénérée quand on a extrait en totalité sa portion basilaire (Philipeaux, 1866 ; expériences non publiées de Guyénot) ; de même, la queue n'est plus régénérée quand elle a été totalement extirpée (expériences non publiées de Schotté).

Cette notion, qui établit un parallèle entre les localisations germinales de l'œuf et les localisations régénératives de l'adulte, nous paraît devoir jouer un rôle essentiel dans l'interprétation de l'expérience de P. Locatelli. Nous avons disséqué sur une aussi grande longueur que possible le nerf de la patte, de façon à le faire aboutir dans le dos, en dehors du territoire, épaule ou bassin (dont les limites exactes seront ultérieurement précisées). Il se forme alors une petite saillie transitoire qui n'est suivie d'aucune évolution. Par contre, ayant repris le nerf brachial qui avait permis la poussée d'une patte hétérotopique dans l'épaule, nous l'avons fait aboutir juste au niveau de la crête médiane, alors peu développée. Il s'est constitué un bourgeon qui a formé une portion de crête (figure), beaucoup plus saillante que le reste de cette partie. Nous avons, de même, dans 5 cas, fait aboutir le sciatique, en arrière, dans le territoire de la queue, et il s'est développé chaque fois une petite queue aplatie, paraissant d'ailleurs dépourvue de squelette (figure). Dans un cas, la queue a été amputée de manière à ce que la section intéresse le nerf transplanté : il s'est formé deux bourgeons de queue, l'un en face de l'axe caudal, l'autre en face du nerf dévié.

Ces expériences montrent que le nerf dévié agit surtout en tant que facteur d'excitation local à la croissance (nous avons précédemment signalé l'existence, dans ces cas, de foyers de mitoses en face des axones) sur un territoire qui répond spécifiquement, suivant ses propres potentialités. Un même nerf peut provoquer indifféremment la formation d'une patte, d'une queue ou d'une crête, suivant qu'on l'a fait aboutir dans l'un ou l'autre de ces territoires. Il ne saurait donc, en tout cas, s'agir d'une action morphogène spécifique du nerf.

(Station de zoologie expérimentale, Université de Genève.)

Reprinted from *Proc. Regeneration in Animals*, 404–419 (1965)

12

AUTONOMOUS MORPHOGENETIC ACTIVITIES OF THE AMPHIBIAN REGENERATION BLASTEMA

J. FABER

Hubrecht Laboratory, International Embryological Institute, Utrecht, The Netherlands

1. Introduction

Discussions concerning the morphogenesis of the amphibian regenerate have been dominated for a long time by the idea that the blastema constitutes an undetermined mass of cells, and that its pattern of differentiation is entirely due to determinative influences emanating from the differentiated stump tissues. This idea has perhaps been able to survive till the present day as a result of the apparently more or less general experience that explanted or transplanted early blastemas fail to differentiate. So it could happen that one of the best recent review articles on regeneration of vertebrate appendages [Goss 1961] is still pervaded by the notion of the complete passivity of the blastema, and the exclusive determinative role of the stump. To quote from the concluding section of Goss' paper: "When a blastema first forms, there is no conclusive evidence that it is in any way determined ... [the] determining factors reside in the tissues of the stump and communicate their influences to the blastema cells."

One of the principal aims of the present paper will be to show that the facts compel us to modify this view. Already some thirty years ago the literature occasionally reported differentiation of transplanted early blastemas (see e.g. [DAVID 1932]), but really conclusive evidence was brought forward in 1939 in an extensive paper by Mettetal, a student of Guyénot in Geneva. The author used a species of *Triturus* and a species of *Salamandra*. He transplanted blastemas of various well-defined stages to the head or the back of the same animal and was able to show that even the youngest blastemas were capable of self-differentiation in these heterotopic sites, although most of them regressed and disappeared. He took care to transplant only blastematous tissue, excluding the differentiated stump tissues beneath the blastema.

The most important result of Mettetal's experiments was his finding

213

that the youngest (flat conical) blastemas transplanted, although derived from amputations through the upper part of the limb, formed only digits; no carpus or tarsus, and no more proximal limb structures were formed. These latter were formed by the grafts only as progressively older blastemas were transplanted. They appeared in a disto-proximal order, first the carpus or tarsus, then the lower part of the limb, and finally its proximal portion.

Mettetal advanced an explanation for these surprising results, but as this was shown later to be incorrect [FABER 1960], I will not discuss it here. However, his results still stand and have been reproduced using a different animal, the axolotl (*ibid.*).

In the discussion which follows I shall speak mostly of the skeleton, since this component of the regenerate is the only one that shows clearcut regional characteristics, and can be used as a marker for regional differentiation.

2. Transplantation of intact blastemas

Using axolotl larvae of 6–11 cm in length, I transplanted fore limb blastemas of four different stages (Fig. 1) to the back of the same animal halfway between the fore and hind limbs. This transplantation site was chosen because it has been shown not to possess limb-forming potencies of its own (cf. [GUYÉNOT *et al.* 1948]), and consequently could be expected not to exert specific morphogenetic influences on the graft. The blastemas were obtained by amputation through the middle of the upper arm. The results were as follows: transplants of stage I formed 1–2 digits with phalanges and metacarpals, and occasionally one or two carpals; transplants of stage II formed 1–4 digits with occasional carpals; one lower arm element appeared in exceptional cases; transplants of stage III formed essentially the same, but with a higher average number of digits, up to 8 carpals (the normal number), and up to 2 lower arm elements; finally, transplants of stage IV were able to form more or less normal limbs, with the exception of the upper arm skeleton (humerus) which appeared only very rarely.

3. Carbon marking experiments; the apical proliferation centre

The finding that transplants of the youngest stage form almost exclusively digits is the more surprising since in normal regeneration it is not the digits, but the upper arm which differentiates first in the

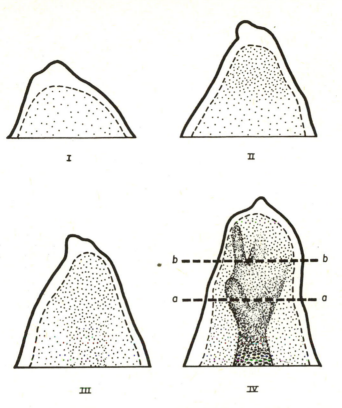

Fig. 1. Blastemal stages used in the transplantation experiments. Mesenchyme and skeletal rudiments stippled. Stages I and II: conical blastemas; stage III: early paddle-shaped blastema; stage IV: paddle-shaped regenerate with rudiments of humerus, radius, ulna and 1st digit (humerus partly precartilaginous). a–a, b–b, explanation in the text.

blastema. In order to provide a satisfactory explanation for this finding it is necessary to know what is the prospective significance of the mesenchyme of the early blastema; in other words, whether in normal regeneration this material is destined to form the tip or the base of the limb. In order to determine this, blastemas of stage I, developing normally on the stump, were provided with a carbon mark, which was introduced with a needle through the epidermis into the mesenchyme at the very tip of the blastema. Fig. 2 shows the fate of such a mark. It is seen that the mark is slightly drawn out in a longitudinal direction but that it stays at about the same distance from the amputation plane, and that new mesenchyme is added distal to the mark by what is no doubt a process of intense proliferation of the mesen-

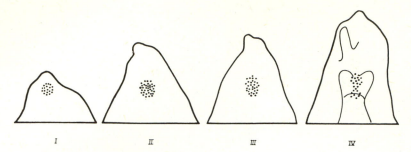

Fig. 2. The fate of a carbon mark inserted into the mesenchyme at the tip of a blastema developing normally on the stump.

chyme at the tip of the blastema. It is significant in this connection that CHALKLEY [1954] has shown by mitotic counts that in blastemas of *Triturus* approximately equivalent to our stages I and III the mesenchymal mitotic index is highest in roughly the distal half of the blastema (Fig. 3). In a stage equivalent to our stage IV the index in the proximal two thirds is even considerably lower than in the distal third. The epidermis seems to expand more uniformly, but this is irrelevant in the present context.

In the differentiated regenerate, which is not shown in the figure, the mark usually lies somewhere in the lower arm, as a rule close to the elbow. This shows that the mesenchyme of the early blastema is destined to form the upper arm and part of the lower arm, and that all the more distal limb parts are laid down later by an "apical proliferation centre".

4. Prospective significance versus self-differentiation of the early blastema

Considering the proximal prospective significance of the mesenchyme of the early blastema it is the more surprising that upon transplantation it should form digits and a few carpal elements, i.e. the most distal limb structures. The obvious conclusion is that the cells of this mesenchyme possess differentiation tendencies* for the formation of distal structures. But is this conclusion justified without further

---

* The term "differentiation tendency" should be used rather than "potency". As RAVEN [1948] has pointed out, the potency of a cell includes both its competence to react to certain inductive stimuli, and its tendency to enter upon and persist in a given pathway of differentiation.

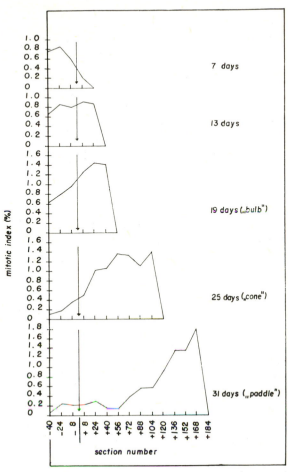

Fig. 3.   Mitotic indices (% cells per section) in stump and blastema in *Triturus* (adapted
from [CHALKLEY 1954]). The stages "bulb", "cone", and "paddle" are approximately
equivalent to our stages I, III, and IV, respectively. Arrows indicate the amputation
plane. Stump section numbers with negative sign, blastemal section numbers with
positive sign.

testing? There are grounds for caution here, since these transplants, and
particularly those of the younger stages, undergo a considerable reduction in size (regression) before they differentiate. This can only
mean that a certain amount of mesenchyme, presumably material located
proximally in the blastema and consequently destined to form proximal
structures, is resorbed or assimilated into the tissues of the transplantation site. On the other hand it could be that the apical prolifer-

ation centre .continues its activity after transplantation, laying down new material for future distal limb structures. If this were true the conclusion that the mesenchyme of the transplanted blastema *as such* possesses distal differentiation tendencies would be false.

This possibility was tested by transplanting blastemas of stage I which had received an apical carbon mark in the mesenchyme just prior to transplantation. Two things were noticed. First, the mark never stayed at the tip of the transplanted blastema. Some new material was always laid down by apical proliferation distal to the mark, and this formed the apical portions of the digits. Second, at the same time the mark usually moved towards the base of the graft, and sometimes was even carried into the tissues of the transplantation site, indicating that the initial regression of the graft indeed involves a loss of material from the proximal portions of the transplanted blastema. However, the most important finding was that in several cases where the mark stayed wholly or partly within the transplant, it extended as far distally as the proximal portion of one or more digits. Fig. 4 shows a recon-

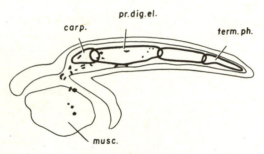

Fig. 4. Graphical reconstruction of a digit differentiated from a transplanted stage I blastema marked with carbon at the tip prior to transplantation. musc. – musculature of the back; carp. – carpus; pr. dig. el. – proximal digital element; term. ph. – terminal phalanx.

struction of one such transplant. In a non-transplanted blastema an apical mesenchymal mark would never come to lie at a level more distal than the lower arm (Fig. 2). This shows that mesenchyme which normally would form proximal structures, after isolation from the stump may take part in the formation of digits. And this can only mean that at the moment of isolation at least the apical mesenchyme of the stage I blastema possesses differentiation tendencies which are *markedly more distal* than would correspond to what this mesenchyme forms in normal regeneration.

5. Transplantation of halved and inverted blastemas; self-organization of blastemal mesenchyme

Before discussing the significance of this finding I first want to describe a different experiment, the outcome of which is even more significant. In this experiment the regenerate of stage IV (Fig. 1) was divided into two halves of about equal volume along the line a–a, and both halves were transplanted side by side to the back. The distal half never produced more than digits and carpal elements, which corresponds roughly to what it would have formed had it been left undisturbed. One would expect the proximal half to form upper and lower arm structures, for which rudiments are already present in the mesenchyme at the moment of transplantation. What happens, however, is something quite different: most of the transplants form exactly the same as the distal half, viz. digits and carpal elements.

A parallel series of transplantations was carried out in which proximal halves were provided with a mesenchymal carbon mark at their extreme distal end (i.e. in the future lower arm region, cf. Fig. 1). It was found that the mark usually moved towards the base of the transplant (proximal regression), and that always new mesenchyme appeared between the mark and the wound-epidermis covering the distal wound. It may be concluded that a new apical proliferation centre was established in the transplanted mesenchyme. This centre laid down new material for the most distal regions of the developing transplant. However, the most important finding again was that in several transplants the carbon mark came to lie in the carpus and the proximal region of one or more digits (Fig. 5). This means that mesenchyme that was already determined to form lower arm structures at the moment of trans-

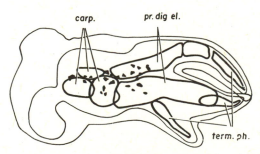

Fig. 5. Graphical reconstruction of a hand differentiated from a transplanted proximal half of a stage IV regenerate which had been marked with carbon prior to transplantation. Abbreviations as in Fig. 4.

plantation, nevertheless was able to take part in the formation of more distal structures after transplantation.

Histological analysis has shown that as a rule the whole proximal half undergoes dedifferentiation as a result of trauma attendant upon transplantation. At the same time, as shown above, at least part of its cells acquire differentiation tendencies for structures which are *markedly more distal* than those they would have formed had the blastema been left undisturbed.

The proximal half thus undergoes upon transplantation a process of internal reorganization which leads to the establishment of a totally unexpected new morphogenetic pattern, and I wish to emphasize here that this whole directive reorganization process takes place *autonomously, i.e. in the absence of any stump influence*.

This aspect is borne out even more strikingly by the experiments which my collaborator Michael carried out (see [MICHAEL and FABER 1961]). He removed the digital plate from regenerates of stage IV (Fig. 1, b–b) and transplanted the proximal portion with reversed proximo-distal polarity into a "pocket" in the muscles of the back, where it became covered by wound epidermis of back origin. These grafts underwent complete dedifferentiation and disorganization. Subsequently apical proliferation was initiated and the grafts produced limb outgrowths of inverted proximo-distal polarity (i.e., with the digits pointing outwards, as in normally oriented transplants). Most of the outgrowths developed into 1 or 2 digits with or without carpal elements. Here self-organization is even more evident than in the case of the non-inverted proximal half. It is significant that in this case it takes place in limb mesenchyme covered by non-limb epidermis. What is again striking in this experiment is that the grafts predominantly formed distal structures (digits) for which no rudiments were present in the material transplanted, since the entire digital plate had been previously removed.

6. The apical organization centre

From all experiments discussed so far a general rule emerges: if the relative numbers of digits and of skeletal elements formed in the various proximo-distal limb regions of the transplants are compared (Fig. 6), it is seen that *the digits and the skeletal elements of more distal regions tend to predominate over the skeletal elements of more proximal regions*. This holds for every series as a whole, as well as for the

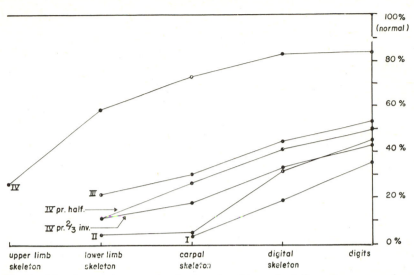

Fig. 6.   Morphogenetic expression in six series of transplantations. Average numbers of digits and skeletal elements expressed as percentages of the numbers of corresponding structures found in the normal fore limb. Note disto-proximal decline in all series. Series I–IV: transplanted intact blastemas of stages I–IV. Series IV pr. half: transplanted proximal halves of regenerates of stage IV. Series IV pr. ²/₃ inv.: stage IV regenerates with digital plate removed, transplanted with inverted proximo-distal polarity.

majority of the individual transplants. In transplanted intact blastemas one might be inclined to explain this by assuming that future proximal limb material disappears selectively from the graft during the regression phase. However, the rule also holds for the more distal limb structures, and for those series where we are sure that the distal limb sections have arisen as a result of a process of self-organization (stage IV, *proximal half* and stage IV, *proximal 2/3 inverted*, see preceding section). This means that distal predominance must be ascribed to the action of the organizing principle that is responsible for self-organization.

Self-organization leads primarily to the formation of one or more digits. If the amount of mesenchyme available is very limited these digits will contain only one or two skeletal elements, and no carpals will be formed. With increasing amounts of mesenchyme more digits will appear, with up to three or four digital elements each, and at the same time carpals will also be formed in increasing numbers. Organization thus proceeds in a disto-proximal sequence, and this has led me to use for the organizing principle the term "apical organization centre" [FABER 1960]. It is evident that its activity is independent of morphogenetic influences of the stump. Finally, there is no reason why an apical organization centre should not be active in the intact blastema just as well as in reorganizing transplants.

7.   The phenomenon of "distalization"

Returning now to our original line of thought, we have seen that at least part of the cells of both the undifferentiated blastema of stage I and the disorganized proximal half of the stage IV regenerate possess differentiation tendencies which are markedly more distal than would correspond to their prospective significance. I propose for this phenomenon the term "distalization". We have seen that distalization is bound up with dedifferentiation, and the most obvious conclusion is that cells dedifferentiating as a result of trauma are always distalized; in other words, that they automatically acquire intrinsic distal differentiation tendencies. Applied to the early blastema, this would mean that all its cells, having arisen from stump tissues by dedifferentiation, would originally possess intrinsic distal differentiation tendencies. The cells formed in the apical proliferation centre, by division of such dedifferentiated cells, would naturally be endowed with similar distal differentiation tendencies. Upon transplantation of the early blastema, the morphological pattern of the developing hand would be determined by the apical organization centre, acting autonomously. The origin and activity of this centre is no doubt bound up with the presence of intrinsic distal differentiation tendencies in the cells of the blastema.

One must also consider the possibility that distalization is not a process going on spontaneously in the mesenchymal cells, but that it is due to the morphogenetic influence of a non-mesenchymal component of the blastema, e.g. the epidermis. Since so far evidence is lacking on this point, in the discussions which follow it will be assumed that distalization is a spontaneous process.

The question when proximal differentiation tendencies appear in the blastema in addition to the intrinsic distal tendencies, and the question of the role played by the stump in the appearance of proximal tendencies, will be discussed by my collaborator De Both later in this symposion (see page 420).

8.   The nature of the apical organization centre

The assumption seems plausible that the apical organization centre is identical with the apical proliferation centre. First of all, several in stances are known of morphogenetically dominant regions which are characterized by high mitotic activity (e.g. in *Hydra*, see [BURNETT 1961]). Secondly, in our work the self-organization of transplanted blastemal mesenchyme was always bound up with the occurrence of apical proliferation. This in its turn seems to depend upon the presence of wound epidermis lacking a dermal layer (cf. [MICHAEL and FABER 1961, pp. 325/26]). There is an obvious connection here with the work of Thornton on the apical epidermal cap of the regenerating limb, although his work essentially deals with an earlier phase of regeneration, i.e. the phase of accumulation of mesenchyme which precedes the phase of growth by apical proliferation. I should like to suggest that wound epidermis in some way conditions or facilitates the appearance of a centre of proliferation and organization in mesenchyme accumulated or transplanted beneath it, in much the same way as the "apical epidermal ridge" of embryonic amphibian and chick limb buds seems to condition apical proliferation and organization in the limb bud mesenchyme (see [ZWILLING 1961] for a review). The information imparted by the epidermis cannot be very specific, since it is known that certain kinds of non-limb epidermis can support essentially normal limb regeneration (inverted transplants, see above; cf. [THORNTON 1960, 1962]).

9. Limited role of the stump in regulation of stump abnormalities

After all I have said so far about the self-organizing capacities of the blastema it will not come as a surprise when I say now that I attribute to the stump a much more modest role in the determination of the regenerate than others have done before me. It is particularly in connection with phenomena of regulation in regenerates arising on experimentally altered stumps, that previous authors have taken recourse to postulating the existence of a "limb field", capable of regulation, which was conceived as being primarily *localized in the stump* and from there extending into the blastema (cf. [WEISS 1939]). Now, in my opinion there are serious theoretical objections against ascribing morphogenetic field properties to *the differentiated stump or its tissues*. A morphogenetic field by definition exists and acts in and through a group of undifferentiated or differentiating cells. Apart from this argument, however, I feel that in view of the demonstrated extensive intrinsic morphogenetic capacities of the blastema it is altogether unnecessary to ascribe field properties to the stump. All regulative phenomena can be attributed to *field properties of the blastema itself,* particularly those associated with the apical organization centre. As has been pointed out repeatedly (see particularly [FABER 1962]) the apical organization centre possesses an inherent tendency to determine the formation of a complete, normal hand pattern in the absence of any stump influence, provided sufficient mesenchyme is available. We may assume that the centre acts likewise in cases of regulative development of the blastema *in situ* on an experimentally altered stump. But it can do so only if the change in composition of the amputation surface does not affect the pattern of morphogenesis in the more apical regions of the blastema.

One type of experiment in particular has yielded results that substantiate this view. Several authors (most recently [Goss 1956]) have studied the regeneration of experimentally doubled stumps, and have often obtained from such stumps regenerates which were internally double proximally, but whose structures converged distally into a unified, single hand. Goss regards such regenerates as potentially double, but I wish to point out that according to his descriptions the blastemas which were to give rise to such regenerates were consistently single in external appearance, and not perceptibly larger than normal blastemas. The interpretation I should like to suggest here is that the distal convergence of structures expresses the proximo-distal decline of the morphogenetic influence of the (abnormal) stump, and that the single hand arises as a result of the action of one single apical organization centre, acting independently of morphogenetic stump influences.

Experiments conducted with regenerating axolotl tails [FABER 1964] suggest that also in the tail alterations in the composition of the amputation surface do not necessarily affect the morphogenesis of the apical regions of the blastema.

10.   The decremental "basal field" and the autonomous "apical field"

We cannot deny that the stump exerts a morphogenetic influence on the basal regions of the blastema adjacent to it, but apparently this influence declines in the blastema, and at a certain distance from the level of amputation is no longer discernible. We may provisionally envisage this influence as being humoral in nature, and taking the form of a diffusion field of limited extension in the blastema. This diffusion field would then lead to the establishment of a morphogenetic field in the proximal portion of the blastema. I should like to call this field the "basal morphogenetic field". On the other hand one must assume that in the apical regions of the elongating blastema an "apical morphogenetic field" arises, which is independent of the stump and has its "high point" in the apical organization centre. These two fields interact and together produce the integrated proximo-distal limb pattern.

In the case of the double stump there must be two basal fields proximally, but these apparently tend to merge into a single field as the distance from the abnormal amputation surface increases. Beyond the distal limit of the basal field the morphogenesis of the regenerate is governed by the single, autonomous apical field.

A longitudinally halved limb stump may give rise to a whole hand (cf. [Goss 1957]), provided, probably, that the amount of mesenchyme produced by the defective stump is sufficient for the apical morphogenetic field to express itself fully. In cases where regulation does not occur or is not complete on such stumps, we must reckon with the possibility that the size of the blastema has been subnormal.

11. Conclusions

The experiments reviewed above have shown that strong intrinsic distal differentiation tendencies are present from the beginning in the blastema, that they persist at least in the apical mesenchyme during its further proliferation, and that they take the lead in the subsequent development of the blastema. The appearance of distal differentiation tendencies accounts for what we have called "distalization". The intrinsic distal tendencies are partly converted later into more proximal tendencies under the morphogenetic influence of the tissues present in the amputation surface ("basal field"). The picture presented here differs markedly from the old concept which holds that the early blastema is morphogenetically neutral, and that its differentiation is dictated entirely by the stump.

The differences between the two concepts are illustrated in the diagram of Fig. 7. In the old concept (top row) the idea is implicit that the limb pattern arises as a result of a chain of determinative steps proceeding from the stump distalwards. To most authors this idea seemed to be a logical inference from the fact that differentiation in the regenerate proceeds in a roughly proximo-distal order. However, it is in sharp contrast with the sequence of events suggested by our experiments (bottom row), in which the proximal differentiation ten-

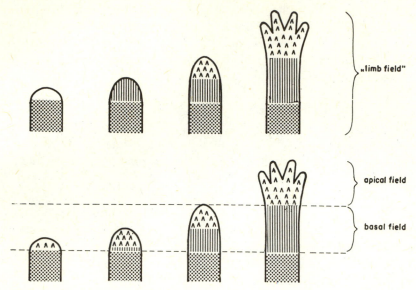

Fig. 7. Diagram illustrating two different concepts of regional organization in the limb regenerate. Four successive stages of regeneration are shown from left to right. In the top row the youngest blastema is depicted as being morphogenetically neutral. Stump stippled. Vertical hatching = proximal differentiation tendencies. ∧ ∧ ∧ = distal differentiation tendencies. Further explanation in the text.

dencies appear later than the distal ones. In our opinion the sequence of actual differentiation does not necessarily reflect the sequence of appearance of differentiation tendencies. A parallel case exists in the regeneration of *Tubularia* (see [FABER 1960] p. 65).

Data on hydroid regeneration [ROSE 1957] and on (anterior) planarian regeneration [WOLFF 1962] indicate that also in these invertebrates dedifferentiated or undifferentiated cells activated by the trauma of amputation tend to form the apical structures of a morphological pattern rather than the more proximal structures. In connection with our results, this can be interpreted to mean that such cells acquire intrinsic apical (distal) differentiation tendencies which subsequently take the lead in the further differentiation of the regenerate.

References

BURNETT, A. L. (1961) *J. Exper. Zool.* **146,** 21
CHALKLEY, D. T. (1954) *J. Morphol.* **94,** 21
DAVID, L. (1932) *Roux' Arch. Entw. Mech.* **126,** 457
FABER, J. (1960) *Arch. Biol.* **71,** 1
————— (1962) *Arch. Biol.* **73,** 379
————— (1964) *Acta Embryol. Morph. Exper.* **7,** 1
GOSS, R. J. (1956) *J. Exper. Zool.* **133,** 191
————— (1957) *J. Morphol.* **100,** 547
————— (1961) *Adv. in Morphogen.* **1,** 103
GUYÉNOT, E., J. DINICHERT-FAVARGER and M. GALLAND (1948) *Rev. Suisse Zool.* **55,** (suppl. 2), 1
METTETAL, CHR. (1939) *Arch. Anat. Histol. Embryol.* **28,** 1
MICHAEL, M. I. and J. FABER (1961) *Arch. Biol.* **72,** 301
RAVEN, CHR. P. (1948) *Fol. Biotheor.* **3,** 1
ROSE, S. M. (1957) *J. Morphol.* **100,** 187
THORNTON, CH. S. (1960) *Developm. Biol.* **2,** 551
————— (1962) *J. Exper. Zool.* **150,** 5
WEISS, P. (1939) Principles of Development. New York
WOLFF, E. (1962) In: "Regeneration" (20th Growth Symp.) ed D. Rudnick, Ronald Press Cy. New York
ZWILLING, E. (1961) *Adv. in Morphogen.* **1,** 301

Discussion

Chairman: Dr. C. S. THORNTON *(USA)*

NEEDHAM *(Britain)* complimented Faber on his work, but wondered whether there was not too great a tendency to swing too far in rejecting old views for newer ones. Perhaps, since a regenerate never produces more than is removed, there may well be an interaction between a "stump factor" and the "apical organisation centre" of the distal region of the blastema. Furthermore, we should be careful in our use of terms. The phrase "limb field", for example, has been used in this symposium in two senses; first as a region in which a particular structure can be evoked; and second, as an organisation field within the regenerate. These terms should be carefully redefined. FABER replied that he was convinced that the stump does have an influence on the regenerate, since otherwise there would not be continuity between the old tissues and the new ones. He pointed out, however, that the stump influence does not extend throughout the blastema. Even without a stump, a blastema can exhibit self-organisation and it is this mechanism which is controlled by the apical centre. As to use of terms he wished to distinguish clearly between the "regeneration territory" which is a district of competence in the Waddington sense, and the "morphogenetic field" which Waddington calls the individuation field. The hypothetical "apical field" is thought to be an individuation field.

TRAMPUSCH *(Netherlands)* cautioned against using terms as a cloak for our ignorance of essential mechanisms of morphogenesis. FABER agreed that concepts must have the positive value of describing effectively and concisely the parameters of phenomena and that they must not be interpreted as explanations.

NEEDHAM commented that in the case of blastemata grafted to a neutral region of the back there might be a possible diffusion of inhibitory chemical substances into the blastema from the neutral region, and that presumably there would be a gradient of concentration of this substance from basal to distal levels of the blastema. Furthermore, when the distal tip of the blastema is excised the resulting distal regeneration of the proximal area might be due to re-regeneration of this distal tip so that therefore one would expect to have the distal structures best represented.

FABER believed that it is difficult to speak of re-regeneration of a mesenchymal mass of cells such as one has in the blastema. He preferred to use the term re-organisation or self-organisation. He also agreed with a suggestion which Needham had made earlier, that some basal factor might have diffused out of the grafted blastema into the neighbouring host tissues and been lost. De Both's results would seem to support this particularly since added amounts of mesenchyme allow a greater extent of basal differentiation to occur. The increased amount of mesenchyme could perhaps act to conserve more of the basal factor. LEHMANN *(Switzerland)* asked how Faber's results could be correlated with those of Tschumi on the morphogenesis of the hind-limb bud of *Xenopus*. According to Faber, it was Tschumi's work which first gave him the idea of carbon marking as a tool for analyzing blastema development. FABER felt that his results and Tschumi's fitted well together, particularly when one includes the results which De Both has reported at this symposion.

JAMES *(Britain)* asked if the more peripheral parts of the blastema had been marked with carbon since one could explain an apparent stasis of the central axis, or even a proximal regression of carbon in the central axis, if cell migration had occurred around the periphery. FABER replied that he had indeed marked the peripheral regions of the blastema but obtained results similar to those already reported. SINGER *(USA)* then asked Faber to describe just how he introduced his carbon marker. Since Faber indicated that the carbon was extracellular, Singer expressed some doubts as to its use as an indicator of cell movements. He then described in detail marking experiments of his own in which nile blue sulphate was introduced into the base of the early blastema by means of a micro-infusion needle. This vital dye is taken up by mesenchymal cells. Various sectors of the blastema were marked, but cells moved mostly distally, as shown by the vital dye, along ventral and posterior sectors of the regenerate – which is just where the major limb nerves lie. Singer thus suggested that nerves constitute an important pathway along which the mesenchymal cells enter the regenerate.

Reprinted from *Develop. Biol.*, **18**, 457–480 (1968)

13

The Urodele Limb Regeneration Blastema: A Self-Organizing System

II. Morphogenesis and Differentiation of Autografted Whole and Fractional Blastemas[1]

DAVID L. STOCUM[2]

*Department of Biology, University of Pennsylvania,
Philadelphia, Pennsylvania 19104*

Accepted July 2, 1968

INTRODUCTION

Regeneration blastemas of larval urodele limbs fail to undergo normal morphogenesis when cultured *in vitro* without their epidermis (Stocum, 1968b), but will develop more normally if transplanted within their epithelium to another site on the animal body. However, the extent to which a grafted blastema is able to develop seems to depend on the presence or absence of limb stump in the graft. In the absence of stump, histologically undifferentiated cone stage blastema transplants usually form only hand structures, whereas essentially complete limbs are formed if a segment of stump is included in the graft (David, 1932; Mettetal, 1939; Pietsch, 1961). Indeed, Faber (1960, 1965) has presented evidence that the morphogenesis of the hand in a regenerating limb is autonomous, while that of more proximal limb structures is dependent on stump induction, a hypothesis

[1] Part of a dissertation presented to the faculty of the Graduate School of Arts and Sciences of the University of Pennsylvania in partial fulfillment of the requirements for the degree of Doctor of Philosophy, 1968. Research supported by U.S.P.H.S. DE-02047, a program project grant under the direction of Dr. C. E. Wilde, Jr.

[2] Predoctoral Fellow, U.S.P.H.S. Developmental Biology Training Grant 5 T1-GM-849-05 and U.S.P.H.S. To 1 DE-00001-11.

Present address: Department of Zoology, University of Illinois, Urbana, Illinois.

457

first suggested by Mettetal in 1952. The examination of this hypothesis is the central concern of the present report.

Faber's hypothesis is based on the following experiments. By tracing the fate of carbon marks inserted into the tips of cone stage blastemas obtained by amputation through the humerus of axolotl larvae, Faber found that the material destined to form the hand arose by proliferation of cells distal to the mark. The marked mesenchyme formed the upper and lower arm of the regenerate. However, a cone blastema marked in the same way formed only hand structures ("distalized") when grafted to the back. In most of these cases, the carbon marks were pulled to the base of the transplant by resorption of cells. But in a few cases, carbon was found in the hand itself, leading to the conclusion that, at the cone stage, cells that normally would form upper and lower arm actually possess differentiation tendencies which are markedly more distal.

In a second experiment, Faber transplanted the distal and proximal halves of palette stage blastemas separately. The proximal halves were provided with a distal carbon mark. The distal half transplants, which contained the rudiments of carpals and digits, formed just those structures. The proximal half transplants, which contained the rudiments of upper and lower arm skeleton, were covered with fresh wound epithelium and dedifferentiated. Subsequently, most of these grafts distalized and formed carpals and digits also, although a few differentiated lower arm skeleton in addition. In those cases which formed the lower arm, the carbon marks were found within this region, as expected. In the majority of those grafts which formed only carpals and digits, the carbon particles were pulled to the base of the transplant by resorption. But in several cases, the carbon was found in the carpals and digits themselves. This result was interpreted to mean that distal differentiation tendencies are acquired as a result of dedifferentiation. Since distalization was expressed only in a transplanted blastema, Faber proposed that, during regeneration *in situ*, the distal differentiation tendencies of the cone stage blastema cells are gradually converted to proximal tendencies under the influence of inductive factors emanating from the stump. The concentration of stump factors grades off distally and the hand forms autonomously from cells proliferated at the tip of the blastema under the influence of an apical organization center which arises there.

That young blastema cells somehow acquire distal morphogenetic tendencies is certainly a reality, since the progeny of such cells derived from the humeral region give rise to all structures distal to that region. But the hypothesis that young blastema cells are all initially distalized to the extent described by Faber is questionable. His conclusions revolve around the few cases in which carbon marks remained within the transplants. In the majority of his cases, however, lack of proximal development was associated with considerable resorption of blastema material. It cannot be ruled out that distalization is an artifact stemming from loss of cells which would have formed proximal skeleton had they remained within the transplant. Therefore, the possibility remains that the capacity of the young blastema for autonomous development has been underestimated.

The critical test of Faber's hypothesis would be to transplant the blastema or fractions thereof in a manner which holds dedifferentiation of the grafts to a minimum. The transplantation experiments of the present study were designed to test the extent of the self-organization capacity of the blastema as a whole, and to determine the morphogenetic behavior of blastema fractions under conditions which either maximized or minimized their dedifferentiation and resorption. The results indicate that the cells of the cone stage blastema are not distalized to the extent that they can form only hand structures in the absence of the stump. Rather, it appears that they are programmed with morphogenetic and differentiative information specifying the more proximal limb components as well and are able to express this information independently of the stump.

MATERIALS AND METHODS

All experiments were carried out on the larvae of *Ambystoma maculatum* (Shaw) raised and maintained as described previously (Stocum, 1968b). Regeneration blastemas were obtained by bilateral amputation of the forelimbs, just proximal to the elbow.

Larvae with regenerating limbs were first anesthetized in 1:1000 MS:222 (Sandoz) in spring water and then transferred to petri dishes containing MS:222 in full-strength Holtfreter solution at a concentration of 1:3000, for operation under continuous low-level anesthesia. The left forelimb was removed and the blastema carefully separated from the stump with fine iris scissors or knives (Heiss, Germany). In some cases, the regenerate of the right forelimb was left intact as an isochronous control. In other cases, it was removed as was the left regenerate and immediately fixed as an operational control. All experimental regenerates were autografted onto wound beds made in the dorsal fin by removing a small patch of skin with watchmaker's forceps. Holtfreter solution promotes wound-healing in urodeles, and healing together of the graft and fin wound borders was complete within 2–3 hours. Upon completion of wound healing, the operated animals were transferred to large finger bowls containing spring water. Transplants were allowed to develop at 21°C for 20–25 days before fixation. For each type of experiment, all three stages of regenerate (cone, palette, and notch) were transplanted. The following types of transplants, totaling 369 cases, were carried out.

Whole Blastemas Transplanted with and without Stump

Whole blastemas were transplanted with and without stump in order to determine the extent to which the blastema can undergo self-organization and to determine the effect of stump on this process. The transplants were positioned in a plane perpendicular to that of the fin, in normal proximodistal orientation. All blastemas were grafted without regard to their *in situ* dorsoventral orientation.

Blastema Fractions Transplanted without Stump

Distal half blastemas. Blastemas were halved transversely, and the distal halves were transplanted onto wound beds in a plane perpendicular to that of the fin, in normal proximodistal orientation. All grafts were made without regard to their *in situ* dorsoventral orientation.

Maximally dedifferentiated proximal half blastemas. Blastemas were halved transversely, and the proximal halves were transplanted onto wound beds in a plane perpendicular to that of the fin, in normal proximodistal orientation. The distal cut ends of such grafts became covered with fresh wound epithelium, and dedifferentiation of the transplants was maximized (see Faber, 1960, 1965). All grafts were made without regard to their *in situ* dorsoventral orientation.

Minimally dedifferentiated proximal half blastemas. A method of transplantation was desired for proximal half blastemas in which conditions favorable to migration of wound epithelium and subsequent dedifferentiation would be minimized. To this end, the following procedure was adopted. The epithelium was stripped from the ventral side of the blastema, after which the latter was halved transversely. The proximal halves were then transplanted onto wound beds with their denuded ventral surfaces facing the fin and with their distal ends oriented toward the dorsal border of the fin. The longitudinal and anterior-posterior axes of such grafts lay in the plane of the fin and the transplant remained covered externally with its dorsal epithelium. Migration of new wound epithelium was restricted to the edges of the transplant.

Minimally dedifferentiated longitudinal blastema fractions. Blastemas were cut longitudinally into anterior and posterior fractions, and the fractions were transplanted onto separate wound beds with their longitudinal cut surfaces facing the fin and with their distal ends oriented toward the dorsal border of the fin. The longitudinal axis of such a graft lay in the plane of the fin, while its anteroposterior axis lay in a plane perpendicular to the fin. Migration of new wound epithelium was again restricted to the edges of the transplant and the latter remained covered with its old epithelium.

Whole blastema transplants were either (a) fixed in Gregg's fixative and stained *in toto* for cartilage with methylene blue by the Van Wijhe method, as modified by Gregg and Butler (published in Hamburger, 1960) or (b) fixed in Bouin's solution, sectioned at 10 μ, and stained with iron hematoxylin and light green for examination of muscle. In addition, all sectioned transplants were examined for the presence or absence of lower arm skeleton. All fractional transplants were fixed in Gregg's and stained *in toto* for cartilage as described above. To gain some idea of the extent to which transplants became reinnervated by nerves from the dorsal fin, five additional whole blastema grafts were fixed in Bouin's after 5 days of development and another five after full development. Ten-micron frontal sections were prepared, stained according to the Holmes silver nitrate method, and counterstained with luxol fast blue (Margolis and Pickett, 1956).

The characteristic skeletal formula of the normal limb is 1:2:8:4:9, signifying the number of skeletal elements in the upper arm, lower arm, carpal, metacarpal, and phalangeal regions, respectively. Individual graft cartilages were easily identifiable by their distinctive morphologies and positions in relation to other cartilages. The degree of development of isochronous control and transplanted regenerates was assessed by counting the number of skeletal elements in each anatomical region of the final regenerate, and dividing by the number of elements found in the corresponding region of the normal limb.

The values obtained are percentages of the numbers of elements in the normal limb regions and will be called "indices of development." For example, if a transplanted regenerate formed six elements in the carpal region, the index of development for that region was 6/8, or 0.75 (75%).

Statistical significance of difference between mean indices of development was determined by t test for small sample size. Significance of difference was tested between mean indices of the different regions of a given graft type, between mean indices of corresponding regions of different graft types, and between mean indices of corresponding regions of isochronous control and experimental regenerates. All data were programmed professionally, and t values were computed by the appropriate method on an IBM 360/40 computer at the University of Pennsylvania Computer Center. The assistance of Mr. Peter Kuner and Computer Associates in preparing the computer program is gratefully acknowledged. The criterion for statistical significance was a $p \leqslant 0.01$. Means, standard deviations, and values for p have been recorded in a doctoral dissertation (Stocum, 1968a). The appropriate t tests and computer programs are on file at the University of Pennsylvania Computer Center.

RESULTS

Operated animals recovered from the anesthesia within half an hour after transfer to spring water and displayed normal behavior up to the time of sacrifice. Autografted regenerates became revascularized within 2–5 days after operation. Transplants of all stages developed somewhat more slowly than their isochronous controls and usually did not attain the size of the latter. Variable degrees of resorption were observed to occur in most transplants. The rate of resorption varied in individual cases, but usually ceased within 2–5 days, after which transplants increased rapidly in size. Sections of operational controls indicated that the surgical procedures employed in isolating the blastemas were effective in completely excluding stump tissues.

Reinnervation of whole blastema transplants by nerves from the dorsal fin was extremely sparse by 5 days after transplantation, although histogenesis was often well underway within the grafts. None of the five transplants examined for innervation at this time possessed more than 5–12 nerve fibers. Figure 1 illustrates two nerve fibers of a stumpless whole cone stage transplant at 5 days after operation.

The few nerve fibers present in this transplant were found only on one side of the graft and penetrated only a short distance into its basal portion. Any nerves present in a transplant at this time were always found in the form of single fibers. By 25 days after transplantation, however, nerve fibers were found in great numbers throughout individual grafts and could be traced to their terminations in muscle bundles and the epidermis. The fibers were now found both singly and in the form of fascicles.

Of the 369 total transplants, 335 (91%) developed. The remaining 34 cases underwent total resorption. Fifteen of these 34 cases were restricted to the maximally dedifferentiated proximal half category. The other 19 cases were about equally distributed among the other five transplant categories. All quantitative data are based on the 335 cases that underwent development.

The degree of development exhibited by an autografted regenerate depended to a great extent on the conditions imposed upon it by the method of transplantation. The details of development for isochronous controls and for each category of transplant are given separately in the following descriptions.

Isochronous Controls (40 Cases)

Four main skeletal regions were regenerated distal to the amputation plane: the lower arm (consisting of radius and ulna), carpals, metacarpals, and phalanges (the latter three regions collectively called the hand). Most control regenerates developed regional numbers of skeletal elements conforming to the normal formula 2:8:4:9, but in some cases the hand regions failed to form the normal numbers. Therefore, the mean indices of development of the latter regions did not quite equal 100%, but the differences from the latter value were almost negligible. Figure 2 represents an isochronous control regenerate which exhibited regional numbers of skeletal elements conforming to the normal formula.

Whole Blastemas Transplanted with Stump (79 Cases; 28 Cone, 28 Palette, and 23 Notch Stage Transplants)

As illustrated by Fig. 3, all mean regional indices of development of these transplants were essentially no different from those of isochronous controls, no matter what the stage of the regenerate at the time of transplantation. One-hundred percent of the cases examined

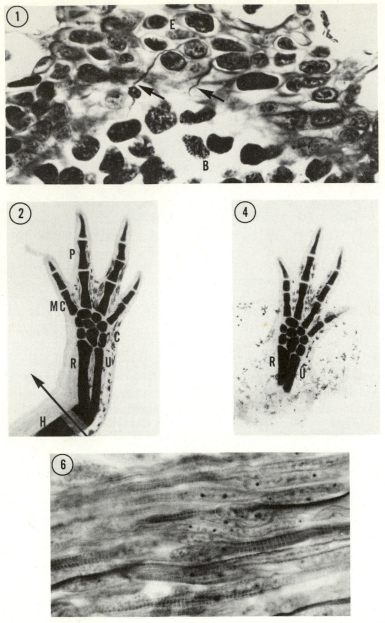

Key to abbreviations: C, carpal; *D,* digit; *H,* humerus; *MC,* metacarpal; *P,* phalange; *R,* radius; *U,* ulna; *Z,* lower arm.

Fig. 1. Nerve fibers (arrows) in epidermis (*E*) of an isolated whole cone

464

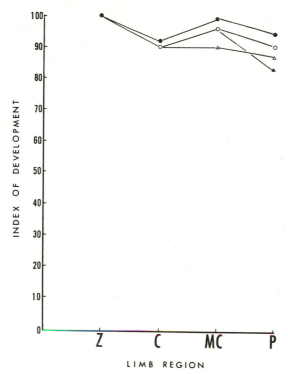

FIG. 3. Comparison of mean regional indices of development of isochronous control regenerates and whole blastemas transplanted with stump. Filled circles, control; open circles, cone stage transplant; open triangles, palette stage transplant; filled triangle, notch stage transplant.

for striated muscle had differentiated this tissue in normal anatomical relationships with the skeleton. However, the quantity of muscle differentiated was always considerably less than that of isochronous controls, and was especially sparse in the hand regions.

---

stage blastema graft 5 days after operation. The section is through a region toward the periphery, and at the base of the transplant. B, blastema cells. × 288.

FIG. 2. Twenty-five-day isochronous control regenerate of normal skeletal formula. Arrow indicates level of amputation. × 18.

FIG. 4. Twenty-five-day regenerate formed from an isolated whole cone stage transplant. Note the entirely normal morphogenesis and differentiation (skeletal formula, 2:8:4:9). × 18.

FIG. 6. Longitudinal section illustrating striated muscle differentiated in an isolated whole cone stage blastema transplant. × 288.

Whole Blastemas Transplanted without Stump (113 Cases; 39 Cone, 36 Palette, and 38 Notch Stage Transplants)

The major finding from these grafts was that numerous cases of each transplant stage were able autonomously to form the skeletal elements of all regions distal to the amputation plane. The morphology of these elements was completely normal, although fusion of cartilages was a common occurrence. Figure 4 illustrates a cone stage transplant that has formed a regenerate of normal skeletal formula.

However, the mean numbers of skeletal elements formed per transplant region never approached the values for corresponding isochronous control regions, and varied with the stage. Figure 5 represents a comparison between corresponding mean regional indices of isoch-

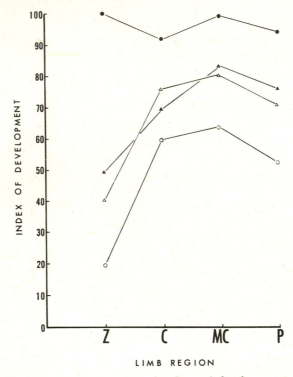

Fig. 5. Comparison of mean regional indices of development of isochronous control regenerates and whole blastemas transplanted without stump. Filled circles, control; open circles, cone stage transplant; open triangles, palette stage transplant; filled triangles, notch stage transplant.

ronous controls and transplants of each stage. No matter what the stage, the mean index of every transplant region was significantly less than that of the corresponding control region. The mean index of each transplant region increased from the cone to the notch stage, but the increase was not significant. However, for every transplant stage, the mean index of each hand region was significantly greater than that of the lower arm, although there were no significant differences among mean indices of the hand regions themselves.

Striated muscle (Fig. 6) was differentiated in normal anatomical relationships to the skeleton in 90% of cone stage transplants and in 100% of palette and notch stage transplants. There appeared to be no stage-dependent difference in the quantity of muscle formed by individual transplants, but the amounts differentiated were always far less than that in isochronous controls and slightly less than that in whole blastemas transplanted with stump.

Blastema Fractions Transplanted without Stump

Distal half blastemas (50 cases; 15 cone, 15 palette, and 20 notch stage transplants). Only skeletal elements of the hand were ever formed by distal half transplants. Figure 7 represents a comparison of mean regional indices of transplants with those of isochronous controls. For all stages, mean indices of transplants were significantly less than those of controls. The frequency of development of metacarpal and phalangeal elements in distal half grafts was comparable to that of whole blastemas transplanted without stump, but that of the carpal region was considerably less. Figure 8 illustrates the degree of development exhibited by a typical distal half graft.

Maximally dedifferentiated proximal half blastemas (18 cases; 6 cone, 7 palette, and 5 notch stage transplants). Unlike distal half transplants, the distal ends of these proximal half grafts became covered with fresh wound epithelium. Subsequently, the grafts began to regress, and eventually assumed a conical configuration by the time regression ceased. Faber (1960) has shown that the cells of partially differentiated proximal half blastema transplants of this type undergo dedifferentiation.

One of the most salient features of this transplant category was the number of grafts that underwent total resorption. Forty-five percent (5/11 total grafts done) of cone transplants, 30% (3/10 total grafts done) of palette transplants, and 58.3% (7/12 total grafts done) of

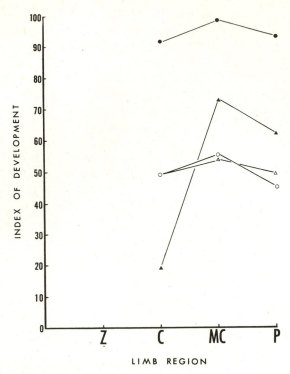

FIG. 7. Comparison of mean regional indices of development of isochronous control regenerates and distal half transplants. Filled circles, control; open circles, cone stage transplant; open triangles, palette stage transplant; closed triangles, notch stage transplant.

notch transplants were totally resorbed. All other surviving cases formed only skeletal elements of the hand, with the exception that one case each was found of cone and palette stage transplants which developed radius and ulna in addition.

Figure 9 represents a comparison of mean regional indices of transplants and isochronous controls. Indices of development were uniformly low in the transplants, and were all significantly less than those of corresponding regions of controls or whole blastemas transplanted without stump. For all stages, mean indices of the hand regions were somewhat higher, though not significantly so, than the mean index of the lower arm. Figure 10 illustrates a typical proximal half transplant which formed only digits.

Minimally dedifferentiated proximal half blastemas (38 cases; 15

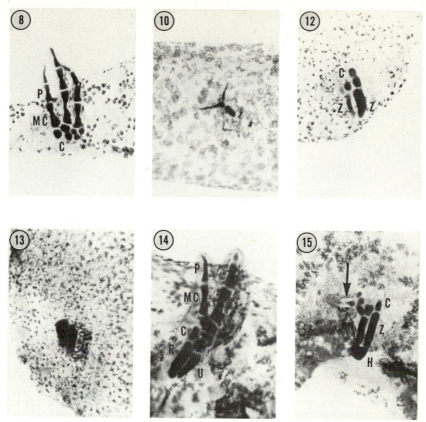

Fig. 8. Palette stage distal half transplant, illustrating exclusive formation of skeletal elements of the hand.

Fig. 10. Cone stage maximally dedifferentiated proximal half transplant. Note the exclusive formation of digital skeleton.

The following six photographs represent 25-day whole mounts of fractional blastema transplants stained for cartilage with methylene blue. All figures, × 17.

Fig. 12. Exclusive development of lower arm and carpal skeleton in a palette stage minimally dedifferentiated proximal half graft. Note that the cartilages are in the plane of the fin.

Fig. 13. Exclusive formation of lower arm skeleton in a notch stage minimally dedifferentiated proximal half transplant. The cartilages are in the plane of the fin.

Fig. 14. Cone stage minimally dedifferentiated proximal half transplant which has formed both lower arm and hand skeleton. The lower arm cartilages are in the plane of the fin, while all other elements are external to the fin.

Fig. 15. Palette stage minimally dedifferentiated proximal half transplant. A short segment of humerus, both lower arm elements, and a few carpals have developed in the plane of the fin. A short digital spike (arrow) has arisen at the edge of the graft, external to the fin.

469

241

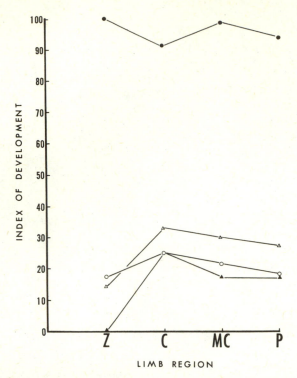

Fig. 9. Comparison of mean regional indices of development of isochronous control regenerates and maximally dedifferentiated proximal half transplants. Filled circles, control; open circles, cone stage transplant; open triangles, palette stage transplant; filled triangles, notch stage transplant.

cone, 12 palette, and 11 notch stage transplants). In contrast to maximally dedifferentiated transplants, minimally dedifferentiated proximal half grafts exhibited a high frequency of development of the lower arm skeleton. Figure 11 represents a comparison of corresponding mean regional indices of transplants and controls. For all transplant stages, the mean index of the lower arm region did not differ significantly from that of the controls, but was significantly greater than that of whole blastemas transplanted without stump.

The majority of cases formed hand skeleton in addition to that of the lower arm. For any given transplant stage, mean indices of the hand regions decreased in a proximodistal direction, and the mean index of any given hand region decreased from the cone to the notch

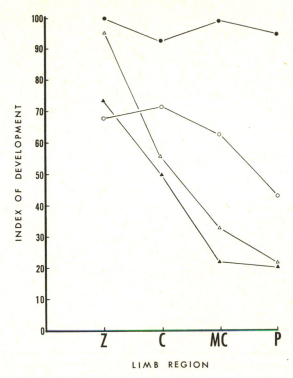

FIG. 11.   Comparison of mean regional indices of development of isochronous control regenerates and minimally dedifferentiated proximal half transplants. Filled circles, control; open circles, cone stage transplant; open triangles, palette stage transplant; filled triangles, notch stage transplant.

stage. Mean indices of all transplant hand regions were significantly less than those of corresponding control regions, but did not differ significantly from corresponding mean indices of whole blastemas transplanted without stump. The mean indices of the metacarpal and phalangeal regions were less than the mean index of the lower arm for all transplant stages. The differences were not significant for the cone stage, but were significant for palette and notch stages. Thus, there appeared to be a significant preference for the formation of lower arm skeleton over hand skeleton in palette and notch transplants of this type.

A finding of major interest was that several cases were found of both palette and notch stage transplants which formed only the radial

472	DAVID L. STOCUM

and ulnar cartilages, or these elements plus a few carpals. Figure 12
illustrates a palette stage transplant from which only the radius, ulna,
and three carpals developed. Figure 13 represents a notch stage graft
which formed only radius and ulna. The lower arm skeleton and
carpals developed in such grafts always lay in the plane of the fin,
in the orientation in which they were grafted. However, in cases
where more distal skeletal elements were formed, they were found external to the fin at an angle to the radius and ulna, and the position
at which they arose was always at the edge of the transplant, where
the wound borders of graft and fin had healed together. Figure 14
illustrates one such cone stage transplant from which radius, ulna,
carpals, and two digits developed. The palette stage transplant represented in Fig. 15 formed five carpals, radius and ulna, and a short
segment of humerus in the plane of the fin. A single digital spike arose
at the edge of the graft.

In summary, restriction of migration of fresh wound epithelium
over proximal half transplants appears to favor the morphogenesis of
proximal skeleton over distal, a result which is in contrast to what is
observed in proximal half grafts were migration of wound epithelium
is maximized.

*Minimally dedifferentiated longitudinal blastema fractions (37 cases;
14 cone, 14 palette, and 9 notch stage transplants).* Transplants of
both anterior and posterior blastema fractions formed the whole
spectrum of missing skeletal elements distal to the plane of amputation in numbers proportional to the size of the fraction. If the blastema was divided equally along its longitudinal axis, each half tended
to form half the skeletal elements of the lower arm, carpal, metacarpal,
and phalangeal regions. Figure 16 illustrates a cone stage transplant
in which each half formed one digit, 2–3 carpals, and one lower arm
element. Since the skeletal elements are obscured by pigment, they
are reproduced in the accompanying drawing. Figure 17 illustrates a
notch stage transplant in which each half formed a single digit, one
or two carpals and a lower arm element.

In the majority of cases the blastema was divided unequally. This
resulted in a corresponding inequality of numbers of elements produced by the fractions. Figure 18 represents a palette stage transplant in which the larger fraction formed seven phalanges, three
metacarpals, the full complement of carpals and both lower arm elements. The smaller fraction produced two phalanges, two carpals

244

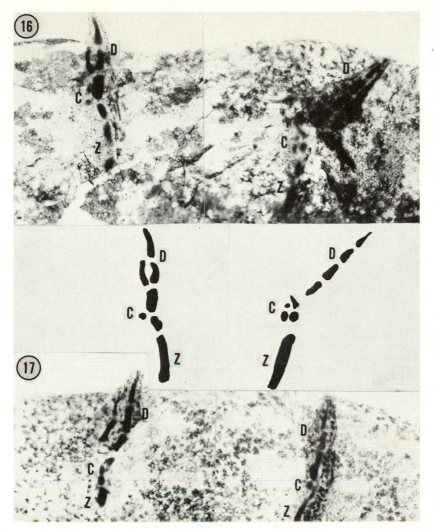

Figs. 16–19. The four photographs represent 25-day whole mounts of fractional blastema transplants stained for cartilage with methylene blue. Figs. 16 and 17, × 18; Figs. 18 and 19, × 16.5.

Fɪɢ. 16. Cone stage minimally dedifferentiated longitudinal blastema fractions of equal size. Each fraction has formed half the total number of skeletal elements produced. The elements are somewhat obscured by pigment and are reproduced in the drawing.

Fɪɢ. 17. Notch stage minimally dedifferentiated longitudinal blastema fractions of equal size. Each fraction has formed about half the total number of skeletal elements produced.

473

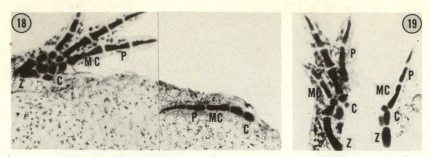

FIG. 18. Palette stage minimally dedifferentiated longitudinal blastema fractions of unequal size. Each fraction has formed numbers of skeletal elements proportionate to its size.

FIG. 19. Notch stage minimally dedifferentiated longitudinal blastema fractions of unequal size. Numbers of skeletal elements formed are in proportion to the size of the fraction.

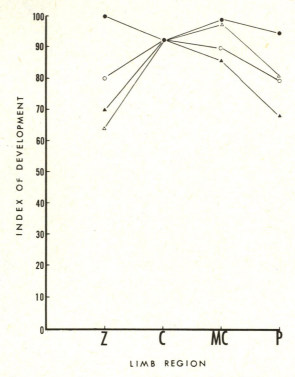

FIG. 20. Comparison of mean regional indices of development of isochronous control regenerates and minimally dedifferentiated longitudinal blastema fractions. Filled circles, control; open circles, cone stage transplant; open triangles, palette stage transplant; filled triangles, notch stage transplant.

474

and a single metacarpal. Figure 19 represents a notch stage transplant in which the larger fraction formed seven phalanges, 5–6 carpals, three metacarpals, and one lower arm element. The smaller fraction produced two phalanges, one metacarpal, three carpals, and a segment of the other lower arm element.

No longitudinal fraction ever tended to regulate toward the formation of a whole regenerate, although in some cases the total number of skeletal elements of a given region formed by both fractions exceeded the number characteristic of the corresponding region of the normal limb. This is considered to be due to the accidental division of the specific anlage by the surgical procedure at the time of transplantation. Figure 20 illustrates a comparison of the mean regional indices of isochronous controls with those obtained by combining the numbers of skeletal elements per region in each transplant fraction. There was no significant differences, either between corresponding regions of control and experimental regenerates, or between different regions of any given stage of experimental regenerates.

DISCUSSION

Autonomy of the Regeneration Blastema

The results of the experiments described herein strongly support the concept that, by the time it has reached the cone stage, the regeneration blastema is capable of self-organization, and suggest that this capacity has been greatly underestimated by previous investigators. Two general conclusions may be drawn from the transplant data. First, transplants of isolated whole blastemas are capable at any stage of autonomously organizing into all the skeletal and muscular elements that would normally regenerate distal to the amputation plane *in situ*. Similar results have also been obtained by Jordan (1960) with conical limb blastemas of larval *Xenopus* grafted to the brain ventricle. Secondly, the various proximal and distal anatomical regions of the missing limb are represented within the blastema by the cone stage as discretely separate portions of a definitive pattern and can independently develop as such when separated in space. Distal half grafts developed into hand structures exclusively. Proximal halves of all stages of regeneration exhibited a distinct preference for the formation of proximal structures when transplanted under conditions that limited their dedifferentiation and resorption. Moreover, each transplanted longitudinal fraction of the blastema tended to autonomously organize the whole proximal-distal sequence of skeletal elements in numbers proportionate to the size of the fraction, a result which again suggests that the young blastema is a patterned structure.

The Influence of Dedifferentiation and Resorption on Development of Transplanted Regenerates

It has been noted that, although stumpless whole blastema transplants were capable at any stage of self-organizing into all the missing limb components distal to the amputation plane, they still exhibited a significant preference for the formation of hand skeleton. Faber (1960, 1965) was unable to obtain any lower arm cartilages in grafts of cone stage blastemas and interpreted this result to mean that cells of the latter possess differentiation tendencies markedly more distal than indicated by their normal fate. Another interpretation, however, would be that the apparent distalization of such grafts was the result of resorption of proximal blastema material which would have formed lower arm skeleton had it remained within the transplant. Applying this interpretation to the present results, those whole blastema transplants which underwent little or no resorption would form all skeletal elements distal to the amputation plane, while those in which large amounts of material were resorbed would form mainly hand skeleton.

However, a second factor in Faber's hypothesis was the observation that cells of a palette stage proximal half transplant, which were already differentiating to form upper and lower arm cartilages, dedifferentiated and subsequently produced mainly hand skeleton. This result has been confirmed in the present investigation with proximal half grafts designed to undergo maximum dedifferentiation. On the other hand, minimally dedifferentiated proximal half transplants exhibited a significant preference for the formation of lower arm skeleton, and the incidence of development of these elements was not significantly different from that in isochronous controls. Moreover, several of these transplants formed only lower arm (or lower arm and carpal) skeleton in the plane of the fin, and any hand cartilages that were formed arose at the edge of the grafts and at an angle to the fin. The only experimental difference between the two types of proximal half transplant was that a maximally dedifferentiated graft became covered by migrating wound epithelium, as opposed to little migration of new wound epithelium over a minimally dedifferentiated graft.

There is a great deal of evidence to indicate that the wound epithelium is involved in the dedifferentiation and accumulation of cells for the blastema (Needham, 1952; Steen and Thornton, 1963; Thornton, 1965). Moreover, the checking of dedifferentiation and the formation of a blastema on an amputated limb depends on the reestablishment of innervation to the area of the blastema (Butler and Schotté, 1941). The blastema fails to grow and resorbs if denervation of the limb is performed at any time prior to the cone stage. The blastema starts to become independent of innervation during the latter stage, since cone regenerates can sometimes proceed with more or less limited morphogenesis in the absence of nerves. Later stages of regeneration are fully independent of innervation for their continued development (Schotté and Butler, 1944; Butler and Schotté, 1949; Singer and Craven, 1948). In the present experiments, all grafts were temporarily denervated, and only a few nerves were just beginning to penetrate them by 5 days after transplantation.

Collectively, these observations suggest that the differential morphogenetic behavior of the two types of proximal half transplant described herein is due to a differential extent of epithelial-induced dedifferentiation of graft cells during the post-transplantation period when the grafts are disconnected from a nerve supply. Maximum dedifferentiation (involving the whole graft) during this period would lead to a high degree of resorption of proximal graft cells, much as a denervated amputated limb resorbs. In effect, a new plane of "amputation" would be established distal to the original amputation plane. The almost exclusive morphogenesis of hand skeleton from a maximally dedifferentiated proximal half transplant would reflect how far distally resorption had progressed.

Conversely, minimal dedifferentiation during the period of nervelessness would involve only a limited region of the graft, similar to the situation in the limb stump during normal regeneration. Those graft cells which underwent dedifferentiation would develop into structures distal to the level to which dedifferentiation had progressed, while the remainder of the graft would continue on its original course, producing more proximal structures. The number of distal skeletal elements regenerated would depend on the extent of dedifferentiation of the transplant. If no dedifferentiation occurred, only those proximal elements represented in the graft would be formed. It might be expected that progressively older graft stages would be more resistant to dedifferentiation than the cone stage and thus form progressively fewer elements of the hand skeleton. Figure 11 indicates that this is actually the case.

Distalization can thus be interpreted as an artifact associated with epithelial-induced dedifferentiation and/or resorption of graft cells in the temporary absence of innervation. Evidence supporting this interpretation has been provided by DeBoth (1965). He demonstrated that if loss of blastema material by resorption was compensated for by transplanting two or more cone blastemas to the same site, or by supplying a single transplant with a deviated nerve, the grafts were able to undergo much more complete development.

In view of the present results, it is apparent that all missing components of an amputated limb can be self-organized by the blastema, at least during the cone stage and thereafter. However, it is still possible that the stump transmits inductive messages to the blastema prior to the cone stage. In this regard, it is worth looking again at the behavior of stumpless proximal half blastemas transplanted in normal proximodistal orientation. Faber (1960, 1965) observed that the partially differentiated cells of a palette stage proximal half dedifferentiated upon transplantation. The same result was obtained when proximal halves were transplanted with reversed proximodistal orientation (Michael and Faber, 1961). In both cases, the dedifferentiated cells subsequently redifferentiated into some of the missing limb structures. The point is, that all the processes of regeneration (dedifferentiation, growth and redifferentiation, and morphogenesis) took place in the complete absence of differentiated stump tissue. Thus, a strong argument can be generated in support of the idea that the mesodermally derived stump tissues play no inductive role in the development of the blastema over its entire history. Although the asymmetry of the future regenerate could possibly be determined by the stump, it is equally possible that development of asymmetry is also an autonomous function of information programmed within the cells of the blastema from their inception. Thus, the role of the stump tissues may only be to ensure a supply of free blastema cells whose developmental information is independently expressed whether in the presence of stump or not.

SUMMARY

Whole limb blastemas of larval *Ambystoma maculatum* (Shaw) were autografted to the dorsal fin, with or without stump, in order to determine the extent of their ability to self-organize. In addition, proximal and distal fractions and longitudinal fractions of blastemas were autografted to the dorsal fin in order to determine whether the separate fractions would develop autonomously as specific parts of the total regenerate which would normally have been formed *in situ*.

Autografts of all stages of whole blastemas were able to self-organize into all the skeletal and muscular components of the lost limb parts distal to the amputation plane in the absence of the stump. Fractional parts of the blastema were able to develop autonomously into separate parts of the total regenerate corresponding to the level of the blastema from which they were obtained, providing their developmental stability was unaltered by the influence of fresh wound epithelium migrating over the graft.

It has been concluded that the blastema, at least from the cone stage on, contains a pattern of discrete and separate parts which make up a wholly self-organizing system.

The author wishes to express his appreciation to Dr. Charles E. Wilde, Jr., for his encouragement and advice through the course of this investigation, and for critical reading of the manuscript.

REFERENCES

BUTLER, E. G., and SCHOTTÉ, O. E. (1941). Histological alterations in denervated non-regenerating limbs of urodele larvae. *J. Exptl. Zool.* **88**, 307–341.

BUTLER, E. G., and SCHOTTÉ, O. E. (1949). Effects of delayed denervation on regenerative capacity in limbs of urodele larvae. *J. Exptl. Zool.* **112**, 361–392.

DAVID, L. (1932). Das Verhalten von Extremitätenregeneraten des weissen und pigmentieren Axolotl bei heteroplastischer, heterotopen und orthotopen Transplantation und sukzessiver Regeneration. *Roux' Arch. Entwicklungsmech. Organ.* **126**, 457–511.

DEBOTH, N. J. (1965). Enhancement of the self-differentiation capacity of the early limb blastema by various experimental procedures. *In* "Regeneration in Animals" (V. Kiortsis and H. A. L. Trampusch, eds.), pp. 420–426. North Holland Publ., Amsterdam.

FABER, J. (1960). An experimental analysis of regional organization in the regenerating forelimb of the axolotl (*Ambystoma mexicanum*). *Arch. Biol.* **71**, 1–67.

FABER, J. (1965). Autonomous morphogenetic activities of the amphibian regeneration blastema. *In* "Regeneration in Animals" (V. Kiortsis and H. A. L. Trampusch, eds.), pp. 404–418. North Holland Publ., Amsterdam.

HAMBURGER, V. (1960). "A Manual of Experimental Embryology," revised ed., p. 196. Univ. of Chicago Press, Chicago, Illinois.

JORDAN, M. (1960). Development of regeneration blastemas implanted into the brain. *Folia Biol.* (*Kracow*) **8**, 41–53.

MARGOLIS, G., and PICKETT, J. P. (1956). New applications of the luxol fast blue myelin stain. *Lab. Invest.* **5**, 459–474.

METTETAL, C. (1939). La régénération des membres chez la salamandre et le Triton. Histologie et détermination. *Arch. Anat. Histol. Embryol.* **28**, 1–214.

METTETAL, C. (1952). Action du support sur la differenciation des segments proximaux dans les régénérats de membre chez les Amphibiens Urodèles. *Compt. Rend. Acad. Sci.* **234**, 675.

MICHAEL, M. I., and FABER, J. (1961). The self-differentiation of the paddle-shaped limb regenerate, transplanted with normal and reversed proximal-distal orientation after removal of the digital plate (*Ambystoma mexicanum*). *Arch. Biol.* (*Liege*) **72**, 301–330.

NEEDHAM, A. (1952). "Regeneration and Wound-Healing," pp. 62–64. Methuen, London.

NEEDHAM, J. (1942). "Biochemistry and Morphogenesis," pp. 430–442. Cambridge Univ. Press, London and New York.

PIETSCH, P. (1961). Differentiation in regeneration. I. The development of muscle and cartilage following deplantation of regenerating limb blastemata of *Ambystoma* larvae. *Develop. Biol.* **3**, 255–264.

SCHOTTÉ, O. E., and BUTLER, E. G. (1944). Phases in regeneration and their dependence on the nervous system. *J. Exptl. Zool.* **97**, 95–122.

SINGER, M., and CRAVEN, L. (1948). The growth and morphogenesis of the regenerating forelimb of adult *Triturus* following denervation at various stages of development. *J. Exptl. Zool.* **108**, 279–308.

STEEN, T. P., and THORNTON, C. S. (1963). Tissue interaction in amputated aneurogenic limbs of *Ambystoma* larvae. *J. Exptl. Zool.* **154**, 207–222.

STOCUM, D. L. (1968a). Doctoral Dissertation, Univ. of Pennsylvania, Philadelphia, Pennsylvania.

STOCUM, D. L. (1968b). The urodele limb regeneration blastema: A self-organizing system. I. Differentiation *in vitro*. *Develop. Biol.* **18**, 441–456.

THORNTON, C. S. (1965). Influence of the wound skin on blastema cell aggregation. *In* "Regeneration in Animals" (V. Kiortsis and H. A. L. Trampusch, eds.), pp. 333–339. North Holland Publ., Amsterdam.

Reprinted from *Develop. Biol.*, **9**, 385–397 (1964)

14

Lens Antigens in a Lens-Regenerating System Studied by the Immunofluorescent Technique

Chinami Takata, Joseph F. Albright, and Tuneo Yamada

Biology Division, Oak Ridge National Laboratory,[1]
Oak Ridge, Tennessee

Accepted March 6, 1964

INTRODUCTION

Transformation of the pigmented iris of an adult newt into the lens upon lens removal is a unique system for analyzing the mechanism controlling tissue specificity. The first approach to understanding the mechanism was through autoradiographic studies of ribonucleic acid (RNA), deoxyribonucleic acid (DNA), and protein synthesis during tissue transformation (Yamada and Karasaki, 1963; Yamada and Takata, 1963; S. Eisenberg, unpublished). These studies indicated enhancement of protein and DNA synthesis, which follows enhancement of RNA synthesis. In order to obtain information concerning tissue specificity of protein synthesized in the system at different stages of regeneration, immunochemical studies are being conducted in our laboratory. This paper describes the results we obtained with the immunofluorescent technique utilizing antisera against the newt lens and with the lens-regenerating cell population derived from the dorsal iris of adult *Triturus viridescens*. Further studies of the same system, in which antisera against specific fractions of newt lens antigens are utilized in immunodiffusion and immunoelectrophoretic analyses, will be reported in separate papers.

MATERIALS AND METHODS

Lens-regenerating system. The lens was removed bilaterally from adult *T. viridescens* through a horizontal cut in the cornea. The animals, which were kept at 21–22°C, were killed 0, 5, 10, 12, 13, 14, 15, 16, 17, 18, 20, and 25 days after the surgery. Lens regeneration

---

[1] Operated by Union Carbide Corporation for the United States Atomic Energy Commission.

was staged according to the scheme proposed by Sato (1940) and adapted to *T. viridescens* by Stone and Steinitz (1953). In the following description, "stage" refers to Sato stage of lens regeneration.

Antisera. Sixty lenses of adult *T. viridescens* were homogenized in 2.0 ml of 0.15 *M* NaCl and incorporated in complete Freund's adjuvant. Five rabbits were used for antibody production; each rabbit was subcutaneously injected with the preparation, including 60 lenses, followed 2 weeks later by a similar injection and 5 weeks later by a single intraperitoneal injection of 50 lenses homogenized in 2.0 ml of 0.15 *M* NaCl. Each rabbit was given 45 mg of lens protein and was bled 9 days after the last injection. Antisera and normal serum from nonimmunized rabbits were collected. Interfacial antiserum titers were of the order 1:4000, and visible precipitates formed with as little as 1.2 μg lens protein per milliliter. Agar diffusion and immunoelectrophoretic analysis of the antisera under various experimental conditions were satisfactory for demonstrating lens antigens.

The globulin fractions were obtained from antisera and normal sera by precipitation with ammonium sulfate at 40% saturation. Salting out was performed twice. The purified globulins were dialyzed exhaustively against 0.15 *M* NaCl buffered at pH 7.0 with 0.01 *M* phosphate buffer (PBS).

Immunofluorescent reagents. Antibody- and normal-globulin fractions were labeled with fluorescein isothiocyanate (Baltimore Biological Laboratory) by the method of Marshall *et al.* (1958) and dialyzed against PBS until the dialyzate was devoid of fluorescence. Excess uncoupled fluorescein was removed by passing the labeled globulins through a column (1.2 × 15 cm) of Sephadex G-50 (Pharmacia, Uppsala). The first-eluted fraction was collected and then concentrated to the original volume by use of Carbowax 20 M (Union Carbide Corporation). After 100 mg of mouse tissue powder (Difco) per milliliter of globulin solution was added, the fraction was stirred occasionally for 2 hours at room temperature, incubated overnight in the cold room, and centrifuged at 12,000 *g* for 60 minutes. The procedure was repeated twice with 50-mg/ml portions of mouse tissue powder. Three absorptions were adequate to remove the weak nonspecific staining of the normal iris that occurred when unabsorbed labeled globulin was applied to tissue sections.

The fluorescein-labeled antibody globulin will be referred to as the "specific immunofluorescent reagent," the labeled normal globulin as

the "control fluorescent reagent," and the unlabeled antibody globulin
as the "specific nonfluorescent reagent."

Tissue sections. The pigmented iris and lens-regenerating cell population were excised together with the covering cornea, fixed in 95%
ethanol at 4°C, embedded in paraffin, and stained with the specific
immunofluorescent reagent according to the method of Sainte-Marie
(1962). Sections 3 μ thick were cut, stained, and then mounted in a
liquid medium of polyvinyl alcohol as described by Rodriguez and
Deinhardt (1960).

Controls. Some control sections were either stained with the control
fluorescent reagent or left unstained, and were mounted as indicated
above. Other control sections were treated with excess specific nonfluorescent reagent and then with specific fluorescent reagent to
demonstrate the specificity of the fluorescent stain localization.

TABLE 1

DETECTION OF LENS ANTIGENS WITH THE FLUORESCENT TECHNIQUE IN THE
LENS-REGENERATING CELL POPULATION

Sato stage[a]	Days after lens removal	Fluorescence[b]	Localization
0	0	− (4/4)	—
I	5	− (4/4)	—
II	10	− (3/3)	—
III	12, 15, 16, 17	− (5/5)	—
IV	12, 13, 14, 15	− (2/8)	
		+ (6/8)	Positive in the cytoplasm of a few cells of the inner wall
V	15, 16	+ (3/3)	Positive in the cytoplasm of several cells of the inner wall
VI	16	+ (7/7)	Same as at stage V
VII	16, 17	+ (4/4)	Positive in the cytoplasm of all cells of lens fiber area and intermediate zone
VIII	17, 18	+ (3/3)	Same as at stage VII
IX	18, 20	+ (6/6)	Same as at stage VII
X	25	+ (4/4)	Positive in the cytoplasm of all cells of regenerate
XI	25	+ (2/2)	Same as at stage X

[a] According to Sato (1940) and Stone and Steinitz (1953). The relation between
stage and time interval after lens removal shown here differs from that reported by
Yamada and Takata (1963), mainly because in the present work the operation was
bilateral rather than unilateral as in the earlier work.

[b] Information in parentheses gives the number of cases corresponding to signs
divided by the total number of cases tested.

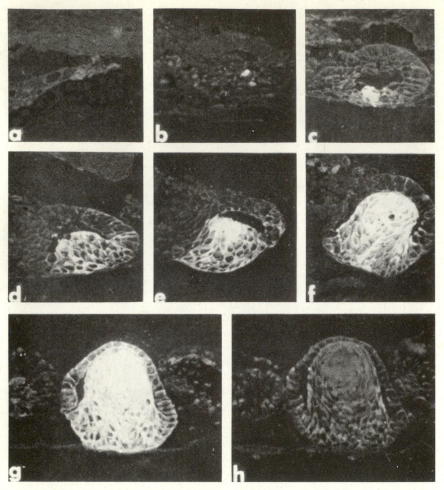

Fig. 1a. Fluorescent photomicrograph of the dorsal margin of normal pigmented iris, stained with the fluorescent globulin fraction of antilens serum. Immunofluorescence is negative in the pigmented iris. Magnification: × 145.

Fig. 1b–g. Fluorescent photomicrographs of various stages of lens regenarate stained with the fluorescent globulin fraction of lens antiserum. Magnification: × 145. (b) Stage IV regenerate. Two cells of the inner wall of lens vesicle are positive. (c) Stage V regenerate. The cytoplasm of several cells of the inner wall is positive. (d) Stage VI regenerate. Fluorescence in the lens fiber cells. (e) Stage VII regenerate. Fluorescence in the lens fiber cells and in the cells of intermediate zone. (f) Stage VIII regenerate. Same as in (e), and in addition, indication of fluorescence in the lens epithelium cells. (g) Stage X regenerate. All cells, including the lens epithelium cells, are positive.

Fig. 1h. A control section for Fig. 1g. Fluorescent photomicrograph of an adjoining section treated with control fluorescent globulin fraction of normal serum. Absence of fluorescence in lens. Magnification: × 145.

388

Fluorescence microscope. A large Zeiss fluorescence microscope equipped with a high-pressure mercury burner HBO-200, exciter filters (Schott) BG 12, UG 5, and UGl, and barrier filter 47 was used. Photographs were made on 35-mm Tri-X Pan Film (Kodak) or High Speed Ektachrome Film (Kodak).

RESULTS

The data obtained from sections treated with the specific immunofluorescent reagent are summarized in Table 1. No fluorescence was indicated in the normal pigmented iris (Fig. 1a) or regenerating irises of stages II and III. At stage IV some of the regenerates showed fluorescence clearly localized in the cytoplasm of a small number of cells present in the internal layer of the lens vesicle (Fig. 1b; see also Fig. 2a). At stages V and VI, fluorescence was observed in all regenerates studied and was localized again in the cytoplasm of cells present in the internal layer of lens vesicle (Fig. 1c and d; see also Fig. 2b). The intensity of fluorescence and the number of positive cells increased between stages IV and VI. According to their location, these positive cells correspond to the prospective primary lens fiber cells.

At stages VII and VIII the area of lens fibers becomes well defined (Fig. 3a). Fluorescence was further intensified in this area and was shared by a larger number of cells (Table 1, Fig. 1e and f). All lens fiber cells displayed cytoplasmic fluorescence. In contrast, most of the cells in the lens epithelium were negative. The positive cells in the zone intermediate between lens epithelium and lens fiber areas were probably the cells at the earliest phase of lens fiber differentiation (Fig. 1e and f). The pattern of fluorescence remained the same at stage IX (Fig. 4).

During stages VIII–IX the lens epithelium of the regenerate is often connected with the dorsal margin of the pigmented iris (Fig. 3a) by a stalk-like tissue consisting of several depigmented cells and probably acting as a source of new cells for the lens epithelium. All cells of the stalk were negative.

At stage X, the lens epithelium is one cell thick and the primary lens fiber cells which make up the core of the lens fiber area are surrounded by younger secondary lens fiber cells (Fig. 3b). Specific immunofluorescence was strongly positive in the primary lens fiber cells, and moderately positive in the secondary lens fiber cells (Fig.

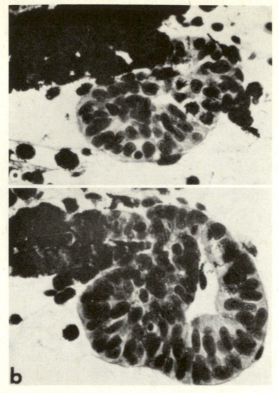

FIGS. 2 and 3. Histological sections of lens regenerate at various stages, as a reference for fluorescent photographs. Stained with Mayer's hemalum and picroni-grosin. Magnification: × 280.

FIG. 2. (a) Stage IV regenerate. Depigmented cells have formed a lens vesicle. Several mitotic figures. Upper portion, external; lower portion, internal side of the eye. (b) Stage VI regenerate. A lens vesicle with the internal (lower) wall made up of many elongated cells (prospective primary lens fiber cells). The central cavity is partly obliterated by a group of cells.

1g). Also, all cells of the lens epithelium indicated weakly positive cytoplasm which formed a thin layer surrounding the oval nucleus.

At no stage was fluorescence detectable in the areas of pigmented iris of the operated eye which do not directly participate in lens formation (Figs. 1b–g), when the specific immunofluorescent reagent was applied to tissue samples.

Besides the lens-forming tissue, fluorescence occurred in corneal

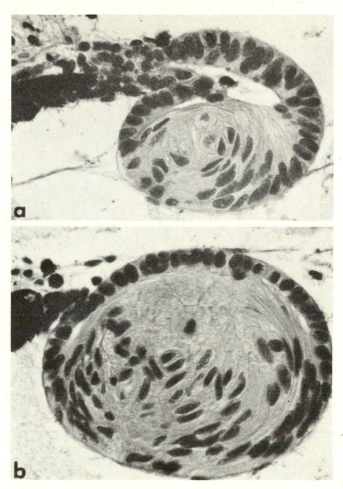

FIG. 3. (a) Stage VIII regenerate. The internal wall of the lens vesicle has
thickened owing to growth of cytoplasm of primary lens fiber cells. Connection
of lens epithelium with the dorsal margin of iris through a stalk. Formation of
secondary lens fiber cells in the zone intermediate between epithelium and fiber
areas. (b) Stage X regenerate. Increase in the area of lens fibers caused by
formation of secondary lens fiber cells and cytoplasmic growth of individual lens
fiber cells. Lens epithelium is now one cell thick.

epithelium, and, in contrast to that consistently occurring in the lens-forming tissue, varied a great deal in intensity and was absent in many cases. At this point we are reserving our decision on whether this occurrence of fluorescence in the corneal epithelium indicates the presence of lens antigens or whether it is an artifact. The question is being studied with other immunochemical techniques.

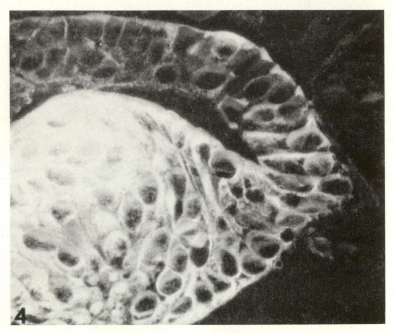

FIG. 4. Fluorescent photomicrograph of a Stage IX regenerate at a higher magnification. Upper left, negative lens epithelium cells. Upper middle, some positive lens epithelium cells. Upper right, positive cells of intermediate zone. Lower half, lens fiber cells with positive cytoplasm. Magnification: × 350.

When the control fluorescent reagent was added to the sections, no fluorescence was noticed in the normal pigmented iris, regenerating pigmented iris, or regenerates of all stages investigated (Fig. 1h). No sign of natural fluorescence was found in those tissues before the immunofluorescent reagent was added. Further, no fluorescence appeared in the sections first treated with excess specific nonfluorescent reagent and then with the specific immunofluorescent reagent.

DISCUSSION

Some information is already available on the immunochemical changes in Wolffian lens regeneration. Titova (1957), who used the anaphylactic technique to study lens antigens in the lens-regenerating system of *Triturus taeniatus*, concluded that they are absent in the 5th-day regenerate but present in the 15th-day regenerate, which, according to an accompanying figure, appears to correspond to stages IX–X. Vyazov and Sazhina (1961) investigated the same lens-regenerating system by the capillary micromethod of double diffusion in agar, testing extracts from the regenerates with rabbit antiserum against the lens extract of larval *Rana temporaria*. The extract from the 5th-day regenerate was negative, but the extract of the 7th-day regenerate gave a diffuse reaction (around stage IV).[2] On the 11th day, when well-defined primary lens fibers are indicated (stages VII–VIII),[2] one clear band appeared. The number of bands increased from 2 to 4 during the 15th to 30th day of regeneration, which may correspond to stages IX–X and XI–XII,[2] respectively.

In his immunofluorescence study of Wolffian lens regeneration in *Triturus pyrrhogaster*, Ogawa (1963) isolated the iris tissue together with the regenerating cell population and treated with the antilens serum conjugated with fluorescein isocyanate and absorbed with iris. He obtained negative results with the control normal iris and 5th-day regenerate, the first sign of positive reaction with the 10th-day regenerate, which appears to correspond to stages II–III, and positive reactions with 15th- to 35th-day regenerate. Since his tests were of unsectioned tissues, he could not elaborate the histological localization of fluorescence. The first sign of reaction he obtained with the immunodiffusion technique was also on the 10th day of regeneration.

We found that in the lens-regenerating system of adult *T. viridescens* the lens-specific antigens become detectable at stages IV–V, in the prospective primary lens fiber cells when they are in the first step of cell elongation. As cell elongation and growth of primary lens fiber cells progressed, lens antigens increased in the cytoplasm. When the secondary lens fiber cells were formed from the intermediate zone between lens epithelium and fiber areas, the same

---

[2] The Sato stages indicated are deduced from the descriptions contained in the papers of Vyazov and Sazhina (1961) and Titova (1957).

sequence of events occurred in their lens antigen composition. All lens fiber cells contained lens antigens in their cytoplasm. In old lens fiber cells, in which the nuclei were losing their basophilic stainability, the cytoplasm showed strong fluorescence. Before stage IX most of the lens epithelium cells did not contain a detectable amount of lens antigens. After stage X, when the lens epithelium becomes one cell thick, all lens epithelium cells contained lens antigens in their cytoplasm. The question of whether or not the lens antigens in the epithelium cells are qualitatively different from those in the fiber cells is under investigation.

To construct the sequence of events occurring during lens differentiation at the subcellular level, we must relate the present data with previous cell biological and cytochemical information obtained in this laboratory:

1. According to autoradiographic studies with H^3-uridine and H^3-cytidine and controlled with ribonuclease, RNA synthesis in the iris cell nucleus is enhanced after lens removal (Yamada and Karasaki, 1963; unpublished data), the enhancement continuing throughout lens regeneration in the cell population which directly participates in lens formation. Transfer of RNA synthesized in the nucleus into the cytoplasm as judged with autoradiography is very slow in stage I but is progressively accelerated during stages II–VI. This acceleration occurs in parallel with the increase in the frequency of ribosomes per unit cytoplasmic area as demonstrated on electron micrographs (Karasaki, 1964). We interpret these observations to indicate establishment of new protein-synthesizing apparatus in the cytoplasm during this phase of regeneration. At a later time in this phase, lens antigens are detected in the cytoplasm, which is in conformity with the idea that the pattern of transcription of genetic information is altered before lens specificity is acquired by the regenerate at the level of antigens.

2. In autoradiographic studies utilizing H^3-thymidine (Eisenberg, unpublished), DNA synthesis is activated soon after lens removal in the iris cells and is evident in the nuclei of all regenerate cells until stage III. At stages IV–V, however, when the lens vesicle is formed, the cells localized in the internal layer of the vesicle stop synthesizing DNA. These are the prospective primary lens fiber cells, in which lens antigens were first detected in our present study. Such observations are consistent with the idea that replication of genetic material precedes transcription of genetic information hitherto not utilized in the cells.

3. A study of protein synthesis with autoradiography suggested two periods in which the level of activity is significantly increased (Yamada and Takata, 1963). During the first period, which corresponds to stages II and III and leads to a lower level of enhancement of protein synthesis, we found no sign of lens antigens in the regenerating system in the present study. The second period is characterized by strong cytoplasmic participation and a higher level of enhancement of the synthetic activity. The period of synthesis of lens antigens as suggested by the present study coincides with the second period. This would indicate that in the early phase of lens regeneration a considerable amount of proteins may be synthesized which lack lens specificity detectable by the present methods.

In the present experiment, the test for lens antigens was negative in all areas of iris tissue of operated eye that do not participate directly in lens formation. This applied for all regeneration stages studied here. On the other hand, it has been shown (Yamada and Takata, 1963) that also in those iris areas negative for lens antigens, protein synthesis is activated by lens removal, as are syntheses of DNA and RNA (Eisenberg, unpublished; Yamada and Karasaki, 1963). This activation temporally corresponds to the first phase of protein enhancement observed in the lens-forming cell population, but not to its second phase, which alone indicates lens specificity.

In various vertebrates, lens antigens in the normal iris, cornea, retina, etc., are reported (Clarke and Fowler, 1960; Clayton, 1954, 1960; Clayton and Feldman, 1957; Flickinger and Stone, 1960; Maisel and Langman, 1961; Maisel and Harmison, 1963; Woerdeman, 1961). It is often asserted that the capacity of iris and other optic tissues to transform into lens depends upon the presence of lens antigens in those tissues. On the basis of their findings in the chick, Maisel and Langman (1961) and Maisel and Harmison (1963) suggest that alpha crystallin is essential for Wolffian lens regenration. In our experiments no lens antigens were detected in the normal adult newt iris. Vyazov and Sazhina (1961) also found no evidence of lens antigens in the normal iris. On the other hand, Ogawa (1963) observed a weakly positive reaction when he tested the normal iris extract with an anti-lens serum. He did not specify, however, whether the reaction was due to a lens specific antigen.

In a similar immunofluorescent study on the developing lens of the embryos of *T. pyrrhogaster* (Takata, Albright, and Yamada, in press), the mode of appearance and the subsequent distribution of lens antigens observed in the lens regenerate appear to correspond very closely to those in the normal embryonic development of the lens.

SUMMARY

Appearance and localization of lens antigens in the cell population engaged in Wolffian lens regeneration in adult *Triturus viridescens* was studied with the immunofluorescent technique. Paraffin sections of the regenerating cell population at various stages of regeneration were treated with the γ-globulin fraction of rabbit antiserum against newt lens, conjugated with fluorescein isothiocyanate.

The normal pigmented iris and regenerates of Sato stages II and III were completely negative. At stage IV some of the regenerates indicated cytoplasmic fluorescence in the prospective primary lens fiber cells. In subsequent stages the cytoplasm of lens fiber cells always showed fluorescence, while the majority of the lens epithelium cells were negative. At stage X, when the lens epithelium becomes one cell thick, the cytoplasm of all lens epithelium cells was positive, as was that of all lens fiber cells.

The data are discussed in connection with previous cell biological data obtained on the same system.

We are indebted to Dr. Toshihiko Sado for his help in fluorescent procedures, and to Mr. Tom W. Evans, Mrs. Lola M. Kyte, and Miss Marion E. Roesel for their technical assistance.

REFERENCES

CLARKE, W. M., and FOWLER, I. (1960). The inhibition of lens-inducing capacity of the optic vesicle with adult lens antisera. *Develop. Biol.* **2**, 155–172.

CLAYTON, R. M. (1954). Localization of embryonic antigens by antisera labelled with fluorescent dyes. *Nature* **174**, 1059.

CLAYTON, R. M. (1960). Labelled antibodies in the study of differentiation. *In* "New Approaches in Cell Biology" (P. M. B. Walker, ed.), pp. 67–88. Academic Press, New York.

CLAYTON, R. M., and FELDMAN, M. (1957). Detection of antigens in the embryo by labelled antisera. *Experientia* **11**, 29–31.

FLICKINGER, R. A., and STONE, G. (1960). Localization of lens antigens in developing frog embryo. *Exptl. Cell Res.* **21**, 541–547.

KARASAKI, S. (1964). An electron microscopic study of Wolffian lens regeneration in the adult newt. *J. Ultrastruct. Res.* in press.

MAISEL, H., and HARMISON, C. (1963). An immunoembryological study of the chick iris. *J. Embryol. Exptl. Morphol.* **11**, 483–491.

MAISEL, H., and LANGMAN, J. (1961). Lens proteins in various tissues of the chick eye and in the lens of animals throughout the vertebrate series. *Anat. Record* **140**, 183–193.

MARSHALL, J. D., EVELAND, W. C., and SMITH, C. W. (1958). Superiority of fluorescein isothiocyanate (Riggs) for fluorescent-antibody technic with a modification of its application. *Proc. Soc. Exptl. Biol. Med.* **98**, 898–900.

OGAWA, T. (1963). Appearance and localization of lens antigens during the lens regeneration in the adult newt. *Embryologia* **7**, 279–284.

RODRIGUEZ, J., and DEINHARDT, F. (1960). Preparation of a semipermanent mounting medium for fluorescent antibody studies. *Virology* **12**, 316–317.

SAINTE-MARIE, G. (1962). A paraffin embedding technique for studies employing immunofluorescence. *J. Histochem. Cytochem.* **10**, 250–256.

SATO, T. (1940). Vergleichende Studien über die Geschwindigkeit der Wolff'schen Linsenregeneration bei *Triton taeniatus* und bei *Diemyctylus pyrrhogaster*. *Arch. Entwicklungsmech. Organ.* **140**, 570–613.

STONE, L. S., and STEINITZ, H. (1953). The regeneration of lenses in eyes with intact and regenerating retina in adult *Triturus v. viridescens*. *J. Exptl. Zool.* **124**, 435–468.

TAKATA, C., ALBRIGHT, J. F., and YAMADA, T. (in press). Study of lens antigen in the developing newt lens with immunofluorescence. *Exptl. Cell Res.*

TITOVA, I. I. (1957). A study of the antigenic properties of the crystalline lens in Wolffian regeneration. *Bull. Exptl. Biol. Med.* (*USSR*) (*English Trans.*) **43**, 715–719.

VYAZOV, O. E., and SAZHINA, M. V. (1961). Immuno-biological study of the process of lens regeneration in *Triton taeniatus*. *Zh. Obsch. Biol.* **4**, 305–310.

WOERDEMAN, M. W. (1961). Eye-lens development and some of its problems. *Koninkl. Ned. Akad. Wetenschap. Proc. Ser. C* **65**, 145–159.

YAMADA, T., and KARASAKI, S. (1963). Nuclear RNA synthesis in newt iris cells engaged in regenerative transformation into lens cells. *Develop. Biol.* **7**, 595–604.

YAMADA, T., and TAKATA, C. (1963). An autoradiographic study of protein synthesis in regenerative tissue transformation of iris into lens in the newt. *Develop. Biol.* **8**, 358–369.

Reprinted from *Develop. Biol.*, **14**, 214–245 (1966)

15

The Influence of Neural Retina and Lens on Lens Regeneration from Dorsal Iris Implants in *Triturus viridescens* Larvae

Randall W. Reyer[1]

Department of Anatomy, School of Medicine, Medical Center, West Virginia University, Morgantown, West Virginia 26506

Accepted March 16, 1966

INTRODUCTION

When the lens is excised from the eye in certain species of urodeles, it is replaced by a lens which develops from the pupillary margin of the dorsal iris. This process is initiated by depigmentation and dedifferentiation involving the loss of the smooth-surfaced endoplasmic reticulum. Synthesis of RNA is enhanced together with an increase in both the number and size of the nucleoli and in the number of ribosomes. DNA synthesis and cell proliferation also take place, resulting in the formation of a depigmented lens vesicle attached to the pupillary margin of the dorsal iris. From this vesicle a new lens develops by the differentiation of elongate lens fibers from those cells facing the neural retina. This is characterized by the synthesis of lens-specific proteins, including γ crystallins (Eguchi, 1963a,b; 1964; Karasaki, 1964; Takata, 1952; Takata *et al.*, 1964, 1965; Yamada and Karasaki, 1963; Yamada and Takata, 1963, 1964; Yamada *et al.*, 1964). The importance of RNA synthesis in these changes is shown by the sensitivity of lens regeneration to the inhibitory action of actinomycin D (Yamada and Roesel, 1964). The initial stimulus for these events comes from the neural retina (evidence in reviews by Reyer, 1954, 1962; Stone, 1959, 1965; Goss, 1964).

[1] This investigation was supported by research grant (NB-01544) from the National Institute of Neurological Diseases and Blindness of the National Institutes of Health, U. S. Public Health Service.

214

Under most conditions, lens regeneration is completely blocked by the presence of a healthy lens of normal size in the pupil of the eye. Thus, the events of regeneration are not initiated until the lens has been removed, and they are prevented if the lens is extirpated and then immediately reimplanted. However, more or less inhibited lens regeneration from the iris has been reported following several experimental procedures where the host lens was displaced so that there was a space between it and the pupillary margin of the dorsal iris (Spirito and Ciaccio, 1931; Ikeda and Kojima, 1940; Ikeda and Amatatu, 1941; Uno, 1943; Zalokar, 1944; Stone, 1953, 1965; Takano et al., 1958; Reyer, 1961, 1962; Eguchi, 1961). Pieces of dorsal iris which will regenerate a lens when implanted into the pupil of the lentectomized eye fail to do so when located in the anterior chamber in the presence of the host lens. When iris is combined in the same eye with a regenerating lens of sufficient size, then lens development from the iris may be partially or completely inhibited (Wachs, 1914; Sato, 1930, 1935; Mikami, 1941; Stone, 1952; Frost, 1961). Attempts to demonstrate that this inhibitory action of the lens is mediated by some factor that is present in the aqueous humor have yielded conflicting results so that no certain conclusion can be reached at this time (Stone and Vultee, 1949; Stone, 1963; Takano et al., 1957; Goss, 1960–1961; reviewed by Goss, 1964, and Stone, 1965). Extracts of the whole lens have shown no inhibitory action (Takano et al., 1957; Goss, 1964), but Smith (1965) observed retardation of lens regeneration following the implantation of a plug of starch gel containing a fraction of an extract of newt lens which had been obtained by starch gel electrophoresis.

Only a few experiments have been performed in which implants of dorsal iris were inserted into the vitreous chamber of an eye still retaining its own normal lens. Uno (1943) reported that an inhibited lentoid developed from 13–14% of implants of dorsal iris into the vitreous chamber with intact host lens in *Triturus pyrrhogaster* larvae, whereas Mikami (1941) observed the regeneration of a lens vesicle in a single case. Takano (1959) cut off the dorsal iris and pushed it into the vitreous chamber dorsal to the lens of the same eye in adults of this species. These implants came to lie close to the remainder of the dorsal iris. Lenses with fibers were found in five out of the 15 cases where the implant was recovered. One of these was attached to the *in situ* dorsal iris, three were attached to the area where the *in situ* and severed iris had healed together, and one lens of stage 11 was in contact only with the implanted iris. For a further discussion of the role of the lens in the inhibition of lens regeneration, see reviews by Stone (1959, 1965) and Reyer (1954, 1962).

The experiments to be reported here represent a study of the differentiation and growth of lenses regenerating from pieces of iris situated in a location within the eye where they can be influenced by the action of both the host neural retina and the host lens. These results are compared with implants of the same tissue to an extraocular site where neither of the above factors is acting and to ocular sites where only one of these tissues is in close proximity to the iris implant. Summaries of the experiments to be described here have been reported during the course of the work in an abstract (Reyer, 1961) and in two symposia (Reyer, 1962, 1965).

MATERIALS AND METHODS

The larvae of *Triturus viridescens* used in these experiments were obtained from eggs laid in the laboratory by females collected locally or in Massachusetts, either in the fall or spring. These adult females were kept in a cold room at 6–10°C until needed. Ovulation was induced by implantation of anterior pituitary glands from *Rana pipiens*. The eggs and larvae were kept at a temperature of 12–15°C or at room temperature. After operation, the experimental animals remained at a room temperature of approximately 21–24°C. Food consisted of brine shrimp (*Artemia*) and white worms (*Enchytraeus*).

Operations were usually performed in full-strength Holtfreter's solution, without bicarbonate, which was buffered with Tris-HCl to a pH of 7.1–7.4. In a few of the early experiments, a bicarbonate buffer was used instead (Reyer, 1948). During the operation, the animals were anesthetized in 0.05% chloretone in Holtfreter's solution. Recovery and wound healing took place overnight in full-strength Holtfreter's solution after which the animals were cultured in aquarium water or in tap water treated with sodium thiosulfate to remove the chlorine. Fixation was in Zenker-acetic or Bouin's fluid at frequent intervals from 10 to 35 days after operation. All cases were studied in 10-μ serial sections stained with hematoxylin and erythrosin.

The donors and hosts were old larvae of *Triturus viridescens* with well developed hindlimbs. The cornea and lens were removed from the donor eye (Fig. 1A), and a sector of middorsal iris was excised

HOST

DONOR

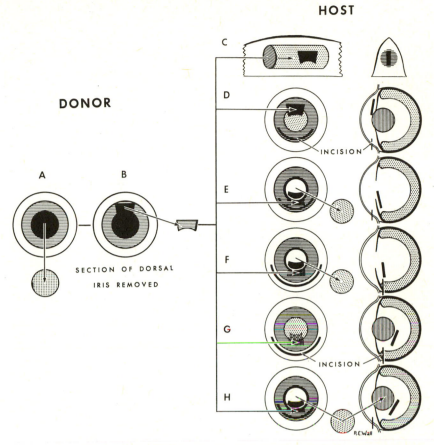

SECTION OF DORSAL
IRIS REMOVED

Fig. 1. Diagram to show the method of operation and the sites of implantation of a sector of dorsal iris from a *Triturus viridescens* larva into a host larva: A, lens removed from donor eye; B, sector of middorsal iris excised for implantation; C, implant in dorsal fin; D, implant in anterior chamber of eye; E, implant in vitreous chamber, host lens removed through cornea; F, implant in vitreous chamber, host lens removed through ventral retina; G, implant in vitreous chamber, host lens *in situ*, incision through ventral retina; H, implant in vitreous chamber, host lens removed and reimplanted, incision through cornea. From Fig. 1, R. W. Reyer (1965).

with iridectomy scissors (Fig. 1B). This piece of iris was then implanted into one of four different sites in host larvae as follows: (1) a pocket in the dorsal fin (Fig. 1C); (2) the anterior chamber of the

eye with the host lens present (Fig. 1D); (3) the vitreous chamber
of the eye after removal of the host lens either through an incision in
the cornea (Fig. 1E) or through an incision in the ventral, or occasionally dorsal, neural and pigmented retina (Fig. 1F); (4) the
vitreous chamber with the host lens *in situ* (Fig. 1G) or after removal
of the host lens through the cornea followed by its reimplantation in
the pupil (Fig. 1H). In order to distinguish any differences in the
response of the implanted iris due to variations between groups of
animals from that due to different implantation sites, the implants into
the eyes were made in pairs. In a particular series of experiments, a
piece of dorsal iris was removed from the right and left eye of each
donor. One piece was implanted into a given site in the right eye of
the host and the other piece was implanted into a different site in the
left eye of the same host. These two sites remained the same for all of
the operations in this series.

RESULTS

The combined data for the six different implantation sites and types
of operation (one dorsal fin and five intraocular) are summarized in
Table 1.

Implants into the Dorsal Fin

Larval dorsal iris usually changed very little when implanted into
a pocket in the dorsal fin of another host larva. Although the epithelial
cells frequently increased in height, they remained completely pigmented. Occasionally, a few partly depigmented cells were observed
while two cases had formed well developed, unpigmented neural
retina which had probably differentiated from cells of the *pars ciliaris
retinae* included with the transplant. No indication of lens formation
from the iris was observed (Fig. 2). These results confirm those of
Stone (1958a,b) and Stone and Gallagher (1958) on adult iris transplants in demonstrating that a factor in the intraocular environment is
necessary for the initiation of lens regeneration from iris tissue.

Implants into the Eye Chamber

The data for the intraocular implants were derived from six series of
operations employing the following paired combinations of implantation sites: *series a:* anterior chamber, lens present *vs.* vitreous chamber,
lens removed through cornea; *series b:* vitreous chamber, lens removed

TABLE 1

SUMMARY: IMPLANTS OF DORSAL IRIS FROM *Triturus viridescens* LARVAE INTO THE DORSAL FIN AND DIFFERENT INTRAOCULAR SITES IN HOST LARVAE

Site of implantation and type of operation	Implants made	Implants recovered	Structures formed by implant										Host iris		
			Pigmented epithelium only	Pigmented and partly depigmented epithelium only	Pigmented and completely depigmented epithelium only	Neural retina	Lens vesicle	Lens with intact fibers	Lens with vacuolated fibers	Lens with herniated fibers	Small fiber lentoid	Lens with degenerating fibers	No change	Depigmentation or lens vesicle	Lens with fibers
1. Dorsal fin	18	18	12	4	0	2	0	0	0	0	0	0	—	—	—
2. Anterior chamber, lens present	33	32	20	9	3	0	0	0	0	0	0	0	31	1	1
3. Vitreous, lentect., cornea	23	23	0	3	3	0	2	13	1	0	1	0	0	3	20
4. Vitreous, lentect., retina	48	48[a]	0	4	5	3	6	26	4	1	2	2	0	4	44
5. Vitreous, lens reimplanted	47	41	0	18	7	1	3	7	4	0	1	0	40	7	0
6. Vitreous, lens *in situ*	68	68[b]	0	10	24	5	4	21	1	0	6	0	68	0	0
Total	237	230													

[a] Three cases with 2 lenses or lentoids; also neural retina in one of these; 1 case with lens vesicle and lens with degenerating fibers.
[b] Three cases with neural retina and lens or lentoid.

through cornea *vs.* vitreous chamber, lens removed and reimplanted; *series c:* anterior chamber, lens present *vs.* vitreous chamber, lens removed and reimplanted; *series d:* anterior chamber, lens *in situ vs.* vitreous chamber, lens *in situ; series e:* vitreous chamber, lens removed through ventral eyeball *vs.* vitreous chamber, lens *in situ; series f:* vitreous chamber, lens removed and reimplanted *vs.* vitreous chamber, lens *in situ.* The variable factors between the operations on the two eyes of the same host for each series are: *series a,* presence or absence of lens and position of implant; *series b and e,* presence or absence of lens; *series c and d,* position of implant; *series f,* treatment of host lens. When the implant was placed in the vitreous chamber, the incision was made through the cornea in series a, b, c, and one-half of f. The host lens was either removed permanently or removed and then reimplanted after the piece of iris from the donor had been inserted into the vitreous body through the pupil. In séries d, e, and one-half of f, the incision was made through the ventral eyeball for implants into the vitreous. The host lens was either removed through this incision or left intact; the piece of iris was inserted into the vitreous body deep to the *in situ* lens or in place of the extirpated lens. In Table 2 are tabulated the number of implants recovered in each of the paired locations for series a–f.

The results from the paired groups of control and experimental operations showed that there were differences in the behavior of the iris implants in the three different intraocular locations: anterior chamber, host lens present; vitreous chamber, host lens absent; vitreous chamber, host lens present. In a particular site, the implants responded similarly in the several different series. Therefore, the results for each site have been pooled for further consideration.

Implants into the Anterior Chamber

When placed in the anterior chamber of a host eye with intact lens, a piece of dorsal iris usually remained as a two-layered, densely pigmented epithelium which was often folded together. Figure 8–39 in Reyer (1962) shows a pigmented iris implant lying in the anterior chamber, 20 days after operation. This resembles the case of the iris isolated in the dorsal fin (Fig. 2). Partial depigmentation of a few cells was observed in nine implants (Fig. 3). In these cells, the pigment had dispersed sufficiently to reveal the nuclei. A group of completely depigmented cells appeared in three animals (Table 1, line 2). However, these depigmented cells had not formed the regular vesicle of columnar cells characteristic of initial stages of lens regeneration. They probably represent an early stage in neural retina differentiation. Therefore, the fate of larval, dorsal iris is the same when implanted either into the dorsal fin or into the anterior chamber of the eye with intact host lens. Differentiation of a lens from the iris does not occur in either location. In a few cases, neural retina regenera-

TABLE 2

NUMBER OF IMPLANTS RECOVERED IN THE SIX SERIES
OF PAIRED OPERATIONS $(a - f)^a$

Fig.	Fig. 1D anterior chamber lens present	Fig. 1E vitreous chamber lentect., cornea	Fig. 1F vitreous chamber lentect., retina	Fig. 1H vitreous chamber lens reimplanted	Fig. 1G vitreous chamber lens in situ
1D, anterior chamber lens present	—	4(a) 4	—	18(c) 18	10(d) 10
1E, vitreous chamber lentect., cornea		—	—	19(b) 16	—
1F, vitreous chamber lentect., retina			—	—	48(e) 47
1H, vitreous chamber lens reimplanted				—	7(f) 11
1G, vitreous chamber lens in situ					—

^{a} Implantation sites are listed at the top and to the left of the table. Each rectangle corresponds to the two implantation sites as indicated by the intersecting columns. The letter refers to the series. For each series, the number at the left of the rectangle gives the cases in the location designated at the left of the table; the number in the right-hand column gives the cases in the location designated at the top of the table.

tion may begin, probably from that part of the iris epithelium next to the *pars ciliaris retinae*.

No difference in the results was observed between the 21 operations where the host lens was left *in situ* and the 12 cases where it was removed and reimplanted. In the latter group, the reimplanted lens remained normal and there was no tendency for lens differentiation to occur from the implant. There were two cases, 13 and 20 days after removal and reimplantation of the host lens, where the pupillary

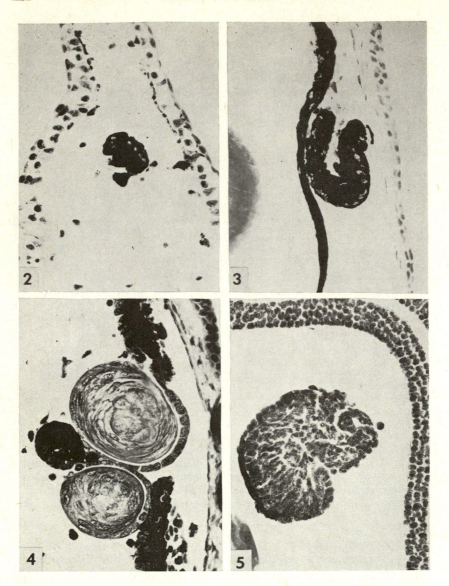

FIG. 2. Implant of dorsal iris into dorsal tail-fin of larva. It remained as pigmented iris epithelium, 20 days after operation. Magnification: × 225.

FIG. 3. Implant of iris into anterior chamber of left eye with intact lens. It formed pigmented iris epithelium a few cells of which were partly depigmented over the nuclei, 21 days after operation. Magnification: × 225.

margin of the host dorsal iris was depigmented. In the older example, two cells had initiated lens fiber differentiation. The host lens was somewhat opaque in the former case; it was clear but displaced a little bit ventrally in the latter one.

Implants into the Vitreous Chamber, Host Lens Removed

It has been known since the work of Wachs (1914) and Sato (1930) that a lens can regenerate not only from the intact dorsal iris of a lentectomized eye but also from an isolated piece of dorsal iris lying within the eye chamber. This has been confirmed by numerous other investigators (see Reyer, 1954, 1962 for a review). Similar operations were performed as controls in the experiments being reported here. Lens or lentoid regeneration from these implants of dorsal iris took place in 63% of the cases (Table 1, lines 3 and 4). Since a lens also regenerated from the dorsal iris of the host, two lenses were present in most of these eyes. Usually, the lens from the host iris was a little larger than that from the implant. In nine cases, the graft and host lenses were fused giving a single or partly double lens with two centers of lens fiber formation within a single lens capsule. In some other cases, there may have been a contribution by the graft to the single lens derived for the most part from the host iris. However, unless there was clear histological evidence of lens formation by the implant, these cases were classified as negative. There was very little difference in the frequency of lens regeneration following lens removal through the cornea or through the retina.

Examples of lens development from both host and grafted iris, following lens extirpation and iris implantation through the cornea in larval *Triturus viridescens*, have been illustrated in two published figures (Fig. 4, Reyer, 1956; Fig. 8-40, Reyer, 1962). Three other cases from *series e* are illustrated in Figs. 4, 10, and 15. In these examples,

---

FIG. 4. Host lens removed and iris implanted into vitreous chamber through ventral retina of left eye. Lens of regeneration stage 10½ from host dorsal iris (average diameter 160 μ) and lens of regeneration stage 10 from iris implant (average diameter 125 μ) lying ventral to it. Both lenses had primary and secondary lens fibers and a simple cuboidal lens epithelium slightly flattened where they were in contact, 20 days after operation. Magnification: × 195.

FIG. 5. Host lens removed through cornea of right eye, iris implanted through pupil into vitreous chamber, and host lens reimplanted in pupil. Lens appeared normal; iris implant formed neural retina but no lens, 27 days after operation. Magnification: × 195.

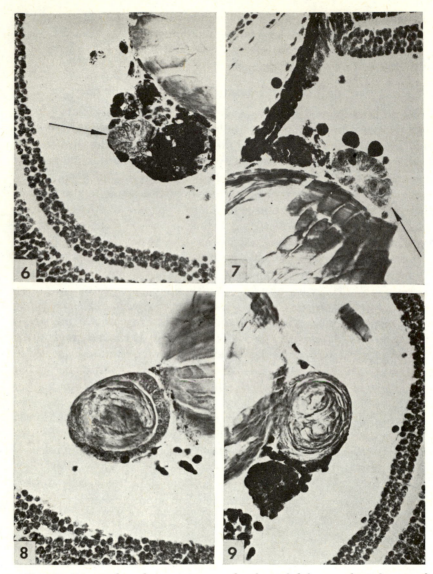

Fig. 6. Iris implanted into vitreous chamber of left eye through ventral retina, host lens intact. Implant formed a depigmented lens vesicle of regeneration stage 5 (indicated by arrow), 10 days after operation. Magnification × 195. (In this and the following cases, the host lens has been damaged during sectioning.)

Fig. 7. Host lens removed from right eye through cornea, piece of iris

the host lens had been removed through an incision in the ventral, pigmented and neural retina. The iris was then implanted through this incision. Therefore, the lens which developed from the implant was frequently located deeper in the vitreous chamber (Fig. 10) than when the iris had been inserted through the cornea.

Implants into the Vitreous Chamber, Host Lens Present

As described under Materials and Methods and in the section on the paired experimental and control series, two different techniques were used in placing the piece of dorsal iris in the vitreous chamber between the host lens and retina. In the first method, the host lens was removed, the iris implant inserted through the pupil into the vitreous body and the host lens then replaced in the pupil (Fig. 1H). An effort was made to orient the reimplanted host lens normally in the pupil. However, this may not have been successful in all instances. A lens with fibers or fiber lentoid regenerated from 29% of the implants retained in the eye (Table 1, line 5). The reimplanted host lens appeared to be clear and normal in all but three cases where there was an opacity in the center of the lens. No changes were observed in the host iris in most specimens. However, depigmentation of a few cells at the pupillary margin of the dorsal iris occurred in seven cases. These represented inhibited attempts at initiating lens regeneration. Five of these were found in eyes where the reimplanted lens had been displaced ventrally leaving a small space between it and the dorsal iris (Fig. 14).

Although most of the host lenses appeared to be completely normal in these experiments, there was a possibility that surgical manipulation

---

implanted into vitreous through pupil and lens reimplanted. Fifteen days after operation, the host lens was slightly less transparent than normal, but no abnormalities were observed in sections. Implant lay dorsal to host lens and formed a lens with initial differentiation of lens fibers (indicated by arrow), regeneration stage 7. Magnification: × 195.

FIG. 8.   Operation as described for Fig. 6. Implant formed a lens with primary and secondary lens fibers and a simple cuboidal lens epithelium; regeneration stage 10 (average diameter 140 μ, 20 days after operation. Magnification: × 195.

FIG. 9.   Operation as described for Fig. 6, on host right eye. Implant formed a lens with primary and secondary lens fibers and a simple squamous lens epithelium which faced away from the retina and was in contact with the host lens where it became obscured by fibers from the shattered host lens; regeneration stage 10 (average diameter 120 μ, 20 days after operation. Magnification: × 195.

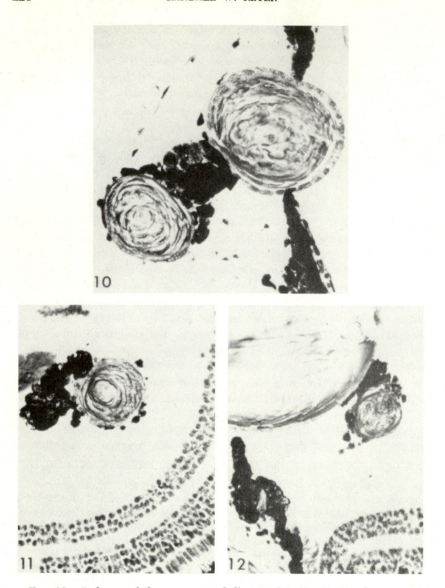

FIG. 10.   Left eye of the same animal illustrated in Fig. 9. Host lens removed and iris implanted into vitreous chamber through ventral retina. A lens of stage 11 regenerated from the host dorsal iris (average diameter 185 μ) and a lens of stage 10 from the implanted iris (average diameter 125 μ). The latter was located quite deep within the vitreous chamber and had a simple squamous lens epithelium facing the pupil. Magnification: $\times$ 195.

might have had some effect on their metabolism sufficient to cause a change in their influence on lens regeneration from the dorsal iris. A different procedure was therefore employed in which the host lens remained undisturbed and the piece of dorsal iris was inserted into the vitreous body through a ventral incision in the pigmented and neural retina (Fig. 1G). Under these conditions, a lens or fiber lentoid regenerated from 41% of the implants recovered (Table 1, line 6). The host lens was normal in all instances and no changes were observed in the host dorsal iris. The results with the two techniques were therefore quite similar. Two examples of lens regeneration from dorsal iris implants in the presence of the host lens have been illustrated in Reyer (1962, Figs. 8-41 and 8-42).

None of the implants of iris remained entirely composed of pigmented cells. Either a few cells showed initial stages in depigmentation so that the nuclei were more clearly visible and the cytoplasm contained discrete melanin granules or else greater or smaller amounts of completely depigmented epithelium developed. In a few cases, recognizable neural retina differentiated from the iris (Fig. 5). The differentiation of a lens from the iris tissue lying in the vitreous body between the host lens and the neural retina showed, for the most part, the same well known series of morphological changes as those observed during lens regeneration from iris tissue implanted after lentectomy. First, there was depigmentation and an increase in height of the cells of the pupillary margin resulting in the formation of a hollow vesicle of depigmented columnar cells (Fig. 6). Then lens fiber differentiation began in the cells of the vesicle facing the neural retina as evidenced by their elongation and increasing acidophilia (Fig. 7) leaving the cells of the vesicle facing away from the neural retina to form the lens epithelium. The cells of the lens epithelium continued to proliferate and, at the equatorial zone, differentiated into secondary lens fibers which soon completely enclosed the first-formed

---

FIG. 11. Iris implanted into vitreous chamber of right eye through ventral retina, host lens intact. Implant formed a lens with primary and secondary lens fibers, but with a deficient lens epithelium composed of only a few scattered, squamous epithelial cells; regeneration stage 10 (average diameter 100 μ), 15 days epithelium facing the pupil. Magnification: $\times$ 195.

FIG. 12. Lens removed from right eye of host through cornea, iris implanted into vitreous chamber through pupil, and lens reimplanted. Host lens appeared clear and normal. Implant formed a small lens of irregular fibers, lacking a lens epithelium (average diameter 70 μ), 20 days after operation. Magnification: $\times$ 195.

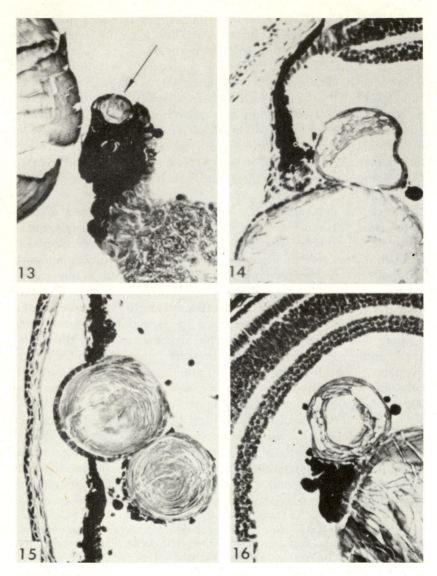

Fig. 13. Iris implanted into vitreous chamber of right eye through ventral retina, host lens intact. Implant formed neural retina which fused with the host neural retina and a tiny fiber lentoid lacking a lens epithelium (indicated by arrow; average diameter of lentoid 38 μ), 25 days after operation. Magnification: $\times$ 225.

primary lens fibers (Fig. 8). As the lens enlarged, the cells of the lens epithelium became flatter, forming a simple cuboidal epithelium in the most normal cases.

A number of deviations from the normal pattern of lens regeneration were found among the lenses regenerating from iris implants. The most common of these involved primarily the lens epithelium and the polarity of the lens. As described above, the regenerating lens regularly attained a normal polarity with respect to the position of the lens epithelium and lens fibers when it had developed within the pupil either from the *in situ* or implanted iris. Regenerating lenses, located deeper in the vitreous body between the host lens and the neural retina, had a more variable and less well defined polarity. There was a graded series of abnormalities characterized by a thinning of the lens epithelium and a gradual reduction in the number of cells which had not started to differentiate into lens fibers. In its mildest expression, this abnormality consisted of a flattening of the lens epithelium so that it was composed of a simple squamous epithelium which still faced away from the neural retina, usually toward the host lens (Fig. 9). This condition was not restricted to eyes having a normal intact host lens but was also found among lenses developing from iris

FIG. 14.   Lens removed from host right eye through cornea, iris implanted into vitreous chamber through pupil, and lens reimplanted. Host lens remained clear and normal but was displaced a little bit ventrally in pupil, leaving a small space next to the pupillary margin of the host dorsal iris. This was occupied by a lens vesicle of stage 4 which was regenerating from the dorsal iris of the host. The implant was located behind the host dorsal iris and dorsal to the host lens. It regenerated a lens with extensive vacuolation and breakdown of lens fibers resulting in the formation of a large cavity, 25 days after operation. Magnification: × 150.

FIG. 15.   Host lens removed and iris implanted into vitreous chamber of right eye through ventral retina. A lens of regeneration stage 11 with normal, simple cuboidal lens epithelium (average diameter 190 μ) has regenerated from the host dorsal iris and a second lens of stage 11 (average diameter 165 μ) from the implant. The latter was located behind the ventral iris and ventral to the host lens. Its lens epithelium was deficient and composed of scattered, squamous epithelial cells. Rostrally, the lens fibers were vacuolated (very slight vacuolation in this section); fixed 25 days after operation. Magnification: × 150.

FIG. 16.   Same animal as illustrated in Fig. 15. Iris implanted into vitreous chamber of left eye through ventral retina, host lens intact. Implant lay deep and dorsal to host lens. It formed a lens with many large vacuoles in the primary and secondary lens fibers and a simple squamous epithelium (average diameter 150 μ). Magnification × 150.

implants which were located equally deep in the vitreous body following removal of the host lens (Fig. 10). A more severe manifestation of this abnormality resulted in a lens having an epithelium composed of scattered, discontinuous, squamous cells on the lateral surface of the secondary lens fibers (Figs. 11 and 15). Finally, other cases were observed where the regenerated lens consisted of a mass of lens fibers completely lacking a covering of lens epithelium (Fig. 12). In a few cases, only a few cells had differentiated into lens fibers, and a lens epithelium was also lacking. These were classified separately as small fiber lentoids (Table 1; Fig. 13). Since this lentoid type of inhibited lens differentiation occurred occasionally when the host lens was absent as well as when it was present, it cannot be attributed directly to an inhibiting effect of the host lens.

Another abnormality of lens differentiation from iris implants, observed both after removal of the host lens and with the host lens present, was the appearance of vacuoles in the lens fibers. These varied from numerous small vacuoles to a few large vacuoles replacing most of the lens fibers (Figs. 14–16). This condition was also found occasionally during lens regeneration from the *in situ* dorsal iris in larvae of *Triturus viridescens, Salamandra salamandra salamandra,* and *Pleurodeles waltlii* (Reyer, 1948, 1963; Reyer and Stone, 1955). Rupture of the lens vesicle and separation of the lens fibers from the lens epithelium by herniation as well as complete degeneration of the lens fibers were very seldom observed in the experiments being reported in this paper.

Growth Rates of Regenerating Lenses

Measurements were made of the size of the regenerating lenses with differentiating lens fibers including those developing from the host, dorsal iris and from iris implants both in the absence of and in the presence of the host lens. Most of the lenses were approximately spherical in shape, and some were spheroidal.

The *average diameter* was used as an index of size. In transverse sections through the center of the lens, the dorsoventral and mediolateral (anteroposterior in human terminology) diameters were measured with an ocular micrometer. The nasotemporal diameter was obtained by counting the number of 10 μ sections passing through the lens and multiplying by 10. These three figures were then averaged.

Figure 17 shows graphically the data obtained for the experiments where the incision was made through the cornea (group A). Average diameter of the regenerating lens in microns is plotted against days after operation. Figure 18 is a similar graph for the experiments where the incision into the eye chamber was made through the ventral retina

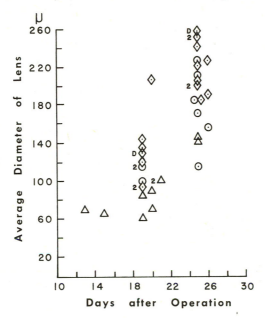

FIG. 17. Graph for group A. Average diameters of regenerating lenses in microns plotted against days after operation. Incision into eye through cornea. Diamond symbol: lens from host iris, host lens removed (treatment 4A). Circle: lens from iris implant into vitreous chamber, host lens removed (treatment 1A). Triangle: lens from iris implant into vitreous chamber, host lens removed and reimplanted (treatment 2A). Numbers to the left of symbols indicate the measurements which fell on a single point. D, a double lens, not included in statistics. (One of the two lenses, 250 μ in diameter, was also double.)

(group B). In order to provide a more accurate comparison of lens size and rate of growth under the three experimental conditions, a statistical analysis was made of the data using the method of covariance as described by Snedecor (1956). In the first analysis, the data for group A and the data for group B were considered separately. Computations were performed at the Computer Center of West Virginia University on an IBM 1620 computer programmed for the above procedure.[2] It was hypothesized that lens diameter (Y) increased with time after operation (X) and that there was a linear relationship between these two variables within the range of observations

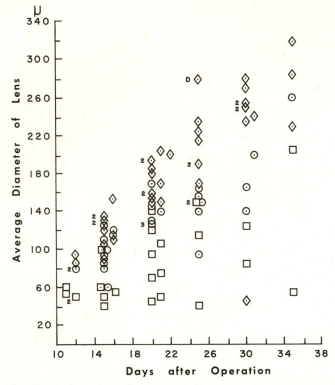

FIG. 18. Graph for group B. Average diameters of regenerating lenses in microns plotted against days after operation. Incision into eye through ventral, pigmented and neural retina. Diamond: lens from host iris, host lens removed (treatment 4B). Circle: lens from iris implant into vitreous chamber, host lens removed (treatment 1B). Square: lens from iris implant into vitreous chamber, host lens *in situ* (treatment 2B).

taken. Therefore, values for the mean lens diameter ($\bar{y}$), mean value for time ($\bar{x}$) and regression coefficient (b) were obtained for each of the three different experimental conditions as follows: *treatment 1*—

[2] The author is indebted to Dr. Albert E. Drake, acting Coordinator of Statistics, Division of Statistics, West Virginia University, who acted as the statistical consultant for these studies.

lenses regenerating from iris implants into the vitreous chamber after extirpation of the host lens; *treatment 2*—lenses regenerating from iris implants into the vitreous chamber with the host lens present; *treatment 4*—lenses regenerating from the host iris after extirpation of the host lens.

In order to determine whether these means and regression coefficients were significantly different or only represented chance deviations between different samples of the same population, the statistical values, obtained with *treatment 4*, were compared with those for *treatment 1* and, in a similar way, the *treatment 1* values were compared with those for *treatment 2*. In this computation, the mean lens size ($\bar{y}$) was first corrected for the difference in the average time after operation at which the measurements were made ($\bar{x}$) in the two treatments being compared, thus giving an adjusted mean of Y, ($\hat{\bar{y}}$). Variance ratios (F) of Fisher were then calculated from the variances of difference between the $\hat{\bar{y}}$'s, b's, and the appropriate error

TABLE 3

COMPARISON OF SIZES AND LINEAR GROWTH RATES OF LENSES REGENERATING FROM IRIS IMPLANTS IN THE PRESENCE AND ABSENCE OF THE HOST LENS AND FROM THE HOST DORSAL IRIS, GROUP A[a]

	Treatments compared		F	P	Conclusion
	4A	1A			
$\bar{x}$	22.8	23.1	—	—	—
$\bar{y}$	181	154	—	—	—
$\hat{\bar{y}}$	182	152	5.28	<0.05	$\hat{\bar{y}}_{4A} \neq \hat{\bar{y}}_{1A}$
b	13.48	10.41	0.47	NS	$b_{4A} = b_{1A}$
n	15	9	—	—	—
	1A	2A			
$\bar{x}$	23.1	19.8	—	—	—
$\bar{y}$	154	93	—	—	—
$\hat{\bar{y}}$	140	105	6.48	<0.025	$\hat{\bar{y}}_{1A} \neq \hat{\bar{y}}_{2A}$
b	10 41	6.66	0.93	NS	$b_{1A} = b_{2A}$
n	9	10	—	—	—

[a] Key to symbols used in Tables 3 to 6. $\bar{x}$, mean of X (days after operation); $\bar{y}$, mean of Y (average diameter of lens in microns); $\hat{\bar{y}}$, adjusted mean of Y; b, regression coefficient, Y on X; n, number of lenses measured; %, percentage of cases which regenerated a lens with fibers or a fiber lentoid; F, variance ratio of Fisher; P, probability that the difference between the statistics being compared is due to chance alone. Conclusion: rejection or acceptance of the hypothesis that the parameters in question are from the same population, hence are equal.

terms for each of the treatment combinations cited above. From these
F tests, probability values (P) were determined. These give the probability that the observed difference between the two statistics being
compared could have been due to chance alone. The pertinent data
for the two comparisons are given in Table 3 for group A and in
Table 4 for group B.

There was a rather small number of cases in group A where the
incision had been made through the cornea (Table 3, Fig. 17). After
lens extirpation in this group, lenses regenerating from the host iris
(treatment 4A) were larger than those forming from the iris implant
(treatment 1A) ($P < 0.05$). Furthermore, the lenses developing from
the implanted iris were larger in the absence of the host lens (treatment 1A) than after removal and reimplantation of the host lens
(treatment 2A) ($P < 0.025$). However, when the regression coefficients were compared in the same way, it was found that there was
no significant difference between the growth rate of the lenses of
treatment 4A and that of treatment 1A; neither was the growth rate of
the lenses of treatment 1A significantly greater than that of treatment
2A. This lack of a real difference may have been due to variability and
the small sample. More data were available for group B where the

TABLE 4

COMPARISON OF SIZES AND LINEAR GROWTH RATES OF LENSES REGENERATING
FROM IRIS IMPLANTS IN THE PRESENCE AND ABSENCE OF THE HOST LENS
AND FROM THE HOST DORSAL IRIS, GROUP B[a]

| | Treatments compared | | F | P | Conclusion |
	4B	1B			
$\bar{x}$	22.7	20.1	—	—	—
$\bar{y}$	184	128	—	—	—
$\hat{\bar{y}}$	177	140	19.15	<0.005	$\hat{\bar{y}}_{4B} \neq \hat{\bar{y}}_{1B}$
b	7.47	6.10	1.02	>0.25	$b_{4B} = b_{1B}$
n	41	26	—	—	—
	1B	2B			
$\bar{x}$	20.1	20.6	—	—	—
$\bar{y}$	128	86	—	—	—
$\hat{\bar{y}}$	130	85	25.87	<0.005	$\hat{\bar{y}}_{1B} \neq \hat{\bar{y}}_{2B}$
b	6.10	3.40	3.96	<0.10	$b_{1B} \stackrel{?}{=} b_{2B}$
n	26	25	—	—	—

[a] See footnote a of Table 3.

incision had been made through the ventral retina (Table 4, Fig. 18). Again the average diameters of the lenses regenerating from the host iris were larger than those of the lenses regenerating from the iris implants following removal of the host lens ($P < 0.005$) and the latter lenses were larger than the lenses developing in the presence of the host lens ($P < 0.005$). The regression coefficients did not differ significantly between the first two treatments (4B and 1B) while, between treatments 1B and 2B, the difference was of doubtful significance ($P < 0.10$).

In order to obtain as much information as possible from the data, it seemed desirable to pool the measurements from group A with those from group B. Therefore, a comparison was made, for each treatment, of the data in group A (operation through the cornea) and that in group B (operation through the ventral retina). This is summarized in Table 5. There was no significant difference in the average sizes of

TABLE 5

COMPARISON OF SIZES AND LINEAR GROWTH RATES OF REGENERATING LENSES
IN GROUP A (OPERATION THROUGH CORNEA) WITH THOSE IN GROUP B
(OPERATION THROUGH VENTRAL RETINA)[a]

	Treatments compared		F	P	Conclusion
	4A	4B			
$\bar{x}$	22.8	22.7	—	—	—
$\bar{y}$	181	184	—	—	—
$\hat{y}$	180	184	0.107	NS	$\hat{y}_{4A} = \hat{y}_{4B}$
b	13.48	7.47	3.46	<0.10	$b_{4A} \overset{?}{=} b_{4B}$
n	15	41	—	—	—
	1A	1B			
$\bar{x}$	23.1	20.1	—	—	—
$\bar{y}$	154	128	—	—	—
$\hat{y}$	140	133	0.411	NS	$\hat{y}_{1A} = \hat{y}_{1B}$
b	10.41	6.10	1.98	>0.10	$b_{1A} = b_{1B}$
n	9	26	—	—	—
	2A	2B			
$\bar{x}$	19.8	20.6	—	—	—
$\bar{y}$	93	86	—	—	—
$\hat{y}$	95	85	0.603	NS	$\hat{y}_{2A} = \hat{y}_{2B}$
b	6.66	3.40	1.11	>0.25	$b_{2A} = b_{2B}$
n	10	25	—	—	—

[a] See footnote a of Table 3.

the lenses between the two groups within each of the three treatment comparisons. In addition, there was no significant difference between the regression coefficients of group A and those of group B for lenses developing from implants in the absence of the host lens ($P > 0.10$) or with the host lens present ($P > 0.25$). The difference between the regression coefficients of the two groups for the lenses regenerating from the host dorsal iris was of questionable significance only ($P < 0.10$).

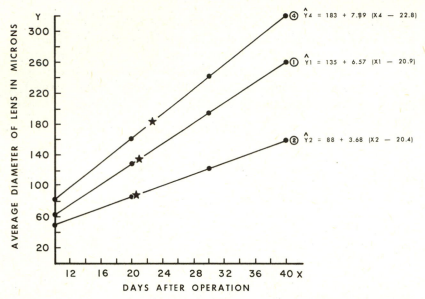

FIG. 19.   Graph showing linear growth rates of lenses regenerating from the host iris (treatment 4) and from iris implants in the absence (treatment 1) or presence (treatment 2) of the host lens. Calculated from the pooled data for the three treatments presented in Figs. 17 and 18 according to the equation for linear regression $\hat{Y} = \bar{y} + b\,(X - \bar{x})$. Filled circle: calculated points. Star: point indicating both $\bar{x}$ and $\bar{y}$.

Since the differences between the results of the two experimental procedures could be attributed to chance alone, the data from groups A and B were pooled for each of the three treatments. The linear regression equation of Y on X, which is given as $\hat{Y} = \bar{y} + b(X - \bar{x})$, was plotted. Four X-values were substituted in the equation and four estimated Y-values ($\check{Y}$) were computed as an aid in plotting the line. The calculated points are indicated by dots and the point designating both $\bar{x}$ (mean of X) and $\bar{y}$ (mean of Y) is marked by a star. In this way, regression lines for Y on X were drawn for each of the three treatments (Fig. 19). The linear regression equation is given for each treatment on the graph. The pooled data are summarized in Table 6. From this, it can be concluded that the lenses, regenerating from the host iris after removal of the host lens were larger than those regenerating from iris implants ($P < 0.005$), but the linear growth rates

TABLE 6

Comparison of Sizes and Linear Growth Rates of Lenses Regenerating from Iris Implants in the Presence and Absence of the Host Lens and from the Host Dorsal Iris, Pooled Data[a]

	Treatments compared		F	P	Conclusion
	4	1			
$\bar{x}$	22.8	20.9	—	—	—
$\bar{y}$	183	135	—	—	—
$\hat{y}$	178	144	22.23	<0.005	$\hat{y}_4 \neq \hat{y}_1$
b	7.89	6.57	1.08	>0.25	$b_4 = b_1$
n	56	35	—	—	—
%	90%	63%	—	—	—
	1	2			
$\bar{x}$	20.9	20.4	—	—	—
$\bar{y}$	135	88	—	—	—
$\hat{y}$	134	89	37.37	<0.005	$\hat{y}_1 \neq \hat{y}_2$
b	6.57	3.68	5.42	<0.025	$b_1 \neq b_2$
n	35	35	—	—	—
%	63%	37%	—	—	—

[a] See footnote a of Table 3

were not significantly different from each other ($P > 0.25$). In addition, lenses which regenerated from iris implants after lens extirpation were also larger than those which regenerated from iris implants in the presence of the host lens ($P < 0.005$). Furthermore, the former lenses had a significantly faster linear rate of growth than did the latter ($P < 0.025$). Therefore, the presence of the host lens had two effects on lens regeneration from iris implants: (1) the percentage of implants which regenerated a lens was considerably reduced (37% as compared to 63%); (2) the growth rate of the regenerating lenses was slower. Possible mechanisms for these results will be considered in the discussion.

DISCUSSION

There are two conditions to be taken into account in the interpretation of the results of these experiments. First, there is the access of the iris implant in a particular site within the eye to the stimulus from the neural retina which is probably mediated by a chemical substance passing through the vitreous body. The availability of this material to the implant would depend on the position of the implant in the eye and on any structures intervening between it and the neural retina. Second, there is the presence or absence of other differentiated lens tissue. In the anterior chamber, the implanted iris is not only in close proximity to the large host lens, but it is also separated by this lens and the suspensory ligament from contact with the vitreous body. It may be that the neural retinal factor is not able to pass readily through this barrier into the aqueous humor. Since virtually nothing is known concerning the chemical nature or method of transport of such a substance, this still remains a conjecture. It has been definitely established, however, that lens regeneration does not take place in the anterior chamber from either *in situ* or implanted iris when a normal lens occupies its normal position within the pupil. When the host lens is removed, the host iris is at once exposed to whatever substance may be passing across the vitreous body from the neural retina. Furthermore, it is no longer subjected to any inhibition that might be exerted by the large lens. Similar conditions also hold for pieces of iris placed in the pupil or within the vitreous body after lens extirpation. For the *in situ* dorsal iris or iris implants into the anterior chamber then, the presence of the lens may somehow block the neural retinal factor while absence of the lens allows this stimulus to reach both the *in situ* iris and iris implants located in the pupil.

Implants of iris into the vitreous chamber with intact host lens are located between the lens and the neural retina. In this position, they have access to any stimulus passing from the neural retina and diffusing through the vitreous body, and at the same time they are in close association with the large host lens. Do the results obtained with these implants provide any further information on the role of the neural retina and lens in lens regeneration? Under these conditions, dorsal iris tissue will regenerate a lens, but this environment is not as favorable as the pupil of the eye after extirpation of the lens. Slightly more than half as many implants of competent dorsal iris regenerate a lens when the host lens is present as when it is absent. These lenses are smaller in size on the average and their growth rate is also slower. However, the morphological abnormalities which were observed here, such as thinning or loss of the lens epithelium and occasional vacuolation of the lens fibers, are also shared with lenses regenerating from iris implants into the lentectomized eye, particularly those lying away from the pupil nearer the neural retina. This type of abnormal lens epithelium was not found among the lenses regenerating from the *in situ* dorsal iris and located within the pupil. Under the latter conditions, the regenerating lens is subjected to a polarized stimulus for lens fiber differentiation which acts most strongly on the side next to the vitreous body and facing the neural retina. Lenses developing away from the pupil within the vitreous body are probably not lying within such a steeply polarized gradient. Although the lens fibers tend to develop from the side facing the neural retina and the lens epithelium from the side facing the host lens, it is likely that there is more uniformity of stimulus within the vitreous body than at its surface within the pupil. Therefore, there is a tendency for cells throughout the lens epithelium to differentiate into lens fibers thereby reducing the number of epithelial cells remaining.

Evidence for the polarizing action of the neural retina was demonstrated by Stone (1954) when he rotated a sector of dorsal iris 180° on its dorsoventral axis and then removed the host lens. The lenses, which regenerated from the inside-out pieces of iris, were normally oriented with the lens epithelium facing the cornea and the lens fiber pole directed toward the neural retina. Coulombre and Coulombre (1963) were also able to reverse the polarity of the young lens in the 5-day-old chick embryo by rotating it 180° on its dorsoventral axis. They observed that lens fiber differentiation was initiated from the lens epithelium, now facing the neural retina, and that there was no further elongation of the previously formed lens fibers, now facing the anterior chamber. Many years ago, Mikami (1941) observed that, in the lentectomized eye of adult *Triturus pyrrhogaster*, iris implants lying deep in the vitreous chamber regenerated a lens less frequently than did those lying in the pupil (32% *vs.* 81%) and that these lenses were also smaller and lacked a lens epithelium. Implants in the middle of the anterior chamber regenerated a small lens in only 40% of the cases. A lens epithelium was present but the lens fibers were poorly differentiated. All this evidence indicates that a lens in the pupil is

under the influence of a sharply polarized stimulus from the neural
retina which is acting most strongly on the internal one-half of the
lens. This, in turn, results in the normal morphology of the lens
epithelium, transitional equatorial zone, and lens fibers.

From the data on the molecular changes occurring during lens
regeneration, outlined in the Introduction, it can be hypothesized
that the neural retinal factor stimulates the activation of the molecular
systems of DNA and RNA in the dorsal iris which provide information
for the synthesis of lens proteins. DNA synthesis and mitotic divisions
as well as RNA synthesis occur during the formation of the lens
vesicle. Then the synthesis of the lens proteins, the α, β, γ crystallins,
takes place in that part of the developing lens facing the source of
the stimulus. The half of the lens facing away from the stimulus
continues to synthesize DNA and RNA as the cells undergo additional
mitotic divisions. Later on, in the equatorial zone, the cells change
from the latter type of synthetic activity to synthesis of the crystallins
as the secondary lens fibers differentiate.[3] Papaconstantinou (1964) and
Papaconstantinou et al. (1964) have found that γ crystallins are present
only in the lens fibers of the bovine lens while α and β crystallins can
be found in both the lens epithelium and the lens fibers. The studies
of Yamada and Roesel (1964) on inhibition of lens regeneration with
actinomycin D have shown that those cells involved in rapid RNA
synthesis are the most sensitive to this drug. Early stages of lens
regeneration, up to the lens vesicle stage, are arrested or retarded in
their further development (depending on duration of treatment)
while, in later stages, the lens epithelium is selectively destroyed,
leaving naked lens fiber bodies of reduced size similar to some of those
described in this paper.

Is there any evidence for a tissue-specific inhibition of lens regeneration by the host lens in the experiments reported here? At first the
answer would appear to be in the affirmative since lens development
from iris implants, under the simultaneous influence of host retina
and lens, was both less frequent and retarded when compared to lens
development from similar implants exposed to the neural retina alone.

[3] Note added in proof: T. Yamada [Am. Zoologist 6, 21–31 (1966)] reviewed
these molecular changes during lens regeneration in the newt. The evidence from
his immunofluorescent studies indicates that the lens fibers synthesize α, β, and γ
crystallins whereas the lens epithelium, from stage 9 on, synthesizes α and β
crystallins, but not γ crystallin.

This inhibition could be due to a competition between the host lens and the implant for the neural retinal factor or to a tissue specific inhibitory action of lens against lens. These possibilities have been discussed by Reyer (1962), Rose (1964), and Smith (1965). However, there is still another condition that could contribute to this retardation. When the large host lens occupies its normal position in the pupil, this region is not available as a site for the implant which is therefore forced to lie deeper within the vitreous chamber. When the host lens is absent, the lens regenerating from the host iris always grows down into the pupil. The implanted iris may also be located here or it may lie behind the iris or deeper in the vitreous. Even in this small larval eye, however, the lens developing from the implanted iris in the lentectomized eye tends to be located nearer to the pupil than the lens developing from an implant into the vitreous chamber of an eye with the host lens *in situ*. As indicated by the early experiments of Mikami noted above, this difference in the average position within the eye could be a factor in the poorer development of lenses in the presence of host lens. It would seem likely that the special conditions in the pupil are optimal for the most normal differentiation and for maximal growth of the lens. Deeper within the vitreous chamber, mitotic division of the cells of the lens epithelium might be decreased in favor of lens fiber differentiation, resulting in the observed flattening of the remaining cells of the lens epithelium, the eventual depletion of cells available for lens fiber formation and therefore decreased growth in size.

Although conclusive evidence for a tissue specific inhibition of the regenerating lens by a more mature, living lens is thus still lacking, the results reported here do show that the regeneration of a lens from an iris implant is possible even though a normal host lens is present provided that the implant has access to the stimulus originating from the neural retina.

SUMMARY

In order to study the role of the neural retina and lens on lens regeneration from the iris, sectors of mid-dorsal iris from larval *Triturus viridescens* donors were implanted into the following sites in larval hosts of the same species: (a) dorsal fin, (b) anterior chamber of the eye, host lens present, (c) vitreous chamber of the eye, host lens absent, and (d) vitreous chamber of the eye, host lens present. In these locations, the implanted iris was subject to the influence of (a) neither neural retina nor lens, (b) probably only lens. (c) neural retina only, or (d) neural retina and lens together.

Lens regeneration failed to take place in the anterior chamber or dorsal fin, thus showing the necessity of the neural retinal stimulus. Lens regeneration occurred from implants into the vitreous chamber not only in the absence of the host lens, but also in its presence. However, both the frequency of lens regeneration and the size and growth rate of the regenerates were decreased when the implant was located between the neural retina and host lens, therefore being subject to the influence of both tissues together. It was suggested that the inhibitory action of the host lens on iris implants in this location could be (1) mechanical, by displacing the implant more deeply into the vitreous chamber or (2) chemical, by competing with the lens developing from the implant for the neural retinal factor or through the production of a specific inhibitor for the homologous tissue.

I should like to thank Dr. Albert E. Drake for his invaluable assistance in the statistical analysis and Mrs. Kathleen Sigler, Mrs. Mary L. Griffith, and Mrs. Evelyn Goldfein for their technical assistance during this research.

Figure 1 is reproduced with modifications from "The Structure of the Eye, II. Symposium" by courtesy of the publisher, F. Schattauer, Stuttgart.

REFERENCES

COULOMBRE, J. L., and COULOMBRE, A. J. (1963). Lens development: Fiber elongation and lens orientation. *Science* **142**, 1489–1490.

EGUCHI, G. (1961). The inhibitory effect of the injured and the displaced lens on the lens-formation in *Triturus* larvae. *Embryologia* (*Nagoya*) **6**, 13–35.

EGUCHI, G. (1963a). Electron microscopic studies on lens regeneration. I. Mechanism of depigmentation of the iris. *Embryologia* (*Nagoya*) **8**, 45–62.

EGUCHI, G. (1963b). An electron microscopic study on differentiation of the lens fiber of the regenerating lens in adult *Triturus pyrrhogaster*. *Bull. Marine Biol. Sta. Asamushi, Tohoku Univ.* **11**, 223–228.

EGUCHI, G. (1964). Electron microscopic studies on lens regeneration. II. Formation and growth of lens vesicle and differentiation of lens fibers. *Embryologia* (*Nagoya*) **8**, 247–287.

FROST, D. (1961). Inhibition of lens regeneration by implanted lenses in the eyes of the adult newt *Diemictylus* (=*Triturus*) *viridescens*. *Develop. Biol.* **3**, 516–531.

Goss, R. J. (1960–1961). Factors affecting lens regeneration in the newt. *Carnegie Inst. Washington Year Book* **60**, *Ann. Rept. Director Dept. Embryol.*, pp. 430–431.

Goss, R. J. (1964). "Adaptive Growth," pp. 96–117. Logos Press, London, and Academic Press, New York.

IKEDA, Y., and AMATATU, H. (1941). Über den Unterschied der Erhaltungs-möglichkeit der Linse bei zwei Urodelenarten (*Triturus pyrrhogaster* und *Hynobius nebulosus*), die sich bezüglich der Fähigkeit zur Wolffschen Linsen-regeneration voneinander wesentlich verschieden verhalten. *Japan J. Med. Sci. Biol.* (*I. Anat.*) **8**, 205–226.

IKEDA, Y., and KOJIMA, T. (1940). Zur Frage der paralysierenden Wirkung der Linse auf die auslösenden Faktoren für die' Wolffsche Linsenregeneration. *Japan J. Med. Sci. Biol.* (*I. Anat.*) **8**, 51–73.

KARASAKI, S. (1964). An electron microscopic study of Wolffian lens regeneration in the adult newt. *J. Ultrastruct. Res.* **11**, 246–273.

MIKAMI, Y. (1941). Experimental analysis of the Wolffian lens-regeneration in adult newt, *Triturus pyrrhogaster*. *Japan J. Zool.* **9**, 269–302.

PAPACONSTANTINOU, J. (1964). The formation of γ-crystallins during lens fiber differentiation. *Am. Zoologist* **4**, 279.

PAPACONSTANTINOU, J., KOEHN, P. V., and STEWART, J. A. (1964). The effect of actinomycin D on lens protein synthesis. *Am. Zoologist* **4**, 321.

REYER, R. W. (1948). An experimental study of lens regeneration in *Triturus viridescens viridescens*. I. Regeneration of a lens after lens extirpation in embryos and larvae of different ages. *J. Exptl. Zool.* **107**, 217–268.

REYER, R. W. (1954). Regeneration of the lens in the amphibian eye. *Quart. Rev. Biol.* **29**, 1–46.

REYER, R. W. (1956). Lens regeneration from homoplastic and heteroplastic im-plants of dorsal iris into the eye chamber of *Triturus viridescens* and *Ambly-stoma punctatum*. *J. Exptl. Zool.* **133**, 145–190.

REYER, R. W. (1961). Lens regeneration from intra-ocular, iris implants in the presence of the host lens. *Anat. Record* **139**, 267.

REYER, R. W. (1962). Regeneration in the amphibian eye. *Regeneration, Symp. Soc. Study Develop. Growth* **20**, pp. 211–265. Ronald Press, New York.

REYER, R. W. (1963). Studies on lens regeneration in *Amblystoma punctatum* and *Pleurodeles waltlii*. *Anat. Record* **145**, 275.

REYER, R. W. (1965). Stimulation and inhibition of the growth and differentia-tion of embryonic and regenerating lenses. *In* "The Structure of the Eye, II. Symposium" (J. W. Rohen, ed.), pp. 495–513. Schattauer, Stuttgart.

REYER, R. W., and STONE, L. S. 1955). A reinvestigation of lens regeneration in *Salamandra salamandra salamandra*. *J. Exptl. Zool.* **129**, 257–290.

ROSE, S. M. (1964). Regeneration. *In* "Physiology of the Amphibia" (J. A. Moore, ed.), pp. 545–622. Academic Press, New York.

SATO, T. (1930). Beiträge zur Analyse der Wolff'schen Linsenregeneration. I. *Arch. Entwicklungsmech. Organ.* **122**, 451–493.

SATO, T. (1935). Beiträge zur Analyse der Wolff'schen Linsenregeneration. III. *Arch. Entwicklungsmech. Organ.* **133**, 323–348.

SMITH, S. D. (1965). The effects of electrophoretically separated lens proteins on lens regeneration in *Diemyctylus viridescens*. *J. Exptl. Zool.* **159**, 149–166.

SNEDECOR, G. W. (1956). "Statistical Methods," 5th ed. Iowa State Univ. Press, Ames, Iowa.

SPIRITO, A., and CIACCIO, G. (1931). Ricerche causali sulla rigenerazione del cristallino nei tritoni. *Boll. Zool.* (*Napoli*) **2**, 1–7.

STONE, L. S. (1952). An experimental study of the inhibition and release of lens regeneration in adult eyes of *Triturus viridescens viridescens*. *J. Exptl. Zool.* **121**, 181–224.

STONE, L. S. (1953). An experimental analysis of lens regeneration. *Am. J. Ophthalmol.* **36**, No. 6, Pt. II, 31–39.

STONE, L. S. (1954). Further experiments on lens regeneration in eyes of the adult newt *Triturus v. viridescens*. *Anat. Record* **120**, 599–624.

STONE, L. S. (1958a). Inhibition of lens regeneration in newt eyes by isolating the dorsal iris from the neural retina. *Anat. Record* **131**, 151–172.

STONE, L. S. (1958b). Lens regeneration in adult newt eyes related to retina pigment cells and the neural retina factor. *J. Exptl. Zool.* **139**, 69–84.

STONE, L. S. (1959). Regeneration of the retina, iris, and lens. In "Regeneration in Vertebrates" (C. S. Thornton, ed.), pp. 3–14. Univ. of Chicago Press, Chicago, Illinois.

STONE, L. S. (1963). Experiments dealing with the role played by the aqueous humor and retina in lens regeneration of adult newts. *J. Exptl. Zool.* **153**, 197–210.

STONE, L. S. (1965). The regeneration of the crystalline lens. *Invest. Ophthalmol.* **4**, 420–432.

STONE, L. S., and GALLAGHER, S. B. (1958). Lens regeneration restored to iris membranes when grafted to neural retina environment after cultivation *in vitro*. *J. Exptl. Zool.* **139**, 247–262.

STONE, L. S., and VULTEE, J. H. (1949). Inhibition and release of lens regeneration in the dorsal iris of *Triturus v. viridescens*. *Anat. Record* **103**, 560.

TAKANO, K. (1959). On the lens-effect in the Wolffian lens regeneration in *Triturus pyrrhogaster*. *Mie Med. J.* **8**, 385–403.

TAKANO, K., YOSHIDA, Y., OHASHI, T., OGASAWARA, T., TAKEUCHI, A., MASAKI, H., MIYAZAKI, A., and MIKAMI, Y. (1957). Experimental analysis of the effect of lens upon the Wolffian lens regeneration in adults of the newt, *Triturus pyrrhogaster*. *Mie Med. J.* **7**, 257–272.

TAKANO, K. YAMANAKA, G., and MIKAMI, Y. (1958). Wolffian lens-regeneration in the eye containing a full grown lens in *Triturus pyrrhogaster*. *Mie Med. J.* **8**, 177–182.

TAKATA, K. (1952). Ribonucleic acid and lens-regeneration. *Experientia* **8**, 217–218.

TAKATA, C., ALBRIGHT, J. F., and YAMADA, T. (1964). Lens antigens in a lens-regenerating system studied by the immunofluorescent technique. *Develop. Biol.* **9**, 385–397.

TAKATA, C., ALBRIGHT, J. F., and YAMADA, T. (1965). Lens fiber differentiation and gamma crystallins: Immunofluorescent study of Wolffian regeneration. *Science* **147**, 1299–1301.

UNO, M. (1943). Zur Frage des Mechanismus der Wolffschen Linsenregeneration. *Japan. J. Med. Sci. Biol.* (*I. Anat.*) **11**, 75–100.

WACHS, H. (1914). Neue Versuche zur Wolffschen Linsenregeneration. *Arch. Entwicklungsmech. Organ.* **39**, 384–451.

YAMADA, T., and KARASAKI, S. (1963). Nuclear RNA synthesis in newt iris cells engaged in regenerative transformation into lens cells. *Develop. Biol.* **7**, 595–604.

YAMADA, T., and ROESEL, M. (1964). Effects of actinomycin D on the lens regenerating system. *J. Embryol. Exptl. Morphol.* **12**, 713–725.

YAMADA, T., and TAKATA, C. (1963). An autoradiographic study of protein synthesis in regenerative tissue transformation of iris into lens in the newt. *Develop. Biol.* **8**, 358–369.

YAMADA, T., and TAKATA, C. (1964). The pattern of RNA synthesis as related to lens antigen formation during lens regeneration in the newt. *J. Cell. Biol.* **23**, 104A.

YAMADA, T., TAKATA, C., EISENBERG, S., and KOHONEN, J. (1964). The pattern of macromolecular synthesis in Wolffian lens regeneration. *Conf. Lens Differentiation. Oak Ridge Natl. Lab.*, pp. 12–13.

ZALOKAR, M. (1944). Contribution à l'étude de la régénération du cristallin chez le *Triton. Rev. Suisse Zool.* **51**, 443–521.

Reprinted from *J. Morphol.*, **99**, 1–15, 26–39 (1956)

16

THE INDUCTIVE ACTIVITY OF THE SPINAL CORD IN URODELE TAIL REGENERATION [1,2]

SYBIL W. HOLTZER

Department of Anatomy, School of Medicine, University of Pennsylvania, Philadelphia

TWENTY-ONE FIGURES

INTRODUCTION

According to recent reviews (Nicholas, '55; Needham, '52) regeneration of the urodele tail and limb is accomplished by different processes. Tissues of the tail, it is contended, are reconstituted by direct budding from parent tissues of the stump, whereas the tissues of the limb are stated to differentiate from an aggregation of indifferent blastema cells. With respect to tail regeneration spinal cord cells are viewed as forming neural cells, muscle cells as forming muscle cells, cartilage cells as forming cartilage, etc. In contrast, Weiss ('25b) and Thornton ('42), working with salamander limbs and Winnick ('52) and Goss ('53) studying frog limbs, demonstrated that the amphibian limb does not regenerate in this fashion. Amputation through the upper arm following removal of the humerus resulted in a normal regenerated limb possessing the usual complement of limb cartilages even though cartilage was absent in the stump.

In a previous report (Holtzer, Holtzer and Avery, '55, see also for bibliography of older literature on tail regeneration) it was stated that following amputation of the larval

[1] Submitted to the Department of Zoology of the University of Chicago in partial fulfillment for the degree of Doctor of Philosophy.

[2] This investigation was supported in part by a research grant no. B 493 from the National Institute of Neurological Diseases and Blindness of the National Institutes of Health, Public Health Service.

1

urodele tail the notochord failed to regenerate. The skeletogenous structures of the regenerated tail consisted of a series
of segmented cartilages, the centra of the tail vertebrae.
Failure of the notochord to regenerate permitted precocious centra formation as does embryonic notochordectomy
(Holtzer, '52). It was also noted (Holtzer et al., '55) that
the presence of the spinal cord at the amputation plane was
necessary for tail regeneration in much the same manner
that peripheral nerves are essential for regenerative activity
in limb regeneration (Weiss, '25a; Schotté and Butler, '41;
Singer, '52). These results would tend to indicate that differences between tail and limb regenerative processes may have
been over-emphasized.

The purpose of this paper is to stress further the similarity
between tail and limb regeneration. Specifically it will be
demonstrated that: (1) Cartilage cells differentiate from the
blastema cells independently of cartilage in the stump, dis-
proving the concept that all tissues in the regenerated tail
bud part by part from stump tissues. (2) The tail blastema
cells differentiate into either cartilage or connective tissue
but may not contribute significantly to the regenerated muscle.
(3) The formation of the skeletogenous elements in the regenerated tail depend upon an inductive process similar to
that operating in the embryo.

MATERIALS AND METHODS

Amblystoma punctatum and *A. opacum* were used for the
experimental series, *A. punctatum* for studying normal regeneration. All animals ranged from 25 mm to 35 mm in
length. Operations were performed, with MS 222 as the
anesthetic, in full strength Holtfreter's solution, the animals
remaining in this solution until healing had taken place, usually for 24 hours. At that time the operated animals were
transferred to spring water.

Animals for the study of normal regeneration came from
the same clutch of eggs. When they reached a length of 25 mm
their tails were cut on a line perpendicular to the long axis

of the trunk just posterior to the cloaca. Fixed in Bouin's at two day intervals and embedded in paraffin, they were sectioned at 10 µ and stained with Delafield's haematoxylin and eosin.

The details of the various operations will be dealt with in context. All experimental animals were fixed in formol-alcohol and stained *in toto* in methylene blue. The formula for this stain, which selectively colors bone and cartilage, was secured from Dr. E. G. Butler of Princeton University. Observations were made on the whole mounts, after which selected specimens were run up in the conventional fashion, embedded, sectioned and stained with Delafield's haemat-oxylin and eosin. This procedure permitted ready evaluation of the skeletogenous structures with subsequent analysis of their detailed histology.

This report is based on the study of 50 normal regenerates and over 350 experimental animals. More than 100 experi-mental animals were prepared for histological examination.

In the following account, references to the regenerated tail includes spinal cord, muscle, cartilage, etc. Descriptions of the reformation of the connective tissue of the fin is not con-sidered, since it does not seem to play any active role in that process.

RESULTS

Normal regeneration. The important spatial and temporal relationships obtaining among the different tissues of the regenerating tail may be seen in figure 1. In this figure the spinal cord and segmented cartilages are projected on a frontal section of the regenerated tail taken at the mid point of the D-V axis. Five successive stages of regeneration are shown. In the regenerated tail the spinal cord always extends most, and the muscle least, posteriorly. The notochord does not regenerate. Its place in the regenerate is taken by a series of segmented cartilages. The terminal point of the cartilage is located somewhere between the most caudal mus-cle fiber and the end of the spinal cord. The tail blastema,

REGENERATION DAYS

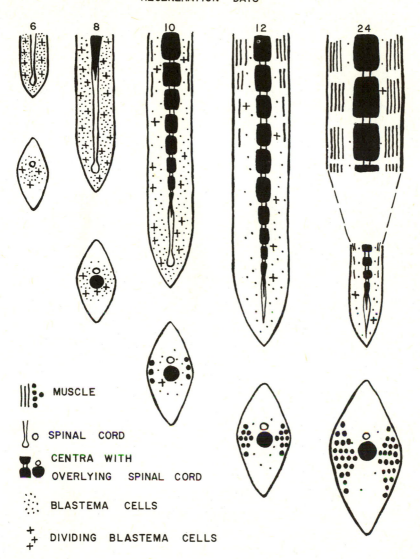

MUSCLE

SPINAL CORD

CENTRA WITH
OVERLYING SPINAL CORD

BLASTEMA CELLS

DIVIDING BLASTEMA CELLS

Fig. 1 Diagrammatic representation of the major tissue relationships during normal regeneration of the tail. The upper row is viewed as a projection of the spinal cord and the notochord on a frontal section of the regenerated tail, taken at the mid-point of the D–V axis. Note the rapid growth of the spinal cord into the blastema. During all stages the neural tissue extends most posteriorly, followed by cartilage and muscle. The lower row represents cross sections taken at the base of the regenerate.

unlike that of the limb, never assumes the form of a tightly packed clump of cells. Instead, the cells of the tail blastema form a diffuse aggregate extending from the region of active cartilage formation posterior to about 200 μ beyond the tip of the spinal cord. During the course of regeneration the blastema maintains a fairly constant position with respect to the spinal cord as the latter elongates. Cross sections reveal that the blastema assumes a bilateral symmetry early in its development. The blastema cells surround the cord, but are most concentrated ventrally and to a lesser extent laterally.

The cells of the blastema deserve special mention here. They most strongly resemble embryonic mesenchyme, having a similar high extra-cellular/cellular ratio. The nuclei are large and vesicular with a finely granulated nucleoplasm. The scanty basophilic cytoplasm has a stellate outline. Cell boundaries are indistinct.

Because tail regeneration involves several tissues in a variety of processes, some degenerative, some regenerative, a brief chronological description of the observed changes will be given first. Each of the tissues will be dealt with in detail in a later section.

24 hours post-amputation. Extravasated blood, both cells and plasma, has coagulated on the wound surface, forming an acidophilic film. Epidermal cells, a single layer in thickness, have migrated over the coagulum from the border of the wound, closing it completely. Degeneration of the muscle is most conspicuous. The fibers have lost their cross striations and fibrils and the cytoplasm has become distinctly more acidophilic than that of normal muscle. Muscle debris is scattered throughout the degenerating area which extends as far anterior as the first myoseptum. The last 100 μ of the spinal cord is greatly disarranged. Nerve cells and fibers have been torn from their locations. Red and white blood cells are found throughout the injured tissue. No changes are observed in staining reaction or cellular appearance of the notochord. The traumatized surface of the notochord lies subjacent to the wound epithelium. There is no invasion of the notochord by leucocytes. No mitoses are seen anywhere in the damaged region.

2–3 days post-amputation. Much of the blood coagulum is still evident, but it has been invaded by many white blood cells. An occasional mitosis can be seen in the epidermis, which is now several cells thick. Degeneration of the muscle has progressed to the point where most of the distal segment is reduced to shreds of glassy acidophilic sarcoplasm sometimes lacking nuclei. Concomitant with this severe degeneration is the appearance in the space between the degenerating muscle and the spinal cord of the first blastema cells. Mitoses are rather frequent in these cells, which are found anterior as far as the degenerating muscle and posterior as far as the end of the spinal cord. Though the morphology of the spinal cord is still disorganized, the neural canal has been closed off. No mitoses are found in the actual wound area of the spinal cord, but they can be found throughout the first few segments immediately anterior. The notochord shows no change.

4 days post-amputation. The extravasated blood has disappeared by this time. The epidermis is still several cells thick, but not conspicuously active mitotically. Degeneration of the muscle has continued. The free, dividing blastema cells have increased in number occupying the space formerly filled by muscle and extend approximately 50 μ posterior to the spinal cord. The latter structure presents much the same appearance as it did two days earlier. The first gross extension in length of the tail, seen at this time, is due to the posterior extension of the epidermis and underlying connective tissue of the dorsal and ventral fins. The notochord remains unchanged.

6 days post-amputation. All degenerated muscle has disappeared at this stage. Small clumps of myoblasts, basophilic, spindle-shaped cells with centrally located nuclei, can be observed at the ends of the most lateral fibers of the surviving muscle. The blastema cells continue to divide at a rapid rate and have arranged themselves so that the greatest number lie ventral to the spinal cord. See figure 9. The long axes of these cells tend to be oriented in the medio-lateral plane. Mitoses are observed in the ependymal layer throughout the length of the regenerated cord, which has grown about 300 μ beyond the amputation plane.' The caudal tip of the neural canal has enlarged to form a terminal bulb (fig. 8). Composed of a single layer of ependymal cells with strikingly enlarged vesicular nuclei (typical ependymal nuclei are elongated ovals, radially arranged about the central canal), this terminal ampula measures approximately 70 μ in diameter and 250 μ in A-P length. This bulb, however, is not particularly rich in mitotic figures.

8 days post-amputation. Muscle regeneration is more evident but still confined to within the stump. In their extension posteriorly, the immature myoblasts keep within 20 μ of the skin at all times. A slender cone of procartilage cells, extending over 200 μ beyond the amputation plane, is the only regenerated tissue besides the spinal cord in the regenerate (fig. 10). The regenerating cartilage is continuous with the cut surface of the notochord anteriorly. This junction between the cut surface of the notochord and the proximal border of the cartilage serves as a permanent marker of the original amputation plane (fig. 21). The procartilage cells, similar in appearance to the blastema cells, are distinguished from the latter by virtue of the small amount of basophilic matrix they have deposited around themselves, by their position immediately below the spinal cord and by their polarization. In frontal sections the procartilage cells present their narrow aspect and are concentrated in the mid-region of the tail. The long axes of the nuclei are perpendicular to the long axis of the tail. In cross section the central cells present the broad aspect of their nuclei while the outer cells apply themselves tangentially to the inner mass. Very few mitoses are observed in the pro-cartilage cells, but the blastema cells continue to divide rapidly.

10 days post-amputation. Myoblasts near the stump, still maintaining their position subjacent to the skin, have arranged themselves in A-P groups which are suggestive of segments. In the same region the myoblasts have increased in number, both medio-laterally and dorso-ventrally. The cartilagenous rod has lengthened and shows the beginnings of segmentation in the A-P direction. Proximally, where the cartilage is approaching full differentiation, the blastema cells are reduced in number and dot not divide so frequently.

12–14 days post-operation. Differentiation in the proximal regenerated muscle has progressed to the point where myofibrils and cross striations are visible. Associated with fibril formation is an increased basophilia in the cytoplasm. Muscle nuclei still remain in the center of the fibers. In proximal regions where cartilage is well differentiated, the number of blastema cells as well as their mitotic activity has greatly decreased. See figure 11.

16 days post-operation. The development of myofibrils and cross striations has proceeded posteriorly. Most nuclei still remain in the center of the fibers however. Small somite growth centers (Holtzer and Detwiler, '53) are seen in the proximal regenerate. The older cartilage shows the beginnings of lacunae formation. Blastema cells have virtually disappeared except from the distal regions where cartilage is forming under the rapidly growing spinal cord.

18–20 days post-operation. The proximal hyaline cartilage has achieved terminal differentiation. Sharply defined capsules are to be seen around large lacunae. The cells within the lacunae are shrunken. Their nuclei stain more darkly and are more granular. The proximal muscle fibers have achieved complete histological differentiation. The fibers have begun to organize themselves into distinct muscle masses. The sarcoplasm exhibits the normal acidophilia and the nuclei, for the most part, have assumed their peripheral position within the fibers. Also in this region of the regenerate a few spinal ganglia and motor roots are now evident.

24–30 days post-amputation. Although presenting much the same picture as the preceding stage, a further step in the morphogenesis of muscle has taken place. The proximal fibers are now arranged in a definite segment. Only a remnant of the blastema remains and that is confined to the region from the tip of the spinal cord to 150 mμ beyond it. See figure 12.

DETAILED HISTOLOGY

Spinal cord. Casual inspection revealed no concentration of mitotic activity at any A-P level of any given stage nor any concentration at any stage after amputation. Mitotic counts were made to see if this were truly the case. Inasmuch as the diameter of the neural canal remains fairly constant except for the enlargement at the tip, and extraependymal mitoses were rarely seen, counts were based on the numbers of mitoses per cross section of ependyma. Every third section was counted, giving the ratio of mitoses per segment for the various stages. This was roughly the same from the 6th day post-amputation through the 30th. During all stages mitotic figures averaged a frequency of about one for every two cross sections of ependyma. In order to see if there were any concentration of mitotic activity along the A-P axis at any given stage, the counts of the total length of the spinal cord were divided into 4 equal parts, each representing 25% of the total length. Figure 2 shows the ratio of mitoses per section per quarter of total length for each of the stages studied. It can be seen that the variation among the quarters is random.

These counts give no indication of the meristem-like growth of the apical bulb such as has been described by Stefanelli and Capriata ('44). The fact that the terminal bulb disappears by the end of the second week and the fact that the cord subsequently continues its elongation at approximately the same rate, make the function of the bulb difficult to comprehend. However, the alteration of its cells from highly columnar ependymal cells with oval nuclei to cuboidal cells with large spherical nuclei suggests a cytoplasmic and nuclear reorganization which may be a prerequisite to subsequent mitotic activity. The absence of a focal point of mitotic activity would indicate that the spinal cord grows equally throughout its length during the whole course of regeneration.

Blastema. The blastema, which occupies the space between the spinal cord and the skin laterally and extends posteriorly a short distance beyond the regenerating spinal cord, first appears as muscle degeneration gets under way. Cells from

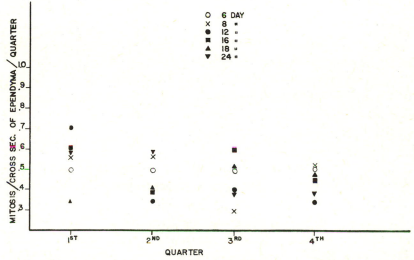

Fig. 2   Figure illustrating the random distribution of mitoses with respect to the A–P axis of the regenerating spinal cord.

the blastema collect under the ventral half of the spinal cord and begin to secrete a matrix, thereby indicating that they are now immature cartilage cells. Later other cells from the blastema apply themselves tangentially to this rod of cartilage. The salient feature of the whole process is that the blastema cells divide rapidly while the chondrifying cells hardly divide at all. The regenerated portion of the tail was divided into 4 longitudinal quarters again, as in the calculation of mitotic activity in the spinal cord. Mitotic indices (number of mitoses per 1000 cells) were calculated for the blastema in each of the quarters of the 6, 8, 10, 12, and 24 day regenerates. The highest mitotic index per quarter was determined for each stage. Quarters having a mitotic index not less than one-half of the highest mitotic index were arbitrarily regarded as having a high mitotic index. Quarters having a mitotic index of less than one-half of the highest mitotic index were regarded as having a low mitotic index. It is to be emphasized that the designations ''high mitotic index'' and ''low mitotic index'' for a given stage have a numerical value only in reference to that stage. The high mitotic index of the 8 day blastema does not have the same absolute value as the high mitotic index of the 12 day blastema.

Two main sources of error in counting the blastema cells are (a) inability to distinguish sharply between the most posterior blastema cells and the connective tissue cells of the fin and (b) difficulty in distinguishing between blastema cells and the definitive connective tissue cells of the tail proper. This error probably begins to appear in the proximal region of the 12 day regenerate. By 24 days, most of the cells counted as blastema cells are most likely differentiated connective tissue cells, except in the region of active cartilage formation.

The data obtained by this method indicate that on day 6 high mitotic activity is confined to the proximal half of the newly established blastema. From day 8 through day 10 mitotic activity is high through the greater extent of the blastema. By day 12 the definitive cartilage has been laid down proximally and the mitotic activity has dropped in that region. High mitotic activity is confined to the distal portion of the regenerate, where cartilage is actively forming. Subsequently the region of high mitotic activity in the blastema moves posteriorly as the new cartilage is forming, as might be expected if the blastema cells are converted chiefly into cartilage. See figure 1.

Mitotic indices for the cartilage were calculated with reference to the boundaries dividing the blastema into 4 parts of equal A-P extent. The mitotic index of the differentiated

cartilage at any stage never rises above 20–40% of the highest
mitotic index found in the blastema of the same stage. Furthermore, the region of *forming* cartilage does not contain
any mitotic figures at all.

The average mitotic index for the entire blastema at each of
the 5 stages studied is seen in figure 3. The 6 day blastema is
dividing rapidly and by the 8th day the rate has increased
further. Subsequently mitotic activity decreases, sharply
at first, gradually after day 12.

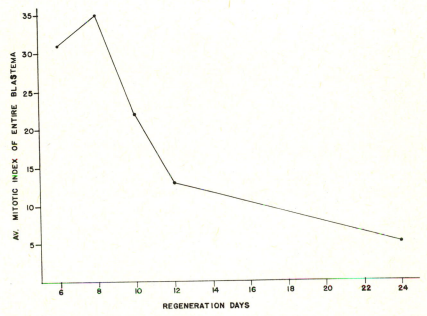

Fig. 3   Graph of the changes in the average mitotic index (number of mitoses
per 1000 cells) of the whole blastema with time.

A parallel phenomenon involving cellular density of the
blastema is shown in figure 4. There the average number of
blastema cells per cross section is plotted against each of
the 4 quarters of the 5 chronological stages studied. A general
decline of cell density with time at comparable levels in the
blastema is seen. When one considers that the average cross-
sectional area of the blastema increase directly with age, the

decrease in cell density is even more marked. See figures 9–12.

Taken together, the graphs and histological description suggests the following temporal sequence of events leading to the formation of cartilage from the blastema: (1) rapid multiplication of the blastema cells, originally formed by migration of single nucleated cells from the ends of muscle

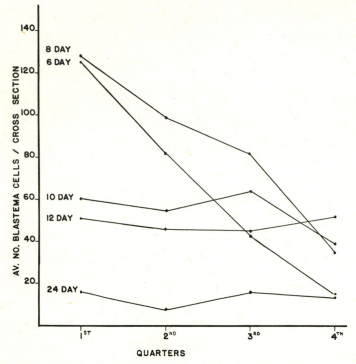

Fig. 4   Graph indicating a progressive decrease in cell density with time at comparable levels of the blastema.

fibers, (2) aggregation of blastema cells to a position adjacent to the motor surface of the spinal cord where the cells form cartilage, (3) decrease in the mitotic rate of the blastema cells following aggregation of cells under the spinal cord, (4) continued deployment of blastema cells into cartilage, depleting the blastema faster than it is formed, (5) formation of the definitive amount of cartilage with the concomitant disappearance of the blastema, (6) low grade mitotic activity in the differentiated cartilage contributing to further growth. The sequence first comes into play in the region of the amputation plane and sweeps posteriorly as the regenerating tail increases in length.

Muscle. After amputation, the most severely injured fibers are seen to degenerate, the pattern being loss of striations and fibrils, increase of sarcoplasmic acidophilia, and dissolution of the fibers. Less severely injured fibers may be seen to liberate single nuclei, surrounded by a scanty rim of cytoplasm from the cut ends of the fibers. All stages of this process may be observed. A nucleus well within a muscle fiber is an elongate oval just under the sarcolemma. Nuclei found closer to the open end are more round and those seen in the process of emergence bulge out into the connective tissue. Once beyond the confines of the differentiated portion of the fiber, the nucleus is quite round and it is indistinguishable from the mesenchyme cells of the blastema. These cells with round nuclei have never been observed to form myofibrils or otherwise give indications of contributing to the regenerating muscle. There is no direct proof, however, that they do not.

Formation of new muscle is seen only in association with the terminal portions of healthy, pre-existing muscle fibers, especially those in the lateral regions of the stump. The regenerating muscle is characterized by a strongly basophilic sarcoplasm. As they grow posteriorly, the pioneering myoblasts always keep within 20 μ of the skin.

The young myoblasts continue to grow in length and to a lesser extent in thickness, but the multiplication of nuclei does not keep pace with the increase of sarcoplasm. As in ontogeny, this process reduces the high nuclear-cytoplasmic ratio of the early myoblasts to the lower level of mature fibers. As the myoblasts grow, they also differentiate. At a much later stage the sarcoplasm assumes its definitive staining properties. The final step in the maturation process is the movement of the nuclei from the center of the fiber to their final position immediately under the sarcolemma.

Mitotic divisions of myoblast nuclei are found infrequently. This may be due to any of several factors: (a) the dense packing of nuclei makes detection of mitotic figures difficult (b) muscle nuclei may require much less time than mesenchymal cells to complete a mitotic cycle (Lewis and Lewis, '17) (c) much nuclear division may be done amitotically (Volkmann, 1893, Le Gros Clark, '46).

Editor's Note: The following is a condensation of the material on pages 15 through 26.

Experimental

Role of the spinal cord in the induction of cartilage

A segment of spinal cord was grafted to the side of the tail of another animal, which subsequently induced segmented cartilage and reorganized muscle near it. Regenerates of such structures formed recognizable tail elements.

Role of the spinal cord in tail axiation

The spinal cord was deflected into the lateral musculature of the tail and amputated. The axis of the subsequent regenerate was determined by the deflected cord.

Ability of anterior regions of the spinal cord to support tail regeneration

Trunk or brachial segments of spinal cord were transplanted to the area normally occupied by sacral and anterior tail segments of the cord. After amputation, normal regenerates were produced even when the anterior–posterior axis of the cord was reversed. Reversed medulla did not give good regenerates.

Inability of the white matter alone to support tail regeneration

When the grey matter was dissected from a segment of spinal cord and amputation subsequently made through that area, tails were not regenerated.

Localization of the cartilage-promoting factor in the motor half of the spinal cord

When a segment of the sacral spinal cord was rotated in the dorso-ventral axis by 180 degrees, formation of cartilage in subsequent regenerates was also rotated.

Independence of tail regeneration from the presence of skeleton in the stump

When spinal cord and associated muscle, but not cartilage, was depleted to the dorsal musculature, subsequent regenerates contained segmented cartilage units.

Source of muscle in the regenerated tail

When muscle was excised unilaterally from a segment of tail and amputation made through that area, regenerates had a concommitant unilateral deficiency of muscle.

DISCUSSION

The results of the foregoing experiments indicate at least two functions of the spinal cord in tail regeneration: (a) It induces cartilage from the blastema cells. (b) It serves as a locus of organization for the structure of the whole tail, all other tissues being oriented with respect to it.

The role of the spinal cord in determining the axis of the regenerate is still obscure, but tentatively may be divided into two parts, one the induction of cartilage, the other a purely mechanical factor produced by the growing spinal cord. Harrison (1898) has shown that in the rapidly growing tail bud of the frog embryo the skin of the tail is under tension due to the more rapid growth of the underlying tissues. Such a tension may be postulated for the regenerating tail, not only on the skin but, more importantly, also on the ground substance of the connective tissue below, giving its ultra-structure an A-P orientation. Such an oriented ultra-structure would provide a preferential pathway for the advance of the amoeboid tips of the muscle buds. In situations where a regenerating spinal cord and its associated tensions are absent, such as deplants of muscle or of muscle and notochord, muscle regeneration is quite limited. Comparable results

313

were obtained by Thornton ('42). This investigator transplanted tail muscle to the forearm from which the humerus had previously been removed. After the graft had healed, the limb was amputated through the upper arm. The transplanted tail muscle regenerated only to a limited degree, forming a few slender strands under the skin of a regenerate otherwise composed of connective tissue similar to that of the fin.

Induction may be defined as a reaction between two tissues such that one, the inductor, causes the second to differentiate into a cell type other than it would have, had the inductor not been present. Theories of induction based on the two most thoroughly analysed situations, the chorda-mesoderm-neural ectoderm and the eye cup-lens formation, have emphasized the necessity for intimate contact between the inducing and reacting tissues. For instance, McKeehan ('51) has shown that the induction of the lens by the eye cup is prevented if a thin membrane is interposed between the two. (However, see Grobstein, '53.) In the regenerating tail the relationship between the tissues is quite different. Instead of two well-organized epithelial layers in apposition, there is a well defined spinal cord surrounded by, but separate from, a loose aggregation of blastema cells. Whether the cartilage-promoting factor of the cord operated directly on the blastema cells, causing them to aggregate as pro-cartilage cells next to the motor surface of the cord, or whether the blastema cells aggregate under some other influence and then become transformed into cartilage cells is not known. In either case induction is not predicated upon surface contact between the involved cells.

As stated earlier, muscle has never been observed to regenerate without the presence of well-formed, healthy muscle in the stump. This is similar to the findings of the pathologists Volkmann (1893) and Waldeyer (1865), of Naville ('22) on histological study of regenerating tadpole tails, and of Speidel ('38) on direct *in vivo* observations of regenerating muscle fibers, also in the tadpole tail. Likewise Le Gros Clark ('46) has shown that in the rabbits, buds from healthy muscle will grow down the old endomysial tubes of muscle which had previously been killed. The persistence in the regenerate of muscle deficiencies in the stump argues for an active role by the stump musculature in the regeneration of muscle. If the stump muscle is not the direct source of the myoblasts, then it may serve as a pattern for the alignment of myoblasts derived from the blastema.

The origin of blastema cells seems dependent upon the presence of muscle or some cells associated with muscle, e.g. connective tissue cells. Spinal cord deplanted to the dorsal fin, either by itself or with notochord, will produce neither muscle nor cartilage. The connective tissue cells of the fin do not form a blastema, either through being too few in number or by being unable to respond to the spinal cord. Whenever tail muscle, together with its associated connective tissue cells, is present with the spinal cord, either in the fin or in the body of the tail, complete regeneration of all tail tissue can take place.

The study of normal regeneration indicates that muscle fibers contribute directly to the blastema. The muscle begins to degenerate within 24 hours after amputation. Within 48 hours the blastema has begun to form. During this time nuclei, surrounded by a small amount of cytoplasm, were seen to leave the ends of the cut muscle fibers and become indistinguishable from the mesenchyme cells of the blastema. Maurer ('39) has observed the transformation of muscle cells into connective tissue cells in mouse and chick material, both in histological section and *in vitro*. In the former case, injured muscle fibers produce connective tissue cells by two methods: (a) by emigration of single nuclei with a bit of cytoplasm from the ends of the muscle fibers and (b) disintegration of whole muscle fibers into uninucleate cells. The transformation of muscle cells into connective tissue cells was followed by regeneration of muscle by means of terminal buds from

healthy fibers. In tissue culture the transformation of muscle
into connective tissue cells was complete, no regeneration of
muscle occurring. The connective tissue cells so formed do
not stain like muscle cells, but assume specific connective
tissue staining properties as soon as they become free cells.
Chèvremont ('48) also reports the *in vitro* transformation
of muscle fibers into connective tissue cells by means of application of choline.

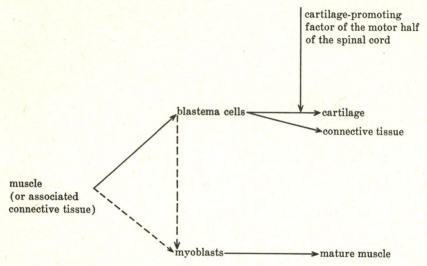

Fig. 7   Figure illustrating the tissue relationships postulated for the process of
regeneration.

These experiments stress the fact that terminally differentiated muscle cells can give rise to connective tissue cells
of a fibroblast nature. The modulation of fibroblasts into the
various specialized connective tissues, e.g. cartilage, has been
stated many times (Fischer, '46; La Croix, '51; Urist and
MacLean, '51). It is worth stressing that this transformation
is undirectional. The converse — fibroblasts into muscle —
has not yet been established. (However, see Levander, '49.)

Thus we are left with the following interpretation. (See
fig. 7.) Injured muscle gives rise to blastema cells which,
under the influence of the cartilage-promoting factor of the

motor half of the spinal cord, forms a rod of segmented cartilage. The blastema cells probably also form the connective tissue of the regenerate. The regenerated muscle may arise either from myoblasts budded off from the muscle of the stump or from cells of the blastema. In the latter case their differentiation into muscle is dependent upon some influence from the muscle of the stump. This relationship — cartilage coming from the blastema and muscle forming in contact with the muscle of the stump — probably exists in limb regeneration as well. It is more apparent in tail regeneration because the greater length of the tissues to be regenerated separates these two processes in time and space. Normal regenerated limbs are often deficient in muscle (David, '34). Preliminary experiments involving deplants of groups of 3 to 6 eight day limb blastemata to the dorsal fin have shown that the resulting limbs have developed all the typical cartilages, but have produced no muscle (Holtzer, Holtzer and Avery, '54). That the blastema cells are converted primarily into skeletal elements received further corroboration from Fritsch's ('11) experiments. He performed a radical extirpation on limb tissues including the humerus, limb girdle, and limb musculature. Though he does not specifically discuss muscle in connection with limb regeneration, his figures clearly show that the regenerated limb consists of skeletal elements only. No muscle has developed from the blastema.

The dissociation of the regeneration of tail muscle and the induction of cartilage can now be made. In the experimental section, mention was made of the limited muscle regeneration that took place in the absence of spinal cord. In such situations cartilage failed to appear. The mere fact that muscle can regenerate does not indicate the presence of conditions proper for the formation of cartilage. Similar findings were obtained by Thornton ('42) when he amputated through the upper arm after tail muscle had replaced the humerus. Muscle regenerated slightly, but no cartilage appeared at all. (However, see Liosner and Voronzova, '35.)

In conclusion, it can be said that the blastema cells are neither strictly determined nor totipotent. Absence of the notochord in the regenerate demonstrates the inability of blastema cells to differentiate into this cell type. On the other hand blastema cells will differentiate into cartilage only by virtue of the cartilage-promoting factor in the motor half of the spinal cord.

SUMMARY

1. The axial skeleton of the regenerated tail is a segmented cartilagenous rod. The larval urodele notochord does not regenerate when the tail is amputated. Each segment represents a centrum of a vertebra and will, in time, develop neural and haemal arches.

2. The cartilagenous rod is induced from the tail blastema by the regenerating spinal cord. This induction takes place regardless of the presence or absence of notochord or cartilage in the stump.

3. Cartilage and connective tissue are probably the only tissue types formed by the blastema.

4. Though the cartilage-promoting factor responsible for the tail vertebrae is found throughout the A-P extent of the spinal cord, it is confined to the motor portion.

5. The spinal cord is also responsible for the axiation of the regenerated tail — the position of the centra, the structuring of the muscle, the angle the regenerated tail makes with the stump.

6. When the spinal cord is absent from the amputation plane, the tail does not regenerate. Spinal ganglia by themselves are unable to support tail regeneration.

7. Regeneration of muscle is not seen unless muscle is present in the stump. Regenerated muscle is the product either of (a) myoblasts budded from muscle fibers in the stump, or (b) of myoblasts derived from blastema cells. In the latter case, the muscle of the stump influences in some way the formation of myoblasts from blastema cells.

LITERATURE CITED

CHÈVREMONT, M. 1948 Le système histiocytaire ou réticulo-endothélial. Biol. Rev., *23*: 267–295.

CLARK, W. E. LE GROS 1946 An experimental study of the regeneration of mammalian striped muscle. J. Anat., *80*: 24–36.

DAVID, A. 1934 La contribution du material cartilagineux at osseux au blastèmes de régénération des membres chez les amphibiens urodèles. Arch. Anat. micro., *30*: 217–234.

DETWILER, S. R., AND R. CARPENTER 1929 An experimental study of the mechanism of coordinated movements in heterotopic limbs. J. Comp. Neur., *47*: 427–441.

FISCHER, A. 1946 Biology of Tissue Cells. Cambridge University Press, Cambridge.

FRITSCH, C. 1911 Experimentelle Studien über Regenerationsvorgänge des Gliedmassenskelets der Amphibien. Zool. Jahrb., *30*: 377–471.

GOSS, R. 1953 Regeneration in anuran forelimb following removal of the radioulna. Anat. Rec., *115*: 311.

GROBSTEIN, C. 1953 Morphogenetic interaction between embryonic mouse tissues separated by a membrane filter. Nature, *172*: 869.

GROBSTEIN, C., AND G. PARKER 1954 *In vitro* induction of cartilage in mouse somite mesoderm by embryonic spinal cord. Proc. Soc. Exp. Biol. Med., *85*: 477–481.

GROBSTEIN, C., AND H. HOLTZER 1955 *In vitro* studies of cartilage induction in mouse somite mesoderm. J. Exp. Zool., *128*: 333–356.

HARRISON, R. G. 1898 The growth and regeneration of the tail of the frog larva. Arch. f. Entwmech. d. Org., *7*: 430–485.

HOLTZER, H. 1950 Differentiation of the regional action systems in the urodele spinal cord. Anat. Rec., *108*: 615–616.

——————— 1952 An experimental analysis of the development of the spinal column. II. The dispensability of the notochord. J. Exp. Zool., *121*: 573–592.

HOLTZER, H., AND S. R. DETWILER 1953 An experimental analysis of the development of the spinal cord column. III. Induction of skeletogenous cells. J. Exp. Zool., *123*: 335–370.

HOLTZER, H., G. AVERY AND S. HOLTZER 1954 Some properties of limb blastema cells of salamanders. Biol. Bull., *107*: 313.

HOLTZER, H., S. HOLTZER AND G. AVERY 1955 An experimental analysis of the development of the spinal column. IV. Morphogenesis of tail vertebrae during regeneration. J. Morph., *96*: 145–168.

LA CROIX, P. 1951 The organization of bones. Blakiston Co., New York.

LEVANDER, G. 1949 On tissue induction. Acta Path. et Microbiol. Scand., *26*: 113–141.

LEWIS, W. H., AND M. R. LEWIS 1917 The duration of the various phases of mitosis in the mesenchyme cells of tissue cultures. Anat. Rec., *13*: 359–367.

LIOSNER, L. D., AND M. N. VORONZOVA 1935 Die Regeneration des Organs mit transplantierten ortsfremden Muskeln. 1. Mitteilung. Zool. Anzeig., *110*: 286–290.

LOCATELLI, P. 1929 Der Einfluss des Nervensystem auf die Regeneration beteiligen Gewebe in Schwanz der Xenopus Larvae. Rev. Suisse Zool., *53*: 683–734.

MCKEEHAN, M. 1951 Cytological aspects of embryonic lens induction in the chick. J. Exp. Zool., *117*: 31–64.

MAPP, F. 1950 Descriptive and experimental studies on the regeneration of the anuran notochord. Ph.D. thesis, University of Chicago.

MAUER, G. 1939 Die histologischen Vorgänge im verletzten Gewebe quergestreiften Muskulatur und ihre Klärung mit Hilfe der Gewebekulture. Arch. Exp. Zellforsch., *23*: 125–168.

NAVILLE, A. 1922 Histogènèse et régénération du muscle chez les anoures. Arch. de Biol., *32*: 37–171.

NEEDHAM, A. E. 1952 Regeneration and wound healing. Menthuen and Co., Ltd. London.

NICHOLAS, J. 1955 Analysis of Development. Ed. by B. H. Willier, P. Weiss and V. Hamburger. W. B. Saunders Co., Philadelphia and London. Section XIII, Chap. 2, 674–698.

SPEIDEL, C. C. 1938 Studies of living muscles. I. Growth, injury, and repair of striated muscle as revealed by prolonged observation of individual fibers in living frog tadoples. Am. J. Anat., *62*: 179–235.

STEFANELLI, A., AND A. CAPRIATA 1944 La rigenerazione del midello spinale nella coda rigenerata dei Tritoni *(Triton cristatus e Triton taeniatus)*. Ricerche di Morfol.

THORNTON, C. 1942 Studies on the origin of the regeneration blastema in *Triturus viridescens*. J. Exp. Zool., *89*: 375–390.

URIST, M., AND F. MACLEAN 1951 Osteogenetic potency and osteogenetic inductor substances of periosteum, bone marrow, bone grafts, fracture callus, and hyaline cartilage transferred to the anterior chamber of the eye. Metabolic Interrelations 3rd Conference. Josiah Macy Jr. Foundation, pp. 55–89.

VOLKMANN, R. 1893 Über die Regeneration des quergestreiften Muskelgewebes beim Menschen und Säugethier. Beiträge path. Anat. und Path., *12*: 233–278.

WATTERSON, R., I. FOWLER AND B. FOWLER 1954 The role of the neural tube and notochord in development of the axial skeleton of the chick. Am. J. Anat., *95*: 337–400.

WALDEYER, W. 1865 Über die Veränderung der quergestreiften Muskulatur bei der Entzündung und dem Typhusprozess sowie über die Regeneration derselben nach Substanz defecten. Virchow's Arch. f. path. Anat. u. physiol. u. f. Klin. Med., *34*: 473–513.

WEISS, P. 1924 Die Funktion transplantierter Amphibienextremitäten. Arch. f. Entwmech. der Org., *102*: 635–672.

———— 1925a Abhängigkeit der Regeneration entwickelter Amphibienextremitäten vom Nervenssytem. Arch. f. Entwmech. der Org., *104*: 317–358.

WEISS, P. 1925b Unabhängigkeit der Extremitäten-regeneration vom Skelett. Arch. f. Entwmech. der Org., *104*: 359–394.

————— 1950 The deplantation of fragments of nervous system in amphibians. I. Central reorganization and the formation of nerves. J. Exp. Zool., *113*: 397–462.

WINICK, M. 1952 The antagonistic action of nerve and bone in a balance of tissue existing in the limbs of adult newts during the process of regeneration. Anat. Rec., *113*: 92.

PLATE 1

EXPLANATION OF FIGURES

8 Cross section through the ampulla in the posterior portion of the spinal cord of a six day regenerate. Compare the diameter of the neural canal in this figure with that in figure 9.

9 Cross section through a six day regenerate just posterior to the amputation plane. Blastemal density and mitotic activity are high at this time. The spinal cord is the only formed structure present in the regenerate. Most of the blastemal cells are found ventral to the spinal cord.

10 Cross section through an eight day regenerate just posterior to the amputation plane. The density and mitotic activity of the blastema are at their peak. The cartilagenous rod has just begun to form under the spinal cord. Muscle has not yet regenerated this far posteriorly.

11 Cross section through a twelve day regenerate just posterior to the amputation plane. Here dense clumps of myoblasts may be seen under the skin. The cartilage rod is larger, though not yet terminally differentiated. Several blastemal cells are still to be seen scattered between the cartilage and spinal cord and the myoblasts. The density and mitotic activity of these blastema cells are much less than those of the eight day regenerate.

12 Cross section through a 24 day regenerate just posterior to the amputation plane. The cartilage has increased in size. The matrix evidences capsule formation, while the cells in the lacunae have shrunken. At this stage and level the muscle has terminally differentiated. In most cases the nuclei occupy the periphery of the fibers, fibrillae may be seen in cross section, and the sarcoplasm has regained its normal staining properties. Such cells that remain in the connective tissue matrix are probably connective tissue cells, not blastema cells.

13 Cross section through a specimen in which a spinal cord from another animal had been grafted into the tail muscle one month previously. The tail of this animal was not amputated. Host spinal cord, notochord, and intact muscle are seen on the left, the ectopic cord on the right. Observe the cartilagenous rod, similar to that found in regenerating tails, lying below the grafted spinal cord. Also well demonstrated is the reorganization of the host musculature in response to the presence of the grafted cord. Bilateral muscle masses flank the graft and the induced cartilagenous centra. It is to be emphasized that the only tissue introduced into the host was a piece of spinal cord. The cartilage present was induced from host mesoderm.

14 Cross section through a regenerate with a unilateral graft of spinal cord. The host cord is seen in the center, the graft on the right. Due to the taper of the tail, the cords have come to lie next to each other. Further posteriorly they have fused completely. The amount of cartilage present is greater than that found in a normal regenerated tail at this level.

36

PLATE 1

INDUCTION IN TAIL REGENERATION
SYBIL W. HOLTZER

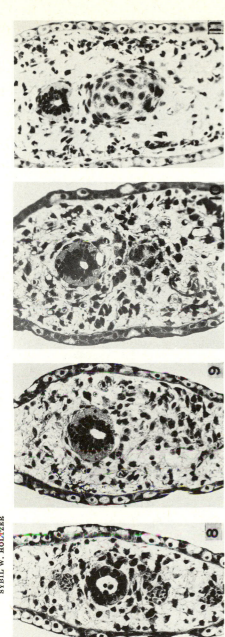

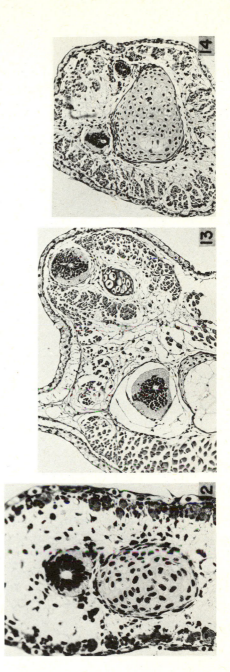

37

PLATE 2

15   Cross section through the stump of an animal two months after unilateral muscle ablation (on the left side). Though total muscle ablation was never attained, a conspicuous deficiency on the left side may be observed.

16   Cross section through the regenerated tail of the same specimen shown in the previous figure. Note the preservation of the asymmetrical muscle distribution in the regenerate. In total area the muscle of the right side exceeds that of the left by over 75%. In all other respects the regenerated tail was comparable to unoperated regenerates.

17   Cross section through the stump of the tail of an animal whose spinal cord was rotated 180° about its long axis. Note that the cartilage has formed in association with the motor surface, which is now dorsal. Below the spinal cord is the notochord.

18   Cross section through the regenerated tail of an animal whose spinal cord was rotated through 180° before amputation. The spinal cord regenerated in such a fashion as to maintain its motor surface dorsad and its sensory surface ventrad. The distribution of the white matter, the presence of ventral roots, and the location of the spinal canal clearly reveal the orientation of the neural tissue. In these specimens, as in all others, the cartilage forms in association, with the motor surface. Thus with respect to the neural tissue, the cartilage forms in the proper location, while with respect to the rest of the animal it is upside down. Neural arches, also upside down, have also been seen in these specimens.

19   Cross section through the regenerated portion of a deplant originally consisting of a piece of spinal cord and trunk muscle introduced into the dorsal fin. In the proximal portion of the regenerate the D–V axis of the spinal cord reflects the 90° rotation of the cord the deplant suffered when it was introduced into the tunnel in the dorsal fin. Observe the induced cartilage associated with the motor surface of the graft.

20   Cross section taken more posteriorly through the same specimen as the preceding one. The spinal cord, having spiraled through 90°, now presents its motor surface dorsally. The induced cartilage reflects this rotation, always being present along the motor surface. Note also the inverted position of the ganglia.

21   A frontal section of a 10 day regenerate. Note the area of juncture between the transected notochord and the forming cartilage.

38

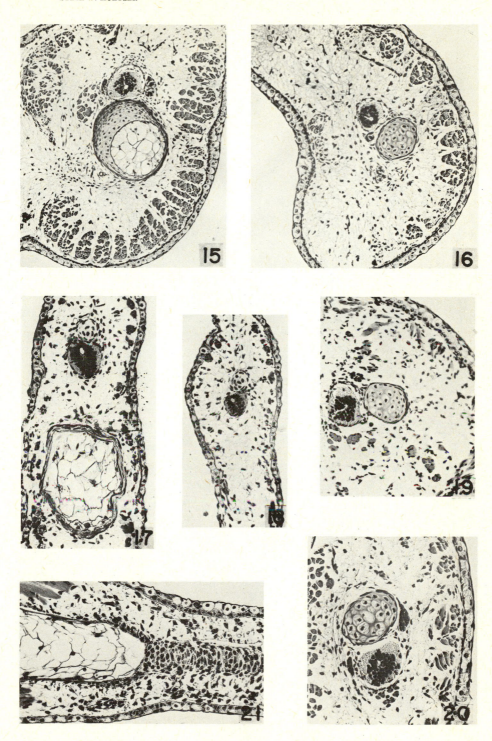

Systemic Factors in Regeneration

III

Editors' Comments on Papers 17 Through 26

The central questions about limb regeneration focus on the manner in which nerves are able to support it. Marcus Singer, originally trained as an anatomist, has done more to elucidate various facets of nerve action in this respect than any other individual. In earlier papers (1946, 1947) he demonstrates that a minimum number of nerve fibers is required if regeneration in newt limbs is to occur and that either motor or sensory fibers are competent (1943, 1945). The first of the papers included here shows that the number of fibers per cross section of a newt limb necessary for regeneration is very similar along the length of the limb, except for the hand, in which the threshold is lower. Thus the permissive role of nerves can be associated with the number of fibers interrupted by the area of the amputation plane. In a series of observations (indicated in Singer's bibliographic references), he shows that organisms that can regenerate limbs have a number of nerve fibers per cross section of limb above a certain threshold; those that cannot have fewer nerves. However, one anuran, *Xenopus,* has a value below the threshold and can still regenerate limbs. In the paper written with Rzehak and Maier, Singer shows that while the number of fibers in *Xenopus* is fewer, the average cross-sectional area per fiber is many times greater. *Rana* also has fibers larger than those in newts (*Triturus*), but if one compares the total cross-sectional area of axoplasm to that of the limb, *Xenopus* and *Triturus*, which both regenerate, have values higher than the permissive threshold, while *Rana*, which does not regenerate, does not. Thus in seeking the neurotrophic factor, one's attention must be directed to some aspect of the cross-sectional area of the axoplasm, probably the total value of axoplasm transected by an amputation plane. To date no one has demonstrated recuperation of regeneration in a denervated limb by any cell-free substance, nervous in origin or not. The causes underlying this failure are as perplexing both theoretically and medically, as they are significant.

In another investigation with Ellen Mutterperl, Singer shows that threshold requirements are not constant; regeneration occurs in transplanted *Triturus* limbs although innervation is far less than that ordinarily necessary. As seen in earlier papers in this volume, responsiveness of the tissues actually undergoing the transformation is variable, and in this case, plastic as well.

Another rather outstanding exception to the notion of a specific nerve threshold for limb regeneration was discovered by Charles Yntema. If two salamander larvae are parabiosed, and the developing central nervous system is removed from one of them, the limbs that form on the denervated individual are very sparsely supplied with nerves or have none at all. Such limbs are able to regenerate. However, if these aneurogenic limbs are allowed to become innervated and are then denervated and amputated, regeneration is prevented, as it is in denervated limbs of conventional ontology. Singer has suggested that limbs become "addicted" to nerves once they are exposed to them, and that a neural factor is apparently produced by the limb cells themselves if they have never been innervated. Once that occurs, the ability of the limb cells to produce that factor is quenched and the function assumed by the invading nerves.

At any rate, it is clear that some cells of aneurogenic limbs differ from those of denervated ones. An ingenious investigation by Steen and Thornton provides insight into these differences. If aneurogenic internal limb tissues are replaced by those from normally innervated limbs, regeneration will occur. But if skin from normal limbs is transplanted to denuded aneurogenics and amputation is made through the transplanted surface, regeneration is inhibited. Thus, the tissue normally affected by nerves would appear to be the skin, regardless of the fact that regeneration will occur in asensory limbs in which no nerves invade the distal epidermis (Thornton, 1960). One must then postulate either that the epidermal–neural interaction is indirect or that the neural factor is able to be transported to the skin by nonneural mechanisms.

An interesting sequel to the inability of reinnervated aneurogenics to regenerate after denervation was discovered by Thornton and Thornton. If aneurogenic limbs are transplanted to a normal host, allowed to become innervated, and then maintained in a denervated condition for several weeks, they gradually lose their nerve dependence. This situation seems to parallel, in some respects, the reduced neural thresholds for normal transplanted limbs; evidently it is a general property of amphibian limb tissue to be able to slowly rehabilitate itself from a previous state of neural addiction. A good deal of effort has been expended in looking for the neurotropic factor without much success; an approach which must be at least as fruitful might be to examine mechanisms that enable these exceptional limbs to free themselves of the necessity for its presence, even during early regenerative stages.

Amputation and regeneration of a limb presumably involves stress responses and certainly involves considerable growth of skeleton and musculature. Endocrine factors can reasonably be expected to be associated with these activities, as indeed they are. Oscar E. Schotté, in a long series of experiments (in summary, Schotté, 1961) has implicated the role of adrenal products in limb regeneration. The paper included here, with J. F. Wilber, illustrates an astonishing amount of limb regeneration in postmetamorphic frogs as a result of implanting adrenal tissue into animals normally unable to regenerate. As Schotté points out, the grafts contained some kidney tissue, which has unspecified stimulatory properties, and there are other considerations involving feedback regulation of the secretory activity of the glands, but the regenerates, heteromorphic though they are, are still the best yet induced in frogs by any means.

A permissive role of pituitary hormones was strongly implicated by Wilkerson (1963), who demonstrated essentially normal regeneration of hypophysectomized newts after chronic administration of a preparation containing mostly growth hormone; Berman, Bern, Nicoll, and Strohman (1964) suggested that prolactin in the preparation could have given the response. Connelly, Tassava, and Thornton (1968) showed that prolactin did indeed give a response in hypophysectomized animals similar to that observed by Wilkerson. Roy Tassava, in his Ph.D. thesis, showed that hypophysectomy does not completely inhibit the ability of newts to regenerate and that either a prolactin–thyroxin combination or a preparation containing growth hormone and prolactin was sufficient to render survival and limb regeneration in the treated animals essentially normal.

From the endocrine work to date, one can make general assumptions about the role of the endocrine system in normal regeneration; it certainly does play a role. Unfortunately, we are still ignorant regarding the identity of normal hormone secretions (the protein hormones mentioned above were of ungulate origin) and the specific roles played by those unknown molecular species. This difficulty has been implicit, given the available endocrinological methodology. It is likely that this obstacle will be overcome in the next few years because of the increasing availability of methods for dealing with small samples with precision and sensitivity.

In any event, we now seem to have an acquaintance with the basic permissive agents of limb regeneration, such as trauma, neural and endocrine factors, and varying cellular response. What we conspicuously lack is a good, testable, unified theory of limb regeneration, or even an indifferent one, for that matter.

Another possible approach to the problem of neural action is simply to do a classical deletion experiment and then monitor the effects of this procedure upon the affected cellular machinery. Marc Dresden used this approach to measure changes upon denervation in synthetic rates of protein, RNA, and DNA of blastemal cells. The general pattern is a reduced ability to synthesize all of these entities.

Lizards are unique among higher vertebrate taxa in their ability to regenerate body appendages, in this case the tail. While the new tail lacks vertebrae and is otherwise easily distinguished from the original tail, it is a reasonable facsimile. S. B. Simpson discovered that the neural requirements for regeneration of this appendage are strikingly different from those of newt limbs; the presence of neurons is not required. However, if ependyma is not present in the stump, regeneration will not ensue. Surprisingly little has been done to determine other properties of ependyma since the appearance of Simpson's paper, but the unusual form of the regenerate, and the immediate survival value probably associated with autotomy, may indicate completely different regeneration mechanisms from those utilized by amphibians.

References

Berman, R., H. A. Bern, C. S. Nicoll, and R. C. Strohman, 1964. Growth promoting effects of mammalian prolactin and growth hormone in tadpoles of *Rana catesbeiana*. J. Exp. Zool., *156*: 353–360.

Connelly, T. G., R. A. Tassava, and C. S. Thornton, 1968. Limb regeneration and survival of prolactin treated hypophysectomized adult newts. J. Morph., *126*: 365–371.

Schotté, O. E., 1961. Systemic factors in initiation of regenerative processes in limbs of larval and adult amphibians. *19th Symposium of the Society for the Study of Development and Growth* (D. Rudnick, ed.), pp. 161–192. Ronald Press, New York.

Singer, M., 1943. The nervous system and regeneration of the forelimb of adult *Triturus*. II. The role of the sensory supply. J. Exp. Zool., *92*: 297–315.

———, 1945. The nervous system and regeneration of the forelimb of adult *Triturus*. III. The role of the motor supply. J. Exp. Zool., *98*: 1–21.

———, 1946. The nervous system and regeneration of the forelimb of adult *Triturus*. IV. The stimulating action of a regenerated motor supply. J. Exp. Zool., *101*: 221–239.

———, 1947. The nervous system and regeneration of the forelimb of adult *Triturus*. VI. A further study of the importance of nerve number, including quantitative measurements of limb innervation. J. Exp. Zool., *104*: 223–250.

Thornton, C. S., 1960. Regeneration of asensory limbs of *Ambystoma* larvae. Copeia, *4*: 371–373.

Wilkerson, J. A., 1963. The role of growth hormone in regeneration of the forelimb of the hypophysectomized newt. J. Exp. Zool., *154:* 223–230.

Reprinted from *J. Exptl. Zool.*, **104**, 251–265 (1947)

17

THE NERVOUS SYSTEM AND REGENERATION OF THE FORELIMB OF ADULT TRITURUS

VII. THE RELATION BETWEEN NUMBER OF NERVE FIBERS AND SURFACE AREA OF AMPUTATION

MARCUS SINGER

Department of Anatomy, Harvard Medical School, Boston, Massachusetts

ONE FIGURE

In preceding studies (Singer, '46a; '47) which have sought to analyze the influence which nerves exert on regeneration of the forelimb of the newt, the importance of number of nerve fibers was established. Each amputation level of the limb was found to require the presence of an adequate quantity of fibers, expressed as a threshold range, in order that regeneration of the limb proceed. Above the threshold range regeneration always occurred, whereas below that range it was invariably absent; within the range, regeneration appeared in some instances and not in others. As far as could be ascertained, quantitative requirements had to be met quite irrespective of the origin of the nerve fibers. Neither the spinal nerve nor the component (motor or sensory) of origin of the nerve quantity appeared to play a decisive role.

Although the influence of the nerve on the regeneration, following amputation of the limb at various levels, is expressed by the collective activity of many nerve fibers, the quantitative requirements for regeneration to occur differ for each amputation level (Singer, '47). The underlying cause of the different requirements at the several amputation levels was discussed in the preceding work (Singer, '47) where a number of possible explanations were considered. The possibility that

251

the size of the amputation stump or, conversely, the amount of limb amputated determines the number of nerve fibers needed for regeneration was early discounted as unlikely. If such were the case, one would expect the nerve requirements to decrease progressively along the length of the limb. Actually, similar fiber requirements were observed following amputation at lower and upper arm levels, although a considerable sector of the limb separates the 2 regions. Nor can regional variations in quantitative requirements be attributed to some progressive change in "intrinsic quality" of the nerve fibers as they traverse the limb. Conceivably such a qualitative difference might exist, but here again one would expect it to be expressed in a sequence along the length of the limb, which is not the case in the experimental observations.

Since the nerve influence expresses itself especially in the early phases of regeneration (Schotté and Butler, '44), it is reasonable to search for variations in the tissues of origin of the regenerate or of the very early regenerate itself which might possibly explain the individuality in fiber demands of each region of the limb. Two variables have appeared as most likely to influence the nerve requirements. For one, there may be variations in the competence of the tissues to respond to the neural influence; other things being equal, the tissues of one region of the forelimb may be more refractory and thus require a greater number of nerve fibers. A second variable would appear to reside in the varying quantity of tissue subjected to the regeneration process. There is a reasonable possibility that the tissue upon which the nerve fibers act varies in amount after transection of the limb at different levels, and, consequently, that the number of nerve fibers required for regeneration depends upon the quantity of regenerating tissue. The present study concerns itself with an investigation of nerve fiber requirements in relation to the quantity of regenerating tissue as well as with a comparison and evaluation of the data thus obtained for the different regions of the limb. For the purpose of arriving at these correlations, it is necessary to find a suitable means of measuring the amount of the regenerating tissue. The early regenerative processes to which the nerve regeneration is related (Schotté and Butler, '44) occurs at and immediately subjacent to the surface of amputation (Hertwig, '27; Thornton, '38; Butler and O'Brien, '42). This involves the mobilization of the blastemal cells from among the old tissues of the stumps and their accumulation at the cut surface. By virtue of this close relation between the surface of injury and the underlying regenerative responses, the area of the surface may be taken as a measure of the amount of tissue subjected to regeneration. It is evident that a larger surface will require a greater fiber number for the expression of its regenerative capacities than would a smaller one. The overall results of previous studies (Singer, '46a; '47) apparently bear out such a relation between fiber quantity and surface area of the amputation site.

MATERIALS AND METHODS

The procedures devised here were directed toward determining the surface area of amputation and the expression of the fiber requirements for regeneration as a function of the area. The measurements involved the areas of cross-section of the normal forelimbs in 6 different regions. The areas comprised the base of fingers 2, 3, and 4; the mid-hand region immediately distal to the fifth digital rudiment; the mid-lower arm level; and, finally, the upper arm about one-third of the distance proximal to the elbow. The normal limbs utilized for the estimation of cross-sectional areas were obtained from animals employed in previous studies. It must be emphasized here that area determinations of fixed and sectioned material does not in the strictest sense represent true areas of the amputation surface of the limbs of living animals since there is unquestionably considerable shrinkage of tissue as a result of histological procedures. Furthermore, the amputation surface is never absolutely flat and is evidently much larger than the histological material because of the swelling due to retraction and lumping of tissues attendant upon amputation. However, these limitations in the method hardly affect the conclusions derived from the experiments, since the conclusions are based on relative values for the different regions, each of which is subject to the same errors.

Eighteen normal limbs were employed in the study. In most cases, area determinations were made for the 6 regions of each individual limb. In some instances, however, only one or a few regions are recorded for a particular limb, the other levels being omitted because of faulty sectioning or staining. Sixteen area determinations were made for the upper arm and 12 each for the remaining regions. When the average of the area determinations for any region was divided by the number of fibers required for regeneration of that region as estimated in previous papers (Singer, '46a; '47), the nerve requirements for regeneration could be expressed as number of fibers per unit area.

In addition to determining the total area of the surface of amputation, the area of the exposed bone included in the cross-section was also determined. Subtraction of the bony area from the total area yielded the value for the soft tissues with which the nerve is presumably concerned in the process of regeneration. From these values and the fiber values, the corresponding number of fibers per unit area of soft tissues could be calculated. Both the fiber value per total area and that per area of soft tissues are presented in the description of results. From either value similar conclusions are derived, but we believe that the fiber value based on the unit area of soft tissues is more pertinent.

As for the method of determining the area, a cross-section was selected from serial sections of the normal limb (the mode of selection was the same as employed for nerve counts, see Singer, '46a; '47) and was projected on paper at a magnification of 59 times. The outer margins of the section and of the contained skeletal elements were outlined in pencil. The outlines were then traced with a planimeter and the readings recorded in square centimeters. Rather than presenting the determinations and subsequent calculations in the magnified units, we have converted these figures into absolute ones by dividing the planimetric readings by the square of the linear magnification. Finally, the decimal value so obtained was converted into the smallest whole number by expressing it in tenths of a millimeter squared, $(100\,\mu)^2$, instead of square centimeters, by multiplying by 10^4.

RESULTS

The surface area of amputation. The areas of the amputated surfaces as well as those of the transected bones are recorded in table 1 where the averages and standard deviations are also noted. Also noted in table 1 is the area of the soft tissues which was obtained by subtracting the average area of bone from that of the entire section. The largest area of cross-section, with or without the bone, occurs at the mid-lower arm level. The next in size is the upper arm, followed successively by the hand and digits. In decreasing order of size, the digits are 3, 2, 4. Because of inadequate information on regeneration in relation to fiber number, the first finger was not subjected to detailed planimetric study. Casual information, however, shows that it has the smallest area of cross-section. The greatest amount of skeletal tissue, compared on an absolute or relative basis, is found in the hand where the osseous elements constitute one-fourth to one-third of the area of cross-section.

The nerve requirements for regeneration of the limb. In previous studies (Singer, '46a; '47) the fiber quantity needed to evoke a regenerative response from each level of the forelimb was determined. It was observed that above a certain value the nerve number was sufficient to produce regeneration in all instances, whereas at other lesser values regeneration appeared in some instances and not in others. At still lower values, there was no regenerative response. The fiber values between the respective limits, which yielded regeneration in all instances or in none, denoted the threshold requirement. Consequently, the threshold value appeared as a range of fiber values. Within this range more instances of regenera-

TABLE 1

Area of the surface of amputation in (100 μ)².

ANIMAL	UPPER ARM Total area	UPPER ARM Area of bone	LOWER ARM Total area	LOWER ARM Area of bone	HAND Total area	HAND Area of bone	DIGIT 2 Total area	DIGIT 2 Area of bone	DIGIT 3 Total area	DIGIT 3 Area of bone	DIGIT 4 Total area	DIGIT 4 Area of bone
1	51.4	8.9	53.8	6.3	80.2	19.8	14.1	2.6	17.5	2.9	8.0	2.9
2	58.3	8.3			70.7	20.4	12.4	2.9	17.0	2.3	9.8	3.4
3	131.9	11.2	141.1	7.8	140.8	23.3	21.6	6.0	34.8	11.5	20.4	8.0
4	123.0	8.9	102.3	6.0	110.3	23.6	15.8	2.9	23.3	5.2	12.1	2.9
5	56.0	10.6	63.5	5.2	69.2	21.0	12.9	2.9	14.7	3.4	8.3	2.9
6	140.8	15.2	127.6	6.6	113.2	25.9	20.1	6.0	29.3	3.4	12.6	5.2
7	153.7	20.4	162.3	6.9	140.8	31.0						
8	106.3	11.5										
9	166.6	14.1	163.2	6.3	141.3	28.4	25.6	5.7	35.3	5.5	17.2	3.2
10	149.1	16.7	138.2	6.0	127.0	33.0	25.6	6.3	30.2	5.7	13.8	3.2
11	114.6	13.5	98.3	6.3								
12	102.8	6.0										
13	106.3	11.2										
14	58.3	8.9										
15	83.3	17.5	79.9	6.3	91.9	32.2	19.8	3.4	26.7	3.2	14.1	2.9
16	132.4	11.8	135.0	6.0	104.0	23.0	18.7	6.0	23.3	8.3	9.5	2.3
17			114.6	6.0	120.7	26.1	20.4	5.7	27.9	3.2	17.5	3.7
18							22.1	6.3	25.6	6.3	12.9	5.5
Average	108.4	12.2	115.8	6.3	109.2	25.6	19.1	4.7	25.5	5.1	13.0	3.8
Standard deviation	± 36.4	± 3.6	± 33.1	± 0.6	± 25.4	± 4.4	± 4.3	± 1.6	± 6.4	± 2.6	± 3.7	± 1.6
Area of soft tissues	96.2		109.5		82.6		14.4		20.4		9.2	

tion were observed the higher the fiber value. It would be possible to express each of the fiber values within the threshold range, extending from minimal to maximal per unit of the surface area. Such a procedure proves, however, to be overly cumbersome and of little use for comparison of different levels. Moreover, in the case of maximal and minimal quantities, it is hardly of value since these limits are ill-defined. Instead, the procedure employed here has been to express the fiber requirements for each level of the limb as a value representing, according to the overall data, enough fibers to induce regeneration in 50% of the animals. The number of nerve fibers needed to induce regeneration in half of the instances was interpolated from the data of the threshold ranges in the following manner.

Interpolation was performed by graphical means (fig. 1); the fiber number for each series of amputations within the threshold range (see table 9, Singer, '46a; tables 3, 6 and 9, Singer, '47) was plotted against the number of positive instances of regeneration expressed as a percentage. In this way a graph was obtained of the entire threshold range for each of the 6 regions of the limb and the probable number of fibers for the 50% value could be interpolated from the curve. As shown in figure 1, the per cent positive instances of each group do not follow strictly in sequence the progressive variation in fiber number. Because of this limitation in the data, the expected fiber quantity for regeneration in half the animals was interpreted broadly as a probable range which would reasonably include this value. The graphs of figure 1 were drawn with such a liberal interpretation in mind. Each solid curve represents a reasonable average of the threshold percentages. Two additional curves (dotted lines) are included in each graph to encompass reasonable maximum and minimum values as determined by the points which deviate most from the solid curve. The horizontal line drawn between the dotted lines represents the range of fiber values which could be expected to produce regeneration in one-half of the instances.

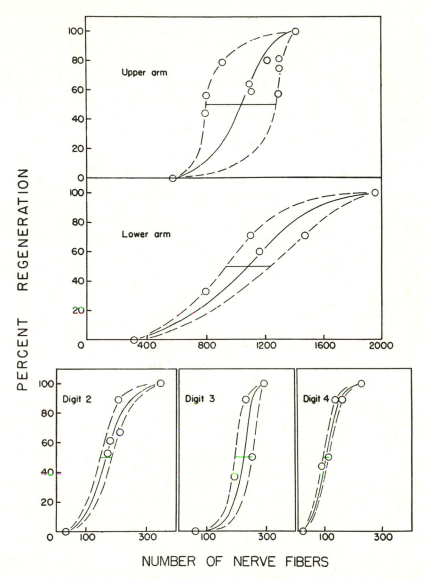

Fig. 1 Statistical determination of the fiber quantity which can be expected to evoke a regenerative response of the forelimb in 50% of the animals.

The point at which the line crosses the solid curve is the median value of this fiber range.

Although the above treatment provides a reasonable interpolation of the fiber quantity needed for regeneration in half of the cases, further extension of the limits of this range may be made without altering fundamentally the conclusions derived below. The 50% range and median value for each of the limb levels derived from figure 1 are recorded in table 2 (columns 1 and 2, respectively). In the case of the hand, no graphical representation was used since the threshold range drops from a fiber value of 506 which provides for 90% regeneration to 196 where regeneration is always absent. The 50% region thus falls within the range of 196 to 506 fibers; this range is listed in table 2 as the probable 50% fiber value and the median value is arbitrarily taken as the halfway point.

The nerve requirements for regeneration expressed in terms of unit area. The nerve requirements for regeneration in one-half of the instances, as interpolated from previous experimental data, were converted into number of fibers per unit area of soft tissue by dividing the range and mid-point of the range by the average area of the surface of amputation, both total and soft tissues, determined in the planimetric studies (refer to table 1). The results of these calculations are recorded in the appropriate columns of table 2.

TABLE 2

Fiber requirements per unit area of amputation.

	TOTAL FIBER REQUIREMENTS AT AMPUTATION SURFACE (SEE FIG. 1 AND TEXT)		FIBER REQUIREMENTS PER $(100 \mu)^2$ OF SURFACE AREA OF SOFT TISSUES		FIBER REQUIREMENTS PER $(100 \mu)^2$ OF SURFACE AREA INCLUDING BONE	
	Expressed as a range	Expressed as a single median value	Expressed as a range	Expressed as a single median value	Expressed as a range	Expressed as a single median value
Upper arm	800–1280	1040	8.3–13.3	10.8	7.4–11.8	9.6
Lower arm	940–1230	1085	8.6–11.2	9.9	8.1–10.6	9.4
Hand	196–506	351	2.4– 6.1	4.2	1.8– 4.6	3.2
Digit 2	145–183	164	10.1–12.7	11.4	7.6– 9.6	8.6
Digit 3	195–255	225	9.6–12.5	11.0	7.6–10.0	8.8
Digit 4	95–115	105	10.3–12.5	11.4	7.3– 8.8	8.1

With but one exception, the hand, the results show that the soft tissues of widely separated regions of the limb have remarkably similar nerve requirements. A similar range of fibers describes the nerve requirements of regeneration for regions as different in bulk, position and anatomy as the upper arm and fourth finger. Indeed, the values 8.3 to 13.3 fibers per $(100 \mu)^2$ cover the needs of the lower and upper arm and the 3 digits.

The notable similarity in the nerve demands of diverse regions of the forelimb is especially apparent when one compares the median value of the range (table 2). The fingers are most alike in fiber needs for unit quantity of surface tissues although the quantities required for the whole amputation area (see first and second columns of table 2) show wide differences. Digits 2 and 4 have the same nerve requirements per unit area of soft tissue and digit 3 deviates from this value only by approximately 3% (see columns 3 and 4). The nerve requirements per unit area of the upper arm, a region farthest removed from the digits, deviate by approximately 5% from digits 2 and 4. The largest deviation from the digit value is seen in the lower arm, but even here the deviation is relatively small when one considers that the total fiber and total area values of the digits and lower arm differ by many hundred per cent. Indeed, the deviation is still less if one should compare each of the median values to an average of them all. Thus it is apparent that when expressed in terms of the size of the surface area of amputation, the nerve requirements for regeneration prove to be similar in diverse regions of the limb.

In addition to fiber values for the unit area of soft tissue, others were calculated for the total area including the area of the bones and are noted in the fifth and sixth columns of table 2. The latter values in no way alter the conclusions derived from consideration of the soft tissues alone. The total area values are included here since there is no absolute certainty at present that the important area is that representing the tissues of the stump from which the blastema is arising rather than the total surface area over which these blastemal cells are accumulating.

The hand provides a notable exception to the general similarity of various regions of the limb in their nerve requirements per unit area of amputation. The fiber requirements are less than one-half that of any other region. Thus, the hand is approximately one-half as refractory, or, stated otherwise, twice as responsive to the nerve influence. Fiber quantities which cannot evoke a regenerative response in the arm or fingers will stimulate development in the hand. The basis for this difference in response of the hand to the nerve influence is not known; various possible explanations are presented below.

DISCUSSION

When a limb is amputated, the tissues at the surface of amputation are stimulated to regenerate by virtue of this injury. From part of these tissues mesenchymatous cells arise which accumulate and proliferate at the amputation surface forming the blastema of regeneration. It is apparent that the amount of tissue subjected to the process of regeneration and consequently the amount of mesenchymal tissue to appear at the amputation surface will vary according to the region of the limb sectioned. Transecting the arm exposes much more tissue than does sectioning a digit as will be appreciated by comparing the cross-sectional areas of these regions. Since the nerve fibers which are concerned with these early regenerative processes must be present in adequate quantity in order that regeneration proceed, it is logical to assume that the number of nerve fibers needed to effect the growth will depend on the quantity of regenerating tissue whether this tissue be that of origin of the blastema or of the blastema itself. This view is supported by the results of the present study in which the nerve requirements were expressed as a function of the surface area of amputation. Indeed, when the nerve requirements are treated in this fashion, they are observed to be relatively constant in widely separated regions such as the arm and the digits.

An exception to the notable similarity in nerve requirements of different regions of the limb was observed for the hand. Less than half of the number of fibers per unit area required elsewhere were needed for regeneration of the hand. Such a difference need not imply that the nerve requirements for regeneration of the hand are independent of the quantity of regenerating tissue. Considering the importance of quantity of regenerating tissue in establishing nerve needs in other parts of the limb, it is logical to assume that such a factor is operating here, too. However, it is quite evident that another factor, not present or not so pronounced in other parts of the limb, serves to increase the responsiveness of the hand. The nature of this factor is unknown, but it may comprise physiological or anatomical variations in the tissues of the hand. For example, there may be a greater abundance of the tissue of origin of the regenerate. Or the mesenchymal cells may be mobilized from greater depths than elsewhere in the limb. Or, simply, the tissues of the hand may be physiologically more competent to respond to the nerve. It should be noted, as has been done previously (Singer, '47), that, perhaps, a comparison of the hand with other regions of the limb is not justified since regeneration of the hand is somewhat atypical in its response to varying quantities of nerve fibers. Whereas, in the limb generally, a progressive decrease in number of regenerates was observed with decrease in fiber quantity, in the hand regeneration tended to occur either in the vast majority of instances or not at all. Observations showed that in many cases with a reduced nerve supply the regenerate appeared initially as a growth restricted to a small area of the surface, but then spread laterally to encompass all or a greater part of the area. In a few instances the regenerate remained restricted to a very small area developing into a localized small growth. These observations, being unlike those for other regions, suggest that conditions of regeneration obtain for the hand that are not apparent in the remainder of the limb.

Disregarding the peculiarities in the case of the hand, the ensemble of experiments shows that the nerve requirements for regeneration of each level of the limb are not influenced in any notable fashion by the extent of amputation, by adjacent levels of the limb, or by any distant region of the body. On the other hand, the nerve needs depend greatly upon the quantity of tissue subjected to the regeneration process as expressed in terms of the surface area of amputation. Thus, each region of the limb demands so much nerve activity dependent upon the amount of its responding tissue.

It is well at this point to summarize briefly some salient facts that have emerged from this series of studies of the action of the nerves on regeneration of the extremity. (Singer, '42–'47). As has been repeatedly pointed out in the past, and amply confirmed here, regeneration of the limb of the urodele depends on the presence of a nerve supply. Although the nature of this dependence is not known, it is not confined to a specific nerve component or spinal nerve. Apparently, all fibers, at least this is true for somatic motor and sensory ones, have the necessary quality to induce limb regeneration. This quality arises within the individual neurons and apparently bears no relation to impulse conduction since it occurs in centrifugal fashion in sensory neurons whereas the nerve impulse runs centripetally. Nor do the neurones depend on connections with the central nervous system for their activity in regeneration. Although each nerve fiber is active in the regeneration process, a certain quantity is required in order that regeneration occur. Thus, the activity of the individual nerve fiber cannot receive expression in the process of regeneration unless it is performed in conjunction with that of many other neurones. The number of fibers required for regeneration at any level is prescribed by that level itself and not by the nervous system, a distant level, or by the limb as a whole. The nerve requirement which any level makes is a function of the quantity of tissue at the amputation surface. This is shown by the fact that widely separated regions of the limb, having entirely different quantities of tissue at the amputation surface, have similar nerve requirements per unit amount.

Precisely what part the nerves play in the process of regeneration or how they act is not known. It appears that the influence provides one of many realizing factors for regeneration. Otherwise, it is non-specific, determining neither the nature of the new structure nor the amount of tissue formed (Weiss, '25; Schotté, '26).

SUMMARY

In the present work the quantity of nerve fibers required for regeneration of the forelimb of the newt has been expressed in terms of number of nerve fibers per unit area of amputation surface. When denoted in this manner, the requirements are seen to be constant in widely separated regions such as the digits and the upper arm. It is evident, therefore, that the fiber requirements of each level of the limb depend on the quantity of tissue present at the amputation surface, and not on the extent of injury, the size of the stump, other limb levels, or the number of nerve fibers available there or at another region of the limb.

The hand, unlike other parts of the limb, does not obey the rule of the constant of fibers per unit area. The hand requires for regeneration less than half the quantity of nerve fibers per unit area than are needed elsewhere. Consequently, it appears to be more sensitive to the influence of the nerves than other regenerating parts of the limb. Factors which might explain the divergence of the hand are discussed in the text, where a limited summary of our present knowledge of the influence of the nerve on regeneration is also presented.

LITERATURE CITED

BUTLER, E. G., AND J. P. O'BRIEN 1942 Effects of localized x-radiation on regeneration of the urodele limb. Anat. Rec., *84:* 407–413.

HERTWIG, G. 1927 Experimentelle Untersuchungen über die Herkunft des Regenerationsblastems. Sitzgsber. u. Abh. naturforsch. Ges. Rostock., *2:* 1–2.

SCHOTTÉ, O. E., AND E. G. BUTLER 1944 Phases in regeneration of the urodele limb and their dependence upon the nervous system. J. Exp. Zool., *97:* 95–121.

SINGER, M. 1942 The nervous system and regeneration of the forelimb of adult Triturus. I. The role of the sympathetics. J. Exp. Zool., *90:* 377–399.

———— 1943 The nervous system and regeneration of the forelimb of adult Triturus. II. The role of the sensory supply. J. Exp. Zool., *92:* 297–315.

———— 1945 The nervous system and regeneration of the forelimb of adult Triturus. III. The role of the motor supply. J. Exp. Zool., *98:* 1–21.

———— 1946 The nervous system and regeneration of the forelimb of adult Triturus. IV. The stimulating action of a regenerated motor supply. J. Exp. Zool., *101:* 221–239.

———— 1946a The nervous system and regeneration of the forelimb of adult Triturus. V. The influence of number of nerve fibers, including a quantitative study of limb innervation. J. Exp. Zool., *101:* 229–337.

———— 1947 .The nervous system and regeneration of the forelimb of adult Triturus. VI. A further study of the importance of nerve number, including quantitative measurements of limb innervation. J. Exp. Zool., *104:* 223–250.

THORNTON, C. S. 1942 Studies on the origin of the regeneration blastema in Triturus viridescens. J. Exp. Zool., *89:* 375–389.

WEISS, P. 1925 Abhängigkeit der Regeneration entwickelter Amphibien — extremitäten vom Nervensystem. Roux' Arch. f. Entw.-Mech., *104:* 317–358.

Reprinted from *Develop. Biol.*, **7**, 180–191 (1963)

18

Nerve Fiber Requirements for Regeneration In Forelimb Transplants of the Newt *Triturus*[1]

Marcus Singer and Ellen Mutterperl

Department of Anatomy, School of Medicine and Developmental Biology Center, Western Reserve University, Cleveland, Ohio

Accepted October 10, 1962

INTRODUCTION

In regeneration of body parts in the newt *Triturus* and in other low vertebrates, the nerves exert a controlling influence without which regeneration is not initiated (reviewed by Singer, 1952). All nerve fibers apparently have the quality to evoke growth, but for growth to occur the number of nerve fibers at the amputation surface must satisfy threshold quantitative demands. In the upper arm of the newt's forelimb approximately one-third to one-half of the fibers ordinarily present at the amputation surface are required to sustain regeneration of that part. If the number falls below this level, regeneration does not occur and the available fibers although individually active constitute an ineffective stimulus. The concept of threshold nerve requirements in regeneration of the limb has been extended to regeneration in frog and salamander larvae (van Stone, 1955; Deck, 1961). It also served as the theoretical basis for successful experiments in which regeneration of the limb was induced in nonregenerating forms, at first in the postmetamorphic frog (Singer, 1954) and later in the lizard (Singer, 1961; Simpson, 1961; Kudokotsev, 1962).

The quantitative demand which the amputation stump makes upon the nervous system is apparently a function of at least two factors. One is the quality and effectiveness of the constituent fibers; fibers may not be equal in their contributions. For example, finer ones may be less effective than larger; the effectiveness of the individual fiber may also be related to the number of times the parent fiber has

[1] Aided in part by grants from the American Cancer Society and the National Institutes of Health, Public Health Service.

180

divided; and, finally, the motor, sensory, and other fibers may not be equally effective. Consequently, the threshold number of fibers needed to evoke regeneration should vary according to the nature of the constituent fibers available at the amputation surface.

A second factor which may also operate in establishing the threshold fiber needs for regrowth is the capacity of the wound tissues to respond to the nerve. The lack of regeneration with a subthreshold number simply means that the tissues are not competent to respond to a low number of fibers even though the available fibers are presumably active. The relation between competence of tissues to respond to the nerve and the quantitative adequacy of the nerve supply is also illustrated in experiments upon the limb of adult frogs and lizards, ordinarily incapable of regenerating. In these instances the full nerve supply normally available at the amputation surface does not evoke regrowth but, when the nerve number is increased by rerouting nerves from another limb to the amputated one, then regeneration does occur. The normal nervous complement has the quality to evoke growth but the tissues could not respond to it unless the fiber number was augmented substantially from an additional source. The tissues of the frog and lizard limb in contrast to the salamander or tadpole limb are, therefore, refractory to the normal nerve content.

In the adult frog, regeneration of the limb was also induced by repeated irritation of the fresh amputation stump with hypertonic solutions of sodium chloride or other noxious agents according to the method of Rose (1945). However, regrowth does not occur if, in these instances, the stump is denervated; and so, induction of regeneration by tissue irritation occurs only against a background of nerve activity (Singer *et al.*, 1957). The results of repeated wounding were interpreted to mean that the capacity of the wound tissues to respond to the nerve was heightened, or, stated in another way, that the threshold fiber requirement for regrowth was decreased with prolonged insult of the wound. A consequence of this interpretation of the effect of injury is that the threshold nerve needs are not absolutely fixed but may vary according to the condition, and therefore the demands, of the wound tissues. This consequence is analyzed in the present work, where it is demonstrated that the quantitative nerve requirements for regeneration in the newt may be lowered substantially through appropriate experimental circumstances. The results enrich the concept of threshold nerve requirements, focus upon the importance of the response of wound tissues to the nerve, and allow us to assess the results of experiments by Yntema (1959) on regeneration in nerveless larval limbs in the light of the new findings.

A sector of the upper arm of the newt *Triturus* was transplanted to the back after removal of a square of skin. In time it was innervated by local fibers. In some instances regeneration of the sector occurred and a limb was formed. In these cases counts were made of the number of fibers innervating the transplant. The number was substantially less than the threshold demands that the normal limb makes upon the nerve.

PROCEDURES

The adult newt *Triturus viridescens*, collected in central Massachusetts, was used in all instances. A sector of the upper arm, approximately 2 mm long, was removed. The bone and soft tissues on each surface were trimmed smooth. Then a square of skin, slightly larger than the diameter of the sector, was removed from the back of the same animal a short distance caudal to the shoulder girdle without injury to the underlying muscle. The proximal surface of the sector was placed in contact with the exposed tissue and the animal was immobilized for 24 hours at a temperature of 5°C. Sixty-eight of 147 animals, or approximately 46%, retained the grafts after they were returned to water at 25°C and survived for a period long enough to determine whether regeneration would occur.

The stumps that regenerated were then removed along with some proximal tissue and sectioned transversely. The sections were stained for nerve fibers according to Bodian's silver method. The nerve fibers in a section taken at the level of the original distal surface were counted and the number expressed as a function of the area of the amputation surface including the bone, according to the method one of us employed in a previous study (Singer, 1947). The quantity could then be compared directly with that reported in previous studies on normal limb regeneration from our laboratory.

In addition, in a series of 11 animals the brachial spinal nerves were dissected to the elbow and deviated to the site of transplantation of the limb sector. In these instances a complement of fibers in addition to those available at the amputation site, was provided to the graft.

RESULTS

All the grafts regenerated which received a deviated supply of the brachial nerves; but, only 6 of the 68 grafts regrew whose innervation was derived only from the locus of transplantation. Obvious signs of regeneration in the positive instances of both series occurred 3–5 weeks after the operation. In normal regeneration of the same level of the forelimb, only about 10 days to 2 weeks elapse before the first visible signs of growth (see stages of regeneration in Singer, 1952). The delay in onset of regeneration in the grafts was probably due to the time required for nerve fibers to invade the transplant and reach the distal surface. The difference in number of positive instances in the two series was undoubtedly due to the difference in number of fibers that invaded the stump, not to the source of the fibers, since it is now well known that nerves are nonspecific in their action on regeneration and that they probably have similar trophic qualities (review, Singer, 1952).

Since it is known that denervated amputation stumps that are adequately innervated can regrow if the wound is freshened by reamputation, amputation was performed a second time 3–7 weeks after transplantation in 27 of the 62 transplants that showed no signs of growth. Six of these subsequently regenerated. The first signs of growth appeared within 2 weeks after reamputation, rather than the longer interval observed above, presumably because of the presence of increased numbers of regenerated nerve fibers at the wound surface. It is likely that if the remaining 35 of the 62 nonregenerating transplants had been reamputated, there would have been additional instances of regeneration. As it is, a total of 12 of the 68 transplants, or 18%, regenerated.

The grafts innervated only by local fibers were of special interest because the number of nerve fibers within the transplant was obviously closer to the minimum needs for regeneration since some grafts regenerated and others did not. Study of the serial cross sections of the sector graft and regenerate showed that there were no obviously large bundles of nerve fibers such as one observes for normal nerve. Instead, the fibers were disposed singly or in small clusters. The fibers were all of fine caliber compared to the larger sizes frequently observed in nerves of the normal limb. It is likely that the number and the caliber of the fibers were even less at the time of onset of regeneration.

The number of nerve fibers at the selected level of the regenerated graft was counted in all the transplants supplied with a brachial nerve and in 8 of the 12 transplants without this complement. The results are depicted in Table 1, in which the absolute values and the

TABLE 1

THE QUANTITATIVE NERVE REQUIREMENTS OF LIMB TRANSPLANTS

With deviated nerve			Without deviated nerve		
	Fiber number	Number per (100 μ)² of amputation surface		Fiber number	Number per (100 μ)² of amputation surface
1.	116	1.2	TRS 2	417	2.9
2.	723	6.3	TRS 12	396	2.8
3.	875	4.2	TRS 17	113	1.3
4.	394	2.7	TRS 32	304	4.1
5.	375	4.9	TRS 41	198	3.6
6.	186	2.5	TRS 43	401	4.2
7.	751	4.4	TRS 61	164	2.6
8.	354	3.3	TRS 64	74	1.4
9.	632	3.3	—		—
10.	1113	4.4	Average	258	2.9
11.	1385	8.3			
Average	628	4.1			

values expressed per unit area of amputation are listed. A comparison of the two series shows that grafts supplied by a brachial source as well as local fibers contained, on the average, many more fibers than those supplied only from the local source. The average number of fibers per 100 μ² of wound surface is, respectively, 4.1 and 2.9. Although there were a number of instances of regeneration in the former series, the fiber values of which overlapped those of the latter series, nevertheless the generally higher count undoubtedly explains the more frequent incidence of regeneration observed in the cases of nerve deviation.

These values are obviously much less than the number of fibers required to evoke regeneration of the normal upper limb stump (Singer, 1947). In the latter instance the threshold needs averaged

9.6 fibers per 100 μ^2, a value which is more than twice that of transplants with deviated nerves and more than three times that of grafts innervated only by fibers from the site of transplantation. In order to emphasize the difference in nerve needs of normal limb and transplanted limb a further comparison may be drawn. Nerve 5, the smallest of the three brachial spinal nerves, contributes on the average 583 fibers to the amputation level of the arm under consideration in the present work (Singer, 1946). This number is greater than any of the values for transplants without nerve deviation. Yet, in various experiments that we have performed in the past in which nerve 5 was isolated in the limb after destruction of nerves 3 and 4, we have never observed regeneration in more than 100 cases studied.

As for the nature of the growth, the regenerates resembled in their sequence of successive stages normal regrowth including digit and hand formation. Examples of regenerate transplants, all taken from the series without added brachial nerves, are shown in Figs. 1–5. Histologically, they showed parts typical of the limb including skeleton and muscle. The size, however, tended to be deficient in comparison to the normal regenerate, and heteromorphosis was seen in some (Fig. 1). However, in terms of the nerve effect on regeneration, these latter faults are of much less significance than the fact that regeneration did occur. Heteromorphosis and size differentials are common outcomes of a variety of harmful conditions, physical or physiological, imposed upon the growing regenerate, and alone are not characteristic of alterations in the nervous supply (Singer and Egloff, 1949).

DISCUSSION

Previous studies (Singer, 1947) showed that the normal forelimb when amputated in the upper arm requires a well-defined threshold number of fibers per unit area of amputation surface before regeneration occurs. Fiber values below this threshold value do not satisfy the need of the wound tissues.

Results of the present study show that the threshold for a given level of the limb is not fixed but can be altered by experimental means. When a limb sector was transplanted to a heterotopic site, instances of regeneration were observed in the presence of a greatly reduced nerve supply. The number of fibers that could evoke regrowth was less than one-third that required for regeneration of the normal limb. Indeed, the number may be considered even smaller if

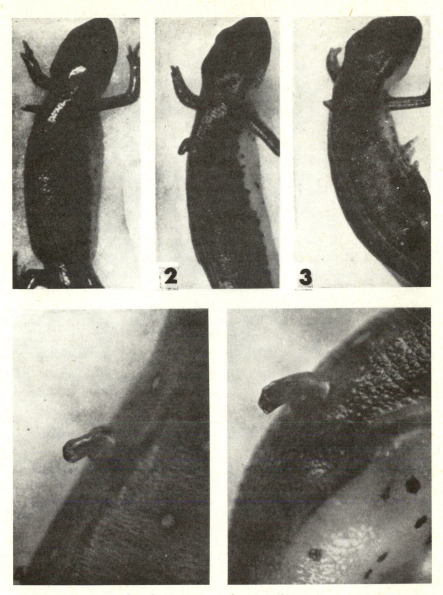

FIGS. 1–5.   Examples of regeneration after transplantation of sectors of the left upper arm to the back of the same side. No innervation other than that which the transplant derived from the locus of transplantation was supplied to the graft. The number of fibers that grew into the graft and stimulated regeneration was

one remembers that the fibers that supply the regenerating surface of
the transplant are themselves regenerating fibers of very small caliber
compared to the spectrum of sizes one finds in the normal limb; and,
assuming that the activity of the individual fiber is related to its
diameter, as is the case for speed of conduction of the impulse, then
the regenerated fibers may be a less potent source of the neuronal influence on limb regeneration than are the larger adult fibers of the
normal limb. Moreover, the fibers which invaded the transplant from
the locus of transplantation have, presumably, branched repeatedly
to supply the enlarged periphery; thus the contribution of the parent
neuron is diffused into a large number of branches. Such being the
case, the fibers counted may represent fewer neurons and therefore
may constitute a smaller fountain of neuronal influence than their
numbers would suggest.

The reason for the decrease in the threshold nerve demands of the
sector transplant is not known. There are a number of possible explanations; for example, that the fibers of the new site have greater
powers of stimulating growth than those of the limb, or that the new
blood supply has special qualities. However, at the moment we prefer
to emphasize another possibility which seems to fit better the available facts. It relates the growth forces within the wound tissues to the
trophic qualities of the neuron. The wound tissues give rise to the new

---

very small compared to the number ordinarily required to evoke regeneration in
the normal limb.

FIG. 1.   Animal TRS 44; 49 days after transplantation. Only two digits formed.
Visible regrowth started on about the twenty-first day. Approximate magnification:
×2.5.

FIG. 2.   Animal TRS 43; 46 days after transplantation. No regeneration before
thirty-first day, at which time transplant was reamputated. Growth started about
10 days later and was very rapid afterward. Regenerate in palette stage showing
elbow and hand formation. Approximate magnification: ×2.5.

FIG. 3.   Animal TRS 17; 55 days after transplantation. No regrowth before
thirty-eighth day, at which time reamputated. A regenerate was visible about 10
days later; it reached the photographed condition after 7 more days. Approximate
magnification: ×2.5.

FIG. 4.   Animal TRS 2; 41 days after transplantation. Visible regenerate enlargement about 21 days after grafting. Regenerate in very early digital stage.
Approximate magnification: ×7.4.

FIG. 5.   Animal TRS 12; 45 days after transplantation. First signs of growth
appeared 23 days after grafting. Regenerate is in early digital stage. Approximate
magnification: ×7.4.

growth. They determine the nature and growth characteristics of the regenerate, whereas the role of the neuron is a relatively nonspecific one, providing some factor important for the growth. These wound tissues must establish, according to their capacity to respond, the quantitative level of nerve activity needed to evoke a new growth. Should the responsiveness of the tissues change, then the demands upon the nerve should vary accordingly. If the capacity to respond decreases, then more fibers should be needed for regrowth of the limb; and, conversely, if the power to respond increases, then fewer are required. The competence of wound tissues to respond to the nerve was explored in the analysis of regenerative capacity of the frog during its life cycle (Singer *et al.*, 1957); it served in part as a working hypothesis for inducing the nonregenerating frog and the lizard to regrow a limb by increasing the number of fibers at the amputation surface and thus satisfying threshold needs that were not met by the normal nerve supply (Singer, 1961; Simpson, 1961; Kudokotsev, 1962). In these animals, in contrast to the urodele amphibian, the limb tissues are less competent to respond to the nerve and consequently a hyperinnervation was required to evoke regrowth. In the present experiments the stump of the newt was rendered even more competent than normal. Apparently, transplantation, with its consequences of prolonged trauma while the graft is revascularized and reinnervated, is one way which causes the responding tissues of the stump to become more sensitive to the neuronal influence. The thought that transplantation causes the wound tissues to become more sensitive to the nerve has been previously advanced by Maron (1956) to explain the regenerative powers of transplanted limbs of *Xenopus* tadpoles.

It is also possible that the increased competence of the wound tissues signifies that wound tissues do not depend entirely upon the nerve for trophic stimulation, but that they produce a trophic factor themselves, perhaps related to that of the neuron; and that the level of production may be altered according to the conditions of the experiment. The dramatic influence of the nerve upon regrowth tends to obscure the possibility that the trophic quality, so-called, or a related growth stimulatory quality, may not be confined to the neuron alone of all cells. It is quite possible that it is a property of cells in general but that the neuron has specialized in its production because of the heavy burden placed upon the cell to maintain the large volume of peripheral cytoplasm. The trophic substance, assuming it is chemical in nature, must originate in the cell body of the neuron and pass thence into the processes, because the processes degenerate immediately when separated from it. In addition to the primary effect upon the neuronal processes, there is also the growth and maintaining influence upon other tissues. And so, the trophic contribution must spill out of the axis cylinder into surrounding tissues. Other cells than the neuron have less exacting problems of maintenance. Their output of trophic substance would therefore be much less but not great enough to meet the high requirements for development of a new body part. However, the generous supplement pouring from the neurons may satisfy the requirements, as apparently is the case for regeneration in urodeles and other forms. The burden placed upon the nerve would be a function of the contribution of nonnervous tissues. In instances in which the tissue contribution is slight or in which the tissues are more refractory for some other reason to the nerve, the nervous supply may not be adequate to the task. Such may be the case in frogs and higher forms that lose the power to regrow. If, however, the tissue contribution itself is threshold adequate, then the neuronal contribution would be superfluous. Such a situation is observed in the amphibian embryo, where the presence of the nerve is not required for the part to develop (Harrison, 1904; Hamburger, 1928). In the present experiments the trauma of transplantation may have stimulated the trophic mechanism of the wound tissues so that less of a supplement from the nerve is now required. A previously inadequate number of fibers now calls forth growth.

If, then, the nerve needs of the regenerating limb are not fixed but are labile according to the history and environment of the regenerating tissues and the extent of their own trophic contribution to growth, it might be possible to create circumstances experimentally in which even fewer fibers, or indeed no fibers, are needed to evoke regeneration. Such a circumstance in which nerves are not required has already been reported by Yntema (1959) for regeneration in the larval salamander. He found that when nerves were prevented from growing into the limb bud during the course of embryonic development, the nerveless larval limb that resulted, nevertheless, could regenerate. It is conceivable in these instances that, in the absence of nerves, the other tissues of the limb retained a high level of production of the trophic quality. Perhaps, the persistence during postembryonic life of the enzymatic mechanisms necessary for production of trophic substances is determined by the availability of such substances in the tissue milieu. When there is an abundance of the trophic substance continuously spewed from nerve fibers, then its production by other cells would decline; in the relative absence of the product, the enzyme mechanism would maintain a high level of activity. Assuming that other cells besides the neuron do, indeed, produce the trophic factor necessary for regeneration, then the present results show that cells of the adult animal can be induced to increase their output of the trophic substance under appropriate experimental circumstances.

SUMMARY

The dependence of limb regeneration on the number of nerve fibers at the amputation surface was examined further in studies of the nerve requirements for regeneration in transplants of the upper arm to the back. In time, the grafts were reinnervated by local fibers and in some of the experiments by brachial nerves that were deviated to the site of transplantation. Counts of nerve fibers at the original regenerating surface of the graft showed that far fewer fibers can induce growth than are required for regeneration of the normal limb. Therefore, the threshold nerve requirements for regeneration are not fixed but vary with the capacity of the wound tissue to respond to nervous stimulation. The ability to respond is increased in the wound tissues of the graft presumably by the trauma of transplantation. The role of the wound tissues in establishing the threshold nerve requirements and possibly in contributing substances important for regeneration are discussed.

REFERENCES

DECK, J. D. (1961). Morphological effects of partial denervation on regeneration of the larval salamander forelimb. *J. Exptl. Zool.* **147,** 299–307.

HAMBURGER, V. (1928). Die Entwicklung experimentelle erzeugter, nervenloser und schwach innervierter Extremitäten von Anuren. *Wilhelm Roux' Arch. Entwicklungsmech. Organ.* **114,** 272–363.

HARRISON, R. G. (1904). An experimental study of the relation of the nervous system to the developing musculature in the embryo of the frog. *Am. J. Anat.* **3,** 197–220.

KUDOKOTSEV, V. P. (1962). Stimulation of the regeneration process in the extremities of lizards by the method of supplementary innervation. *Doklady Akad. Nauk S.S.S.R.* **142,** 233–236.

MARON, K. (1956). The innervation of regenerating grafted limbs in Xaenopis laevis tadpoles. *Folia Biol.* **4,** 317–326 (in Polish).

Rose, S. M. (1945). The effect of NaCl in stimulating regeneration of limbs of frogs. *J. Morphol.* **77**, 119–139.

Simpson, S. B. (1961). Induction of limb regeneration in the lizard, Lygosoma laterale, by augmentation of the nerve supply. *Proc. Soc. Exptl. Biol. Med.* **107**, 108–111.

Singer, M. (1946). The nervous system and regeneration of the forelimb of adult *Triturus*. V. The influence of number of nerve fibers, including a quantitative study of limb innervation. *J. Exptl. Zool.* **101**, 299–337.

Singer, M. (1947). The nervous system and regeneration of the forelimbs of adult *Triturus*. VII. The relation between number of nerve fibers and surface area of amputation. *J. Exptl. Zool.* **104**, 251–265.

Singer, M. (1952). The influence of the nerve in regeneration of the amphibian extremity. *Quart. Rev. Biol.* **27**, 169–200.

Singer, M. (1954). Induction of regeneration of the forelimb of the postmetamorphic frog by augmentation of the nerve supply. *J. Exptl. Zool.* **126**, 419–471.

Singer, M. (1961). Induction of regeneration of body parts in the lizard, *Anolis*. *Proc. Soc. Exptl. Biol. Med.* **107**, 106–108.

Singer, M., and Egloff, F. R. L. (1949). The nervous system and regeneration of the forelimb of adult *Triturus*. VIII. The effect of limited nerve quantities on regeneration. *J. Exptl. Zool.* **111**, 295–314.

Singer, M., Kamrin, R. P., and Ashbaugh, A. (1957). The influence of denervation upon trauma-induced regenerates of the forelimb of the postmetamorphic frog. *J. Exptl. Zool.* **136**, 35–52.

van Stone, J. M. (1955). The relationship between innervation and regenerative capacity in hind limbs of *Rana sylvatica*. *J. Morphol.* **97**, 345–392.

Yntema, C. L. (1959). Regeneration in sparsely innervated and aneurogenic forelimbs of *Amblystoma* larvae. *J. Exptl. Zool.* **140**, 101–124.

Reprinted from *J. Exptl. Zool.*, **166**, 89–97 (1967)

The Relation between the Caliber of the Axon and the Trophic Activity of Nerves in Limb Regeneration [1,2]

19

MARCUS SINGER,[3] KAROL RZEHAK,[4] AND CHARLES S. MAIER
Department of Anatomy, School of Medicine, and Developmental Biology Center, Western Reserve University, Cleveland, Ohio

ABSTRACT   Previous work from this laboratory demonstrated a relation between ability to regenerate a limb and number of nerve fibers available per unit area of amputation wound. Animals with fewer fibers than the newt, Triturus, (for example, Rana, Anolis, and Mus) do not regrow the limb. An exception appeared in Xenopus whose limb is sparsely innervated and yet which regenerates. The present work demonstrates that the nerve fibers of Xenopus make up in individual size what they lack in number. When the average cross-sectional area of the axons was multiplied by the number of fibers per unit area, the results showed that the Xenopus limb is about as well supplied with axoplasm as that of Triturus. This was not true for Rana. The results also demonstrate for the first time a direct relation between caliber and trophic effectiveness of the axon, the evolutionary significance of which is discussed.

In previous studies from this laboratory on the dependence of limb regeneration upon the nerve, a direct relation was drawn between the number of nerve fibers available at the amputation surface of the forelimb and the ability of the stump to regrow a new limb (see review, Singer, '52). In the newt, Triturus, the number of nerve fibers normally available per unit area of amputation surface is more than sufficient to cause regrowth. However, if this number is reduced below a critical threshold, by selective destruction of the brachial nerves or nerve roots, regeneration does not ensue. In forms which do not regenerate the limb, for example the frog, Rana, the lizard, Anolis, and the mouse, the number of nerve fibers per unit amputation area is substantially less than that found in the limb of the newt; indeed, it is within or is less than the range of nerve fibers required for regeneration of the newt's limb (Rzehak and Singer, '66, '66a). One of us (Singer, '51. '54, '61) also found for Rana and Anolis that, if the number of nerve fibers in the limb is augmented by deviating an extraneous large nerve into the limb, regeneration occurs presumably because the nerve fiber threshold for limb regeneration is now satisfied (see also Simpson, '61; Kudokotsev, '62).

An exception appeared to the rule that quantity of innervation is directly related to ability of the limb to regenerate (Rzehak and Singer, '66a). We observed that the number of fibers normally available at the amputation of the upper arm is much less in the adult African clawed frog, Xenopus, than in Triturus; and, indeed, substantially less than the threshold number required for regeneration in Triturus (see fig. 2, left). Yet, the limb of Xenopus did regenerate. Initial histological observations on the nerves of Xenopus suggested an answer to this paradox. The peripheral fibers are in general much larger than those of any other form we studied. This being the case, it must follow that the amount of axoplasm per nerve fiber is much greater at the amputation surface in Xenopus than in Triturus. If trophic activity of the axon is directly related to the quantity of axoplasm available at the amputation wound, the axons of Xenopus should be more effective agents of limb regrowth than the smaller ones of Triturus. Assuming all of this to be true, number of fibers is not by itself a completely adequate measure of the nervous influence on regeneration. In-

[1] The term trophic, according to our usage here, is defined as that quality of the neuron, probably represented in a secretory discharge at the axonal ending, responsible for the important action of the nerve in the growth and maintenance of the organs and organelles upon which it ends.
[2] Supported by grants from the American Cancer Society and the National Institutes of Health.
[3] Guggenheim Foundation Fellow during 1967 in the Istituto Anatomia, Università Cattolica, Rome, Italy.
[4] Visiting scientist during the year 1964–65 from the Department of Biology and Embryology, Medical Academy, Krakow, Poland.

stead, a factor of axon size must also be introduced and nerve activity expressed as amount of axoplasm available per unit area of amputation surface. The present study attempts such an evaluation on three forms—two regenerating ones with different numbers of nerve fibers, Xenopus and Triturus, and a nonregenerating one, Rana.

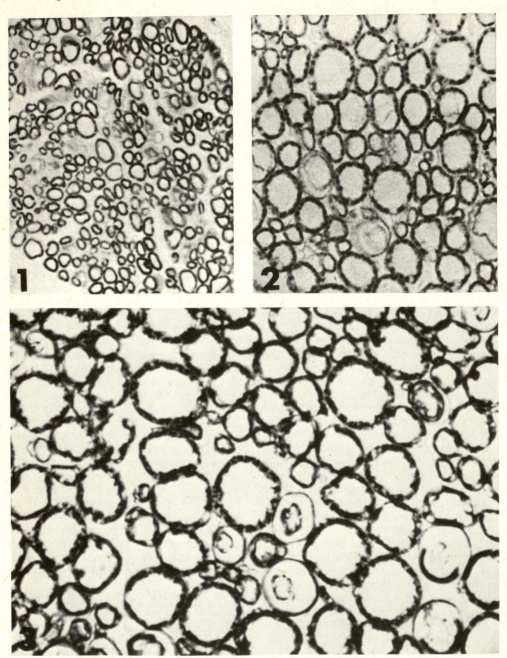

Figs. 1–3   Cross-section of a brachial nerve sectioned at $5\,\mu$ after fixing in 10% formalin and staining according to the Luxol-fast blue technique. Figure 1, Triturus; 2, Rana; 3, Xenopus. All photographs enlarged to 1000 times original size. Note the great difference in caliber of the fibers.

Axon diameters were measured and cross-sectional areas calculated; the average values of the areas multiplied by the number of fibers per unit area (previously reported: Singer, '47, for Triturus; Rzehak and Singer, '66a, for Xenopus and Rana) established the amount of axonal substance available in the wound area. When expressed in this way, the results show that Xenopus and Triturus limbs are about equally innervated. These results explain how a numerically sparse supply of nerve fibers yields regeneration of the Xenopus limb; they also assert for the first time a direct relation between the diameter and the trophic activity of the axon.

MATERIALS AND METHODS

The major nerve trunks of the upper arm of three adult Amphibia, *Triturus viridescens*, *Rana pipiens*, and *Xenopus laevis*, were removed, stretched on a piece of card, and fixed in 10% formalin solution for one hour. The nerves were embedded, sectioned at 5 μ, and stained according to the Luxol-fast blue procedure. A section was selected at the level of the nerve corresponding approximately to the region of the upper arm through which we amputate ordinarily to test for regenerative capacity,—namely, a short distance proximal to the elbow. Photographs were taken of the sections and printed at a magnification of 1000 times. Figures 1–3 show a representative photograph from a brachial nerve of each animal. The axon diameter was determined by measuring the diameter internal to the myelin sheath with a caliper and then by correcting for linear magnification. In sheaths of irregular contour, the smallest and largest diameters were averaged. Three hundred fibers were measured for Xenopus and Rana; and 400 for Triturus. The results were fed into a computer programmed to yield average diameters, average cross-sectional areas (πr^2), distribution of classes of fiber diameters and surface areas, and the standard deviation of the mean.

RESULTS

1. *The diameter of axons.*

Figures 1–3 compare stained sections of the nerves of the three amphibians, showing the obvious difference in caliber of the fibers. Table 1 lists the diameter of the axons for the three species. The fiber diameters are arranged in frequency groups with an increasing order of 1 μ; for example in the case of Triturus, the smallest being 0–1 μ and the largest 5–6 μ. The

TABLE 1

Diameters and cross-sectional areas of the axons of brachial nerves

Diameter	Frequency	Per cent frequency	Cross-sectional area (average per frequency)
μ			μ²
Triturus viridescens (400 fibers)			
– 1.0	3	0.75	0.59
1.0– 2.0	149	37.25	1.89
2.0– 3.0	152	38.00	4.30
3.0– 4.0	64	16.00	8.72
4.0– 5.0	26	6.50	14.85
5.0– 6.0	6	1.50	20.16

Average, 2.35   Average, 5.00
Stand. dev., ±0.93   Stand. dev., ± 4.14

Rana pipiens (300 fibers)			
1.0– 2.0	10	3.33	1.97
2.0– 3.0	20	6.67	4.96
3.0– 4.0	46	15.33	9.52
4.0– 5.0	58	19.33	15.56
5.0– 6.0	40	13.33	22.44
6.0– 7.0	40	13.33	31.93
7.0– 8.0	30	10.00	41.02
8.0– 9.0	20	6.67	54.50
9.0–10.0	14	4.67	68.38
10.0–11.0	10	3.33	83.89
11.0–12.0	8	2.67	100.33
12.0–13.0	3	1.00	121.49
13.0–14.0	1	0.33	147.41

Average, 5.70   Average, 30.22
Stand. dev., ±2.45   Stand. dev., ±25.88

Xenopus laevis (300 fibers)			
1.0– 2.0	4	1.33	1.79
2.0– 3.0	17	5.67	5.01
3.0– 4.0	19	6.33	9.14
4.0– 5.0	33	11.00	15.10
5.0– 6.0	44	14.67	22.26
6.0– 7.0	42	14.00	30.60
7.0– 8.0	26	8.67	41.87
8.0– 9.0	25	8.33	54.74
9.0–10.0	30	10.00	69.36
10.0–11.0	25	8.33	83.78
11.0–12.0	9	3.00	103.09
12.0–13.0	10	3.33	118.88
13.0–14.0	6	2.00	144.26
14.0–15.0	5	1.67	158.40
15.0–16.0	4	1.33	185.09
16.0–17.0	1	0.33	219.03

Average, 7.16   Average, 47.98
Stand. dev., ±3.14   Stand. dev., ±40.73

number of fibers in each class is also listed as well as the per cent of the total fiber number making up a class. The average for all the fibers and the standard deviation are also given.

The frequency of the size groups is plotted in figure 4 for each animal; and the average value is denoted by paired arrows. This figure and the tabulation show that the axons of Xenopus surpass in size those of Rana and Triturus, the averages being respectively 7.16, 5.69, and 2.35. The diameter of the average and the largest fibers of Xenopus are about three times

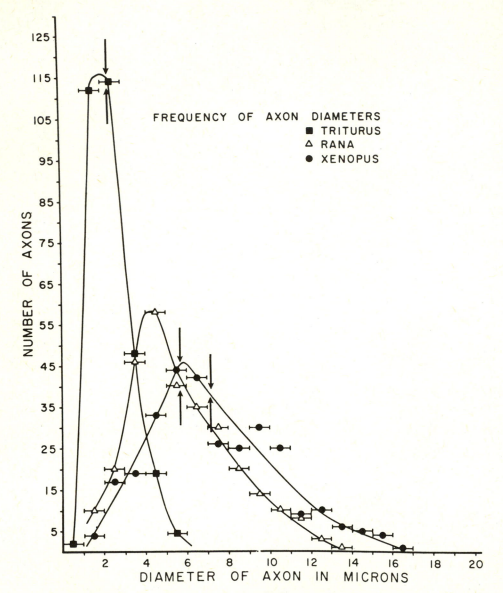

Fig. 4 Population of axon diameters in brachial nerves of three amphibians. Each curve represents 300 fibers. In the case of Triturus, 400 fibers were measured (see table 1); however, in graphing figure 4 the number was reduced by decreasing each listed range of table 1 by one fourth, thus providing for strict comparison of the curves. The paired arrows point to the mean values.

those for Triturus. The axons of Rana occupy intermediate values but are substantially thicker than those of Triturus.

2. The area of cross-section of axons

Table 1 also lists the average cross-sectional areas in square microns of axons for the selected diameter ranges. The tabulation shows that the cross-sectional area is greatest in Xenopus, followed in order by Rana and Triturus. Indeed, the average axonal area in Xenopus is approximately ten times that of Triturus. That of Rana is about two-thirds the average area of Xenopus but approximately six times that of Triturus.

3. The amount of nervous tissue at the wound of the amputated forelimb

It is now possible to estimate the amount of axonal material available at the amputation surface of the upper arm for the three animals. In previous papers, we reported that the number of fibers per unit area of amputation surface for Triturus, Xenopus and Rana, was respectively 24.5, 2.6, and 1.8 fibers per $(100 \mu)^2$. If we multiply this number by the average cross-sectional area in microns of the fibers (Triturus, 5.0; Xenopus, 48.0; and Rana, 30.2), we obtain an estimate of the total amount of neuroplasm available per unit area of amputation surface. The values are, in the same order, 122.8, 114.8, and 54.4 μ^2 per $(100 \mu)^2$ of wound tissue. These results are depicted in the right half of figure 5. For comparison the number of fibers per unit area is shown on the left half of the diagram; this part of the figure is taken from Rzehak and Singer ('66a). The threshold range for regeneration of the Triturus limb is also shown in each diagram as a striped parallel horizontal bar. In the case of the right diagram the threshold range is recalculated as axonal cross-sectional area. The results show that the amount of axoplasm available at the amputation surface per unit area of wound

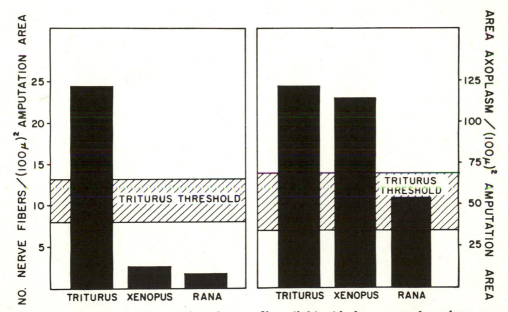

Fig. 5   A comparison of the number of nerve fibers (left) with the amount of axoplasm per unit cross-sectional area (right) of the upper arm in the three amphibians. The values on the left were taken from Rzehak and Singer ('66); the ones on the right were calculated by multiplying the average cross-sectional area in square microns (see table 1) times the number of fibers per unit area taken from the left graph. The horizontal striped bars represent the threshold of nerve fibers required for regeneration in the Triturus limb; on the left the number of fibers, on the right the amount of axoplasm. The axoplasmic area also may be taken as a measure of the axonal volume for a given length of nerve. Note that the two regenerating forms, Xenopus and Triturus, have comparable neuroplasmic volumes but that non-regenerating Rana has substantially less.

tissue is very similar in the two regenerating forms, Xenopus and Triturus. Indeed, they also show that the value for Xenopus does exceed the threshold range established for Triturus. However, in the case of Rana the amount of axoplasm available at the amputation surface does not exceed, but falls within, the threshold range established for Triturus.

DISCUSSION

The individual neuron is the source of the important action which nerves exert upon regeneration of a body part. It impresses its influence upon the tissues of the wound by way of its axonal endings (Singer, '64). The combined action of many neurons is required to accomplish regeneration; a single neuron is in itself an ineffective stimulus of regrowth. The number of axons required to evoke growth constitutes a threshold. In the case of the newt, threshold needs are more than satisfied by the available nerve supply. This is apparently not so for some higher forms that we studied. The frog, Rana; the lizard, Anolis, and the mouse have much fewer nerve fibers per unit volume of forelimb; and the forelimb does not regrow after amputation (Zika and Singer, '65; Rzehak and Singer, '66). However, hyperinnervation of the stump by rerouting additional nerves into it yields regeneration in Rana and Anolis. In contradiction to these facts supporting a direct relation between quantity of nerve fibers and ability to regenerate a limb, even fewer fibers were found in the limb of Xenopus than in those of all other forms we studied including the mouse; yet, the Xenopus limb does regenerate. The resolution of this contradiction apparently lies in the fact that the caliber of the axons in the peripheral nerves of Xenopus is greater than that of the other vertebrate forms. We interpreted this fact to mean that the axons of Xenopus are individually more effective agents of growth than those of Rana, Anolis, and the mouse; and that by virtue of axonal size their combined activities could satisfy the nervous threshold required for regeneration. Indeed, our present calculations support this conclusion. Instead of comparing number of nerve fibers per unit wound area in these animals, we compared the amount of neu-

ronal substance available at the amputation surface by multiplying the average cross-sectional area of the axons, as a measure of axonal volume, with the number of axons; the values showed that the Xenopus limb is about as well innervated as the Triturus one. In this way, we have defined trophic innervation to mean the total amount of neuronal substance rather than the number of axons. The limb of Rana is much less innervated, which thus accounts for its inability to regrow without an augmented nerve supply.

Some time ago, one of us (Singer, '52, pp. 182–183) speculated that axons are not equal in their trophic effectiveness. There was then reason to believe that the contribution varied according to a number of factors, a salient one being axonal volume. And, it was presumed that fibers of large diameter are more effective agents of limb regeneration than smaller ones. However, we found no way to test this possibility in Triturus since we were unable to isolate the activity of an individual axon from the combined activity of an axonal population of mixed caliber. The fortunate circumstance which emerged in our quantitative analysis of the regenerative ability of the frog, Xenopus, has allowed us to assess the problem of the relation between size of fiber and its trophic ability and to provide a reasonable answer; namely, that large fibers are on the average more effective agents of limb regrowth than small ones.

There are other important implications to the present findings. Physiologists have repeatedly emphasized that speed of conduction is in general directly related to fiber caliber; large fibers conduct more rapidly. The present work shows that there is another significance to size, namely trophic effectiveness of the individual fiber; larger axons deliver more of the trophic agent than smaller ones. Consequently, considerations of the physiological significance of the caliber of axons can no longer be confined to conduction velocities. Size implies trophic effectiveness as well; and there may be still other implications unknown at present. For example, the rate of axoplasmic flow may be greater in large fibers because of the smaller frictional resistance of the axolemma.

In the evolution of the neuron, there was probably a number of physiological and morphological factors of survival value which affected the caliber of the neuron; for example, trophic contribution, speed of conduction and central integration of neuronal activity. The survival value of each of these factors may have been in part supplementary and in part contradictory. We do not have enough information to evaluate each of these factors. However, our findings against the background of our present knowledge of the neuron allow us to make some suggestions which may be useful. First of all, we must ask ourselves the question: what is the survival value of trophic neuronal activity for regeneration and morphological maintenance of a body part? The survival value of regenerative capacity has been argued for a long time (see illuminating discussion by Morgan, '01; and arguments in the revival of the debate by Needham, '61; Barr, '64). A cardinal principle of nervous function is integration of body activity into one functional whole. If a lost external body part is able to regrow, then regeneration must have survival value because integration of body activity also implies integrity of body structure, for a mutilated and structurally deficient body cannot function as a whole. Since it assists in maintaining and reconstituting body structure, the trophic function of the neuron must also have survival value. But, conduction speed and, particularly, central integration of neuronal action are also important in the preservation of body structure. Indeed, it seems more reasonable to suppose that morphological integrity is better preserved by more effective response and use of body parts than by regeneration of these parts after their loss, requiring as it does weeks and months for completion. Moreover, the completed part is seldom structurally and functionally as good as the original. Indeed, the speed of sensory response and the extent of motility in the appendage of animals capable of regeneration is limited and stereotyped compared to the intricacy of motile expressions seen in the non-regenerating limb of higher forms.

The most favorable evolutionary circumstance might have been supplementary elaboration of the trophic effectiveness of nerves and increased integration of limb functions. As it is, however, the trophic capacities of the nerve diminished with elaboration of integrative mechanisms. In part, the decline was due to decrease in the quantity of peripheral innervation and, in part, to limitation in size of nerve fibers. The mammal has much fewer fibers per unit volume of limb structure than does the newt and the diameter of the axons is of about the same order as that of the newt (see, for example, Cragg, '55). Yet, the motility and purposive use of the mammalian limb reflects a higher order of evolution of the nervous system. Stated otherwise, the limb of the mammal is a much more effective organ of motile exploration and defense because of the elaboration of the central nervous system although the sensory input and, perhaps, the motor saturation is much less than in that of the lower vertebrate.

Amphibians seem to have "experimented" with the evolutionary possibilities of trophic and integrative activities of the nervous system. In the case of the urodele, Triturus, the peripheral nerve fibers are small in caliber, of the order of mammalian ones. However, the number of fibers per unit limb volume far exceeds that of higher forms. We do not know the extent of innervation in the fish appendage but it is conceivable that saturation of the periphery with nerve fibers is the rule in forms still lower than the urodeles. In the uniform aqueous environment of the fish, saturation of the periphery is advantageous for it allows maximum peripheral sensory input. At the same time, it satisfies trophic needs in maintenance and regrowth of peripheral tissues. In the case of the anuran, two trends are observed. One of these anticipates the trend in higher forms, a striking decline in the number of fibers per unit volume. There is a lesser numerical saturation of the periphery; but, at the same time, advanced integrative action of the central nervous system and more effective use of the limb. The second trend is a striking increase in the diameter of the axons. This is shown particularly in the primitive frog, Xenopus, in which the axons surpass those of Triturus and the mouse by many times. Such an increase served to augment the trophic flow and to make up in individual axonal effectiveness what was lacking in combined axonal activities. However, in Rana, the increased trophic flow of the individual axons failed to compensate for the decrease in fiber number; and, therefore, the total contribution falls short of regenerative needs. Yet, even in this animal, regeneration only hovers on complete extinction because one often observes limited or aborted signs of regeneration (Thornton and Shields, '45; Singer, '54). In forms higher than Amphibia, the nerve fibers are relatively small in caliber and sparse in number, a pattern upon which evolution finally settled.

The above speculations emerged from study of the trophic activity of the neuron and they may be of significance in analysis of certain developmental phenomena, including widespread destruction in motor nuclei and sensory ganglia that attends embryogenesis of these parts. It may, however, be important to stress at this moment that, although there is a decline in trophic activity of the nervous system in regeneration of a body part, there is not a complete loss. Indeed, it is appropriate to assert that all neurons in whatever animal are trophic. One of us has discussed this point repeatedly in previous papers (see for example, Singer, '64). In brief, the trophic function is primary for the neuronal processes and is required for maintenance and growth of the tremendous axonal volume of cytoplasm that the neuron supports. The excess of trophic agent, whatever its nature, spills onto the peripheral end organ and has thereby a salutary effect. This function remains, the only difference being that the amount contributed peripherally is insufficient to evoke limb regrowth in higher forms. However, it is enough to maintain muscle, taste organs, and other structures in all animals including man.

Another point which needs emphasis at this time· is the role of the peripheral tissues in establishing the trophic threshold. We have stressed this point repeatedly in past publications and have presented evidence to show that the extent of response of peripheral tissues to the nerve establishes the nerve requirements (Singer and Mutterperl, '63). There is no reason to assume that the limb wound tissue of all vertebrates responds equally to the trophic agent of the nerve. Indeed, there is more reason to anticipate a difference in response. The threshold requirement for regrowth in Triturus probably differs from that in Rana and Xenopus. It may be lower or higher. In the case of the lizard and mammal, it is conceivable that the tissues are relatively refractory to the nervous agent establishing therewith a higher threshold. We have also pointed out before and would like to re-emphasize here the possibility that the limb of other urodeles, for example Amblystoma, may be more responsive to the nerve and thus require much fewer fibers than does that of Triturus. A sparse innervation in a highly responsive limb may be as effective a trophic source as the rich one of the newt.

LITERATURE CITED

Barr, H. J. 1964 Regeneration and natural selection. Am. Naturalist, 48: 183–186.

Cragg, B. G. 1955 A physical theory of the growth of axons. J. Cell. and Comp. Physiol., 45: 33–59.

Kudokotsev, V. P. 1962 Stimulation of the regenerative process in the extremities of lizards by the method of supplementing innervation. Dokl. Akad. Nauk SSSR, 142: 233–236.

Morgan, T. H. 1901 Regeneration. MacMillan, New York.

Needham, A. E. 1961 Evolution of regeneration, and its possible bearing on philosophy. Nature, 192: 1255–1256.

Rzehak, K., and M. Singer 1966 The number of nerve fibers in the limb of the mouse and its relation to regenerative capacity. Anat. Rec., 155: 537–540.

———— 1966a Limb regeneration and nerve fiber number in Rana sylvatica and Xenopus laevis. J. Exp. Zool., 162: 15–21.

Simpson, S. B. 1961 Induction of limb regeneration in the lizard, Lygosoma laterale, by augmentation of the nerve supply. Proc. Soc. Exp. Biol. Med., 107: 108–111.

Singer, M. 1947 The nervous system and regeneration of the forelimb of adult Triturus. VII. The relation between number of nerve fibers and surface area of amputation. J. Exp. Zool., 104: 251–265.

———— 1951 Induction of regeneration of the forelimb of the frog by augmentation of the nerve supply. Proc. Soc. Exp. Biol. Med., 76: 413–416.

———— 1952 The influence of the nerve in regeneration of the amphibian extremity. Quart. Rev. Biol., 27: 169–200.

———— 1954 Induction of regeneration of the forelimb of the frog by augmentation of the nerve supply. J. Exp. Zool., 126: 419–472.

———— 1961 Induction of regeneration of body parts in the lizard, Anolis. Proc. Soc. Exp. Biol. Med., 107: 106–108.

——— 1963 Nervous control of the regrowth of body parts in vertebrates. In: The Effect of Use and Disuse on Neuromuscular Functions. E. Gutmann and P. Hnik, eds., Publishing House of the Czechoslovak Academy of Sciences, Prague, pp. 88–94.

——— 1964 The trophic quality of the neuron: Some theoretical considerations. In: Progress in Brain Research, Mechanisms of Neural Regeneration. M. Singer and J. P. Schadé, eds., Elsevier, Amsterdam, 13: 228–232.

Singer, M., and E. Mutterperl 1963 Nerve fiber requirements for regeneration in forelimb transplants of the newt, Triturus. Dev. Biol., 7: 180–191.

Thornton, C. S., and T. W. Shields 1945 Five cases of atypical regeneration in the adult frog. Copeia, 1: 40–42.

Zika, J., and M. Singer 1965 The relation between nerve fiber number and limb regenerative capacity in the lizard, Anolis. Anat. Rec., 152: 137–140.

Reprinted from *J. Exptl. Zool.*, **142**, 423–439 (1959)

20

BLASTEMA FORMATION IN SPARSELY INNERVATED AND ANEUROGENIC FORELIMBS OF AMBLYSTOMA LARVAE

C. L. YNTEMA

State University of New York, Upstate Medical Center and Marine Biological Laboratory, Woods Hole, Massachusetts

THIRTEEN FIGURES

INTRODUCTION

Regeneration in amphibian limbs is closely associated with innervation under many experimental conditions (Singer, '52). Formation of the blastema after amputation is prevented if the limb is denervated (Schotté and Butler, '41). The number of nerve fibers needed to sustain regeneration is considerably less than the normal complement but if depletion is too great regeneration is prevented (Singer, '46).

A peculiarity of an innervated regenerating limb is the early and rich nerve supply of the blastemal epidermis (Singer, '49; Taban, '49). A close association between regeneration and innervation of the epidermis of the stump has been described by Thornton ('54, '56, '56a) who has suggested that the nerve fibers are needed for the formation of the thickening of the distal epidermis to form the apical cap which in turn is necessary for formation of the blastema.

Once the blastema is well formed, differentiation can proceed without nerves in *Amblystoma* larvae (Butler and Schotté, '49). In the work of Skowron and Komala on *Xenopus* ('57) it has been found that regrowth of tissues following amputation of a regenerating limb occurs without nerves. The cartilaginous rod of the taste barbel *Ameiurus* regenerates following transection or removal of its proximal portion in the absence of nerves (Goss, '56). The findings indi-

423

cate that the crux of the question regarding the influence of nerves on limb regeneration has to do with the formation of the blastema rather than subsequent differentiation of tissues.

The experiments which have been noted above deal with regeneration in denervated structures. The generalities arising from these findings do not necessarily apply to regeneration of limbs which developed from embryonic stages with few or no nerves. Such sparsely innervated or aneurogenic limbs can regenerate readily in *Amblystoma* larvae; some characteristics of these regenerates have been described in an earlier paper (Yntema, '59). In this report, stages in the regeneration of these limbs will be described. Study of such material leads to the conclusion that the morphology of regeneration in absence of nerves is essentially like that which occurs after amputation of the normal limb except that nerve fibers are missing.

PROCEDURE

Aneurogenic or sparsely innervated forelimbs of *Amblystoma* larvae were obtained by procedures described in an earlier report (Yntema, '59). In brief, pairs of embryos in the head process stage were joined in parabiosis and the neural tube was extirpated from the levels of the hindbrain and the first 12 somites. The latter operation was done on the right member at a tailbud stage. Subsequently the ectoderm of the postauditory dorsolateral placodes and that of the more posterior epibranchial placodes was removed from the right side of the same member. These operations produced forelimbs in the parasitic twin which usually were sparsely innervated; in some cases no nerve fibers occurred and these limbs were aneurogenic.

Two sets of experiments were performed. In the first, L80 to L87, the limbs were amputated through the distal brachium and the animals were kept at room temperature following amputation. The animals were preserved at 5 to 12 days following amputation. In each case the two forelimbs of the parasite were amputated as well as the right forelimb of the host; the latter serves as a control. In the second series, L99, the animals were reared at 20°C following amputation through the proximal antibrachium and preserved 9 and 11 days later. There were 31 twins in the first series and 20 in the second. Thus 102 experimental limbs were available for study along with 51 control limbs.

The animals were preserved in Bouin's fluid and before embedding were divided so that sections could be made through the long axis of the limb in each case. A modification of the staining methods of Bodian ('37) and Ungewitter ('51) was used to impregnate the nerve fibers in sections. The stain in each of the 51 sets of limbs was satisfactory.

RESULTS OF EXPERIMENTS

Rates of regeneration in experimental and control limbs

The experimental limbs of the parasite regenerate more slowly than the control limbs. In the previous report (Yntema, '59) the delay in digit formation was found to average two days and the mode of delay was one day. In the present series, the regenerating limbs can be classified according to stage of the blastema by study of sections. Stages in the development of the blastema have been described by Butler ('33) and others; the classification of Hay ('52) was found useful. The observations are summarized in table 1. The rates of regeneration were different in the two groups of experiments, one conducted at room temperature, the other at 20°C. The 9-day control at 20°C is similar to the 7- or 8-day control at room temperature; the 11-day control at 20°C like the 9- or 10-day control at room temperature.

Lag in regeneration of sparsely innervated or aneurogenic limbs is shown by comparison of the data for the experimental limbs with those for the control limbs in the room temperature series. The shift for the control limbs from early to proliferative blastema occurs at 6 days; that for the experimental limbs at 8 days. Similarly the shift from

C. L. YNTEMA

TABLE 1

Stages of experimental and control blastemata fixed 5 to 12 days after amputation

TEMP.	DAYS	LIMB TYPE	NO. CASES	EARLY	STAGE OF BLASTEMA		
					Prolif.	Conden.	Digit
					%	%	%
Room	5	Control	4	100			
	6		4	50	50		
	7		4		100		
	8		3		100		
	9		4		50	50	
	10		4			100	
	11		4			100	
	12		4				100
	5	Exper.	8	100			
	6		8	100			
	7		8	100			
	8		6	50	50		
	9		8	25	75		
	10		8		100		
	11		8	13	87		
	12		8			75	25
20°C	9	Control	8		100		
	11		12			100	
	9	Exper.	16	63	37		
	11		24	50	50		

proliferative to condensing blastema for controls is at 9 days; for experimentals between 11 and 12 days. The average delay is about two days.

Incidence of nerve fibers in experimental limbs

Serial sections of the 102 experimental limbs stained by an adequate silver method were examined for presence and number of nerve fibers. The findings are summarized in table 2. About one third of the limbs had one to 10 nerve fibers; in 7 cases no nerve fibers could be found. The latter are most probably aneurogenic. In even the least satisfactory experiments the incidence of nerve fibers is much less than normal; hence all limbs in which nerve fibers are present are considered to be sparsely innervated.

TABLE 2

Incidence of nerve fibers in experimental limbs

TEMP.	DAYS	NO. CASES	NO. OF NERVE FIBERS		
			0	1–10	11+
Room	5	8		4	4
	6	8		5	3
	7	8		2	6
	8	6	1	2	3
	9	8		4	4
	10	8	1	2	5
	11	8	1	4	3
	12	8			8
20°C	9	16	1	4	11
	11	24	3	5	16
Totals		102	7	32	63

Relation between incidence of nerve fibers and rate of regeneration

Three groupings of experimental limbs are set up in table 3 based on the number of nerve fibers present in the limbs: 0, 1–10, 11+. Four sets of experiments in which one or more aneurogenic limbs occurred are included in the table. No correlation between numbers of fibers in the experimental limbs and the stages of the blastemata is found in these data. The small numbers in many of the categories do not carry much statistical significance but the observations as they stand are not a source of encouragement to pursue the problem by adding more cases. The observations lead to the conclusion that variations in fiber numbers in the experimental limbs does not influence rate of regeneration.

Resorption in control and experimental limbs

Observations on the living animals and on sections of the limb regenerates showed a surprising number of instances in which considerable resorption of the brachium occurred. This was true of both control and experimental limbs.

The incidence of the resorption is given in table 4. In the room temperature series the amputation was through the

TABLE 3

Relation between incidence of nerve fibers and stage of blastema

TEMP.	DAYS	NO. FIBERS	NO. CASES	STAGE OF BLASTEMA		
				Early	Prolif.	Conden.
				%	%	%
Room	8	0	1	100		
		1–10	2	50	50	
		11+	3	33	67	
		Control	3		100	
Room	10	0	1		100	
		1–10	2		100	
		11+	5		100	
		Control	4			100
20°C	9	0	1		100	
		1–10	4	50	50	
		11+	11	73	27	
		Control	8		100	
20°C	11	0	3	67	33	
		1–10	5	60	40	
		11+	16	44	56	
		Control	12			100

TABLE 4

Incidence of resorption in experimental and control limbs

TEMP.	DAYS	LIMB TYPE	NO. CASES	CONDITION OF HUMERUS	
				More than half resorbed	Half or more present
				%	%
Room	10, 11, 12	Control	12	59	41
	11, 12	Exp.	16	38	62
20°C	11	Control	12	33	67
	11	Exp.	24	42	58

distal brachium. The shortening was obvious by 10 days if it were to occur in the controls and by 11 days in the experimental limbs. Among the 12 controls, 7 showed marked resorption; in the 16 experimental limbs, 6 showed such changes. In the group of animals kept at 20°C, the amputation was done through the proximal antibrachium. The regenerates of the animals reared for 11 days were advanced enough to make a reasonable estimate of the extent of resorption which would occur. These observations summarized in table 4 indicate that extensive resorption may occur in both control and experimental limbs following amputation through the proximal antibrachium.

Examples of varying amounts of resorption following amputation are illustrated in plate 1; the length and condition of the humerus indicates the amount in each case. Three of these figures (figs. 1–3) show sections of control limbs which were amputated through the distal brachium and preserved after 9 or 10 days at room temperature. In the first of these three the length of the humerus indicates that little was resorbed after amputation. In the next figure only the proximal part of the shaft of the humerus persisted. In figure 3 a small part of the humeral shaft remains next to the more proximal tuberosity and head; the position of the shaft is marked by two rows of osteoclasts. In the next example (fig. 4) a control limb was amputated through the proximal antibrachium and preserved after 9 days at 20°C. The humerus remained intact and a small portion of the ulna is seen as a cartilage distal to the humeral condyle. Resorption of the elbow and part of the humerus occurred in some experiments of this type as indicated in table 4.

Similar responses to amputation occur in the experimental cases; examples are shown in sections of three aneurogenic limbs illustrated in figures 5, 6 and 7. The first two were amputated through the distal brachium and fixed after 8 and 10 days at room temperature. In the first, most of the shaft of the humerus had been resorbed; this occurred in the normally innervated stump illustrated in figure 3. In the second of the group (fig. 6) the length of the humerus indicates little resorption; this situation is like that in the first of the control limbs (fig. 1). The third limb (fig. 7) was amputated through the proximal antibrachium and preserved after 11 days at 20°C. In this nerveless stump the elbow has disappeared and the distal end of the humeral shaft is open to the surrounding medium. From observations on similar cases one would expect that the distal part of the bony shaft of the humerus would have sloughed off and the proximal part resorbed if the animal had been carried longer. The head of the humerus would have persisted and a stump like one of those represented in figures 2 and 3 would be expected.

According to the observations given above, extensive resorption after amputation may occur in both control and experimental limbs. If it were noted in the latter only, it would be reasonable to associate it with fewness or absence of nerve fibers. However, conditions in the controls are essentially like those in the experimental limbs; presumably the same cause or causes underlie the changes in both types of limbs.

Structure of the regenerating limbs

The development and differentiation of the limb following amputation in the larval salamander has been described by Butler ('33) and others. Descriptions of the morphology are essentially in agreement although there is some discussion as to the origins of the blastemal components. In this report two features of the nerveless regeneration of special interest are the formation of a blastema and the presence of an apical cap of epidermis.

The experiments of Butler and Schotté ('49) have shown that the critical period for support of regeneration by nerves is during the formation of the blastema. The demonstration of nerveless blastemata then is crucial for the thesis of aneural regeneration and depends upon the reliability of the nerve fiber stain. To the best of my judgment the stain used made individual nerve fibers detectable and small groupings of fibers obvious.

Sections through 4 of the 7 available aneurogenic blastemata are illustrated (figs. 5–7, 9, 10, 12 and 13); two normally innervated blastemata are also shown (figs. 8, 11). The animal L83-2R which was the source of the section shown in figures 5 and 10 was kept at room temperature for 8 days after amputation through the distal brachium. Shortening of the humerus occurred as shown by presence of only a small part of the shaft next to the head. Two osteoclasts are evident in the central part of the bud; the one on the right is related to a remnant of the bony portion of the shaft. The blastema is in an early stage and not vigorous. In the distal portion the inner layer of epidermis is disoriented. This area is small and bounded by the edge of intact basement membrane; Leydig cells with clear cytoplasm are nearby. The distal thickened epidermis probably is in an early stage in the formation of the apical cap. No nerve fibers were found in the stump.

Two similar blastemata are shown in figures 6, 9 and 12. In L99-16R amputation was through the proximal antibrachium; the elbow resorbed but the distal portion of the humeral shaft was maintained (fig. 12). In L85-2R following amputation through the distal brachium the resorption was slight since much of the humerus is present (figs. 6, 9). Both of the blastemata were active when fixed: L99-16R after 11 days at 20°C, L85-2R after 10 days at room temperature. The basement membrane and Leydig cells are absent distally. The apical epidermis is three or 4 cells thick in contrast to the two layered condition proximally. The considerable thickening of the apical cap represented in the low power illustration of L85-2R (fig. 6) is probably due to a tangential plane of section. No nerve fibers were found in these two regenerating limbs.

In L99-20L illustrated in figures 7 and 13, the amputation was done through the proximal antibrachium. As pointed out in a previous section (p. 429) resorption in this stump is not yet complete. However, an accumulation of cells occurs above the humerus in the distal part of the stump and a smaller group is below the humerus. This cell group rings the shaft of the humerus and is held to be part of an early blastema. The thickening of epidermis around the humerus is associated with retraction of soft tissues; it may represent the apical cap. No nerve fibers were found in this stump.

A marked increase of nerve fibers in the blastemal epidermis of regenerating limbs was described by Singer ('49) and Taban ('49) and investigated further by Thornton ('54, '56, '56a). Two examples of these fibers in the apical caps of innervated regenerates are shown in figures 8 and 11. These apical caps were among the most striking cases seen in the 51 control limbs available. In figure 8 the blastema is in an early stage. The epidermis is papillate and many fibers occur which run directly to the apical cap or through the more proximal epidermis to reach to the tip. In figure 11 the thickening is still present at the proliferative stage; several nerve fibers occur in the apical region of the section shown.

These control apical thickenings have been selected as conspicuous examples which are richly innervated. Apical epidermis in the absence of nerve fibers is illustrated in figures 9, 10 and 12. In figure 10 the thickening is in an early stage and due to changes in the inner layer of epidermis. In the proliferative stage shown in figures 9 and 12 the apical thickenings are moderate; they form parts of essentially typical blastemata except for the absence of nerve fibers. Thickenings such as seen in the apical epidermis of the aneurogenic limbs can be found on control limbs. The two examples given for the controls (figs. 8 and 11) are among the most striking available in my material; less conspicuous ones also occur. The conclusion indicated is that the apical caps of the experimental limbs fall within the range of types seen in the control limbs.

DISCUSSION

The finding that aneurogenic limbs of larval salamanders regenerate was presented in an earlier paper (Yntema, '59) which dealt with the characteristics of these regenerates and the incidence and rates of their regeneration. Stages in this process of nerveless regeneration have been presented above; they are remarkably like what is found in regenerating stumps of normally innervated limbs except for their nervelessness. According to these findings, tissues in the aneurogenic limb differ from those in the innervated and denervated limbs. The latter types have come to rely on presence of nerve fibers for the initiation of regeneration. If this point of view is taken, there is no basic contradiction among the observations on the ability of innervated and aneurogenic limbs to regenerate and the inability of denervated limbs to do so.

The relations in time and position between innervation of the distal epidermis following amputation of the normal limb and the thickening to form the apical cap have been carefully studied by Thornton ('54, '56, '56a). Fibers from the abundant cutaneous nerve supply which occurs in the epidermis following amputation (Singer, '49, Taban, '49) appear in the distal epidermis before it forms the characteristic cap with its increased thickness. Thornton indicates that this change is a crucial event in the formation of the blastema. The transformation of the epidermis may be dependent upon the influence of the penetrating nerve fibers.

In aneurogenic limbs apical caps of modest thickness formed without demonstrable nerve fibers. It is possible that the correlation between epidermal innervation and formation of the apical cap in the innervated regenerate may be temporal only and that the epidermal changes are not dependent upon presence of nerves. However, it is also possible that the epidermis is the tissue which is basically different in innervated and denervated limbs as opposed to aneurogenic limbs. If the latter is the case, aneurogenic epidermis on a denervated limb might initiate regeneration.

Resorption following amputation of aneurogenic and sparsely innervated limbs was reported in an earlier paper (Yntema, '59). It was found that the resorption was not related to presence or number of nerve fibers. In the present report, resorption was described in normally innervated limbs as well as in the experimental cases. The incidence and extent of the resorption showed no relation to presence of nerve fibers. The retrogression was not so extensive as that following denervation in the experiments of Schotté and Butler ('41). In each instance the blastema was maintained and a conspicuous stump persists even though it may be considerably foreshortened. This resorption is due to factors other than amount of innervation. A number of possibilities might be suggested: method of amputation, temperature or other characteristics of the water in which the animals were kept, diet of the animals, the fact that all the limbs are parts of animals in parabiosis. Whatever the reason, similar incidence of resorption in both control and experimental limbs indicates that it is not related to amount of innervation.

SUMMARY

1. Stages in regeneration of aneurogenic and sparsely innervated forelimbs were described.

2. Regeneration in these limbs is delayed an average of about two days. This delay is not related to incidence of nerve fibers.

3. Considerable resorption may follow amputation of both normal and experimental limbs.

4. The structure of blastemata on aneurogenic or sparsely innervated limb stumps is essentially like that on normally innervated stumps except for incidence of nerve fibers.

5. The apical epidermis can differentiate into an apical cap of moderate thickness without nerve fibers under conditions of the experiments.

LITERATURE CITED

BODIAN, D. 1937 The staining of paraffin sections of nervous tissues with activated protargol. The role of fixatives. Anat. Rec., 69: 153–162.

BUTLER, E. G. 1933 The effects of x-irradiation on the regeneration of the forelimb of *Amblystoma* larvae. J. Exp. Zool., 65: 271–315.

BUTLER, E. G., AND O. E. SCHOTTÉ 1949 Effects of delayed denervation on regenerative activity in limbs of urodele larvae. Ibid., 112: 361–392.

GOSS, R. J. 1956 An experimental analysis of taste barbel regeneration in the catfish. Ibid., 131: 27–49.

HAY, E. D. 1952 The role of epithelium in amphibian limb regeneration studied by haploid and triploid transplants. Am. J. Anat., 91: 447-481.

SCHOTTÉ, O. E., AND E. G. BUTLER 1941 Morphological effects of denervation and amputation of limbs in urodele larvae. J. Exp. Zool., 87: 453–493.

SINGER, M. 1946 The nervous system and regeneration of the forelimb of adult *Triturus*. V. The influence of number of fibers, including a quantitative study of limb innervation. Ibid., *101:* 299–337.

———— 1949 The invasion of the epidermis of the regenerating forelimb of the urodele, *Triturus*, by nerve fibers. Ibid., *111:* 189–210.

———— 1952 The influence of the nerve in regeneration of the amphibian extremity. Quart. Rev. Biol., *27:* 169–200.

SKOWRON, S., AND Z. KOMALA 1957 Limb regeneration in post-metamorphic Xenopus laevis. Folia biol. Kraków, *5:* 53–73 (English summary).

TABAN, C. 1949 Les fibres nerveuses et l'épithélium dans l'édification des régénérates de pattes (in situ ou induites) chez le Triton. Arch. Sci., *2:* 553–561.

THORNTON, C. S. 1954 The relation of epidermal innervation to limb regeneration. J. Exp. Zool., *127:* 577–601.

———— 1956 Epidermal modifications in regenerating and in non-regenerating limbs of anuran larvae. Ibid., *131:* 373–393.

———— 1956a The relation of epidermal innervation to the regeneration of limb deplants in *Amblystoma* larvae. Ibid., *133:* 281–299.

UNGEWITTER, L. H. 1951 A urea silver nitrate method for nerve fibers and nerve endings. Stain Tech., *26:* 73–76.

YNTEMA, C. L. 1959 Regeneration in sparsely innervated and aneurogenic forelimbs of *Amblystoma* larvae. J. Exp. Zool., *140:* 101–123.

1 Photomicrograph of longitudinal section of right forelimb of L84–4N Amputation was through the distal brachium and the animal was kept for 9 days at room temperature. In this normally innervated limb, the blastema is proliferative and little resorption of the brachium followed amputation. (Since sections in this case were made from posterior to anterior, the limb points in the direction opposite to that of the right limbs sectioned from anterior to posterior which are illustrated below.) $\times$ 40.

2 Photomicrograph of longitudinal section of right forelimb of L85-2N. Amputation was through the distal brachium and the animal was kept for 10 days at room temperature. In this normally innervated limb, the blastema is in the condensing stage. Considerable resorption of the brachium occurred as indicated by the shortened residuum of the humerus. $\times$ 40.

3 Photomicrograph of longitudinal section of right forelimb of L84–2N. Amputation was through the distal brachium and the animal was kept for 9 days at room temperature. In this normally innervated limb, the blastema is in the condensing stage. As in L83-2R (fig. 5) the head of the humerus and a short portion of the shaft escaped resorption following amputation. $\times$ 40.

4 Photomicrograph of longitudinal section of right forelimb of L99–3N. Amputation was through the proximal antibrachium and the animal was kept for 9 days at 20°C. This animal did not eat well and remained relatively small. In this normally innervated limb the resorption following amputation was confined to the antibrachium. $\times$ 40.

5 Photomicrograph of longitudinal section of right forelimb of L83-2R. Amputation was through the distal brachium and the animal was kept for 8 days at room temperature. No nerve fibers were found in this blastema which is in the accumulative stage. The head of the humerus and a short portion of the shaft escaped resorption following amputation. $\times$ 40.

6 Photomicrograph of longitudinal section of right forelimb of L85–2R. Amputation was through the distal brachium and the animal was kept for 10 days at room temperature. The blastema of this aneurogenic limb is in the proliferative stage. As in L84–4N (fig. 1), little resorption of the brachium followed amputation. $\times$ 40.

7 Photomicrograph of longitudinal section of left forelimb of L99–20L. Amputation was through the proximal antibrachium and the animal was kept for 11 days at 20°C. In this aneurogenic limb, the resorption would probably have continued until the stump resembled those illustrated in figures 2 and 3. The bony collar of the humerus has not resorbed as yet. Accumulation of cells to form an early blastema has started in the distal part of the stump. $\times$ 40.

436

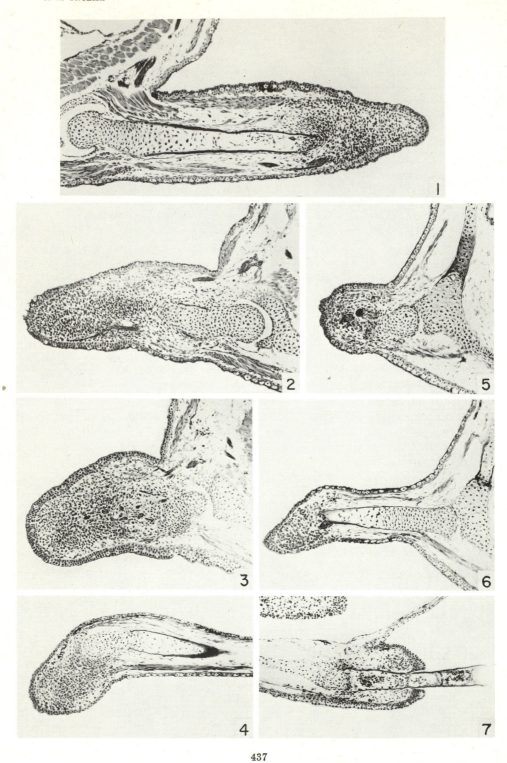

PLATE 2

8   Photomicrograph of longitudinal section of the innervated early blastema in right limb of L82–2N. Amputation was through the distal brachium and the animal was kept for 7 days at room temperature. Nerve fibers are seen traversing the core and epidermis of the blastema to reach the distal portion. The apical epidermis is basically 3 to 4 cells thick and has localized areas of greater thickness forming projections. × 145.

9   Photomicrograph of longitudinal section of the aneurogenic proliferative blastema in the right limb of L85-2R. Amputation was through the distal brachium and the animal was kept for 10 days at room temperature. The apical epidermis is three cells thick. × 145.

10   Photomicrograph of longitudinal section of the aneurogenic early blastema in the right limb of L85–2R. Amputation was through the distal brachium and the animal was kept for 8 days at room temperature. The outer layer of epidermis in the apex of the blastema has developed ridges. The inner layer has become ill-defined and the basement membrane in the area has disappeared. × 145.

11   Photomicrograph of longitudinal section of the innervated proliferative blastema in right limb of L84-3N. Amputation was through the distal brachium and the animal was kept 9 days at room temperature. Nerve fibers can be seen in the apical epidermis which is 5 to 8 cells thick. × 145.

12   Photomicrograph of longitudinal section of the aneurogenic proliferative blastema in the right limb of L99–16R. Amputation was through the proximal antibrachium and the animal was kept for 11 days at 20°C. The apical epidermis is three cells thick. × 145.

13   Photomicrograph of longitudinal section of the aneurogenic early blastema in the left limb of L–99–20L. Amputation was through the proximal antibrachium and the animal was kept for 11 days at 20°C. Resorption in this case is incomplete. The shaft of the humerus would have broken down and only the proximal end of the humerus would have survived. Accumulations of blastemal cells occur above and below the humerus. × 145.

438

383

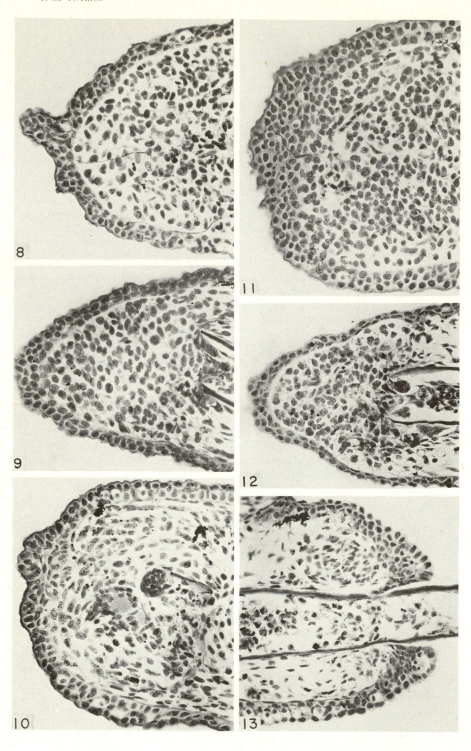

384

Reprinted from *J. Exptl. Zool.*, **145**, 207–221 (1963)

Tissue Interaction in Amputated Aneurogenic Limbs of *Ambystoma* Larvae[1]

21

TRYGVE P. STEEN AND CHARLES S. THORNTON
*Department of Biology, Yale University, New Haven, Connecticut and
Department of Zoology, Michigan State University, East Lansing, Michigan*

In recent years evidence has accumulated which clearly indicates that neural and epidermal tissues exert significant influences on the process of regeneration in amphibian limbs (Schotté and Butler, '41; Singer, '52; Rose, '44; Thornton, '60). The presence of nerves in the limb stump is essential if those post-amputation phases of regeneration which are concerned with the limitation of tissue regression and with the formation of a blastema are to occur at all (Schotté and Butler, '44; Butler and Schotté, '49). Although the nature of the neural control of the early phases of regeneration is still unknown (Singer, '60), a considerable body of evidence (Singer, '52; Van Stone, '55) has demonstrated that the neural influence is expressed quantitatively. The role of the skin in amphibian limb regeneration has been extensively reviewed by Singer and Salpeter ('61). In addition to an important function in removing cellular debris resulting from the trauma of amputation (Bodemer, '58; Singer and Salpeter, '61), the wound epithelium of the limb stump possesses important but still unknown properties involved in stimulating the onset of regeneration (Effimov, '31, '33; Rose, '44; Thornton and Steen, '62). Indeed, there is evidence that the wound epidermis, apparently as a function of the activities of the epidermal cap (Thornton, '62), influences the aggregation of the mesenchymatous cells to form a regeneration blastema.

Important aspects of tissue interactions in regenerating limbs have been disclosed in a variety of ways. For example, Butler and Puckett ('40) prevented limb regeneration in *Ambystoma* larvae by exposing the limb tissues to x-rays. When irradiation prevented the formation of the regeneration blastema, the limb tissues underwent a great exaggeration of the regression which normally occurs after amputation so that the greater part of the limb was resorbed. Limb irradiation at the blastema stage also prevented regeneration but excessive limb regression did not occur. Similarly, Butler and Schotté ('41, '49) produced excessive regression in forelimbs of *Ambystoma* larvae by sectioning the brachial spinal nerves to the limb. However, if denervation of the limb was delayed for nine days after amputation, when a proliferating blastema had formed, no excessive regression of the stump tissues occurred. Butler has concluded, therefore, that the regeneration blastema interacts with the injured stump tissues to restrict the extent of their regression.

Dependence of limb regeneration on neural influences is apparently acquired during the development and differentiation of the limb. Van Stone ('57, '60), for example, has found that the thigh tissues of the early stages of limb development in *Rana sylvatica* tadpoles have a lower nerve requirement for regeneration than do the thigh tissues of older stages of limb development. Most remarkable, however, is the discovery by Yntema ('59a, '59b) that limbs of *Ambystoma* larvae which have never been innervated do not require the presence of nerves for successful regeneration. Furthermore, these aneurogenic as well as sparsely innervated limbs regenerate in typical fashion. After amputation they pass successively through stages of wound healing, limited regression, epidermal cap formation, and accumulation as well as differentiation of blas-

[1] Supported by a PHS research grant from the National Institute of Neurological Diseases and Blindness. This work was begun at Kenyon College, Gambier, Ohio.

tema cells. Epidermal cap influence on the aggregation of blastema cells of aneurogenic limbs has been demonstrated by Thornton and Steen ('62). Thus, regenerating limbs which have never experienced innervation still exhibit a typical pattern of interaction between the wound epidermis and the internal limb tissues.

The striking differences in the reactions of *innervated, denervated,* and *aneurogenic* limb tissues to the stimulus of amputation provide a basis for analyzing the specific functions of the different limb tissues in regeneration. In the present paper attention has been focused primarily on the part the skin plays in the aggregation of mesenchymatous cells to form a regeneration blastema in aneurogenic limbs.

MATERIAL AND METHODS

Embryos and larvae of *Ambystoma maculatum* were used in all experiments. A complete account of the procedures for producing aneurogenic forelimbs has previously been given (Thornton and Steen, '62), and no further description is necessary here. Great care was taken to maintain the embryos and larvae in strictly sterile conditions. Feeding provided problems in maintaining aseptic conditions. Thus, the Artermia eggs were irradiated with ultra-violet light for 12 hours before hatching in sterile salt water, and the hatched larvae were washed several times in sterile tap water containing Fulvicin[2] before they were offered to the *Ambystoma* larvae. Infections, both fungal and bacterial, were kept to a minimum.

In preparation for limb transplantations, the *Ambystoma* larvae were first anesthetized in a 1/10,000 solution of MS222, and then placed on a pad of sterile gauze in a petri dish containing sterile Steinberg's solution and enough MS222 to bring the anesthetic to a dilution of 1/50,000. At the end of the transplantation operation the larvae were transferred to fresh Steinberg's solution and kept at approximately 5°C for five to eight hours. This procedure kept the larvae immobile during the healing period. After the healing period the larvae were transferred to Fulvicin-saturated, autoclaved tap water

and maintained at a temperature of 20°C for the duration of the experiment.

All transplantations of skin between limbs of different larvae were accomplished as follows: An incision was made through the skin around the proximal end of the forelimb. With watch makers forceps, ground to fine points, the skin was pulled distally and carefully freed from the underlying tissues as far distally as the mid-forearm. Under a magnification of 40 × to 45 × the inner surface of the skin, turned inside out in the process of removal, was carefully cleaned of bits of tissue and then pushed back again to occupy its former position on the limb. Amputation of the limb was then effected just proximal to the elbow. The loosened cylinder of skin was then easily pulled from the limb, examined for contaminating tissues, and then slipped into place on the host limb which had similarly been denuded of its own skin. In many cases the humerus of the host limb projected slightly beyond the distal border of the skin graft. In such cases the humerus was clipped back to the level of the skin. As will be reported in Section C, a variation of the skin transplantation described above involved placing skin sleeves from *two* donor limbs onto the host limb. In these latter cases, skin from an innervated limb was transplanted to an aneurogenic limb and trimmed to cover only the proximal two-thirds of the host limb stump. The skin which had been removed from the host aneurogenic limb was trimmed carefully and transplanted back to the host limb so that it covered the distal third of the limb stump where it subsequently provided the wound epithelium.

At the end of each experiment limbs were fixed in Bouin's fluid, sectioned longitudinally at 10 μ and stained for nerves according to the Bodian silver impregnation technique. Normally innervated larval tissues were included on each slide in order to serve as a control for the quality of nerve impregnation.

In three series of experiments the internal mesodermal tissues of innervated limb stumps were transplanted to aneurogenic

2 We are indebted to Dr. J. S. Sickles of Schering Corporation for kindly supplying us with Fulvicin tablets.

limb stumps whose mesodermal tissues had been previously extirpated. In these experiments, the aneurogenic limbs were first amputated just proximal to the elbow. The humerus was then grasped with watch makers forceps and pulled out. Most of the sparse limb musculature came out with the humerus. The interior of the remaining cylinder of aneurogenic skin was then cleaned, under a magnification of 40 × to 45 ×, as carefully as possible of all mesodermal tissues. Mesodermal tissues taken from innervated limbs were then packed into the aneurogenic skin sleeve.

In two series of experiments tritiated thymidine labeled mesodermal tissues from normally innervated forelimbs were implanted into aneurogenic limbs whose mesodermal tissues had previously been removed. Two methods were used for obtaining labeled mesodermal tissues. (1) Larvae with regenerating limbs at the late mound blastema stage, when dedifferentiating stump tissues are preparing for mitoses (Hay and Fischman, '61) received single intraperitoneal injections of 2.25 microcuries of tritiated thymidine per gram of body weight. The humerus and musculature of the distal third of the limb stump, including the mound blastema, was removed from the donor limb 6 to 12 hours after the thymidine injections and transplanted to an aneurogenic limb whose mesodermal tissues had been extirpated. Commonly the mesodermal implant consisted of the distal segments of two donor limbs. (2) A second group of larvae received intraperitoneal injections of 2.25 microcuries of tritiated thymidine per gram body weight on each of three successive days. At the time of thymidine injections the forelimbs of these larvae possessed cone blastemata. Morphogenesis of the blastemata was allowed to proceed until the regenerates had reached the 2-digit stage, when the mesodermal tissues of the regenerates were removed and implanted into aneurogenic limb stumps prepared, as described above, to receive them.

Aneurogenic limbs which received unlabled mesodermal implants were fixed at late blastema and two digit stages, and their histological treatment was identical to that of the skin transplant limbs.

Aneurogenic limbs which received labeled mesoderm tissues were fixed in Bouin's fluid at mound and cone blastema stages, sectioned at 10 μ, stained in Delafield's hematoxylin, air dried and dipped for radioautography, using Kodak bulk emulsion NTB-3 (diluted two parts emulsion to one part water) according to the technique of Messier and Leblond ('57). Some modifications of their photographic processing technique were used to preserve hematoxylin stain in the sections. Development of the exposed sections was for one and one-quarter minutes in D-72 solution containing half the standard amount of sodium sulfite and one-quarter the usual amount of potassium bromide. The emulsion was hardened in Kodak SB-5 and fixed in 24% hypo.

We wish to express our thanks to Mrs. Mary Thornton for her expert technical assistance.

RESULTS

A total of 105 aneurogenic limbs was used in the various experiments described in this paper, and nerve counts were made on 45 of these limbs. However, the total number of cases relevant to this paper in which careful counts of nerve fibers have been made is 77, if the 32 aneurogenic and sparsely innervated limbs with deviated epidermal caps described by Thornton and Steen ('62) are included. Since these 32 limbs were from larvae which underwent central nerve tissue excision during the same period of time and under the same laboratory conditions as the larvae used for the experiments in the present report, it is assumed that the data for the number of nerve fibers in all 77 available counted limbs reflects the general standard of technique obtaining for the group as a whole. Although actual nerve counts have been made on slightly more than half (77 of 137) the aneurogenic limbs used in these studies, it should be noted from table 1 that in only 3 of 77 limbs counted were there more than 15 nerve fibers and in one were there more than 40. Indeed, 74 of the 77 limbs possessed fewer than 11 nerve fibers and 69 of these had fewer than six nerves. It should be noted that the term aneurogenic may be applied reasonably to somewhat less than half the limbs (30 of

TABLE 1

Nerve distribution in experimental limbs

Number of nerves per limb	Number of limbs counted in:				Totals
	Section B	Section C	Section D	Deviated cap group	
0	11	3	1	15	30
1–5	16	2	8	13	39
6–10	1	–	2	2	5
15–20	—	–	–	2	2
30–40	1	–	–	—	1
					77

77), and that the remaining limbs whose nerve fiber counts have been recorded are "sparsely innervated" in accordance with the definition of Yntema ('59a).

An additional general point is that after simple amputation and also after extensive experimental treatment as detailed in this paper, the internal stump tissues of aneurogenic and sparsely innervated limbs became pigmented. In most cases the pigment was particularly concentrated at the distal end of the limb stump, but often following the transplantations it was more generally distributed. Histological observation revealed that the pigment consisted of brown intracellular granules. The pigment was argyrophylic, staining intensely with the Bodian protargol technique, and was not stained with Delafield's hematoxylin. Since treatment for five to seven hours with a 10% hydrogen peroxide solution completely bleached the pigment, it is neither a melanin nor a chromaffin (Pearse, '60). It should be noted that after five to seven hours of bleaching a brown pigment found in the intestinal wall of these larvae was completely removed but pigment found in the skin was not fully bleached. We have therefore tentatively concluded that this pigment is pseudomelanin, which is regarded by Pearse ('60) as belonging to the class of lipfuscin pigments but containing additional elements from lipid-soluble phenolytic oxidation products.

A. Autotransplantation of aneurogenic skin

The complete exchange of limb skin would seem to involve considerable surgical trauma. It was thought necessary, therefore, to provide a control series of operations in which the effect of surgical trauma on the survival of skin grafts and on the subsequent regenerative activities of aneurogenic limbs could be assessed. A series of 12 aneurgoenic limbs underwent removal of the skin of the upper arm, and using techniques similar to those described on page 208, the skin was autoplastically grafted back to its former location on the upper arm. Gross observation revealed that regeneration proceeded typically in all 12 limbs (Yntema, '59a; Thornton and Steen, '62). There was, however, a noticeable contraction of melanophores in the skin during the first three days after the operation. Healing was completed by 12 hours. From the fourth to the eighth day, the internal limb tissues, especially those at the tip, could be observed to become pigmented. The epidermis of the graft appeared slightly thickened during days five through eight, and during this period, the tip of humerus of each limb stump became round and more opaque, indicating that processes of tissue dedifferentiation were in progress. A mound blastema developed by the eighth day; the palette blastema stage occurred by the eleventh day; and two digits developed by the fifteenth day. The rate and degree of morphogenesis of the blastema, therefore, were clearly within the ranges observed in aneurogenic limb regeneration after simple amputation. It is concluded, therefore, that the trauma accompanying skin transplantation does not significantly influence regeneration.

B. Skin of innervated limbs transplanted to aneurogenic limbs

We turn now to the question of the possible effect on regeneration of skin that is taken from normally innervated limbs and grafted to aneurogenic limb stumps. It is known that skin exerts an essential stimulating influence on limb regeneration (Rose, '44; Thornton, '62) and that during the normal course of development once an amphibian limb has become innervated, the skin's participation in regeneration is conditioned by the presence of nerves in the limb stump (Thornton, '54).

An interchange of limb skin between parabiotic partners provided evidence for the analysis of this problem. The skin of 53 aneurogenic limbs was removed and replaced by skin taken from the forelimbs of their fully innervated parabiotic partners. Five larvae died during the course of the experiment and eight were preserved at too early a stage to yield useful data. Therefore, these results are based on a total of 40 cases. In 29 of the 40 cases no signs of regenerative activity were ever observed. In 28 of these, limb regression was excessive and resulted in the reduction of the limb to a small mound at the shoulder or to a tiny stub. In the remaining non-regenerative limb, regression was limited and at the end of 36 days there remained a long stump, very nearly as long as at the time of amputation. However, 11 limbs underwent delayed regeneration after a period of either excessive, or only slight, regression. Table 2 summarizes the gross data obtained in this experiment.

It should be emphasized that 90% of the aneurogenic limbs covered with skin from innervated limbs regressed excessively during a period of time sufficient for typically regenerating aneurogenic limbs to form cone or palette blastemata. Limbs were considered to have regressed excessively when 25%, or less, of the original limb stump remained. It should also be noted (table 2) that the rate of regression varied considerably, a phenomenon also observed in denervated injured limbs (Thornton and Kraemer, '51). Nevertheless, a maximal degree of regression occurred within a period of three weeks after amputation in 32 of the 40 cases. Those limb stumps (table 2) which lost no more than 30% of their original length, as measured at amputation, were considered to have undergone only slight regression since this degree of reduction in length is within the range observed for normally regenerating limbs whether they are aneurogenic or fully innervated.

Histological observations of all limbs which underwent excessive regression revealed a common pattern of tissue dedifferentiation, remarkably similar to that seen many times in denervated larval limb stumps (Butler and Schotté, '41, '49; Thornton, '53). The cartilaginous matrix of the humerus and the organized strands of atrophic musculature disintegrated and released mesenchymatous cells which, in contrast to control limbs, did not aggregate at the limb tip to form blastemata. Continued rapid regression in 36 of the 40 cases reduced the limb stumps to tiny stubs. Excessively regressed, non-

TABLE 2

The effect of transplanting limb skin from innervated larvae to aneurogenic limb stumps

Extent of regression	Number of cases	% of Total	Days needed to regress maximally	Number of limbs
Excessive, with no subsequent regeneration	28	70	7–12	8
			13–18	9
			19–21	7
			25–30	4
Excessive, with subsequent regeneration	8	20	7–12	2
			13–18	5
			19–21	1
Slight, with subsequent regeneration	3	7.5	22–25	3
None, with no subsequent regeneration	1	2.5	—	—

regenerating limbs are shown in figures 1, 2 and 3, which are representative of this group. The regressed limb shown in figure 1 (case 2GRA) possessed two nerves, 21 days after amputation when the larva was fixed. Although little external evidence of the limb is observable, the head and a small portion of the shaft of the humerus still persist. In figure 2 is shown a similar limb (case 9CRA) which possessed no nerves 12 days after amputation when this larva was fixed. Case 9CRA, therefore, illustrates the great rapidity with which some of the limbs in this series regressed. In figure 3 (case 2MRA) is shown a limb fixed 19 days following amputation. There are no nerves in this limb and regression can be seen to have spread to the shoulder girdle which is in the process of dedifferentiation. The head of the humerus remains, but evidence of its continued dedifferentiation is clear.

In all the above regressed limbs there is little evidence of an accumulation of mesenchymatous cells. As with denervated, regressed limbs (Butler and Schotté, '41), the present observations do not provide any clues as to the manner in which dedifferentiating cells are removed from the limb. In figure 4, however, there is clear evidence of a beginning accumulation of mesenchymatous cells to form an early blastema. This limb (case 30RA) is representative of those limbs which underwent excessive regression with subsequent indication of regeneration, although it is atypical of the group in exhibiting regenerative activity as early as 12 days after amputation. The great rapidity of regression in case 30RA has nevertheless resulted in the reduction of the limb skeleton to a remnant of the head of the humerus, while approximately half the shoulder girdle has undergone dedifferentiation. A characteristic of all regressed limbs possessing blastemata is the thickened epidermis of the limb tip, which is shown in figure 4 and is strikingly similar to the epidermal cap of the aneurogenic limbs described by Yntema ('59b) and Thornton and Steen ('62). Examination of the epidermis of the shoulder areas bordering the regressed but subsequently regenerating limbs has disclosed the interesting fact that this bordering epidermis is excessively thin. Indeed, in all of these cases the shoulder epidermis at the base of the regressed limbs consists of only a single layer of cells whose nuclei, instead of exhibiting the generally rounded shape characteristic of epidermal nuclei, are instead elongate and turned so that their long axes are parallel to the skin's surface. The outer cell membranes of these epidermal cells show irregular wavy outlines similar to those of epidermal cells bordering the limb amputation surface during the process of wound healing. It is suggested that during excessive regression the transplanted skin covering the limb stumps is reduced so extensively in area that bordering aneurogenic epidermis of the shoulder region migrates onto the limb stump to replace the skin graft. Regeneration, therefore, would occur subsequent to an invasion of the limb area by bordering aneurogenic epidermis. Although eight excessively regressed limbs subsequently regenerated, 28 did not. In analyzing the data of this experiment, a factor which may be of significance in accounting for the discrepancy in regenerative behavior of the two regressed groups of limbs was uncovered. Fixation of the 28 excessively regressed limbs, which failed to exhibit signs of regeneration, followed the day of maximal regression by an average for the group of 1.5 days. On the other hand, fixation, on the average, followed the day of maximal regression by 10.4 days in the eight regressed limbs which did exhibit regenerative activity. Perhaps more cases of regeneration of excessively regressed limbs might have occurred had the fixation date been delayed in the group of 28 nonregenerating limbs; the assumption here being that this would have allowed time for a possible replacement of the small amount of remaining graft skin by host aneurogenic skin bordering the limb base.

The explanation for regeneration in the group of three larvae with only slight limb regression is not clear. Figure 5 (case 6GRA) illustrates a typical limb of this group fixed 26 days after amputation. Regression apparently involved only the distal third of the remaining humerus. Beginning on approximately the twentieth day, the mesenchymatous cells derived from the regressing stump tissues began to accumulate to form a blastema. By the twenty-sixth day a long, atypical cone blastema had formed within which there are some indications of cartilage differentiation. The epidermis at the limb tip is thickened, while the epidermis of the pre-axial border of the limb, particularly where it is continuous with the epidermis of the shoulder, is only a single layer thick and resembles closely the thin epidermis bordering the excessively regressed limbs described above. Whether or not this thinning of the pre-axial epidermis indicates that a partial replacement of transplanted skin by aneurogenic skin took place is impossible to say. Gross observations, although limited by relatively low powers of magnification, provide no evidence of such a skin replacement. It is interesting, however, that the three limbs of this group are alike in possessing a thin epidermis at the base of the limb. Nerve counts indicated that these limbs each possessed from four to five nerve fibers.

C. Skin from innervated limbs transplanted together with aneurogenic skin to aneurogenic limbs

Skin from innervated limbs transplanted to aneurogenic limb stumps inhibits or greatly delays regeneration in the great majority of cases, as has been described above. The question arises as to the nature of the inhibition. Is the skin derived from innervated limbs incapable of taking part in the early phases of regeneration in the continued absence of nerves, or does the presence of the previously innervated skin induce a chemical imbalance in the aneurogenic limb stump which is incompatible with regeneration? Since much previous experience (Thornton, '54, '57, '58, '60) had emphasized the importance of the wound epidermis for initiating limb regeneration, an experiment was designed which allowed for the proximal two-thirds of the aneurogenic limb stump to be covered with skin from an innervated limb, but for the distal third, which would provide the wound epithelium, to be covered by the host's own aneurogenic skin. The operation (see page 208) was performed on 16 limbs, of which 13 survived to the end of the experiment. Typical regeneration occurred in all cases. Six limbs

were fixed at mound and cone blastema stages and seven limbs were allowed to regenerate to the two-digit stage. In two cases the rate of blastema formation was unusually slow, although after they formed blastemata the limbs presented a typical histological picture. It should be emphasized, however, that in these and all the other cases there was no excessive regression. Thus, in figure 6 is illustrated a limb (case 11CRA) with a late mound blastema, fixed 21 days after limb amputation. The major part of the regenerative delay of this limb was apparently a direct consequence of the extremely late onset of tissue dedifferentiation, since protocol records show that gross signs (rounding of the humerus tip with accompanying opacity) of dedifferentiation were not noted until the twelfth day. This and other cases of delayed onset of tissue dedifferentiation exhibited a general edema beneath the skin transplant during approximately the first five days after the operation. These events may possibly be causally connected. In spite of the delay, however, limb 11CRA is quite representative of mound stage blastemata of this series of larvae and histological study discloses a typical blastematous aggregation of cells beneath a thickened tip epidermis.

It is concluded, therefore, that when skin derived from innervated limbs is grafted to aneurogenic limb stumps it fails to form a wound epithelium adequate for supporting regeneration, and that the mere presence of previously innervated skin on an aneurogenic limb is not of itself inhibitory to regeneration.

D. Mesodermal tissues of innervated limbs transplanted to aneurogenic limbs

The cells comprising the blastema of a regenerating limb are derived from the mesodermal tissues of the limb stump (Hay, '62). But on a denervated limb stump no cells gather to form a blastema (Butler and Schotté, '41), and in the absence of innervation no thickened wound epithelium is observed (Thornton, '54). However, in the absence of nerves the skin of an aneurogenic limb stump forms a thickened wound epithelium able to stimulate blastema formation as well as regeneration (Yntema, '59b; Thornton and Steen, '62). Thus one is led to ask whether internal tissues from an innervated limb would participate in blastema formation in the absence of nerves when subjected to the influence of the wound epithelium formed on an aneurogenic limb stump.

As a first approach to this problem, the humerus and musculature of 15 aneurogenic limb stumps were removed as completely as possible under magnification of $40 \times$ to $45 \times$. Muscle and cartilage from the upper arm of the normally innervated parabiotic partner were removed and carefully packed into the empty cylinder of skin, which was the prepared aneurogenic limb stump. Six of the limbs were subsequently fixed at cone or early palette stages, and the remaining nine limbs regenerated completely. Figure 7 (case 30LA) illustrates a palette blastema fixed 12 days after implantation of mesodermal tissues from an innervated limb and is quite typical of this experimental group. All limbs that received implants regenerated with no delay. Aneurogenic control limbs in this experiment regenerated to the late cone-early palette blastema stage by 12 days after amputation. The aneurogenic limbs of this series, regenerating with previously innervated mesodermal tissues, *all* formed palette blastemata by 12 days after tissue implantation (which was also the day of amputation).

Since the experimental aneurogenic limbs described in this section had undergone extensive extirpation of their own mesodermal tissues, the fact that both the rate of regeneration and the size of the regenerating blastemata of these limb stumps were equivalent to those of control limbs would seem to attest to the likelihood that the blastema cells of the experimental limbs were derived from the implanted mesodermal tissues. Such a conclusion, however, must be accepted with caution, since there is no direct evidence that the blastema cells were in fact derived from the mesodermal implants.

An attempt was made, therefore, to label the tissues being implanted into aneurogenic limbs in the hope that marked blastema cells would be produced whose ancestry could be traced. Labeling of larvae with regenerating forelimbs (see page 209) was done in two ways: (A) In 30 cases tritiated thymidine injections of innervated larvae occurred at a time when the regenerating forelimbs were in the mound blastema stage, a period when the dedifferentiating mesodermal tissues of the stump and the first regeneration blastema cells are actively synthesizing DNA (Hay and Fischman, '61). These labeled limb mesodermal tissues were implanted into ten aneurogenic limbs whose mesodermal tissues had been extirpated. In 35 cases the tritiated thymidine injections of innervated larvae were so timed that the label would be taken up by the regenerating limb during the late cone blastema stage when blastema cells are rapidly proliferating but approaching the phase of differentiation. The labeled cells were allowed to differentiate into muscle and cartilage tissues which were then transplanted to 15 aneurogenic forelimb stumps whose own mesodermal tissues had been extirpated.

In both groups larvae were fixed when the experimental limbs had mound and cone blastemata (days 6 through 11). Labeled nuclei were found in the blastemata of all 25 limbs. It should be noted that the ten limbs which received tissues labeled at the mound stage consistently showed lower numbers of labeled nuclei and also that variations in the intensity of the label were noted in all limbs. Two regenerating aneurogenic limbs of group B are illustrated as typical examples of the nature of the labeling. In figure 8 (case 11GLA) is illustrated an eight-day regenerate in the long cone blastema stage. Labeled cells have accumulated prominently at the blastema tip. Scattered throughout the blastema are cells containing pigment granules which are readily distinguished from the radioautographic silver grains during direct observation, for they are brownish in color and distributed throughout the cell. The grains of silver due to the radioactive label are confined to areas over the nuclei. In figure 10, a higher magnification of some of the labeled blastema cells shown in figure 8, the nature of the label is shown more clearly. Figure 9 is a longitudinal section through the early cone blastema of case 12HLA, seven days after the operation.

As was illustrated in figure 8, the labeled blastema cells of case 12HLA form a compact group near the tip of the blastema. Some pigment granules are present but, as figure 11 indicates under higher magnification, the nuclei of many of the blastema cells are clearly labeled. It should also be noted that the label shows dilution in the nuclei near the blastema tip, although one nucleus in the central area of the blastema shows little dilution of the label. It is assumed that dilution of the label resulted from the mitotic division of the mesenchymatous cells which were originally labeled with tritiated thymidine. It should be noted that the duration of regeneration in both of these aneurogenic larvae was seven and eight days. Hay and Fischman ('61) present evidence that when tritiated thymidine is injected into the adult newt with a ten-day limb regenerate, the label is first diluted at three days post-injection and that a second dilution may begin at about the sixth day. Evidence which we have obtained (unpublished) for larval regenerating limbs indicates that a similar interval of sequential mitotic proliferation occurs in these limb stumps. Therefore, it is probable that the labeled cells shown in figure 11 have divided at least once since being implanted and possibly a second time as well.

It would appear that, unlike skin, mesodermal tissues of innervated limbs can take part in regeneration when implanted into aneurogenic limbs. The possibility must be considered, however, that tritiated thymidine label could be transferred following the death and destruction of a labeled cell. In such an event the origin of a labeled cell would be difficult to determine. Although label transfer can not be completely eliminated as a possibility in these experiments, some observations of the limbs of this series indicate that if a transfer of label did take place it must have been at a very low level. A small epidermal blister, apparently involved in the elimination of cell debris and containing cells with pseudomelanin pigment, was observed on several limbs but in no case was there any evidence of radioactive label in the material within, or associated with, the blisters. Except for these few small blisters, very little evidence of cellu-

lar destruction has been observed in these limb stumps. It would be reasonable to assume, if there were a significant transfer of label due to cell destruction as observed in the adult newt by Riddiford ('60), that labeled nuclei would be found in the wound epidermis. Since mitotic activity was observed in the epidermis of these experimental limbs, epidermal cells must have been synthesizing DNA and could have thus been able to take up released label. In the 25 limbs which constitute this experiment not a single epidermal nucleus was found to be labeled. This provides some indication that any transfer of label in these limbs was apparently at a level lower than was detectable by our radioautographs. Therefore, it would seem reasonable to assume that the observed labeled blastema cells were derived from the dedifferentiation of the implanted mesodermal tissues.

DISCUSSION

The experiments described in the preceding pages provide data which emphasize the importance to the initiation of regeneration that a specific tissue, the skin, of the aneurogenic limb stump of larval *Ambystoma* possesses. Thus, most aneurogenic limbs whose skin has been replaced by limb skin from a fully innervated larva will, after amputation, fail to initiate the early phases of regeneration and, indeed, in 36 out of 40 cases (see table 2, page 211) will undergo extreme regression with loss of most of the formed structures of the limb stump. However, previously innervated skin is the only tissue which will produce this drastic effect. For, when mesodermal tissues of innervated limbs replace the mesodermal tissues of aneurogenic limb stumps, regression is limited and subsequent regeneration is typical, both in rate and degree of morphogenesis, when compared with that of regenerating control aneurogenic limbs. No inhibitory effect on regeneration by implants of previously innervated mesodermal tissues has been observed in a total of 40 aneurogenic limb stumps. This evidence, therefore, provides a basis for understanding the observed contrast between the presence of regeneration in aneurogenic limbs (Yntema, '59a) and the failure of regeneration in limbs whose nerves have been transected (Singer, '52).

In recent years an increasing amount of attention has been devoted to the importance of the part taken by limb skin in the process of regeneration. An important advance in analyzing the function of the wound epithelium in limb regeneration was made by Rose ('44) and Gidge and Rose ('44) who demonstrated that adult frog limbs can be induced to regenerate if dermis is prevented from taking part in the wound healing process. The resulting wound epidermis maintains an unimpeded interaction with the internal tissues of the stump, a circumstance which both Rose ('44) and Polejaiev and Faworina ('35) believe to be essential for the onset of regeneration. Wound epidermis derived from head skin of *Ambystoma talpoideum* which has been transplanted to the forelimb will not support regeneration (Thornton, '62). In this case no epidermal cap develops, and instead of a regeneration blastema, a connective tissue pad and cartilaginous callus form at the limb stump tip, much as in the non-regenerating limbs of adult frogs. Further, it has been shown that in denervated limbs of larval salamanders wound healing occurs but no epidermal cap develops and blastema cell aggregation does not take place (Thornton, '54). Such limbs exhibit excessive regression (Butler and Schotté, '41).

The failure of regeneration and the occurrence of extreme regression in the aneurogenic limbs covered with previously innervated limb skin, as described in preceding pages, is not a result of either a mechanical or a chemical effect on the stump tissues. When only the proximal two-thirds of the aneurogenic limb stump is covered with skin derived from innervated limbs and the distal third of the limb stump retains its own aneurogenic skin, regeneration is similar to that of control aneurogenic limb stumps. In these cases, the limb stump tissues have an extensive contact with the transplanted, previously innervated skin, yet regeneration proceeds without delay. The significant variable in these experiments is the presence of the aneurogenic skin at the tip of the limb stump where it provides the wound epithelium which migrates over the amputation surface. Beneath this wound epithelium mesenchymatous cells accumulate to form a regeneration blastema. It is interesting to note that the presence of previously innervated limb skin covering the greater part of the aneurogenic limb stump has not influenced in any perceptible way the activities of either the internal aneurogenic limb tissues or the distal aneurogenic skin. Conversely, the internal aneurogenic mesodermal tissues have not, apparently, affected the competence of the transplanted previously innervated limb skin, since inhibition of regeneration and excessive regression occur in the great majority of cases where the skin transplant forms the entire skin covering of the aneurogenic limb stump. In such cases the skin graft, derived from innervated limbs, reacts as though it were a part of a denervated limb. Since mesodermal tissue implants, derived from innervated limbs, apparently take part in the regeneration of limbs whose skin is aneurogenic, it would seem that failure of blastema formation and the production of excessive limb regression may be causally related to a change in character of limb skin that developed on an innervated limb. It would be interesting, therefore, to know if the excessive regression of denervated limbs of larval *Ambystoma* would be halted by transplants of skin from aneurogenic limbs.

SUMMARY

1. The skin of the limb stumps of 40 aneurogenic larvae was removed and replaced with limb skin from innervated larvae. Thirty-six of these aneurogenic limb stumps subsequently regressed excessively, and eight of the 36 limb stumps formed blastemata after maximal regression. Evidence indicates the possibility that aneurogenic skin migrated onto the excessively regressed limb stumps of these eight cases. Three of the 40 limbs regressed only slightly, but subsequently formed blastemata. Possibly some aneurogenic skin migration to these limb stumps occurred, but evidence is not conclusive. One limb stump neither regressed nor regenerated.

2. When skin from an innervated limb covered only the proximal two thirds of the limb stump, while the distal third retained its own aneurogenic skin, regeneration occurred.

3. Regeneration of aneurogenic limb stumps, whose own mesodermal tissues had been replaced by mesodermal tissues of innervated limbs, proceeded typically in rate and in morphogenesis. Innervated mesodermal tissues labeled with tritiated thymidine produced labeled blastema cells within the aneurogenic limb site.

4. The results of this investigation emphasize the importance which the limb skin possesses for regeneration of aneurogenic limbs of *Ambystoma* larvae.

LITERATURE CITED

Bodemer, C. W. 1958 The development of nerve-induced supernumerary limbs in the adult newt, *Triturus viridescens*. J. Morph., *102:* 555–582.

Butler, E. G., and W. O. Puckett 1940 Studies on cellular interaction during limb regeneration in *Amblystoma*. J. Exp. Zool., *84:* 223–238.

Butler, E. G., and O. E. Schotté 1941 Histological alterations in denervated non-regenerating limbs of urodele larvae. J. Exp. Zool., *88:* 307–341.

———— 1949 Effects of delayed denervation on regenerative activity in limbs of urodele larvae. J. Exp. Zool., *112:* 361–392.

Effimov, I. M. 1931 Die materialen zur Erlernung der Gesetzmässigkeit in den Erscheinungen der Regeneration. J. Exp. Biol. (Russ.), *7:* (summarized by Polejaiev and Faworina ('35).)

———— 1933 Die Rolle der Haupt in Prozess der Regeneration eines Organs beim Axolotl. Zhur. Biol. (Russ.), *2:* (summarized by Polejaiev and Faworina ('35).)

Gidge, N. M., and S. M. Rose 1944 The role of larval skin in promoting limb regeneration in adult anura. J. Exp. Zool., *97:* 71–93.

Hay, E. D. 1962 Cytological studies of dedifferentiation and differentiation in regenerating amphibian limbs. In: Regeneration (D. Rudnick, ed.). Ronald Press, New York. Pp. 177–210.

Hay, E. D., and D. A. Fischman 1961 Origin of the blastema in regenerating limbs of the newt, *Triturus viridescens*. Devel. Biol., *3:* 26–59.

Messier, B., and C. P. LeBlond 1957 Preparation of coated radioautographs by dipping sections in fluid emulsion. Proc. Soc. Exp. Biol. Med., *96:* 7–10.

Pearse, A. G. E. 1960 Histochemistry, theoretical and applied. Second edition, Little, Brown and Co., Boston.

Polejaiev, L. W., and W. N. Faworina 1935 Über die Role des Epithels in den anfänglichen Entwicklungsstadien einer Regenerationsanlage der Extremität beim Axolotl. Arch. Entwmech., *133:* 701–727.

Riddiford, L. M. 1960 Autoradiographic studies of tritiated thymidine infused into the blastema of the early regenerate in the adult newt, *Triturus*. J. Exp. Zool., *144:* 25–32.

Rose, S. M. 1944 Methods of initiating limb regeneration in adult anura. J. Exp. Zool., *95:* 149–170.

Schotté, O. E., and E. G. Butler 1941 Morphological effects of denervation and amputation of limbs in urodele larvae. J. Exp. Zool., *87:* 279–322.

———— 1944 Phases in regeneration of the urodele limb and their dependence upon the nervous system. J. Exp. Zool., *97:* 95–121.

Singer, M. 1952 The influence of the nerve in regeneration of the amphibian extremity. Quart. Rev. Biol., *27:* 169–200.

———— 1960 Nervous mechanisms in the regeneration of body parts in vertebrates. In: Developing Cell Systems and Their Control. (D. Rudnick, ed.), Ronald Press, New York. Pp. 115–133.

Singer, M., and M. M. Salpeter 1961 Regeneration in vertebrates: The role of the wound epithelium. In: Growth in Living Systems. (M. X. Zarrow, ed.), Basic Books, New York. Pp. 277–311.

Thornton, C. S. 1953 Histological modifications in denervated injured forelimbs of *Amblystoma* larvae. J. Exp. Zool., *122:* 119–150.

———— 1954 The relation of epidermal innervation to limb regeneration in *Amblystoma* larvae. J. Exp. Zool., *127:* 577–602.

———— 1957 The effect of apical cap removal on limb regeneration in *Amblystoma* larvae. J. Exp. Zool., *134:* 357–381.

———— 1958 The inhibition of limb regeneration in urodele larvae by localized irradiation with ultra-violet light. J. Exp. Zool., *137:* 153–180.

———— 1960 Influence of an eccentric epidermal cap on limb regeneration in *Amblystoma* larvae. Devel. Biol., *2:* 551–569.

———— 1962 Influence of head skin on limb regeneration in urodele amphibians. J. Exp. Zool., *150:* 5–16.

Thornton, C. S., and D. W. Kraemer 1951 The effect of injury on denervated unamputated forelimbs of *Amblystoma* larvae. J. Exp. Zool., *117:* 415–440.

Thornton, C. S., and T. P. Steen 1962 Eccentric blastema formation in aneurogenic limbs of *Ambystoma* larvae following epidermal cap deviation. Devel. Biol., *5:* 328–343.

Van Stone, J. M. 1955 The relationship between innervation and regenerative capacity in hind limbs of *Rana sylvatica*. J. Morph., *97:* 345–392.

———— 1957 The quantitative relationship of nerves to the loss of regenerative capacity in developing hind limbs of *Rana sylvatica*. Anat. Rec., *128:* 635–636.

———— 1960 The influence of sodium-1-thyroxine upon anuran hind limb regeneration. Anat. Rec., *136:* 295.

Yntema, C. L. 1959a Regeneration in sparsely innervated and aneurogenic forelimbs of *Amblystoma* larvae. J. Exp. Zool., *140:* 101–124.

———— 1959b Blastema formation in sparsely innervated and aneurogenic forelimbs of *Amblystoma* larvae. J. Exp. Zool., *142:* 423–440.

PLATE 1

EXPLANATION OF FIGURES

1   Photomicrograph of a longitudinal section through the right forelimb 2GRA, which contains two nerves. Twenty-one days after amputation and complete replacement of its skin by skin from an innervated limb, this case regressed to the extent that there was little external evidence of the limb. Note that only the head and a small bit of the shaft of the humerus persist. × 180.

2   Photomicrograph of a longitudinal section through limb 9CRA, a nerveless right forelimb, fixed 12 days after amputation and transplantation of innervated skin over the entire limb stump. This case illustrates the great rapidity of regression that took place in some limbs of this series. × 180.

3   Photomicrograph of a longitudinal section through limb 2MRA, a nerveless right forelimb, which was fixed 19 days after amputation and transplantation of innervated skin over the entire limb stump. The excessive regression in this limb resulted in the loss of the shaft of the humerus and parts of its head as well as part of the shoulder girdle. × 180.

4   Photomicrograph of a longitudinal section through limb 3ORA, a nerveless right forelimb. Following amputation and transplantation of innervated skin to it, this limb underwent excessive regression, which reduced the skeleton to a remnant of the head of the humerus and only about half of the shoulder girdle. It is representative of those limbs that regressed and then showed signs of regeneration, for it has a clearly thickened epidermis at its tip and an early blastema. × 180.

5   Photomicrograph of a longitudinal section through limb 6GRA, a sparsely innervated right forelimb containing four nerves. Of the 40 aneurogenic and sparsely innervated forelimbs in the experimental group with innervated skin completely replacing their own skin, three cases represented by this limb regressed only slightly and then regenerated. Fixed 26 days after amputation and skin transplantation, this limb shows signs of regression only on the distal third of the remaining humerus and has formed a long, atypical cone blastema. × 90.

6   Photomicrograph of a longitudinal section through limb 11CRA, an aneurogenic or sparsely innervated right forelimb on which no nerve count could be made. This case was fixed 21 days after amputation and transplantation of skin from an innervated limb to its basal portion followed by immediate replacement of its own skin to the distal part of the limb stump. The aneurogenic skin of this limb has formed a typically thickened apical epidermis, and a mound stage blastema has formed under the epidermal thickening. × 225.

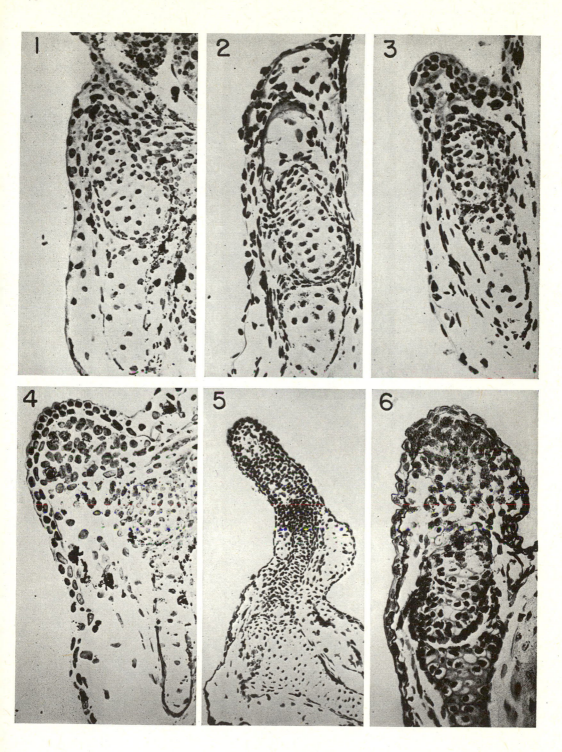

PLATE 2

7   Photomicrograph of a longitudinal section through limb 3OLA, a nerveless left forelimb stump, which had its mesodermal tissues replaced by internal tissues from innervated limbs. Twelve days after amputation and tissue transplantation, this limb has formed a typical palette blastema. The rate of regeneration and the morphology of this regenerate are both closely equivalent to those of control aneurogenic limbs. $\times$ 180.

8   Photomicrograph of a longitudinal section through limb 11GLA, an eight day regenerate in the long cone blastema stage. This right aneurogenic or sparsely innervated forelimb stump had its mesodermal tissues replaced by tritiated thymidine labeled tissues from innervated limbs, and labeled nuclei have accumulated prominently at the tip of the blastema. Since a black and white photomicrograph does not allow one to distinguish between the silver grains of the radioautograph and the pigment present in these limbs, the pigment cells (P) in the distal portion of this regenerate are marked. No nerve count could be made on this limb. $\times$ 180.

9   Photomicrograph of a longitudinal section through limb 12HLA, a left aneurogenic or sparsely innervated forelimb stump that had its mesodermal tissues replaced by tritiated thymidine labeled tissues from innervated limbs. This early cone blastema formed seven days after the operation, and labeled nuclei again identify blastema cells derived from innervated tissues. No nerve count could be done on this limb. $\times$ 180.

10   This photomicrograph of the same section as appears in figure 8 illustrates more clearly the labeled nuclei near the tip of this blastema as well as one of the pigment localizations (P). Arrows from L indicate nuclei which are representative of the range of labeling levels observed. $\times$ 720.

11   This photomicrograph illustrates the central area of the blastema shown in figure 9. It includes the many lightly labeled nuclei positioned near the tip of the blastema as well as one heavily labeled nucleus from the central area of the blastema. $\times$ 720.

220

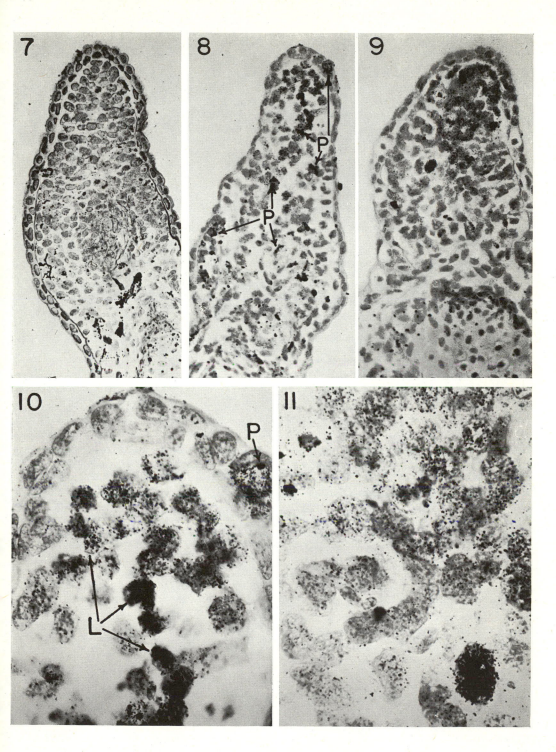

Reprinted from *J. Exptl. Zool.*, **173**, 293–301 (1970)

Recuperation of Regeneration in Denervated Limbs of Ambystoma Larvae [1]

22

CHARLES S. THORNTON AND **MARY T. THORNTON**
Department of Zoology, Michigan State University,
East Lansing, Michigan 48823

ABSTRACT   The current neurotrophic theory of amphibian limb regeneration allows for a non-neural tissue contribution of trophic substance (TS) which during ontogeny is suppressed by the production of large amounts of neural TS. A difficulty has been to account for the inability of denervated limbs to initiate regeneration in the absence of neural TS when non-neural tissue TS synthesis, no longer "quenched," might be expected to function. This difficulty has now been overcome in the aneurogenic limb system. Aneurogenic forelimbs of A. maculatum larvae were transplanted orthotopically and homoplastically in place of left forelimbs of normally innervated larvae. The brachial nerves of the host larva were allowed to grow into the grafted aneurogenic limb. When tested by subsequent denervation and amputation, the grafted limbs became progressively nerve-dependent, for regeneration purposes, between 10 and 13 days post-transplantation, dependence on nerves being complete at 13 days. Grafted, aneurogenic limbs were allowed to become innervated for periods of 2–3 weeks. Then for at least 30 days the limbs were maintained in a nerveless state by denervations, repeated at five day intervals, with histological checks on effectiveness of denervation. Amputation through the forearm was performed at 30 days with continued denervations to prevent reinnervation. Regeneration occurred in 49% of cases (in 16 of 33 limbs) with histological verification of nerveless or sparsely innervated condition. Thus, regeneration-competent aneurogenic limbs may become nerve-dependent after transplantation and then, after prolonged denervation, recover ability to regenerate under essentially nerveless conditions.

In a series of elegant experiments, Singer ('52, for review) established a quantitative relationship between the presence of peripheral nerve fibers in a salamander limb and the ability of that limb to regenerate after amputation. Between one-third and one-half of the normal limb nerve complement was required for regeneration to proceed. Singer ('65) proposed that a "trophic substance" (TS) is produced in great abundance in the neuron primarily to maintain its great mass of axonal neuroplasm, but that significant amounts "spill over" on to other tissues which then come to depend on the nerve for their own regenerative activity. Sensory nerves are as competent as motor nerves in supporting limb regeneration so that impulse transmission mechanisms seem not to be important for this regeneration-supporting action of the nerve.

Lately it has been necessary to modify the neurotrophic theory of regeneration to accommodate recent evidence that nerves are not needed for limb regeneration under certain conditions. Thus Yntema ('59) developed aneurogenic limbs in *Ambystoma* larvae by excising the neural tube of tailbud embryos. Amputation of the nerveless limbs at the larval stage was followed by typical regeneration, results which Thornton and Steen ('62) have extended. Polejaieff ('39), Bodemer ('60) and Singer and Mutterperl ('63) find that transplantation trauma will lower the neural requirement for limb regeneration. Singer ('65), in modifying his neurotrophic theory to accommodate these newer data, now proposes that the TS which the nerve contributes to the growth process is not unique for the neuron but is also synthesized by all other cells in the embryo. The outgrowth of nerves, with their massive addition of TS, quenches this nonneural synthesis of TS. Under conditions such as trauma, however, he proposes that nonneural production of TS might be resumed so that despite a reduced neural contribution enough TS is present in the limb to meet regeneration requirements.

[1] Supported by grants-in-aid from the National Science Foundation (GB-2618; GB-7748) and the National Institutes of Health (NB-04128).

A problem with this theory is to account for the inability of *denervated* limbs to initiate regeneration in the absence of neural TS when nonneural tissue synthesis of TS, no longer quenched, might be expected to function. When one denervates a limb and simply amputates at a later period, no matter how long delayed, regeneration of the denervated limb never occurs. Perhaps the long-continued functioning of limb nerves during ontogeny and larval development produces such a strong inhibition that interventions in addition to simple nerve withdrawal are needed to reactivate the nonneural tissues. The experiments of Bodemer ('60) and Singer and Mutterperl ('63), for example, would seem to point to this possibility. If this is the case, then perhaps a shorter term of innervation will allow limbs, subsequently denervated, to recuperate the ability to regenerate after simple amputation.

The aneurogenic limb system provides an excellent means of examining this possibility. During ontogeny nerves are absent in the limb, yet nerves can be introduced naturally by orthotopic transplantation of the aneurogenic limb to normal larvae, where brachial nerves may then invade the graft. This new innervation can then be withdrawn at will and the effect of this on regeneration observed. Experiments to be described below clearly demonstrate that transplanted aneurogenic limbs do become nerve-dependent after innervation but that in approximately half these cases prolonged withdrawal of innervation can later result in regeneration in the absence of nerves.

MATERIALS AND METHODS

Only larvae of *Ambystoma maculatum* were used in these experiments. Aneurogenic limbs were produced by methods previously described (Thornton and Steen, '62; and especially Thornton and Tassava, '69). Ten larvae with 4-digit aneurogenic limbs underwent amputation of the left forelimb at the mid-forearm level in order to determine regeneration rates of aneurogenic limbs *in situ* (table 1). The remaining 192 aneurogenic 4-digit limbs were transplanted orthotopically to normally innervated larvae 20–25 mm long as has previously been described (Thornton and Tas-

TABLE 1

Regeneration of aneurogenic limbs in situ

Stage	Day	No. at stage
Mound blastema	6	10/10
Cone blastema	8	7/10
Paddle blastema	10	9/10
2 — Digits	12	9/10
3 — Digits	16	8/10
4 — Digits	20	7/10

Temp. 18° ± 1°C
N = 10

sava, '69). Those limbs which were sectioned (7 μ) were stained with the Bodian Protargol method and since the host was always an innervated larva, and was sectioned with the attached limb, each section was its own control for effectiveness of silver impregnation of limb nerves. Additional details, where appropriate, are included in the descriptions of the results of the experiments.

RESULTS

Regeneration of aneurogenic limbs

Although regeneration rates of aneurogenic limbs of *A. maculatum* have been described by Yntema ('59) and also by Steen and Thornton ('63), it is nevertheless important to know the norm of regeneration in the aneurogenic limbs used in the present experiments. Therefore, ten aneurogenic larvae with 4-digit limbs underwent amputation of the left forelimbs at the midforearm level. Regeneration rates and morphogenetic development of the blastemata were close to those recorded earlier by Yntema ('59) and Steen and Thornton ('63). Table 1 summarizes these data.

Since a major emphasis of the present work is devoted to analysing the response of transplanted aneurogenic limbs to innervation from the host larva, it is important to know if there is an influence of the host on the aneurogenic limb which is *not* mediated by local nerve action. Therefore 20 aneurogenic forelimbs were transplanted orthotopically in place of left forelimbs of normally innervated larvae. Removal of the host's forelimb severed the brachial nerves. To maintain a continued nerveless state of the aneurogenic limb transplant, the host's brachial nerves were severed at the shoulder seven days following limb transplantation and again at intervals of six days thereafter until the end of the experiment on the twenty-fifth day (see Thornton and Tassava, '69). At each denervation sample limbs were fixed to check on the effectiveness of the previous denervation operation and, therefore, on the status of innervation of the experimental limbs during the course of their regeneration. Amputation was performed through the mid-forearm of the limb graft seven days following transplantation. As inspection of table 2 will demonstrate, regeneration rates and morphogenesis of the aneurogenic limb transplants in a continued nerveless or sparsely innervated condition is within the range of those of aneurogenic limbs regenerating *in situ* (table 1). For example, there is no significant difference statistically in the percentage of limbs reaching the 2-digit stage at 12 days in the two groups (Chi Square Test, 10% level). It is concluded that in the absence of the host's brachial nerves there is no unusual influence of the host, through its circulatory system or otherwise, on rate and morphogenesis of the transplanted regenerating aneurogenic limb.

Development of nerve dependence of innervated limb grafts

In studying the influence of the host's brachial nerves on the limb transplant, attention was first focused on the rate of the growth of brachial nerves into the aneurogenic limb. Relatively little consideration has been given to the pattern of the resulting innervation of the limb graft since Piatt ('42, '57) has shown that nerve patterns of aneurogenic forelimbs grafted orthotopically to innervated hosts are approx-imately normal. Our own general observations of nerve patterns in forelimb grafts do not indicate any important variance with Piatt's data. Furthermore, Singer ('46) has clearly demonstrated that increasing the motor innervation to an atypical degree will allow regeneration of asensory newt limbs to proceed. Thus, quantity of innervation, rather than its type and pattern, would seem to be the prime prerequisite for successful limb regeneration. A similar conclusion can be deduced from the successful regeneration of *Ambystoma* larval limb deplants whose sole innervation was derived from intracentral nerve fibers (Thornton, '56).

A total of 25 grafted aneurogenic forelimbs were used to study the rate of the growth of the host larvae's brachial nerves into the aneurogenic limbs (summarized in table 3). Although nerve counts were not made, each symbol, X, represents an approximately normal number of nerve bundles at the limb level indicated. Thus, for example, at ten days post-transplantation the grafted forelimbs have become fully innervated and show considerable movement.

Since this innervation is the first that the grafted limbs have ever experienced, it becomes of great interest to determine the influence of these nerves on the response of the limb grafts to amputation. Aneurogenic limbs will regenerate. Will the addition of nerves to aneurogenic limbs in any way influence their regeneration? To determine this, 20 aneurogenic forelimbs were transplanted in place of the left forelimbs of normal host larvae. After three weeks, when the limb grafts had been fully innervated for at least 11 days (see table 3), they underwent amputation through

TABLE 2

Regeneration of aneurogenic limbs transplanted orthotopically to normal host larvae and kept denervated

Denervation	Day	No. cases	Stage regen.	No. at stage	No. nerves [1]
Transplantation	0	20	—		—
1st (Amputation)	7	20	—		0;5
2nd	13	18	Cone	16/18	0;0;8
3rd	19	15	2–Digits	14/15	0;2
Fixed	25	13	3–Digits	6/13	0;4;3
			4–Digits	7/13	

Temp. 18° ± 1°C
[1] Ten limbs were fixed for nerve counts.

TABLE 3

Rate of brachial nerve growth into grafted aneurogenic forelimbs

Day post-transplant	No. cases	Level of limb reached by nerves			
		Head humerus	Elbow	Midforearm	Carpus
5	1	X			
6	3	X	X		
7	2	X	X	X	
9	4	X	X	X	X [1]
10	8	X	X	X	X
11	2	X	X	X	X
12	5	X	X	X	X

Temperature, 20° ± 1°C
[1] Indicates one nerve bundle only has reached this level.

the distal third of the upperarm. Two types of control series were studied at the same time: (a) normally innervated limbs amputated *in situ;* (b) normally innervated limbs transplanted in place of left forelimbs of normally innervated hosts and amputated on the twenty-first day post-transplantation. The results of the experiment are shown in figure 1. Although there are slight variations in the per cent of limbs reaching the 2-digit and 4-digit stages of regeneration these are not significantly different statistically (Chi Square Test, 10% level). Thus, there is no evidence that addition of nerves affects either morphogenesis or rate of regeneration in aneurogenic limb grafts.

Now it is relevant to ask: After acquiring brachial nerves do transplanted aneurogenic limbs become nerve-dependent for successful regeneration? If so, when does this occur? To answer these questions, 65

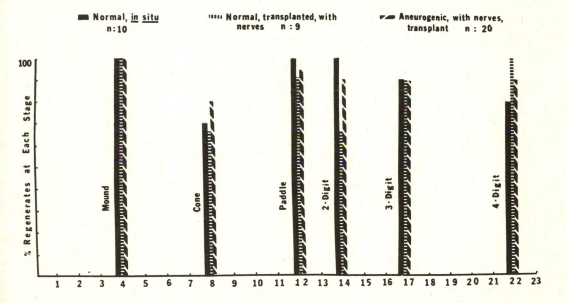

Fig. 1 Comparison of regeneration and morphogenesis of innervated limbs of *Ambystoma maculatum* larvae after: (a) orthotopic transplantation of aneurogenic limbs to normal hosts; (b) orthotopic transplantation of normal limbs to normal hosts; (c) no transplantation, normal limbs maintained *in situ*. It should be noted that these larvae are from a 1967 hatch, while those of tables 1 and 2 are from a 1968 hatch. Comparisons of rate of regeneration are made within larvae of single egg clusters only. Temperature 20°±1°C.

aneurogenic forelimbs were transplanted orthotopically to normally innervated hosts. The transplanted limbs were amputated through the midforearm and denervated at 10 days (n = 20); 11 days (n = 10); 12 days (n = 20); and at 13 days (n = 15). Subsequent denervations were performed at successive five day intervals over a period of 19 days. Table 4 summarizes the results obtained from the surviving larvae. From 10 days to 13 days post-transplantation, the appearance of successful limb regenerates in the *absence* of nerves is reduced from 74% to 0%. During this 3 to 4 day period, therefore, the limb grafts have become progressively "nerve-dependent."

A phenomenon associated with the non-regenerating limb stumps of these larvae is of particular interest. *All* non-regenerating ("nerve-dependent") limb stumps underwent excessive regression similar to that first described by Schotté and Butler ('41) for denervated limb stumps of *Ambystoma* larvae. The nerve-dependent limb stumps thus respond to denervation much as do normal *in situ* limb stumps deprived of their nerves. *Nerve-dependence seems to be similar in these two types of limb stumps.* The response to denervation and amputation is quite different, however, in other types of limb transplantation. Thus, for example, Thornton and Tassava ('69) transplanted normally innervated forelimbs orthotopically in place of forelimbs of normal host larvae. The limb grafts were maintained in a denervated or sparsely innervated state by repeated nerve section. Nevertheless, after amputation the resulting limb stumps underwent *limited* regression and, in approximately 50% of cases,

successfully regenerated. The interpretation of these results, as well as those of Polejaieff ('39) and Singer and Mutterperl ('63), was that transplantation trauma stimulated production of threshold levels of nonneural trophic substance in the limbs which regenerated. In the present experiment since all of the denervated non-regenerating limb stumps regressed excessively, it is concluded that the presence of the host brachial nerves in the 13-day series (table 4) had eliminated any effect of transplantation trauma in these limbs.

Recuperation of regeneration

Since the foregoing experiments clearly demonstrate that innervation of aneurogenic limb grafts renders them nerve-dependent (for regeneration) within 13 days post-transplantation, the question now raised is: Will these nerve-dependent limbs recover their ability to regenerate without nerves after a prolonged period of denervation?

In an attempt to answer this question, 60 aneurogenic limbs were transplanted orthotopically to innervated host larvae and allowed to become fully innervated. On day 19, 21 or 33 post-transplantation the grafted limbs were denervated, but *not* amputated. Successive denervations were performed at 5-day intervals over a period of 40 days. On the thirtieth day of denervation 33 of the limb grafts were amputated through the midforearm and maintained in a denervated state to the thirty-sixth to fortieth day when they were fixed. As inspection of table 5 will show, 49% of these denervated limbs formed proliferating or condensing regenerates (figs. 2, 3, 4). Of particular interest is the fact that only 4 of the 17 non-regenerating limb stumps regressed. The remaining cases (13) continued as stable limb stumps. Perhaps failure of regression is related to a minimal non-neural TS activity in the limb stump. Failure of limb stumps of *Ambystoma* larvae to regress when innervation is subthreshold for regeneration has been attributed similarly to minimal TS (neural) activity (Thornton, '54; Deck, '61).

A consideration of great importance to these results is the verification of nerve-lessness *throughout the course of the ex-*

TABLE 4

Rate of development of nerve dependence of innervated limb grafts

Day of nerve section	No. cases	No. regen.	Regen.
			%
10	19	14	74
11	9	4	44
12	16	2	12
13	12	0	0

Temperature 20° ± 1°C.
Amputation at midforearm level at time of denervation.

TABLE 5

Recuperation of regeneration in denervated limbs following transplantation of aneurogenic limbs orthotopically to normal hosts with subsequent innervation and prolonged denervation

First denervation on day post-transplant	No. cases	No. denervations	No. regen.	Regen. %	Amputation on day	Fixed on day
19	10	7	5	50	30	36–40
21	7	7	4	57	30	40
33	16	7	7	43	30	36–40
Total	33	7	16	49	30	36–40

Temperature 20° ± 1°C.
Host length at transplantation, 20–23 mm.
Host length at termination of experiment, 30–35 mm.

periment. Therefore, 27 of the denervated limbs were fixed, as is shown in table 6, for nerve counts during the seven periods of denervation. Sample size was a minimum of three limbs. Limbs were fixed at each period of denervation, but were not themselves denervated at the time of fixation. Therefore, each sample period served as a check on the effectiveness of the previous denervation. As inspection of table 6 will show, denervations were adequate (except at the 4th, when a recognized deficiency in technique resulted in 3 poor denervations). Also, 14 of the 16 regenerates obtained (see table 5) were subjected to nerve counts and they proved to be nerveless or sparsely innervated (table 6). It is concluded, therefore, that the nerve dependence for regeneration was reversed in approximately half the long-term denervated limbs.

DISCUSSION

The theory of a neurotrophic mechanism in support of amphibian limb regeneration, as extensively propounded by Singer (see '52; '65, for pertinent reviews), has been of invaluable aid in stimulating meaningful inquiry into the processes of regeneration. However, comprehensive as this theory is, there are two outstanding problems which must be resolved before it can be considered well established. The first of these is the failure to identify the neurotrophic substance. Despite intensive effort (see Thornton, '68, for review) little progress has been made in isolating a neurotrophic agent. The second, and less difficult, problem has been the subject of the present paper. In essence, it relates to the inability of limbs denervated for prolonged periods to undergo regeneration on subsequent amputation. According to the current neurotrophic theory, the trophic substance is considered not to be unique to the nerve, but may be synthesized by all tissues, at least in the embryo. During growth and development the nerves, in order to satisfy the imperative demands of the enormous volume increase of axonal

TABLE 6

Nerve counts of denervated limbs. Denervations at 5-day intervals

Denervation	No. days	No. cases	No. nerves
1st	5	3	0; 3; 5
2nd	10	3	0; 0; 10
3rd	15	3	0; 2; 23
4th	20	4	0; 3; 4; 4
5th	25	8 [1]	73; 79; 143; 0; 3; 12; 4; 3
6th (Amputation)	30	3	0; 2; 8
7th	35	3	0; 2; 3
Regenerates	36–40	14 [2]	3; 4; 2; 0; 7; 3; 2; 11; 2; 0; 5; 15; 0; 12

[1] At the fourth denervation three larvae were poorly denervated due to excessive bleeeding that obscured the field. Additional cases were denervated and all cases included in the count.
[2] Due to faulty impregnation, two regenerates were judged inadequate for accurate nerve counts.

cytoplasm, are thought to produce quantities of trophic substance (TS) which spill over on to other tissues and suppress their further production of TS by a kind of end-product feedback inhibition. Theoretically, one might expect that removal of the neural TS would eventually result in lifting the inhibition on synthesis of nonneural tissue TS so that limbs, after prolonged denervation, could regenerate without neural assistance. Until now this result has not been obtained.

It has been demonstrated in preceeding pages that aneurogenic limbs provide a model system which can be used to advantage in exploring the interaction between nerves and limb tissues in *Ambystoma* larvae. Aneurogenic limbs regenerate (Yntema, '59), yet when nerves are allowed to enter the limb, after orthotopic transplantation to a neurogenic host larva, the limb graft becomes dependent on the presence of nerves for successful regeneration. It is of particular interest that the process of nerve-dependence is achieved in a period of about 3 to 4 days, from day 10 to day 13 post-transplantation. It is tempting to speculate that during this 4-day interval the newly acquired innervation is acting to quench the synthesis of TS in the nonneural tissues. Certainly the data are in accord with such an interpretation. The relatively short interval of time during which nerve-dependence is acquired favors analysis of this phenomenon with modern electron microscopic and biochemical techniques. Such analyses are now under way in this laboratory.

Nerve-dependence, in this system, can be reversed by the simple expedient of withdrawing the newly acquired innervation. This result was obtained, however, on limb grafts subjected to relatively short-term innervation. Nevertheless, it should be emphasized that nerve withdrawal and *no other intervention* was adequate to reverse the acquired nerve-dependence. Normally innervated limbs have been reported to undergo regeneration with a greatly reduced (Singer and Mutterperl, '63) or sparse (Thornton and Tassava, '69) innervation after transplantation. The trauma of transplantation was, however, a necessary factor in the regeneration of

nerve-deprived limbs (Thornton and Tassava, '69). The experiments reported in the present paper, therefore, represent the first demonstration that nerve-dependent *Ambystoma* limbs can undergo regeneration during periods of prolonged absence of nerves and in the absence of transplantation trauma (see page 297). This aneurogenic limb graft system, therefore, should provide a valuable means for obtaining further insights into the neural:nonneural tissue interaction in the salamander limb and, consequently, the possibility of gaining an understanding of the trophic mechanisms underlying limb regeneration.

LITERATURE CITED

Bodemer, C. 1960 The importance of quantity of nerve fibers in development of nerve-induced supernumerary limbs in *Triturus* and enhancement of the nervous influence by tissue implants. J. Morph., 107: 47–60.

Deck, J. D. 1961 Morphological effects of partial denervation on regeneration of the larval salamander forelimb. J. Exp. Zool., 147: 299–307.

Piatt, J. 1942 Transplantation of aneurogenic forelimbs in Amblystoma punctatum. J. Exp. Zool., 91: 79–101.

——— 1957 Studies on the problem of nerve pattern. II. Innervation of the intact forelimb by different parts of the central nervous system in Amblystoma. J. Exp. Zool., 134: 103–126.

Polejaieff, L. W. 1939 Uber die Bedeutung des Nervensystems bei der Regeneration der Extremitaten bei den Anuren, C. R. (Doklady) Acad. Sci. Nauk., SSSR 25: 543–546.

Schotté, O. E., and E. G. Butler 1941 Morphological effects of denervation and amputation of limbs in urodele larvae. J. Exp. Zool., 87: 279–321.

Singer, M. 1946 The nervous system and regeneration of the forelimb of adult *Triturus*. V. The influence of number of nerve fibers, including a quantitative study of limb innervation. J. Exp. Zool., 101: 299–338.

——— 1952 The influence of the nerve in regeneration of the amphibian extremity. Quart. Rev. Biol., 27: 169–200.

——— 1965 A theory of the trophic nervous control of amphibian limb regeneration, including a re-evaluation of quantitative nerve requirements. In: Proc. Regen. in Animals. V. Kiortsis and H. A. L. Trampusch, eds. North Holland Publ. Co., Amsterdam, pp. 20–32.

Singer, M., and E. Mutterperl 1963 Nerve fiber requirements for regeneration in forelimb transplants of the newt, *Triturus*. Dev. Biol., 7: 180–191.

Steen, T. P., and C. S. Thornton 1963 Tissue interaction in amputated aneurogenic limbs of *Ambystoma* larvae. J. Exp. Zool., 154: 207–221.

Thornton, C. S. 1954 The relation of epidermal innervation to limb regeneration in *Amblystoma* larvae. J. Exp. Zool., 127: 577–602.

———— 1956 The relation of epidermal innervation to the regeneration of limb deplants in *Amblystoma* larvae. J. Exp. Zool., *133:* 281–300.

———— 1968 Amphibian limb regeneration. In: Advances in Morphogenesis. M. Abercrombie, J. Brachet and T. King, eds. 7: 205–249.

Thornton, C. S., and T. P. Steen 1962 Eccentric blastema formation in aneurogenic limbs of *Amblystoma* larvae following epidermal cap deviation. Dev. Biol., *5:* 328–343.

Thornton, C. S., and R. A. Tassava 1969 Regeneration and supernumerary limb formation under sparsely innervated conditions. J. Morph., *127:* 225–232.

Yntema, C. L. 1959 Regeneration of sparsely innervated and aneurogenic forelimbs of *Amblystoma* larvae. J. Exp. Zool., *140:* 101–123.

PLATE 1

EXPLANATION OF FIGURES

2   Larva IFd. Total number of denervations: 7. Number of nerve fibers observed in limb stump after fixation and Bodian stain: 4. Amputation on day 30 of denervation and at time of sixth denervation. Fixation on day 36 of denervation. Note proliferating mound blastema. × 300.

3   Larva IFg. Total number of denervations: 7. Number of nerve fibers observed in limb stump after fixation and Bodian stain: 2. Amputation on day 30 of denervation (at 6th denervation). Fixation on day 38 of denervation. Note proliferating cone blastema. × 300.

4   Larva IFb. Total number of denervations: 7. Number of nerve fibers observed in limb stump after fixation and Bodian stain: 2. Amputation on day 30 of denervation (at 6th denervation). Fixation on day 40 of denervation. Note condensing paddle blastema. × 300.

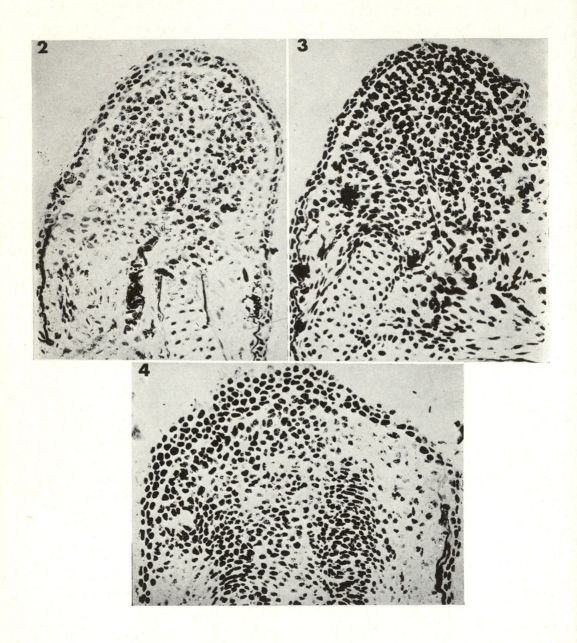

301

Reprinted from *J. Embryol. Exptl. Morphol.* **6**, 247–261 (1958)

Effects of Adrenal Transplants upon Forelimb Regeneration in Normal and in Hypophysectomized Adult Frogs

by O. E. SCHOTTÉ *and* J. F. WILBER[1]

From the Department of Biology, Amherst College

23

WITH FIVE PLATES

INTRODUCTION

THE frog provides among the vertebrates the best opportunity to investigate the factors which determine loss of regeneration within the ontogenetic boundaries of a single organism, for in tadpoles the capacity to regenerate tapers off along the proximo-distal axis of their limbs and in adult frogs limb regeneration is substantially absent.

It was generally accepted that metamorphosed frogs lose this capacity because of progressive complexity of structure in the growing limbs (Marcucci, 1916; Polejaiev, 1936; and Forsyth, 1946). However, the conclusion that it is the nature of the limb tissues which determines presence or absence of regeneration was shown to have only limited validity when regeneration was obtained in metamorphosed frogs. Thus Polezhajev (1945) demonstrated that repeated trauma to the amputational surface of limbs of adult frogs (which according to his own theory had irrevocably lost the powers of regeneration because of progressive differentiation) was sufficient for recuperation of this faculty in a substantial number of cases. By immersing metamorphosed frogs in hypertonic salt solutions, Rose (1944, 1945) obtained limb regeneration; and finally, Singer (1954) (and indirectly also Van Stone, 1955), by increasing the nerve-supply of limbs, obtained regeneration from normally unresponsive amputational levels.

The papers of Walter (1911), Schotté (1926), Richardson (1940–5), Hall & Schotté (1951), and Schotté & Hall (1952) concurred in showing that both the pituitary and the thyroid glands influenced regeneration, but these findings could not explain the cessation of regeneration in Anura (Naville, 1927; Guyénot, 1927; and Schotté & Harland, 1943).

The suspicion that cellular properties did not play so dominant a role in regeneration as the above evidence seemed to suggest, became a near-certainty

[1] *Authors' addresses*: O. E. Schotté, Department of Biology, Amherst College, Amherst, Mass., U.S.A.; J. F. Wilber, Harvard University Medical School, Cambridge, Massachusetts, U.S.A.

[J. Embryol. exp. Morph. Vol. 6, Part 2, pp. 247–269, June 1958]
5584.6

S

when the endocrinological factors in regeneration were re-examined in the light of discoveries made by mammalian endocrinologists in regard to the pituitary-adrenal synergism under stress conditions (Selye's general adaptation syndrome, 1947; Selye & Stone, 1950). In 1952 Schotté & Hall proposed that, in a manner already known from mammalian endocrinology, the role of the pituitary in urodele regeneration was probably confined to stimulation of the adrenal cortex after amputational stress. The reality of a pituitary-adrenal synergism in respect to regeneration was demonstrated when, in newts deprived of their pituitary, replacement therapy with ACTH (Schotté & Chamberlain, 1955) and with cortisone (Schotté & Bierman, 1956) restored the regenerative capacities in these animals. The direct involvement of the adrenals in regeneration of urodeles became a certainty when Schotté & Lindberg (1954) induced regeneration in hypophysectomized newts by transplantations of frog adrenals.

Because of marked similarities between urodele and anuran regeneration, and because of the well-known involvement of endocrines in metamorphosis, a process which seemed to coincide with loss of regenerative processes in Anura, it became imperative to determine whether artificial changes in the endocrine system of frogs could modify regeneration of their limbs. The first attempts to investigate this problem were made a few years ago in this laboratory when it was shown that the transplantation of additional adrenals could lead to restoration of regenerative capacity in premetamorphic tadpoles at normally non-regenerating amputational levels (Lindem, unpublished thesis, Amherst College, 1954).

The realization that lost regenerative ability was regained in tadpoles by inducing a hyperadrenal state offered an irresistible invitation to conduct still further investigations on adult frogs. The purpose of this research, therefore, has been to determine whether or not the normally absent capacity for limb regeneration in adult frogs could be recovered by introduction of additional adrenal glands into these animals.

MATERIALS AND METHODS

The effects of adrenal transplants upon the regeneration of forelimbs were studied in two American species of frogs: *Rana clamitans*, commercially procured from the south of the U.S.A.; and *R. pipiens* (donor of all the adrenal glands), secured from Wisconsin and northern Vermont. These postmetamorphic frogs were force-fed twice weekly on beef-liver with bone-meal supplement before and after the operations. For all the operations the frogs were narcotized by immersion in a solution of MS 222 (1 : 1,000), a meta-amino-benzoic acid ethylester in the form of methanesulphonate (Sandoz, Basle).

Amputations were made at either the mid radius-ulnar level or at the distal humerus level of the frog forelimb. These amputations were performed under a dissecting microscope, and, immediately after the initial severance, the protruding bone elements were re-amputated to adjust them to the level of the retracted soft tissues.

For transplantation experiments both narcotized host and donor frogs were placed in petri dishes upon sterile surgical gauze moistened with sterile physiological salt solution. The adrenal glands in frogs appear as thin ribbons of a golden yellow colour extending along the median ventral surface of each kidney. The donors' kidneys were dissected out and placed upon surgical gauze in another sterile dish and the adrenals were excised from the kidneys with iridectomy scissors and Swiss watchmaker's forceps underneath a binocular microscope. Care was taken to minimize the amount of kidney tissue adhering to the adrenal bodies. All transplantations were made into the lower jaw because this region of the frog is highly vascularized. For the insertion of the prepared adrenals a small opening was made in the skin of the lower jaw with iridectomy scissors. A sterile probe was used to enlarge the subcutaneous pocket into which the pieces of adrenals were pushed as far anteriorly as possible to prevent postoperative extrusion; no transplants were lost.

The pituitary gland in the frog is easily identified through the semi-cartilaginous parasphenoid bone. The operative procedure was adopted from Levinsky & Sawyer (1952), with the difference that we extirpated the exposed pituitary with watchmaker's forceps instead of with a pipette.

After any surgical procedure the animals were kept for one day at 13° C. ($\pm 1°$ C.) in finger-bowls, the bottoms of which were covered with surgical gauze and moistened with sterile physiologic solution to which a small quantity of sodium sulphadiazine was added. After 1 day the animals were transferred into demineralized water, about ten per aquarium, and were maintained at 20° C. ($\pm 1°$ C.) for the duration of the experiment.

This research is based upon a study of sixty-five surviving animals from a much larger number of operated frogs. From these cases forty-seven forelimbs were studied histologically; and in addition, the lower jaws of five frogs were sectioned serially to investigate the condition of the adrenal transplants especially in regard to vascularization; also, the heads of six hypophysectomized animals were examined in sections to verify the completeness of pituitary removal. The tissues were fixed in Bouin's, decalcified in Jenkin's solution, stained with Harris's modification of Delafield's hematoxylin and counterstained with orange G. For the study of slides of adrenal tissues the sections were stained by Mallory's polychrome method.

OBSERVATIONS AND RESULTS

1. *Survival and functional state of adrenal transplants*

At various times the conditions of forty-five heteroplastic or homoplastic adrenal transplants were studied among the fifty-three frogs having received similar transplants. For gross study the whole subcutaneous area of the lower jaws was exposed by dissecting away the skin and in all cases the transplants were found embedded within the richly vascularized sub-dermal tissues. Examination of these transplants *in vivo* under the dissecting microscope and study on slides both concur to show that the transplanted adrenals remained viable, normally pigmented, and fully vascularized for as long as 125 days, the longest period examined.

The colour photograph of Plate 1 illustrates the survival and the state of vascularization of two consecutive heteroplastic transplantations of adrenal glands from *R. pipiens* cut in both instances into four pieces to enlarge the surface area. (See explanation of plates for further details.)

The amount of tissue diminishes with time and gradual histolysis was unmistakable; but fragments fixed for histological study around 30 days after the last transplantation appeared normal and healthy: the frog erythrocytes were easily identified within the blood-vessels connecting the host's tissues with the heterotopic adrenal bodies. The cortical elements appeared as oblong groups of cells and anastamosing cords, separated by blood sinuses, and this aspect complies in all respects with the histological description of anuran cortices given by Jones (1957). Therefore, from both gross and histological evidence it is inferred that the adrenals *ex situ* were alive and received an adequate blood supply during the period considered. These indices of healthy survival of adrenocortical elements suggest, but do not prove, the functional activity of these tissues. Few of the many physiological effects of the hyperadrenal state indicated in the literature (Selye & Stone, 1950) have been detected in this study.

No visible atrophy of the adrenals was observed in normal frogs with the exception of pigmentation losses in cases with large dosages of adrenal transplants. Cameron (1953) described peripheral stasis of small blood-vessels and the presence of mucous exudates as symptomatic of a hyperadrenal state, but neither of these conditions was seen in these experiments. Our observations would suggest that the adrenal transplants were not acting cumulatively.

2. *Morphological and histological effects of forelimb amputations in control frogs* (24 postmetamorphic *R. clamitans*; mean body length from nose tip to cloaca, 4·7 cm., range, 4·2 to 5·7 cm.; 16 cases studied histologically)

(*a*) *Simple amputations* were performed in ten cases to study the histological aspects of non-regeneration in adult frogs, since no such investigations are known to us from the literature. Singer (1954) confines himself to descriptions of the morphological features of early cicatrization in amputated forelimbs, and we concur fully with his descriptions.

Limbs were amputated either through the forearm or the upper arm and fixed at intervals between 15 to 50 days after amputation. No differential rate of cicatrization was noted between the two amputational levels and this is fully confirmed by slides. Two sections (one for each level) are presented: Plate 2, fig. A represents the usual features of non-regeneration in a longitudinal section of a limb amputated through the radius-ulna: the wound surface is completely healed with epidermis and dermis containing newly formed skin glands covering the amputation surface; directly beneath lies the conspicuous and uninterrupted basement membrane. Another feature that defines wound-healing in a non-regenerating limb is the presence of fibrous connective tissue and of muscle elements, oriented perpendicularly to the ends of the cut bone-shafts. These elements have infiltrated distally to the vacuolated regions of the perichondrium and the hemopoietic area of the bone-marrow; they constitute a fibroid wall and in spite of the presence of numerous 'blastematous' cells further progress of regeneration is always prevented in this and in similar cases. As time progresses the stratified layers covering the skeletal structures will become, in cases of more advanced amputation age, much thicker and serve as an impenetrable callus over the cut bones. Another invariable concomitant of inhibited regeneration is the transformation of the periosteal tissue into massive cartilaginous formations.

Plate 2, fig. B is a section from the upper arm of a much younger frog. In contrast to the preceding case, this amputation surface completely lacks identifiable blastematous cells. Moreover, while showing the features of wound-healing enumerated above, this limb has a much thicker basement membrane and a more clearly defined cartilaginous cap immediately distal to the cut humerus at this later stage of cicatrization. It should be noted that this limb, whose humerus is made up exclusively of cartilage, and is therefore less structurally differentiated, demonstrates no greater capacity for dedifferentiation and regeneration than did the former limb.

Among the ten controls examined we have found no regeneration on gross inspection, and on slides we found no free, unencapsulated blastematous masses. Singer (1954) records regenerative response in 15 per cent. of the controls in 'recently metamorphosed frogs' while our own experiments, performed on much older frogs, corroborate the findings of Thornton & Shields (1945).

(b) *Amputations combined with interference with cicatrization processes* were performed in another series of fourteen *R. clamitans*, ten limbs of which were fixed 52 days after amputation and studied histologically.

Since early cicatrization is generally considered coincidental with non-regeneration it was thought advisable to test this proposition by means of surgical removal of the skin which, in frogs, precociously invades the amputation surfaces.

Plate 2, fig. C illustrates a case in which skin circumcision was performed twice, once on the seventh and once on the fourteenth day following amputation. The figure shows that normal wound-healing processes are operating, although they have been retarded. While the callus is well established over the skeletal shafts, and a connective tissue pad has formed distally, the basement membrane and skin glands have not yet been reconstituted over the whole amputation surface. The staining properties of some of the perichondrial elements visible in this section suggests some dedifferentiation and blastema-like cell formations may be discerned anteriorly to the cut bone-shafts. This attempt at regeneration, however, has been halted by the fibrous layer capping them. The generality of this aspect of early regeneration 'interfered with' has been verified in the remaining nine cases of this series. The skin-removal treatment employed twice on each limb was therefore only able to retard wound-healing, not to induce a recovery of lost 'embryonic' properties of cells requisite for regeneration.

3. *Effects of adrenal transplants upon regeneration in hypophysectomized frogs* (25 postmetamorphic frogs—16 *R. clamitans* and 9 *R. pipiens*; mean body length, 4·4 cm., range, 4·1 to 4·9 cm.; 16 limbs investigated histologically)

The success in reawakening lost powers of regeneration in hypophysectomized newts by replacement therapy with cortical hormones (Schotté & Bierman, 1956) and with xenoplastic adrenal transplants (Schotté & Lindberg, 1954) suggested the use of a similar procedure in frogs. Concomitant with hypophysectomy one forelimb was amputated in all cases. The time and number of adrenal transplantations are summarized in Table 1 along with the data concerning the effects the transplants exerted on regeneration.

TABLE 1

Results of heteroplastic transplantations of adrenals after forelimb amputation in hypophysectomized frogs

(Groups marked with * were all *R. pipiens*, the other hosts were *R. clamitans*. All adrenals *R. pipiens*; one adrenal transplanted represents the adrenal tissue from one kidney)

No. of frogs	Level of amp.	Number and order of transplantations of adrenals in respect to amp. age						Macroscopic and histological results				
		Day of administration					Total no. adrenals	Beginning of reg. observed No. days after amp.	Gross regeneration		Fixation after amp. (days)	Verified histo-logically
		I	II	III	IV	V			Pres.	Abs.		
9*	Distal humerus (8) Radius-ulna (1)	7	14	21	..	..	3	40–50	4	5	40–85	6
7	Distal humerus	3	3	3	10	10	5	40–50	3	4	75	3
9	Distal humerus	7	14	21	28	..	4	40–50	4	5	50	8

Macroscopic observation showed that during the 7 days preceding transplantation very rapid skin invasion occurred. Experiments to test the comparative rates of wound-healing in normal and hypophysectomized frogs confirmed that wound-healing was much more rapid in the latter: untreated hypophysectomized frogs at 15 days have a reformed dermis, epidermal skin glands, and a state of muscle-repair not yet present in limbs of normal frogs of the same amputational age.

In spite of rapid healing our data show that after adrenal transplantations, regeneration of some sort has occurred in eleven cases among the twenty-five examined. Excellent regeneration may be observed in the case represented on Plate 3, fig. D. After amputation through the radius-ulna, coincident with hypophysectomy, this frog received a total of three adrenal transplants, administered separately on the seventh, fourteenth, and twenty-first day following amputation. This case afforded the most extensive regeneration of the hypophysectomized hosts, and yet the amount of differentiation and of proliferation is less than in the majority of the normal hosts to be reported below. The section reproduced on the photomicrograph illustrates an abortive attempt at regenerative activity, since the most distal region of the limb remains blastematous and is not engulfed in connective tissue at this stage.

It should be made clear that the kind of regeneration observed in this series was different from that observed in normal hosts to be described below, for only the case just mentioned underwent proliferation, while the remaining limbs did not progress beyond the point of 'accumulation blastema' characteristic of the 'critical' phase of regeneration described by Forsyth (1946) for anuran premetamorphic tadpoles at non-regenerating levels. This is true even for the group of nine cases that received as much as five consecutive adrenal transplants without any significant increase in the frequency of regeneration (Table 1), which indicates that in hypophysectomized frogs the required adrenal threshold to restore regenerative capacity is much higher than in normal frogs. This result is in keeping with the general observation that in the absence of the pituitary cicatrization takes place much more rapidly, and it also suggests that in the absence of adrenocorticotropic stimulation the transplanted adrenals were at a low functional level.

4. *Effect of adrenal transplants upon regeneration in normal frogs* (16 postmetamorphic *R. clamitans*; mean body length, 4·9 cm., body range, 4·7 to 5·3 cm.; 15 limbs studied histologically)

Because previous investigations had shown extreme delays in blastema formation, the amputational surfaces in all cases but four were reopened by skin circumcision simultaneously with the introduction of the first transplant.

All the results from this series are summarized in Table 2. Among the twenty-one limbs (from sixteen individuals) studied only two presented no regeneration. When one considers that among the twenty-four individuals of the previously reported control group only two limbs exhibited any regeneration at all the importance of the adrenal factor introduced in these experiments appears impressive.

Our data show that the first signs of regeneration, in the form of macroscopically detectable blastematous mounds, appeared 30 days after the last transplant, at an amputation age of about 45 days (at 20° C.) in all but one exceptional case (see explanation in Table 2). The second observation which results from our morphological and histological findings is that after an apparently normal start regeneration almost invariably becomes abnormal. A third general conclusion supporting the second centres around the disparity in stages of development at comparable amputation ages between anuran and urodele regeneration. This

TABLE 2

Results of heteroplastic transplantations of adrenals after forelimb amputation in normal R. clamitans

No. of frogs	Level of amp.	Day on which forelimb amp. performed		Day on which adrenal transpl. performed after amp. of right limb			Total no. adrenals	Macroscopic and histological results						
				Day of administration				Gross regeneration					Fixation after amp. days	Verified histologically
		Left	Right	I	II	III		Amp. age in days	Left		Right			
									Pres.	Abs.	Pres.	Abs.		
10	Radius-ulna	45 days later	1st day	7	14	..	2	45–50 (right) 85 (left)	1	0	10	0	65–160	9 (1)
4	Radius-ulna	1st day	33 days later	40	47	..	2	50 (left) 75 (right)	3	1*	4	0	75–160	5
2	Radius-ulna	..	1st day	7	14	21	3	45–50 (right)	..	..	1	1†	130	1 (1)

Special remarks: ( ) Individuals still alive and regenerating after 8 months (July 1957).
* The right limb of this case did regenerate.
† This right limb presents only instance of non-regeneration with amputation performed 7 days before transplantation.

is well illustrated by the figures of Plate 4, in which five morphologically successive normal stages and one abnormal regenerate are assembled. A regenerate 50 days old (Plate 4, fig. G) corresponds to a newt regenerate of about 20 days of amputation age; the regenerate 60 days old (Plate 4, fig. H) corresponds to a newt regenerate of about 25 days; the regenerate 72 days old (Plate 4, fig. I) is comparable to the 'palette' stage of a newt about 35 days old. Finally, Plate 4, fig. J, of an amputation age of 110 days is comparable to the regenerate of a newt at 40 days. This particular regenerate, FJW_{26}, was fixed, and a longitudinal section is represented on Plate 3, fig. E. Our records show that at the time of fixation this regenerate had not ceased to grow in length, an observation supported by the histological aspect of this section: while at its distal portion separate digital phalanges are discernible, the other bones of the forearm are still unindividualized.

The fifth normally regenerating case is represented by Plate 4, fig. K, and it shows clearly visible digital differentiations 130 days after amputation. A similar state of regeneration in an adult newt would have been achieved as early as 50 days after amputation. On slides this case showed clearly separated carpal, metacarpal, and phalangeal cartilaginous islets. However, a certain degree of syndactylism is evidenced by carpal fusion.

Frog regenerates of this type have the essential attributes of morphogenesis, growth, and differentiation that make these formations appear like true urodele regenerates. That this is not always the case has been observed in several instances where, instead of a regular terminal manus formation, a mushroom-like growth results. Such conditions may be discerned at the earliest beginnings of regenerative proliferation, and they are illustrated *in statu nascendi*, as it were, in Plate 3, fig. F. The lack of symmetry in cellular configuration is dramatized by the disorderly alignment of multiple cartilaginous and connective tissue whorls. It becomes understandable how after a long period (up to 8 months after amputation) continuous chaotic growth leads to a regenerate of abnormal appearance, such as has occurred in the left limb of case FJW$_{19}$ (Plate 5, fig. M).

Not always, however, does abnormal regeneration of this type degenerate into a formation devoid of 'order'. In other cases, such as represented by Plate 4, fig. L, axial growth is at first normal, but the distal portions of the regenerate, no doubt due to early morphogenetic disorientation, are abnormal. On slides this case exhibits a normal stem formation, but where the manus should differentiate there is within a large cartilaginous terminal mass only a vague suggestion of subdivision, even as late as 7 months after amputation.

Such cases, where gross observation and histological verification show early differentiation of multiple cartilaginous cores, always degenerate into abnormal appendages, and they have been observed several times in this investigation. Similar results have also been obtained from another research performed at this laboratory following administrations of ACTH and of cortisone to amputated adult frogs (Schotté, unpublished).

That true regeneration, still not perfect but of a type not yet reported, may result from adrenal transplants is exemplified by the corresponding right limb of the same frog, FJW$_{19}$ (Plate 5, figs. M and N), the photograph being taken after 8 months of regeneration. This regenerate has always kept the appearance of a normal blastema, and at the time of writing there is a somewhat syndactylous manus with, however, one individualized digit and another finger (as indicated on the X-ray picture) with separated metacarpals and at least two phalanges fused with the rest of the manus.

To recapitulate the data from this series of experiments, it may be stated that the addition of the equivalent of the adrenal complement from one adult frog (administered within one to two weeks after amputation) has brought about regeneration in all but one post-metamorphic frog. The most surprising new finding was that in three cases the delayed transplantation of adrenals brought about regeneration in limbs which had visibly healed over their amputational surfaces.

DISCUSSION

This research has been undertaken to investigate whether or not the lost regenerative capacity in adult frogs could be reinstated by introducing additional adrenal glands. The results provide an affirmative answer to this question only so far as the positive action of the adrenals is concerned, for an investigation of the separate effects of other tissues, glands, or substances is not yet concluded. We are therefore unable at this time to attribute the restoration of regenerative capacity in frogs to the direct and exclusive action of the transplanted frog adrenals. In this respect it is imperative to test the possible role of kidney tissue, as it was impossible to entirely separate kidney elements from the adrenal islets by ordinary surgical procedures. Moreover, preliminary experiments performed with newts have revealed a complex situation in respect to the possible stressor effect of actively secreting and therefore poisonous and irritating transplanted kidney tissues. Since stress in Selye's sense invariably involves the pituitary-adrenal axis, it would not be surprising to discover that transplanted kidneys also are capable of creating a hyperadrenal state conducive to regeneration (Schotté & Lindberg, 1954).

Before concluding that the adrenal transplants are the sole agents in bringing about regeneration in adult frogs, the importance of skin removal and of amputational stress must first be considered because of the possible stressor role of surgical trauma in newts (Pellman & Schotté, 1955; Lindberg & Schotté, 1955; Schotté & Bonneville, 1955). The amputation alone of another limb must be excluded as a stressor agent capable of promoting regeneration, since it was followed in all cases simply by the normal wound-healing mechanism. Secondly, concerning the effects of skin circumcision, only one of the fourteen forelimbs upon which repeated skin removal around the amputational surface was performed exhibited any regeneration, and such a regenerate did not continue growth beyond the accumulation phase. Thus, this treatment did not restore any regenerative activity comparable to that observed in the transplantation experiments.

These brief considerations will suffice to re-emphasize the belief that the results obtained are attributable to the introduction of additional adrenals. But then to what effect of the transplanted adrenals may one attribute the induced regeneration?

Since experimentation in this laboratory has implicated a pituitary-adrenal synergism in establishing conditions favourable to regeneration in newts, one wonders whether such a mechanism is operative in adult frogs also. Unpublished investigations regarding the effects of stress upon adult frogs and adrenal transplants upon regeneration in tadpoles would intimate that this is so. However, to support this hypothesis, an important piece of information is required. Were the adrenal transplants synthesizing cortico-steroids found to be essential for the initiation of regeneration in urodeles (Schotté & Bierman, 1956)? Besides the gross and microscopical condition of the adrenal tissues described in the experimental part there is additional evidence to indicate that the adrenal transplants were functioning, but at a basal level.

Firstly, in hypophysectomized frogs the bright yellow pigmentation of the adrenals faded in a manner identical to that in normal hosts receiving high adrenal dosages by transplantation. The paleness of the hypophysectomized host's own adrenals was no doubt due to the fact that involution of the adrenal cortex, in the absence of ACTH, produces a lesser concentration of the yellow-coloured lipids in the cortical cells. The atrophy in the normal hosts receiving additional adrenals implies that the transplanted adrenals were functional, since in the endocrine system compensation for more than normal concentrations of a hormone is by atrophy of the organ synthesizing it. Secondly, bioassays of cortical secretions from isolated mammalian adrenal glands through which blood was perfused showed that deprivation of ACTH did not interrupt, but only lowered their secretory activity (Vogt, 1951). This result suggests that the frog cortices would not completely terminate their activity in the absence of the trophic hormone.

The next important problem concerns the duration of activity of the transplants. Both these results and those from a previous study in newts (Schotté & Lindberg, 1954) imply that there is a transitory period of corticoid activity, for in both experiments regeneration was only associated with transplantations made around two weeks after amputation, not when adrenals were introduced earlier. Apparently the period of cortical activity must be integrated with the time, late in wound repair, that is most susceptible to corticoid action. The failure to induce regeneration with smaller dosages in other experiments was probably not the result of inadequate quantitative dosage, but the result of premature administration, before the limb tissues were sensitive to proper inter-action with cortico-steroids like cortisone. For this hormone is known to effect many changes in wound repair, such as diminishing cellular migration and infil-tration, inhibiting fibroblast formation, and fibrin deposition (Cameron, 1953); and all these responses repress normal wound-healing processes which in turn interfere with the mechanisms of regeneration.

If these propositions regarding the delicate hormonal changes at the propi-tious moments when the frogs' limbs are in a state of repair most receptive to their action are valid, why is it that the adrenal transplants are unable to facili-tate regeneration in hypophysectomized frogs to the same degree? It seems that there may be two main reasons. Firstly, the total absence of ACTH in animals deprived of their pituitaries is no doubt responsible for a much lower level of activity of the transplanted adrenal glands in these frogs; in normal hosts, how-ever, after vascular communication has been established, ACTH is present for cortical activation. Secondly, the much higher rate of cicatrization observed in hypophysectomized frogs possibly modifies the time at which the corticoid activity is most instrumental in preventing wound-healing.

The general results of this study, then, together with the still unpublished aforementioned experimental findings, support the proposition that a pituitary-adrenal synergism is operating in adult frogs as in newts; they also suggest that an artificially-induced hyperadrenal state is sufficient to determine recuperation of regenerative capacities in adult frogs. This, of course, does not preclude other factors (nerves, for example) from being just as effective in restoring regenerative potencies.

In conclusion it may be stated with a certain degree of confidence that the normal loss of regenerative potencies in Anura, more or less coincidental with metamorphosis, is attributable to endocrine changes in these organisms, rather than to irreversible modifications in properties of the cellular constituents of their limbs.

SUMMARY

1. Experiments to test the proposition that systemic factors and not fixed properties of the cells within limbs determine loss of regeneration in frogs were performed upon the forelimbs of post-metamorphic *R. clamitans* and *R. pipiens* varying in body length from 4·4 to 5·7 cm.
2. Control experiments consisting of simple amputations through the mid-radius ulnar or distal humeral regions showed that the frogs at this stage were incapable of detectable regeneration. The repeated removal of the cicatricial skin in another control series elicited regeneration, of an abortive type, in only one case among fourteen limbs tested.
3. In twenty-five hypophysectomized *R. clamitans* adrenals from *R. pipiens* were transplanted heterotopically beneath the jaw concomitantly with amputation. Weak regenerative response was observed in eleven cases. It is concluded that the absence of the pituitary in the hosts lowers the activity of the transplanted adrenals.
4. The transplantation of *R. pipiens* adrenals into normal *R. clamitans* hosts 7 and 14 days after amputation proved most effective, since among the twenty-one limbs examined only two did not show any regeneration.
5. It is suggested that: (1) the transplanted adrenals were functional, although at a low level of activity; (2) the time of transplantation of the adrenals is critical, since in order to elicit regeneration it must coincide with a period in amputational wound repair that is most susceptible to corticoid activity; (3) the loss of regenerative potencies after metamorphosis in Anura is attributable to changes in the endocrine system of these organisms, rather than to changes in the properties of the cells of their limbs.

ACKNOWLEDGEMENTS

Thanks are due to the Division of Grants of the U.S. National Institutes of Health (Grant 2236) which has made this research possible. We also wish to thank Mrs. Jean Francis for excellent histological work and Mr. Carl Howard for consummate skill in photography.

REFERENCES

CAMERON, G. R. (1953). A survey of tissue responses to ACTH and cortisone. *The Suprarenal Cortex*. New York: Academic Press.

FORSYTH, J. W. (1946). The histology of anuran limb regeneration. *J. Morph.* **79**, 287–322.

GUYÉNOT, E. (1927). La perte du pouvoir régénérateur des Anoures, étudiée par les hétérogreffes, et la notion de territoires. *Rev. suisse Zool.* **34**, 1–51.

HALL, A. B., & SCHOTTÉ, O. E. (1951). Effects of hypophysectomies upon the initiation of regenerative processes in the limb of *Triturus viridescens*. *J. exp. Zool.* **118**, 363–88.

JONES, I. C. (1957). *The Adrenal Cortex*. London: Cambridge University Press.

LEVINSKY, G., & SAWYER, W. (1952). Influence of the adenohypophysis on the frog water-balance response. *Endocrinology*, **51**, 110–16.

LINDBERG, D. A., & SCHOTTÉ, O. E. (1955). Amputational trauma and regeneration of limbs in normal and in hypophysectomized newts. *Anat. Rec.* **121**, 331.

MARCUCCI, E. (1916). Capacità rigenerativa degli arti nelle larve di Anuri e condizioni che ne determinano la perdita. *Arch. zool. (ital.), Napoli*, **8**, 89–117.

NAVILLE, A. (1927). La perte du pouvoir régénérateur des Anoures étudiée par les homogreffes. *Rev. suisse Zool.* **34**, 269–84.

PELLMAN, C., & SCHOTTÉ, O. E. (1955). Influence of repeated anesthesia upon regeneration of limbs in normal and in hypophysectomized newts (*Triturus viridescens*). *Anat. Rec.* **121**, 351–2.

POLEJAIEV, L. W. (1936). Sur la restauration de la capacité régénérative chez les Anoures. *Arch. Anat. micr.* **32**, 437–63.

POLEZHAJEV, L. V. (1945). Limb regeneration in adult frog. *C.R. Acad. Sci. U.R.S.S.* **49**, 609–12.

RICHARDSON, D. (1940). Thyroid and pituitary hormones in relation to regeneration. 1. The effects of anterior pituitary hormone on regeneration of the hind limb in normal and thyroidectomized newts. *J. exp. Zool.* **83**, 407–25.

—— (1945). Thyroid and pituitary hormones in relation to regeneration of the hind limb of the newt, *Triturus viridescens*, with different combinations of thyroid and pituitary hormones. *J. exp. Zool.* **100**, 417–27.

ROSE, S. M. (1944). Methods of initiating limb regeneration in adult Anura. *J. exp. Zool.* **95**, 149–70.

—— (1945). The effect of NaCl in stimulating regeneration in limbs of frogs. *J. Morph.* **77**, 119–39.

SCHOTTÉ, O. E. (1926). Hypophysectomie et régénération chez les Batraciens urodèles. *C.R. Soc. Phys. Hist. nat. Genève*, **43**, 67–72.

—— & BIERMAN, R. (1956). Effects of cortisone and allied adrenal steroids upon regeneration in hypophysectomized *Triturus viridescens*. *Rev. suisse Zool.* **63**, 353–75.

—— & BONNEVILLE, M. (1955). The systemic effect of injury and of repair within dehumerized limbs of *Triturus viridescens*. *Anat. Rec.* **121**, 364.

—— & CHAMBERLAIN, J. (1955). Effect of ACTH upon limb regeneration in normal and in hypophysectomized *Triturus viridescens*. *Rev. suisse Zool.* **62**, 253–79.

—— & HALL, A. B. (1952). Effects of hypophysectomy upon phases of regeneration in progress. (*Triturus viridescens*.) *J. exp. Zool.* **121**, 521–60.

—— & HARLAND, M. (1943). Amputation level and regeneration in limbs of late *Rana clamitans* tadpoles. *J. Morph.* **73**, 165–80.

—— & LINDBERG, D. (1954). Effect of xenoplastic adrenal transplants upon regeneration in normal and in hypophysectomized newts. (*Triturus viridescens*.) *Proc. Soc. exp. Biol. Med. N.Y.* **87**, 26–29.

SELYE, H. (1947). *Textbook of Endocrinology*. Acta Endocrinologica. Montréal: Université de Montréal.

—— & STONE, H. (1950). *On the experimental morphology of the adrenal cortex*. Springfield, Ill.: Charles E. Thomas.

SINGER, M. (1954). Induction of regeneration of the forelimb of the postmetamorphic frog by augmentation of the nerve supply. *J. exp. Zool.* **126**, 419–72.

THORNTON, C. S., & SHIELDS, T. W. (1945). Five cases of typical regeneration in the adult frog. *Copeia*, **1**, 40–42.

VAN STONE, J. M. (1955). The relationship between innervation and regenerative capacity in hind limbs of *Rana sylvatica*. *J. Morph.* **97**, 345–92.

VOGT, M. (1951). Cortical secretion of the isolated perfused adrenal. *J. Physiol.* **113**, 129–56.

WALTER, F. K. (1911). Schilddrüse und Regeneration. *Arch. EntwMech. Org.* **31**, 91–130.

EXPLANATION OF PLATES

PLATE 1

Ventral view of lower jaw of an adult *R. clamitans* (case FJW_9) after removal of skin showing four prominent adrenal transplants (from *R. pipiens* donors) 125 or 118 days after transplantation. Several smaller islets below derive from other unsuccessful or regressing adrenal transplants. Note vascular connexions of the four main transplants.

PLATE 2

FIG. A. Photomicrograph of a longitudinal section of left forelimb of case FJW_{67} amputated through the radius-ulnar region and fixed after 23 days of amputation age. Note reformation of skin over amputational surface, complete with basal membrane. Also characteristic for arrested regeneration are the masses of procartilage formed on both sides of the periosteum at the expense of periosteal and other connective tissue cells. ($\times 25$)

FIG. B. Photomicrograph of a longitudinal section of left forelimb of case AFC_{82}, amputated through distal humerus and fixed 50 days after amputation. Note lack of blastematous cells and the cartilaginous nature of the humerus. ($\times 25$)

FIG. C. Photomicrograph of a longitudinal section of right forelimb of case FJW_{93} amputated through the radius-ulnar level and fixed 50 days later. Note the incompletely reformed basement membrane, the crescent-like pad of fibrous tissue distal to the cut shafts of the bone collar, and the dense periosteal cartilage. ($\times 25$)

PLATE 3

FIG. D. Photomicrograph of the forelimb of case FJW_{24} fixed 120 days after amputation. The amputation level is situated at the bottom of the figure where the old bony formations are still visible. Within the regenerated procartilaginous mass forming the distal portion of the radius and the general, still not separated mass of the manus, may be distinguished small hemopoietic islets. From the aspect of the distal skeletal formations it is doubtful whether digital formations would ever emerge. ($\times 25$)

FIG. E. FJW_{26}. Photomicrograph of a longitudinal section of left forelimb of case FJW_{26}, at 110 days amputation age. Radius and ulna are well regenerated and elements of the basipodium of the metacarpals and even the phalanges are well indicated on this and on other sections. ($\times 25$)

FIG. F. Photomicrograph of a longitudinal section of the right forelimb of case FJW_9 fixed 60 days after amputation and 46 days after the second adrenal transplantation. The long cartilaginous shaft is the radius and ulna, probably fused, around which periosteum begins to form. From the distal area of these procartilaginous formations several whorls of cartilage are derived which have distinct and irregularly distributed centres. Since there is no regular connective tissue cap precluding further growth, these irregular blastematous centres may diverge into tridimensional, often digit-like formations which eventually form a mushroom-shaped regenerate. ($\times 25$)

PLATE 4

All figures here represented are regenerates from hosts which received two successive *R. pipiens* adrenal transplants 7 and 14 days after amputation of the forelimb.

FIG. G. *R. clamitans* forelimb regenerate at the stage of early blastema, 50 days after amputation.

FIG. H. *R. clamitans* regenerate at the blastema stage, 60 days after amputation.

FIG. I. *R. clamitans* regenerate at the stage of flattened palette, 72 days after amputation.

FIG. J. *R. clamitans* forelimb regenerate 110 days after amputation. (A photomicrograph of a section of that limb is reproduced on Plate 3, fig. E.)

FIG. K. Forelimb regenerate at the stage of prominent digital differentiation, 130 days after amputation, showing clearly defined three digital indentations.

FIG. L. Abnormal regenerate from forelimb amputated through the proximal radius-ulna showing a boxing-glove-like curvature at its tip, 7 months after amputation.

PLATE 5

FIG. M. Ventral view of *R. clamitans*, 5·3 cm. body length, and amputated bilaterally, the left limb, however, amputated 45 days after the right (case FJW$_{19}$). Two adrenal transplantations were performed 7 and 14 days respectively after amputation of the right limb. Note mushroom-like regenerate at left and excellent regeneration with digital differentiations at right.

FIG. N. X-ray photograph of the right limb of the above case taken 8 months after amputation and adrenal transplantations. Note the fused osseus rod of radius-ulna, some carpal formations and particularly phalangeal ossifications within the free digit.

(Manuscript received 19:viii:57)

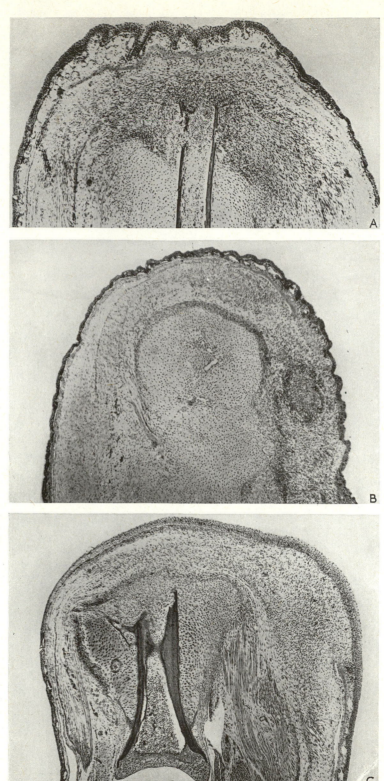

O. E. SCHOTTÉ *and* J. F. WILBER

Plate 2

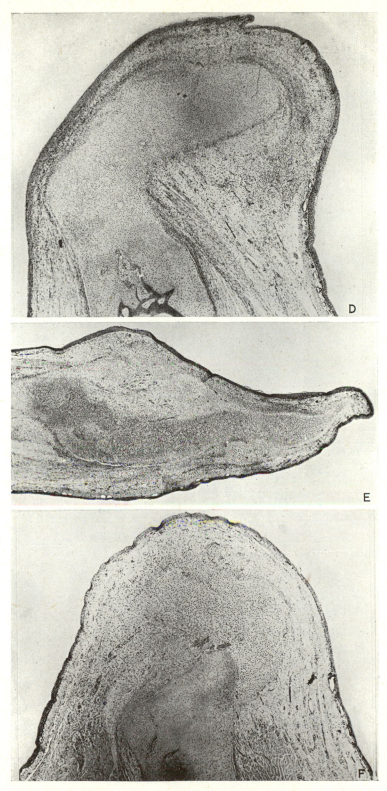

O. E. SCHOTTÉ *and* J. F. WILBER

Plate 3

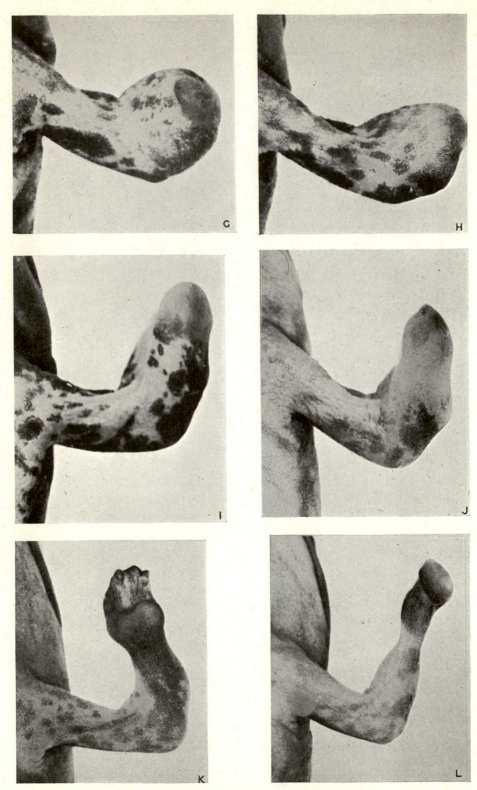

O. E. SCHOTTÉ *and* J. F. WILBER

Plate 4

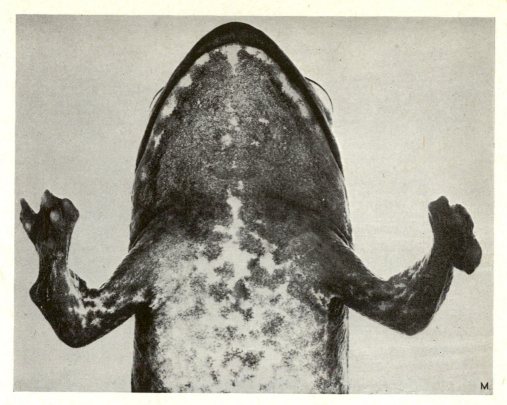

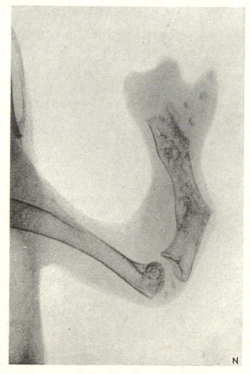

O. E. SCHOTTÉ *and* J. F. WILBER

Plate 5

Reprinted from *J. Exptl. Zool.*, **170**, 33–53 (1969)

Hormonal and Nutritional Requirements for Limb Regeneration and Survival of Adult Newts

24

ROY A. TASSAVA

Department of Zoology, Michigan State University, East Lansing, Michigan

ABSTRACT   Newts which were fed daily for two weeks prior to hypophysectomy survived significantly longer and regenerated limbs better than newts fasted for two weeks prior to hypophysectomy. Since limbs were amputated five days after hypophysectomy, regeneration was initiated in the complete absence of pituitary hormones.

Newts which were hypophysectomized 14 days after limb amputation were found to possess blastemas significantly smaller at eight and also at 16 days after the operation than the blastemas of sham-operated newts. Thus, hypophysectomy, performed after limb regeneration had progressed through the wound healing and dedifferentiation phases, resulted in a significant retardation in the growth of the blastema.

To determine which hormones are essential to normal limb regeneration and survival, newts were hypophysectomized and treated with various combinations of hormones, or grafted with pituitaries from newts or from axolotls (*Ambystoma mexicanum*). A prolactin-thyroxine combination, growth hormone, and ectopic pituitary grafts from newts or axolotls, significantly prolonged the life of hypophysectomized newts. These newts also regenerated limbs in a normal fashion. Thyroxine alone, prolactin alone, thyroxine + ACTH, ACTH, or saline were not effective in restoring the health of hypophysectomized newts and were not effective in restoring normal limb regeneration ability.

The adult newt, *Notophthalmus* (*Triturus*) *viridescens*, requires pituitary hormones for survival and normal limb regeneration (Richardson, '45; Dent, '67; Connelly et al., '68). The exact nature of the essential pituitary hormones and their mechanism of action on survival and limb regeneration are unknown.

Schotté ('61) has suggested that in typical limb regeneration the newt pituitary is stimulated to produce ACTH (adrenocorticotrophic hormone) by the stress of amputation; the ACTH then activates the adrenal gland which produces steroid hormones. The adrenal hormones are thought to be essential for proper wound healing during the first six days after limb amputation and this is prerequisite for initiating regeneration (Schotté and Hall, '52). However, evidence in support of an ACTH-adrenal mechanism in newt limb regeneration and survival is not compelling. Thus, ACTH injections into hypophysectomized adult newts brought about limb regeneration and enhanced survival (Schotté and Chamberlain, '55), but adrenal steroids did not enhance survival or limb regeneration of hypophysectomized adult newts (Schotté and

Bierman, '56). Schotté and Chamberlain ('55) found that ACTH injections into intact newts caused a temporary delay in limb regeneration, yet Bragdon and Dent ('54) observed normal limb regeneration in intact newts injected with ACTH or cortisone.

There is other evidence which makes doubtful the possibility of an ACTH involvement in limb regeneration. A pituitary-adrenal axis apparently exists in the bullfrog (*Rana catesbiana*) (Piper and DeRoos, '67), nevertheless the adrenal of both anuran (Hanke and Weber, '65) and urodele amphibians (Wurster and Miller, '60) displays considerable independence from the pituitary gland. A hypophysectomized adult newt with an ectopic pituitary gland will regenerate amputated limbs in a normal fashion (Schotté and Tallon, '60), yet in other vertebrates, experiments have shown that the ectopically transplanted pituitary (removed from its hypothalamic connections) does not respond well to stress (Van Dongen et al., '66; Mangili et al., '66; Purves and Sirett, '67). Also, it is known in mammals that the response to stress is nearly immediate, oc-

428

curring within seconds of the onset of stimulation and, furthermore, that denervation of a limb or ablation of the spinal cord prevents the ACTH response to stimuli applied to the denervated territory (Fortier, '66). When the lumbo-sacral spinal cord of adult *Ambystoma* is ablated, amputated hind limbs and tail nevertheless regenerate (Liversage, '59). Furthermore, when limb and spinal cord segments are transplanted to the dorsal fin, followed by spinal cord ablation of the host and later amputation of the limb, normal regeneration follows. In these ablation experiments, neural stimulation of endocrine activities (especially the pituitary-adrenal axis) was prevented (Liversage, '59).

Hormones other than ACTH have been investigated in newt limb regeneration. Connelly et al. ('68) found that prolactin, when combined with thyroxine, was very effective in enhancing survival and promoting limb regeneration in hypophysectomized adult newts. Prolactin alone, in larger quantities, was less effective and thyroxine alone was completely ineffective. Richardson ('45) found that Antuitrin G, a crude growth hormone preparation, supported limb regeneration in hypophysectomized newts but not as well as Antuitrin G combined with thyroxine. Wilkerson ('63) obtained excellent limb regeneration in hypophysectomized adult newts with growth hormone (NIH) even when injections were begun 14 days after amputation and hypophysectomy. However, the growth hormone (GH) used by Wilkerson ('63) contained prolactin as a contaminant in an amount equivalent to that used by Berman et al. ('64) which stimulated growth and inhibited metamorphosis in the frog tadpole. The prolactin contamination in the GH was also comparable to the amount of prolactin which will induce the red eft, the land stage of the newt, to migrate to water (Grant and Grant, '58). That prolactin may play a role in newt limb regeneration was further suggested by Niwelinski ('58) and Waterman ('65) who found an enhancement of limb regeneration by prolactin injections into intact newts, and by the report of Chadwick and Jackson ('48) that prolactin increases mitotic activity in newt epidermis.

It has been suggested that pituitary hormones may act during the growth phase of limb regeneration (Hay, '56, '66) instead of only during the wound healing phase (Schotté and Hall, '52). Hay ('56, '66) points out that the best regenerative response in hypophysectomized newts occurred when frog adrenals were implanted *15 days* after amputation (Schotté and Lindberg, '54) and when growth hormone injections were begun 14 days after amputation (Wilkerson, '63). The data of Schotté and Hall ('52) also indicate that pituitary hormones may influence the growth of the blastema. Thus, when hypophysectomy of adult newts was delayed until 14 days after limb amputation, all of the limbs regenerated but in 79% of the cases regeneration was abortive or delayed (Schotté and Hall, '52). At 14 days after amputation, wound healing is essentially complete, as is dedifferentiation (Hall and Schotté, '51), yet hypophysectomy apparently has an adverse effect on the 14 day regenerate (Schotté and Hall, '52).

It is not clear, therefore, which pituitary hormones are essential to newt survival and limb regeneration; whether these pituitary hormones act through the adrenal gland; which stage(s) of limb regeneration is influenced by pituitary hormones; and finally, whether pituitary hormones are absolutely essential to the initiation of limb regeneration. Therefore, experiments described in this paper were designed to determine: (1) whether limb regeneration absolutely requires pituitary hormones, particularly when newts are in good nutritional condition at the time of hypophysectomy; (2) whether the 14 day early bud regenerate of an adult newt will continue normal growth and differentiation in the absence of pituitary hormone; and finally (3) which hormones or combination of hormones are most effective in prolonging survival and promoting limb regeneration of hypophysectomized adult newts.

GENERAL METHODS

Adult newts, *Notophthalmus* (*Triturus*) *viridescens*, were obtained from Lewis Babbitt, Petersham, Massachusetts. Except for some newts in Series I (Nutritional Effects), all newts were fasted for two weeks prior to hypophysectomy and were not fed during the experimental period. Using fine pointed watchmaker's forceps, a small flap of the sphenoid bone was loosened directly ventral to the pituitary. The anterior end of the bone flap was then lifted away while the entire pituitary gland was removed by gentle suction with a fine glass pipette (Connelly et al., '68). The flap of bone was then replaced. If any hypophysectomy was doubtful due to injury of the gland or to excess bleeding, that particular newt was immediately discarded.

Newts which are completely hypophysectomized do not shed the outer layer of their epidermis because of a thyroxine deficiency resulting from the absence of thyroid stimulating hormone (TSH) (Dent, '66). These newts develop a rough, black skin, quite different from the dark, smooth and slippery skin of newts subjected to an excess melanophore stimulating hormone (MSH) secretion. Any hypophysectomized newts which did not exhibit complete lack of molting were discarded.

Heads and limbs were fixed in Bouin's fluid, decalcified, dehydrated, cleared in methyl salicylate, embedded in paraffin, and serially sectioned at 10 μ. Sections of heads were stained by the PAS technique or Herlant's tetrachrome (Pearse, '60). Sections of limbs were stained with hematoxylin-eosin, iron-hematoxylin, or Masson's trichrome (Merchant et al., '63).

EXPERIMENTAL

Series I. Nutritional effects

This series was designed to compare the ability to regenerate limbs and the length of survival of: (1) newts which were in good nutritional condition at the time of complete hypophysectomy; (2) newts which were fasted for two weeks prior to *complete* hypophysectomy; and (3) newts which were fasted for two weeks prior to *incomplete* hypophysectomy.

A. *Methods.* Forty-four adult newts, weighing from 1–2 gm, were randomly selected from planted aquaria and maintained in 2-quart finger bowls at a density of four newts per bowl. All newts in this and the following series were maintained at 22°C ± 2°C. Of the 44 newts, 16 were randomly selected and fed beef liver maxi-

mally each day for two weeks. The remaining 28 newts were fasted for the same two week interval. After the two week interval of feeding or fasting, the individual body weights of the newts fed for two weeks were significantly greater than the individual body weights of the newts which were fasted for the same two week period (Mann-Whitney U Test, 0.01 level of significance). On the average, each fed newt weighed 0.32 gm more than each fasted newt. All 16 of the fed newts and 16 of the fasted newts were then completely hypophysectomized. The remaining 12 fasted newts were partially hypophysectomized. Using the suction technique of hypophysectomy, it is much easier to remove the entire pituitary than to leave a fragment. Therefore, for these partial hypophysectomies, only forceps were used. Care was taken to cut the pituitary gland *in situ* and to remove all but one or two fragments of the anterior lobe.

Both forelimbs of the newts in all three groups were amputated just distal to the elbow at five days post-operation (complete or partial hypophysectomy). None of these newts were fed during the course of the experiment. The day of death of each newt was recorded. Limbs and heads were fixed either immediately or within a few hours after death or were randomly sampled from the surviving newts.

Of the 16 newts which were fed for two weeks prior to hypophysectomy, eight were still surviving 26 days post-hypophysectomy (21 days post-amputation). Both forelimbs and the head of each of these eight newts were fixed at that time (26 days post-hypophysectomy) and examined histologically.

B. *Results.* Of the newts in Series I which were fed maximally for two weeks prior to hypophysectomy, 12 of 16 were alive 20 days post-hypophysectomy. One of the 16 newts which were fasted for two weeks prior to hypophysectomy was alive 20 days post-hypophysectomy. Two of the 12 fasted newts which were incompletely hypophysectomized were alive 20 days post-hypophysectomy. Figure 1 compares the per cent survival of these three groups. The 12 surviving, maximally-fed newts represent a significantly greater number than the one surviving, fasted, hypophy-

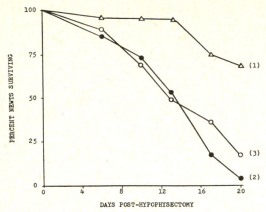

Fig. 1 A comparison of the per cent survival of (1) hypophysectomized newts fed for two weeks prior to hypophysectomy, (2) hypophysectomized newts fasted for two weeks prior to hypophysectomy, and (3) partially hypophysectomized newts fasted two weeks prior to patrial hypophysectomy.

sectomized newt or the two surviving, fasted, partially hypophysectomized newts (X^2 Test, 0.01 level of significance). Of the 16 hypophysectomized newts which were fed for two weeks prior to hypophysectomy, eight survived to day 26 post-hypophysectomy and were sacrificed. The 16 amputated limbs of these eight newts (day 21 post-amputation) revealed regeneration blastemas in every case (fig. 8; see table 1). Examination of serial sections of the heads of these eight surviving newts revealed no pituitary fragments (compare figs. 10, 11). In contrast, 10 of 12 of the heads of the partially hypophysectomized newts contained small pituitary fragments (fig. 12), yet, a significantly smaller number of these newts survived than did newts which were completely hypophysectomized and had been previously well fed. These ten newts with small pituitary fragments nevertheless failed to molt. Thus, the small pituitary fragment may not have contained the essential cell type(s) for newt survival. Histological examination of four limbs of the fasted completely hypophysectomized newts and six limbs of the fasted incompletely hypophysectomized newts which were fixed 20–26 days post-hypophysectomy (15–21 days regeneration) revealed regeneration blastemas although retarded as compared to normal regenerates (table 1; fig. 9).

In summary, 16 of 16 limbs of maximally fed newts which were amputated five days post-hypophysectomy exhibited clear indications of early stages of regeneration when examined histologically 26 days post-hypophysectomy. Thus, limbs of newts which are amputated after hypophysectomy do initiate regeneration in the absence of the pituitary gland. Furthermore, survival depends upon the nutritional state of the newt at the time of hypophysectomy.

Series II. Pituitary influence on growth of the 14 day regenerate

The results of Series I clearly demonstrate that limb regeneration can proceed at least to the medium bud stage in the complete absence of pituitary hormones. This series was designed to determine whether regenerates which have reached the bud stage in the presence of pituitary

TABLE 1

Delay of regeneration produced by hypophysectomy

Days after amputation [1]	Stages of regeneration attained		
	Intact newts [2]	Hypophysectomized previously fed newts	Hypophysectomized previously fasted newts [3]
10–13	Accumulation blastema		
14–15	Early bud	Accumulation blastema	
16–20	Medium bud	Early bud	Accumulation blastema
21–25	Late bud and cartilage differentiation	Medium bud	Early bud
26–29	Palette		Medium bud
30 +	Digit		

[1] Limbs were amputated five days post-hypophysectomy.
[2] Limb regeneration of intact newts injected with ACTH (UpJohn) corresponds to that of intact untreated newts. Regeneration stages after Singer ('52).
[3] Limb regeneration of hypophysectomized newts treated with saline, prolactin alone, ACTH, and/or thyroxine corresponds to that of hypophysectomized, previously fasted newts.

hormones can continue to grow and differentiate in a normal fashion after removal of the pituitary hormones.

A. *Methods*. The right or the left forelimb (chosen randomly) of each of 36 newts, previously fasted for two weeks, was amputated through the distal portion of the humerus. At 16 days post-amputation these newts were randomly divided into two groups. The limbs of two newts were sampled at this time for histology to ascertain the status of regeneration. Both limbs exhibited early bud regenerates. The newts in one group (n = 20) were hypophysectomized while the remaining 16 newts were sham-operated. The sham-operation consisted of removing a portion of the bone over the sella turcica without removing the pituitary. The bone was then replaced. At eight days post-hypophysectomy or post-sham (22 days post-amputation), six of the newts from each group were randomly selected, sacrificed and the amputated limb and the head of each newt were fixed and prepared for histological examination. Limbs were oriented randomly in paraffin, serially sectioned longitudinally at 10 μ, and stained with hematoxylin and eosin. The area and the length of the regeneration blastema were determined for each of five sections of each limb. The image of the section to be measured was projected on to graph paper from a constant height. The image of the blastema of each section sampled was then traced on to the graph paper with a fine pointed pencil. The number of squares representing the blastema was counted for each section sampled (5 sections for each limb). The five sections to be sampled were determined by selecting the approximately largest section (usually in the center of the limb), and measuring the area of the blastema of that section. Two sections, 100 μ apart, on both sides of this largest section were also sampled. Thus the five sections sampled, each 10 μ thick and spaced at 100 μ intervals, covered a total distance through the blastema of 450 μ. The mean blastema area and length for each limb were determined (mean of 5 sections) and the means were compared statistically for the two groups using the non-parametric Mann-Whitney U Test.

In addition to the six newts fixed eight days after hypophysectomy, seven of the 20 hypophysectomized newts died during the course of the experiment. The seven surviving hypophysectomized newts were sacrificed 16 days after hypophysectomy (30 days post-amputation) and their limbs and heads prepared for histology. Area measurements were made of these blastemas as described above. The limbs of the eight surviving sham-operated newts were also fixed at this time and prepared for histology. Area measurements were also made of these blastemas. These mean blastema areas, 16 days post-hypophysectomy or post-sham operation, were compared statistically.

B. *Results*. The results of Series II are summarized in figure 2, figure 3, and table 2. The mean blastema areas of the sham-operated newts were significantly greater (0.01 level) than the mean blastema areas of the hypophysectomized newts at eight days post-hypophysectomy (15.90 squares vs. 9.25 squares) and also at 16 days post-hypophysectomy (22.47 squares vs. 15.22 squares). At eight days post-operation, the blastemas of the sham-operated newts were also significantly longer (0.01 level) than the blastemas of the hypophysectomized newts (17.5 units vs. 11.5 units). Histological and gross observations also revealed that the regenerates of the sham-operated newts were in more advanced stages of differentiation. Digit differentiation was apparent in five of the six regenerates of the sham-operated newts at 16 days post-sham-operation whereas only one of the six regenerates of the hypophysectomized newts exhibited digit differentiation 16 days post-hypophysectomy. Cells in mitosis could be observed in the regenerates of the hypophysectomized newts at eight days post-hypophysectomy; therefore growth was not completely stopped. Furthermore, the mean blastema areas were greater at 16 days than at eight days post-hypophysectomy suggesting that some growth did occur (table 2; fig. 3) in the complete absence of the pituitary.

In summary, the results of Series II demonstrate that the normal growth of the blastema requires pituitary hormones. Hypophysectomy has an adverse effect on limb regeneration even after the wound

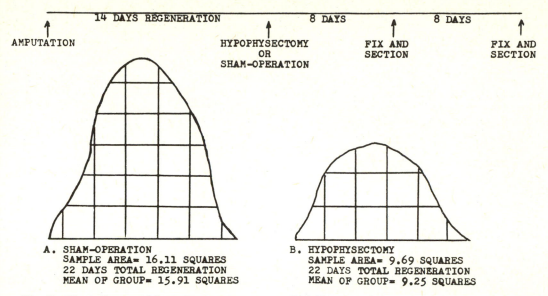

Fig. 2   The influence of hypophysectomy on newt limb blastema area. Fasted newts were hypophysectomized or sham-operated 14 days after limb amputation. At eight days and also at 16 days after hypophysectomy or sham-operation, the limbs were sampled for sectioning and staining. In A and B the section areas were obtained by projecting the blastema image onto graph paper from a constant height and tracing the blastema outline. The mean blastema areas of the sham-operated newts were significantly larger (0.01 level of significance; Mann-Whitney U Test).

healing and dedifferentiation phases are complete.

Series III.   Hormone replacement therapy

The results of Series I and Series II demonstrate that pituitary hormones are more important to the later growth phase of limb regeneration than to the early, wound healing phase. In addition factors which promote survival of hypophysectomized newts also promote limb regeneration (i.e. good nutritional condition). These facts do not lend support to an ACTH-adrenal mechanism but suggest instead that pituitary hormones which promote survival also make it possible for normal limb regeneration to proceed. Therefore, Series III was designed to determine which hormones or combination of hormones effectively enhanced survival and/or limb regeneration.

A. *Methods.*   Fasted newts were hypophysectomized and randomly divided into groups. Different groups (tables 3, 4) of hypophysectomized newts were either not treated or received ectopic pituitary glands from donor newts or donor axo-

lotls, *Ambystoma mexicanum*, or intraperitoneal injections of either water, saline, prolactin (NIH), growth hormone (NIH-GH) or adreno-corticotropic hormone (UpJohn-ACTH). In addition some hypophysectomized newts were treated with thyroxine, with and without prolactin or ACTH. The thyroxine was dissolved in the water in which the newts were maintained at a concentration by weight of one part thyroxine/10 million parts of water (1×10^{-7} conc. in water). The newts were changed to fresh solution every other day. The prolactin and GH were dissolved in 0.9% saline and the ACTH in sterile water at concentrations such that 0.1 cm³ of solution contained the desired amount of hormone. For the pituitary transplants, the host newt was anesthetized in MS 222 and after hypophysectomy of the host, the donor (either newt or axolotl) was decapitated and its entire pituitary (neural, intermediate and anterior lobes) was quickly removed and transplanted to the musculature of the lower jaw of the host.

Plastic disposable syringes with 27 gauge hypodermic needles were used for

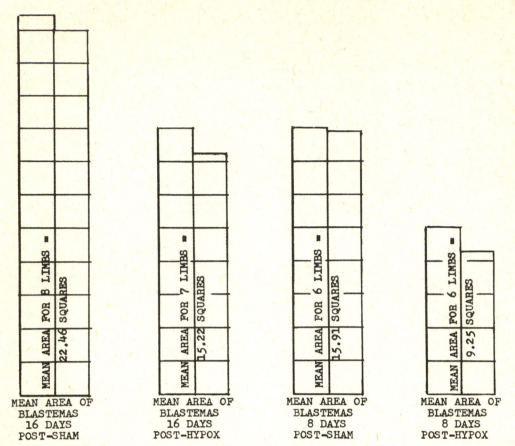

MEAN AREA OF BLASTEMAS 16 DAYS POST-SHAM   MEAN AREA OF BLASTEMAS 16 DAYS POST-HYPOX   MEAN AREA OF BLASTEMAS 8 DAYS POST-SHAM   MEAN AREA OF BLASTEMAS 8 DAYS POST-HYPOX

Fig. 3   A comparison of the mean blastema areas of newts hypophysectomized or sham-operated at 14 days after limb amputation and fixed eight days post-operation or at 16 days post-operation.

TABLE 2

The effect of hypophysectomy on blastema area

Mean area of blastema sections 8 days post-operation (22 days regeneration)				Mean area of blastema sections 16 days post-operation (30 days regeneration)			
Hypophysectomy		Sham [1]		Hypophysectomy		Sham	
Limb	Mean area of 5 sections	Limb	Mean area of 5 sections	Limb	Mean area of 5 sections	Limb	Mean area of 5 sections
1	1.86	1	11.72	1	11.26	1	16.47
2	5.07	2	13.18	2	11.87	2	17.20
3	9.55	3	16.13	3	14.19	3	17.78
4	10.37	4	16.30	4	14.49	4	19.95
5	11.71	5	17.11	5	16.21	5	25.46
6	16.93	6	21.03	6	18.21	6	25.51
				7	20.26	7	27.63
						8	29.68
Mean = 9.25		Mean = 15.91		Mean = 15.22		Mean = 22.46	

[1] The mean blastema areas (5 sections/limb) of the individual limbs of the sham-operated newts are significantly greater than the mean blastema areas (5 sections/limb) of the individual limbs of the hypophysectomized newts, both at eight days post-operation and at 16 days post-operation (Mann-Whitney U Test — 0.01 level of significance).

injections. Injections into hypophysectomized newts were begun either five days after hypophysectomy, at which time both forelimbs were amputated, or at ten days after hypophysectomy (delayed injections) and continued every two days until the end of the experiment.

Limbs and heads of newts from all groups were sampled for histological examination at various times during the experiment. The number of surviving newts of each group was recorded and compared statistically using the X^2 Test on days 20 and 24 post-hypophysectomy.

The hormones and control solutions with which hypophysectomized newts were treated are listed in tables 3 and 4 along with the number of newts treated, and the effect of the treatment on limb regeneration and survival of hypophysectomized newts. The amounts and activities of the contaminating hormones present in the hormone preparations used are listed in table 5. It should be noted that two growth hormone dosages were used, 0.3 mg and 0.03 mg GH/newt/2 days. The growth hormone (NIH) used in these experiments contained ten times more prolactin in each mg of preparation than did the growth hormone (NIH) used by Wilkerson ('63). Thus, the smaller amount of growth hormone (0.03 mg) still contained an amount of prolactin (0.016 U) equivalent to that given with the larger amount of GH (0.3 mg) by Wilkerson ('63).

Hypophysectomized euryhaline fish will survive when maintained in dilute saline and, in these fish, prolactin plays a role in water balance (Ball and Ensor, '67). Therefore, to test whether deaths of hypophysectomized newts were due to loss of sodium, control, untreated hypophysectomized newts, and hypophysectomized newts treated with prolactin, were provided NaCl in the aquarium water (table 3).

ACTH (1U/newt every other day) dissolved in water, was also administered to eight *intact* newts for a period of 24 days. Water (0.1 cm³/newt) was administered to another eight intact newts. Four days after beginning the injections, both forelimbs of all 16 newts were amputated through the elbow. The status of limb regeneration of these newts was observed grossly until day 30 post-amputation.

B. *Results*.   The results of Series III are summarized in tables 3 and 4 and figures 4, 5, 6, and 7. Untreated or saline-treated hypophysectomized adult newts begin to die approximately one week after hypophysectomy and few survive beyond 24 days post-hypophysectomy. A similar survival pattern is observed when hypophysectomized newts are treated with (1) prolactin alone (0.015 U/newt) (fig. 4), (2) prolactin + NaCl (fig. 5), (3) thyroxine alone (fig. 6), (4) ACTH (fig. 7), and (5) ACTH + thyroxine (tables 3, 4). The thyroxine and ACTH were no more effective when treatment was delayed until ten days post-hypophysectomy (table 4).

Hypophysectomized adult newts treated with ACTH do not molt and their skin becomes rough and turns uniformly dark. In addition, one observes the typical melanophore response of ACTH (Geschwind, '67) so that these ACTH treated newts become extremely black by two weeks post-hypophysectomy. The skin of hypophysectomized newts treated with thyroxine is smooth and has a normal olive-green appearance. Hypophysectomized newts treated with saline, ACTH (UpJohn) and/or thyroxine are sluggish and will not eat, even when repeatedly offered fresh beef liver. It should be recalled that none of the newts in these investigations were fed during the course of the experiments.

The treatments effective in enhancing survival (0.01 level of significance, X^2 Test) and limb regeneration of hypophysectomized newts were: (1) prolactin + thyroxine, (2) growth hormone, (3) ectopic pituitary grafts of newts and (4) ectopic pituitary grafts of larval axolotls (*Ambystoma mexicanum*). A prolactin-thyroxine combination significantly enhanced survival of hypophysectomized newts even when treatment was delayed until ten days post-hypophysectomy. Both quantities of growth hormone given (table 3) were effective and it should be noted that the smaller quantity (0.03 mg/newt) nevertheless contained a significant prolactin contamination (table 5). The growth hormone also contained more TSH as a contaminant than did the prolactin (table 5). The lower thyroxine concentration (1×10^{-8}), when

TABLE 3

The effect of hormone treatments on limb regeneration and survival of hypophysectomized newts when treatment is begun five days post-hypophysectomy

Treatment	No. of newts	Limb regeneration [1]	Survival [2]
Prolactin (0.015 U/newt/2 days) and Thyroxine (1×10^{-7} conc. in water) [3]	23	Normal	+
Growth Hormone (0.03 mg/newt/2 days)	15	Normal	+
Growth Hormone (0.3 mg/newt/2 days)	6	Normal	+
Ectopic Newt Pituitary (1 pit/newt)	10	Normal	+
Ectopic Axolotl Pituitary (2 pit/newt)	16	Normal	+
Prolactin (0.015 U/newt/2 days)	20	Delayed	−
Thyroxine (1×10^{-7} conc. in water)	17	Delayed	−
ACTH (1 U/newt/2 days)	15	Delayed	−
ACTH (1 U/newt/2 days) and Thyroxine (1×10^{-7} conc. in water)	12	Delayed	−
Saline (0.9% 0.1 cm³/newt/2 days)	30	Delayed	−
Prolactin (0.015 U/newt/2 days) and NaCl (3.6 g/liter of aquarium water)	9	Delayed	−
Prolactin (0.015 U/newt/2 days) and NaCl (0.36 g/liter of aquarium water)	8	Delayed	−
NaCl (3.6 g/liter of aquarium water)	6	Delayed	−

[1] Normal regeneration refers to the amount of blastema cells comparable to that found in a regenerate of a normal (intact) newt at the same amputation age. Delayed regeneration refers to amputated limbs which may exhibit some blastema cells but an amount not comparable to that found in a regenerate of a normal (intact) newt at that same amputation age. Complete absence of regeneration exemplified by dermal pad formation and normal skin differentiation (skin gland formation) over the amputation surface was not observed. Limbs were amputated five days post-hypophysectomy.
[2] Survival classed as positive (+) refers to *over* 80% survival at 21 days post-hypophysectomy. Survival classed as negative (−) refers to *less* than 20% survival at 21 days post-hypophysectomy.
[3] Another eight hypophysectomized newts received prolactin (0.015 U/newt/2 days) + thyroxine (1×10^{-8} conc. in water). This experiment is depicted in figure 5.

TABLE 4

The effect of hormone treatment on limb regeneration and survival of hypophysectomized newts when treatment is begun ten days post-hypophysectomy [1]

Treatment	No. of newts	Limb regeneration	Survival
Prolactin (0.015 U/newt every other day) and Thyroxine (1×10^{-7} conc. in water)	22	Normal	+
Thyroxine (1×10^{-7} conc. in water)	9	Delayed	−
ACTH (1.0 U/newt every other day)	12	Delayed	−
0.9% Saline (0.1 cm³/newt every other day)	9	Delayed	−
No treatment	10	Delayed	−

[1] Limbs were amputated five days post-hypophysectomy.

combined with prolactin, appeared to be equivalent to the higher concentration (1×10^{-7}) in promoting survival and limb regeneration of hypophysectomized newts (fig. 5).

It should be noted that figures 1, 4, 5, 6, and 7 each represents a separate experiment and statistical comparisons were made only within each experiment. This is important because the newts were received in several shipments. Thus, the nutritional state of the newts receiving the experimental treatments probably varied from group to group, even though the newts were

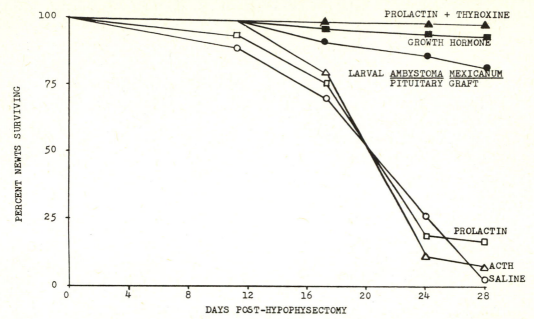

Fig. 4 A comparison of the per cent survival of hypophysectomized newts given growth hormone (0.03 mg/newt/2 days), prolactin (0.015 U/newt/2 days) + thyroxine (1×10^{-7} conc.), prolactin alone (0.015 U/newt/2 days), ACTH (1 U/newt/2 days), 0.9% saline (0.1 cm³/newt/2 days) and ectopic, *Ambystoma mexicanum* pituitary grafts (2 pituitaries/newt).

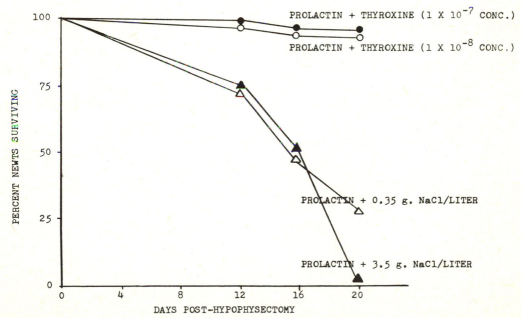

Fig. 5 A comparison of the per cent survival of hypophysectomized newts treated with prolactin (0.015 U/newt/2 days) + thyroxine (1×10^{-7} conc.), prolactin (0.015 U/newt/2 days) + thyroxine (1×10^{-8} conc.), prolactin (0.015 U/newt/2 days) and maintenance in aerated water containing 0.35 gm NaCl/liter, and prolactin (0.015 U/newt/2 days and maintenance in aerated water containing 3.5 gm NaCl/liter.

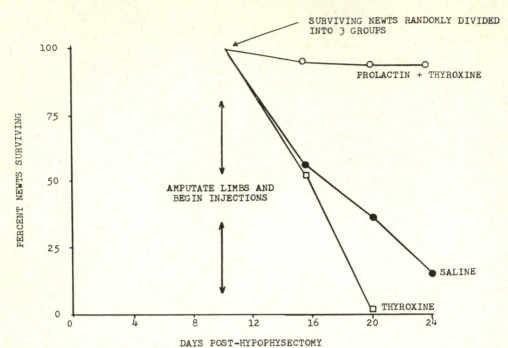

Fig. 6 A comparison of the per cent survival of hypophysectomized newts treated with prolactin (0.015 U/newt/2 days) + thyroxine (1×10^{-7} conc.), thyroxine alone (1×10^{-7} conc.) and 0.9% saline (0.1 cm³/newt/2 days).

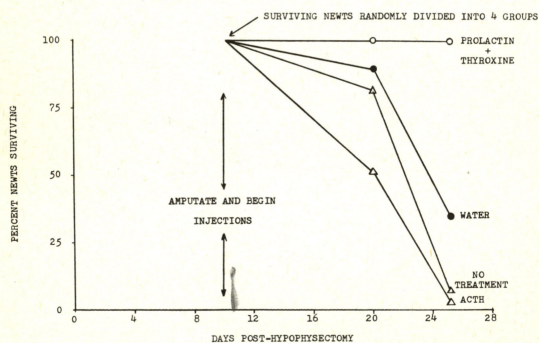

Fig. 7 A comparison of the per cent survival of hypophysectomized newts receiving no treatment, prolactin (0.015 U/newt/2 days) + thyroxine (1×10^{-7} conc.), ACTH (1 U/newt/2 days), and water (0.1 cm³/newt/2 days).

TABLE 5

Hormone quantities injected into hypophysectomized newts

Preparation	Hormone activities present	Amount injected/newt every other day	
		Dose 1	Dose 2
Growth hormone Bovine NIH-GH-B12	Growth hormone (GH) 1.0 U/mg	0.3 mg = 0.291 U	0.03 mg = 0.0291 U
	Thyroid stimulating Hormone (TSH) 0.013 U/mg	0.0039 U	0.00039 U
	Luteinizing hormone (LH) 0.0085 U/mg	0.00255 U	0.000255 U
	Follicle stimulating Hormone (FSH) 0.001 U/mg	0.0003 U	0.00003 U
	Prolactin 0.54 U/mg	0.162 U	0.0162 U
Prolactin Bovine NIH-P-B2	Prolactin 19.9 U/mg	0.015 U	
	TSH 0.0002 U/mg	0.00000015 U	
	GH 0.01 U/mg	0.0000075 U = 0.00075 mg	
	LH 0.0004 U/mg	0.0000003 U	
	FSH 0.01 U/mg	0.0000075 U	
ACTH UpJohn Co.	ACTH 46.3 U/mg (No contaminants listed)	1.0 U	

fasted for two weeks prior to hypophysectomy.

Histological examination of a total of over 40 heads (Series I, II, and III) revealed complete hypophysectomies in over 97% of the cases (fig. 11). In addition, the ectopic pituitary grafts from both newts and axolotls appeared healthy and well vascularized even 35 days after grafting (fig. 13). No neural lobe tissue was observed in these grafts. The skin of the hypophysectomized newts with ectopic pituitary grafts became dark due to excess MSH secretion from the ectopic gland (Tassava et al., '68). The skin of these newts was smooth and slippery and not rough and black as is typical of the skin of untreated hypophysectomized adult newts. In addition, hypophysectomized newts treated with prolactin + thyroxine, GH, or ectopic pituitary grafts were active and readily ate fresh beef liver 30 days post-hypophysectomy. They also underwent occasional molts. Limb regeneration was normal in this group of newts (tables 3, 4) whereas, hypophysectomized newts treated with saline, prolactin alone, ACTH and/or thyroxine exhibited delayed limb regeneration (tables 3, 4).

The *intact* newts injected with ACTH (UpJohn) became dark in color and remained dark during the 24 days of treatment, whereas the eight intact newts injected with water appeared normal. All eight newts in each group survived the experiment and newts of both groups regenerated their amputated limbs in a comparable manner; 100% of the amputated limbs of the 16 newts showed typical regeneration on day 20 post-amputation. On day 30 post-amputation, all 16 limbs had reached the digit stage of regeneration (see footnote, table 1).

In summary, the results of Series III indicate that prolactin + thyroxine, growth hormone, and ectopic pituitary grafts effectively enhance both survival and limb regeneration of hypophysectomized newts. ACTH had no significant effect either on survival or limb regeneration of either hypophysectomized or intact newts.

DISCUSSION

The results described in the preceding pages make it clear that adult newt limb regeneration can be initiated and will proceed to at least the medium bud stage in the complete absence of pituitary hormones. Thus, the conclusion of Schotté ('61), that the stress of limb amputation initiates pituitary ACTH secretion and consequent release of adrenal steroids which are essential to proper wound healing, is not supported by the above data. These data are consistent with the findings of Liversage ('59), however, that amputated limbs will regenerate even though previously isolated from the central nervous system by spinal cord ablation.

It was apparent that regenerates on normal intact newts were more advanced in development than those on hypophysectomized newts of a comparable age, therefore indicating that pituitary hormones insure a more rapid rate of regeneration. These results agree with the observations of DeConinck et al. ('55) who found that hypophysectomized *Triturus alpestris* and *Triturus vulgaris* regenerated limbs which were amputated simultaneously with or four days after hypophysectomy; however, regeneration was considerably delayed. Hay ('56) and DeConinck et al. ('55) suggested that the primary influence of pituitary hormones was on the growth of the regeneration blastema and the results of the present study support this credence. Regenerates which had developed for 14 days in the presence of pituitary hormones nevertheless required these pituitary hormones to continue *normal* development. The 14 day regenerates increased in size after hypophysectomy but to a significantly lesser degree than did those of their sham-operated counterparts. Schotté and Hall ('52) also found that pituitary hormones were necessary for the normal growth of the limb blastema. By delaying hypophysectomy until 14 days after limb amputation, these authors obtained abortive or delayed regenerates in over 75% of the cases. Even more adverse effects were noted when hypophysectomy was delayed for only 7–13 days after limb amputation; 100% of the regenerates were abortive or delayed (Schotté and Hall, '52). The blastema cells begin DNA synthesis as early as the fourth day after amputation (Hay and Fischman, '61) and cell proliferation is an essential part of blastema formation (Chalkley, '54). Therefore it is not surprising that limb regeneration seldom proceeds to completion when hypophysectomy is performed prior to, concomitantly with, or shortly after limb amputation (Hall and Schotté, '51; Schotté and Hall, '52; Connelly et al., '68).

In addition to being able to regenerate lost limbs, the adult newt can readily replace an excised lens. The hormonal requirement is apparently comparable for regeneration of both organs. This supposition is supported by the data of the present investigation and by the experiments of Stone and Steinitz ('53) who found that when newts were lentectomized 5–8 days after hypophysectomy, regeneration was considerably delayed and the final result was often an abnormal lens. Similar delays and abnormalities of lens regeneration were observed in thyroidectomized newts. Thus, in the absence of the pituitary and/or the thyroid, there occurs a retardation in the rate of development of the regenerate of the limb and of the lens, both accompanied by various abnormalities. It is important to note that in the present experiments and in the experiments of Stone and Steinitz ('53), regeneration could not be correlated with small pituitary remnants due to improper hypophysectomies. Thus, thyroid and pituitary hormones are essential to both *normal* lens regeneration (Stone and Steinitz, '53) and *normal* limb regeneration (Connelly et al., '68). Hormones from the *in situ* pituitary, without thyroid hormone, are not enough to promote normal limb or lens regeneration (Richardson, '45; Schotté and Washburn, '54; Stone and Steinitz, '53). Connelly et al. ('68) found that hypophysectomized newts treated with prolactin alone (1.2 U/newt every other day) survived significantly longer than hypophysectomized saline-treated newts but not as long as newts treated with prolactin + thyroxine. The hypophysectomized newts which survived when treated with the larger amount of prolactin all regenerated amputated limbs. The large dose of prolactin (NIH) used may have contained enough TSH as a contaminant to activate the thyroid to a minimum level. It may be that very little thyroxine is needed for the normal health of newts when combined with prolactin, since as the data of the present study demonstrate, even a 1×10^{-8} concentration of thyroxine, with prolactin, significantly enhances survival and limb regeneration of hypophysectomized newts. Of interest in this connection are the experiments of Schmidt ('58) who found that hypothyroid newts regenerated limbs even faster than normal newts. A low thyroxine level may increase prolactin secretion and the higher prolactin quantity may then stimulate regeneration (see also Thornton, '68). Grant and Cooper ('65) have shown that when thyroxine levels are low, as in the adult newt, prolactin secretion is high (aquatic environment), whereas when thyroxine levels are high, as in the eft, prolactin levels are low (land environment). These authors also report that thyroxine treatment will cause aquatic newts to resume a land habitat, again suggesting decreased prolactin secretion under conditions of high thyroxine. High exogenous thyroxine will cause abnormal limb regeneration of intact newts and the effect of the thyroxine appears to be most pronounced during the growth phase, beginning two weeks after amputation (Hay, '56). Since it was shown in this investigation that pituitary hormones, specifically prolactin and TSH, are important to the normal growth of the blastema, it is tempting to speculate that in Hay's ('56) experiments the high thyroxine levels decreased prolactin secretion by the pituitary, thus resulting in a type of *in situ* hypophysectomy.

In the present study, a prolactin-thyroxine combination was found to be very effective in enhancing survival and also limb regeneration of hypophysectomized newts. This hormone combination resulted in survival of almost 100% of the hypophysectomized newts so treated, even when the hormones were not administered until ten days post-hypophysectomy. Thyroxine alone in the concentration used was ineffective in enhancing survival and limb regeneration and prolactin alone (0.015 U/newt every other day) was also ineffective. These findings strongly suggest that prolactin and thyroxine act synergistically in some as yet unknown way and the combination of the two hormones is essential to normal health and limb regeneration of the adult newt. Growth hormone was shown to be as effective as the prolactin-thyroxine combination in enhancing survival and limb regeneration of hypophysectomized newts, even at 1/10 the quantity used by Wilkerson ('63). The effectiveness of this smaller quantity of growth hormone (NIH) used in the present investigation may have been due to the fact that it contained ten times as much prolactin per mg of preparation as the growth hormone (NIH) used by Wilkerson. Thus, the newts given 0.03 mg of growth hormone received 0.016 U prolactin/newt every other day. It was pointed out by Berman et al. ('64) that this amount of prolactin, equivalent to the prolactin contaminant which Wilkerson ('63) administered to newts, will elicit a growth response when injected into frog tadpoles. It cannot be said with certainty whether the quantity of growth hormone administered in this investigation contained enough TSH to activate the thyroid of a hypophysectomized newt. However, occasional molting or partial molting was observed during the treatment period suggesting some thyroid activation. Richardson ('45) and Wilkerson ('63) both suggested that some thyroid activity is important to normal limb regeneration. Additional investigations should determine whether growth hormone, completely free of prolactin and/or TSH would still be effective in supporting limb regeneration and survival of hypophysectomized newts.

The finding that a prolactin-thyroxine combination is important to newt limb regeneration and survival agrees with the abundant evidence that prolactin and TSH are produced by the newt pituitary and that these hormones (TSH acting via the thyroid) may influence newt physiology. The larval newt undergoes a primary metamorphosis to a land form, called an eft, which lives from 1–5 years on land before migrating to water as an aquatic, reproductively mature adult. The "water drive" of the eft has been shown to be induced by prolactin and the migratory behavior, and the structural and physiological changes involved, have been termed "second metamorphosis" (Grant, '61). The induced water drive of the eft has been recommended as an adequate assay for prolactin (Grant, '59) and has been used to demonstrate the presence of prolactin in pituitary tissue of *Bufo, Fundulus, Cyprinus, Natrix* (Chadwick, '41; Grant, '61) and adult newts (Reinke and Chadwick, '39). Antuitrin G will induce water drive (Chadwick, '40) but LH, ACTH, posterior pituitary tissue, TSH, and Antuitrin S have no water drive activity (Grant and Grant, '58). Since the ectopically transplanted eft pituitary will induce a water drive (Masur, '62) it is probable that prolactin secretion is under negative control by the hypothalamus (Grant, '61). The ectopic pituitary of the adult newt has been shown to produce thyroid stimulating hormone (TSH) in near normal amounts (Dent, '66) and ultrastructural studies of the adult newt pituitary suggest that TSH and prolactin are produced by the normal as well as the ectopic gland (Dent and Gupta, '67).

There is no evidence that prolactin and/or thyroxine act through the adrenal gland thereby increasing or maintaining the secretion of adrenal hormones. Adrenal hormones did not enhance survival of hypophysectomized newts (Schotté and Bierman, '56) and in this investigation prolactin and thyroxine, but not ACTH, significantly enhanced survival and limb regeneration of hypophysectomized newts. Furthermore, in mammals prolactin causes a decreased secretion of the adrenal gland (Bates et al., '64). Why then, was the ACTH used by Schotté and Chamberlain ('55) beneficial to survival and limb regeneration of hypophysectomized newts? It is likely that ACTH was *not* the active hormone in the preparation used by Schotté and Chamberlain ('55) for three reasons: (1) their ACTH contained 1.14 U/ACTH/mg (crude Armour ACTH, '55 preparation) whereas it was not until 1962 that Armour produced purified ACTH containing 33 U/ACTH/mg (Evans et al., '66); the ACTH used in this investigation, on the other hand, was essentially pure ACTH (UpJohn Co., '66 preparation). (2) The description given by Schotté and Chamberlain ('55) of the response of the hypophysectomized newts to the crude ACTH — "newts remained active, had good appetites, and slippery skin" — closely resembles the response of the hypophysectomized newts in this investigation to prolactin + thyroxine, growth hormone, or ectopic pituitary grafts. In the present investigation it was found that hypophysec-

tomized newts given pure ACTH (UpJohn) are sluggish, will not feed, and instead of having slippery skin, have course granular skin similar to that of saline-treated hypophysectomized newts. (3) Schotté and Chamberlain ('55) reported that ACTH (1 U/newt/2 days) resulted in inhibition of limb regeneration in intact newts. This effect could also have been due to contaminating hormones. In this investigation, pure ACTH (1 U/newt/2 days) *did not* inhibit limb regeneration in intact newts. Therefore, the evidence implies that prolactin and thyroxine are essential to adult newt survival and normal limb regeneration and furthermore, that adrenal hormones are not limiting after hypophysectomy.

The exact pathway by which prolactin and thyroxine influence survival and limb regeneration is unknown and more work must be done on this problem. However, some possible functions of prolactin and thyroxine in the newt are suggested by observations of the present and other investigations. Prolactin may act on cell proliferation. Inoue ('56) found a diminished mitotic proliferation in epidermal cells of amputated limbs of hypophysectomized adult newts. Waterman ('65) and Niwelinski ('58) increased the rate of *intact* newt limb regeneration by prolactin treatment. Waterman ('65) also found that prolactin increased appetite and body weight of intact newts. Thyroxine will act directly on the skin of adult newts (Clark and Kaltenbach, '61) and also on denervated skin (Taban and Tassava, unpublished) and Grant and Cooper ('65) found that prolactin would maintain newt skin in organ culture but thyroxine alone was ineffective. Prolactin also acts on the skin of lizards by raising the frequency of sloughing (Maderson and Licht, '67) and prolactin treatment will increase the molting frequency and the mitotic rate of red eft skin (Chadwick and Jackson, '48).

Grant ('61) suggested that prolactin may induce the water drive of the red eft because of the role of this hormone in the water balance of the newt. This suggestion is important since it is known that prolactin does influence water balance in eels (Olivereau and Ball, '64), *Fundulus* (Ball and Ensor, '67) and *Tilapia* (Dharmamba et al., '67). In this investigation, survival of hypophysectomized newts was not enhanced by maintaining newts in NaCl solutions; however, measurements of tissue sodium and sodium transport before and after hypophysectomy would be a more worthwhile approach to this problem.

Prolactin may also play a role in metabolism of fat, protein and carbohydrate. Hypophysectomized *Tilapia mossambica* cannot form liver glycogen from amino acid precursors (Swallows and Fleming, '67) which suggests that pituitary hormones are essential for gluconeogenesis in these fish. Prolactin enhances food consumption and body weight gain in both newts (Waterman, '65) and lizards (Licht, '67) and prolactin treated lizards also show a significant weight increase of regenerating tails. It may be that prolactin and thyroxine are involved in normal metabolism in the adult newt and these hormones are required for energy (glucose) production from protein. Thus, hypophysectomized newts which were previously fasted, survived only when given prolactin + thyroxine. Furthermore, hypophysectomized newts which were previously well fed survived significantly longer than fasted hypophysectomized newts. These observations suggest that newts in particularly good nutritional state at the time of hypophysectomy may contain food reserves, such as liver glycogen, which can be utilized for energy in the absence of hormones. The energy from the food reserves could then be used to maintain survival and also for growth of the regenerate. This speculation is supported by the fact that newts in Series I which were fed daily for two weeks weighed significantly more at the time of hypophysectomy and survived significantly longer after hypophysectomy than newts fasted for two weeks prior to hypophysectomy. Whether prolactin and thyroxine act on metabolism, which cells and tissues are acted upon and exactly how these hormones interact in enhancing survival and limb regeneration of adult newts will receive further attention in future investigations.

Schotté ('61) reported that larval *Ambystoma punctatum* and larval newt pituitary grafts did not support limb regeneration in hypophysectomized adult newts. It was therefore surprising to discover in this investigation that larval *Ambystoma mexicanum* (axolotl) pituitaries (2 pituitary grafts/newt) did significantly enhance survival and limb regeneration of hypophysectomized newts. Prolactin has been identified in the pituitary of *Necturus* and adult *Ambystoma tigrinum* (Nicoll and Bern, '68) and the results of the present experiments suggest that larval *Ambystoma mexicanum* pituitary tissue also contains prolactin. The epidermis of hypophysectomized newts with ectopic axolotl pituitary grafts does not build up as in hypophysectomized newts given only prolactin; and, in addition, these newts undergo occasional molts. Thus, the axolotl pituitary, ectopically transplanted to the newt apparently also secretes TSH, thus activating the newt's thyroid. It may be that in the axolotl no TSH releasing factor is present, whereas, grafted in the newt, the axolotl pituitary responds to TSH releasing factor which reaches the pituitary through the blood circulation.

ACKNOWLEDGMENTS

This investigation was conducted while I held a predoctoral fellowship (NDEA) and was included in a dissertation submitted to Michigan State University in partial fulfillment of the requirements for the Ph.D., June, 1968. This work was supported by NIH grant NB-04128 and NSF grants GE-2618 and GB-7748 administered by Dr. C. S. Thornton, Department of Zoology, Michigan State University. I would like to express my appreciation to Dr. Thornton for his helpful suggestions and advice during the course of this investigation and in the preparation of the manuscript.

LITERATURE CITED

Ball, J. N., and D. M. Ensor 1967 Specific action of prolactin on plasma sodium levels in hypophysectomized *Poecilia latipinna* (Telostei). Gen. Comp. Endo., 8(3): 432–440.

Bates, R. W., S. Milkovic and M. M. Garrison 1964 Effects of prolactin, growth hormone, and ACTH, alone and in combination, upon organ weights and adrenal function in normal rats. Endo., 74: 714–723.

Berman, R., H. A. Bern, C. S. Nicoll and R. C. Strohman 1964 Growth promoting effects of mammalian prolactin and growth hormone in tadpoles of *Rana catesbeiana*. J. Exp. Zool., *156*: 353–360.

Bragdon, D. E., and J. N. Dent 1954 Effect of ACTH and cortisone on renal fat and limb regeneration in adult salamanders. Proc. Soc. Exp. Biol. Med., *87*: 460–462.

Chadwick, C. S. 1940 Induction of water drive in *Triturus viridescens* with anterior pituitary extract. Proc. Soc. Exp. Biol. Med., *43*: 509–511.

———— 1941 Further observations on the water drive in *Triturus viridescens*. II. Induction of the water drive with the lactogenic hormone. J. Exp. Zool., *86*: 175–187.

Chadwick, C. S., and H. R. Jackson 1948 Acceleration of skin growth and molting in the red eft of *Triturus viridescens*. Anat. Rec., *101*: 718.

Chalkley, D. T. 1954 A quantitative histological analysis of forelimb regeneration in *Triturus viridescens*. J. Morph., *94*: 21–70.

Clark, N. B., and J. C. Kaltenbach 1961 Direct action of thyroxine on the skin of the adult newt. Gen. Comp. Endo., *1*: 513–518.

Connelly, T. G., R. Tassava and C. S. Thornton 1968 Survival and limb regeneration of prolactin treated hypophysectomized newts. J. Morph. In press.

DeConinck, L., M. Denuce, Fr. Dierckx and M. Janssens 1955 Acides ribonucleiques, hormone somatotrope et regeneration chez *Triturus* (Urodeles). Societe Royale Zoologique De Belgique Annales, LXXXVi: 191–234.

Dent, J. N. 1966 Maintenance of thyroid function in newts with transplanted pituitary glands. Gen. Comp. Endo., *6*: 401–408.

———— 1967 Survival and function in hypophysial homografts in the spotted newt. Amer. Zool., *7(4)*: 714.

Dent, J. N., and B. J. Gupta 1967 Ultrastructural observations on the developmental cytology of the pituitary gland in the spotted newt. Gen. Comp. Endo., *8*: 273–288.

Dharmamba, R., R. I. Handin, J. Nandi and H. A. Bern 1967 Effect of prolactin on fresh water survival and on plasma osmotic pressure of hypophysectomized *Tilapia mossambica*. Gen. Comp. Endo., *9*: 295–302.

Evans, H. M., L. L. Sparks and J. S. Dixen 1966 The physiology and chemistry of adrenocorticotropin. In: The Pituitary Gland. Vol. 1. G. W. Harris and B. T. Donovan, eds. Univ. Calif. Press, Berkeley.

Fortier, C. 1966 Nervous control of ACTH secretion. In: The Pituitary Gland. Vol. 2. G. W. Harris and B. T. Donovan, eds. Univ. Calif. Press, Berkeley.

Geschwind, I. I. 1967 Molecular variation and possible lines of evolution of peptide and protein hormones. Amer. Zool., *7*: 89–108.

Grant, W. C. 1959 A test for prolactin using the hypophysectomized eft stage of *Diemictylus viridescens*. Endo., *64*: 839–841.

———— 1961 Special aspects of the metamorphic process: second metamorphosis. Amer. Zool., *1*: 163–171.

Grant, W. C., and J. A. Grant 1958 Water drive studies on hypophysectomized efts of *Diemictylus viridescens*. Part I. The role of the lactogenic hormone. Biol. Bull., *114(1)*: 1–9.

Grant, W. C., and G. Cooper 1965 Behaviorial and integumentary changes associated with induced metamorphosis in *Diemictylus*. Biol. Bull., *129*: 510–522.

Hall, A. B., and O. E. Schotté 1951 Effects of hypophysectomies upon the initiation of regenerative processes in the limb of *Triturus viridescens*. J. Exp. Zool., *118*: 363–382.

Hanke, W., and K. Weber 1965 Histophysiological investigation on the zonation, activity, and mode of secretion of the adrenal gland of the frog, *Rana temporaria*. Gen. Comp. Endo., *5*: 444–455.

Hay, E. D. 1956 Effects of thyroxine on limb regeneration in the newt *Triturus viridescens*. Bull. Johns Hopkins Hosp., *99*: 262–285.

———— 1966 Regeneration. Holt, Rinehart and Winston, N. Y., N. Y.

Hay, E. D., and D. A. Fischman 1961 Origin of the blastema in regenerating limbs of the newt *Triturus viridescens*. Dev. Biol., *3*: 327–342.

Inoue, S. 1956 Effect of growth hormone and cortisone acetate upon mitotic activity in normal and regenerating tissues of amphibians. Endo. Japan., *3*: 236–239.

Licht, P. 1967 Interaction of prolactin and gonadotropins on appetite, growth and tail regeneration in the lizard, *Anolis carolinensis*. Gen. Comp. Endo., *9(1)*: 49–63.

Liversage, R. A. 1959 The relation of the central and autonomic nervous systems to the regeneration of limbs in adult urodeles. J. Exp. Zool., *141*: 75–118.

Maderson, P. F. A., and P. Licht 1967 Epidermal morphology and sloughing frequency in normal and prolactin treated *Anolis carolinensis*. J. Morph., *123*: 157–172.

Mangili, G., M. Motta and L. Martini 1966 Control of adrenocorticotropic hormone secretion. In: Neuroendocrinology. Vol. 1. L. Martini and W. F. Ganong, eds. Academic Press, N. Y.

Masur, S. 1962 Autotransplantation of the pituitary of the red eft. Amer. Zool., 2: 538.

Merchant, D. J., R. H. Kahn and W. H. Murphy 1964 Handbook of Cell and Organ Culture. Burgess Publishing Co., Minneapolis, Minn.

Nicoll, C. S., and H. A. Bern 1968 Further analysis of the occurrence of pigeon crop-stimulating activity (prolactin) in the vertebrate hypophysis. Gen. Comp. Endo., *11(1)*: 5–20.

Niwelinski, J. 1958 The effect of prolactin and somatotropin on the regeneration of the forelimb in the newt, *Triturus alpestris*. Folia Biol., *6*: 9–36.

Olivereau, M., and J. N. Ball 1964 A contribution to the histophysiology of the hypophysis of teleosteans in particular the cells of *Poecillia*. Gen. Comp. Endo., *4*: 523–532.

Pearse, A. G. E. 1961 Histochemistry: Theoretical and Applied. Little, Brown and Co., Boston.

Piper, G. D., and R. DeRoos 1967 Evidence for a corticoid-pituitary negative feedback mechanism in the American bullfrog (*Rana catesbeiana*). Gen. Comp. Endo., 8: 135–142.

Purves, H. D., and N. E. Sirett 1967 Corticotropin secretion by ectopic pituitary glands. Endocrinology, 30: 962–968.

Reinke, E. E., and C. S. Chadwick 1939 Inducing land stage of *Triturus viridescens* to assume water habitat by pituitary implantation. Proc. Soc. Exp. Biol. Med., 40: 671–693.

Richardson, D. 1945 Thyroid and pituitary hormones in relation to regeneration. II. Regeneration of the hind limb of the newt, *Triturus viridescens*, with different combinations of thyroid and pituitary hormones. J. Exp. Zool., 100: 417–429.

Schmidt, A. J. 1958 Forelimb regeneration of thyroidectomized adult newts. II. Histology. J. Exp. Zool., 139: 95–125.

Schotté, O. E. 1961 Systemic factors in initiation of regenerative processes in limbs of larval and adult amphibians. In: Molecular and Cellular Structures, 19th Growth Symposium. D. Rudnick, ed. The Ronald Press Co., N. Y.

Schotté, O. E., and R. H. Bierman 1956 Effects of cortisone and allied steroids upon limb regeneration in hypophysectomized *Triturus viridescens*. Rev. Suisse Zool., 63: 353–375.

Schotté, O. E., and J. L. Chamberlain 1955 Effects of ACTH upon limb regeneration in normal and in hypophysectomized *Triturus viridescens*. Rev. Suisse Zool., 62: 253–279.

Schotté, O. E., and A. B. Hall 1952 Effect of hypophysectomy upon phases of regeneration in progress, *Triturus viridescens*. J. Exp. Zool., 121: 521–556.

Schotté, O. E., and D. A. Lindberg 1954 Effect of xenoplastic adrenal transplants upon limb regeneration in normal and in hypophysectomized newts (*Triturus viridescens*). Proc. Soc. Exp. Biol. Med., 87: 26–29.

Schotté, O. E., and A. Tallon 1960 The importance of autoplastically transplanted pituitaries for survival and for regeneration of adult *Triturus*. Experimentia, 16: 72–76.

Schotté, O. E., and W. W. Washburn 1954 Effect of thyroidectomy on the regeneration of the forelimb in *Triturus viridescens*. Anat. Rec., 120: 156.

Singer, M. 1952 The influence of the nerves in regeneration of the amphibian extremity. Quart. Rev. Biol., 27(2): 169–200.

Stone, L. S., and H. Steinitz 1953 Effects of hypophysectomy and thyroidectomy on lens and retina regeneration in the adult newt, *Triturus v. viridescens*. J. Exp. Zool., 124: 469–504.

Swallows, R. L., and W. R. Fleming 1967 Effect of hypophysectomy on the metabolism of liver glycogen of *Tilapia mossambica*. Amer. Zool., 7(4): 715.

Tassava, R. A., F. J. Chlapowski and C. S. Thornton 1968 Limb regeneration in *Ambystoma* larvae during and after treatment with adult pituitary hormones. J. Exp. Zool., 167: 157–163.

Thornton, C. S. 1968 Amphibian limb regeneration. In: Advances in Morphogenesis. Vol. 7. Academic Press, N. Y.

Van Dongen, W. J., D. B. Jorgensen, L. O. Larsen, P. Rosenkilde, B. Lofts and P. G. W. J. Van Oordt 1966 Function and cytology of the normal and autotransplanted pars distalis of the hypophysis in the toad, *Bufo bufo*. Gen. Comp. Endo., 6: 491–518.

Waterman, A. J. 1965 Prolactin and regeneration of the forelimbs of the newt. Amer. Zool., 5: 237.

Wilkerson, J. A. 1963 The role of growth hormone in regeneration of the forelimb of the hypophysectomized newt. J. Exp. Zool., 154: 223–230.

Wurster, D. H., and M. R. Miller 1960 Studies on the blood glucose and pancreatic islets of the salamander, *Taricha torosa*. Comp. Biochem. Physiol., 1: 101–109.

PLATE 1

8 A longitudinal section through the regeneration blastema of a hypophysectomized newt which was fed for two weeks prior to hypophysectomy. This limb was amputated five days post-hypophysectomy and
fixed 26 days post-hypophysectomy. Histological examination of the
head of this newt demonstrated complete hypophysectomy. Masson's
Trichrome Stain. × 160.

9 A longitudinal section through the regeneration blastema of a hypophysectomized newt which was fasted for two weeks prior to
hypophysectomy. This limb was amputated five days post-hypophysectomy and fixed 26 days post-hypophysectomy. Hematoxylin-eosin.
× 160. Enlarged × 1.5.

10 A median sagittal section through the pituitary gland of a sham-
operated newt. PAS Stain. × 100.

11 A median sagittal section through the base of the infundibulum of
a completely hypophysectomized newt which was fed for two weeks
prior to hypophysectomy. At the time of fixation, 26 days post-
hypophysectomy, both forelimbs of this newt showed positive regeneration. PAS Stain. × 100.

12 A median sagittal section through the base of the infundibulum of a
partially hypophysectomized newt which was fasted for two weeks
prior to hypophysectomy. The small pituitary fragment (arrow) had
little or no survival value since this newt survived for only 17 days
post-hypophysectomy. PAS Stain. × 100.

13 A median sagittal section through the grafted larval axolotl (*A. mexicanum*) pituitary tissue. This completely hypophysectomized newt
was fixed 35 days post-hypophysectomy. At the time of fixation (30
days post-amputation) the forelimbs of this newt exhibited three
digit regenerates. Herlant's Tetrachrome Stain. × 100.

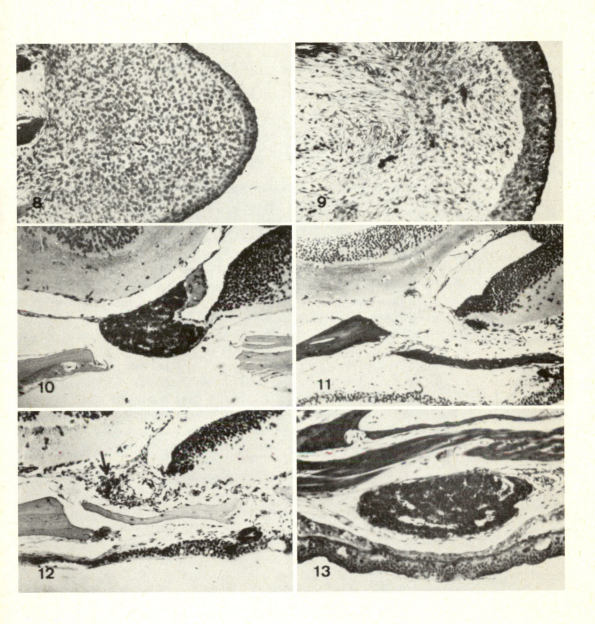

Reprinted from *Develop. Biol.*, **19**, 311–320 (1969)

25

Denervation Effects on Newt Limb Regeneration: DNA, RNA, and Protein Synthesis[1]

MARC H. DRESDEN[2]

The Developmental Biology Laboratory, Department of Medicine, Massachusetts General Hospital and Harvard Medical School, Boston, Massachusetts

Accepted December 9, 1968

INTRODUCTION

The ability of adult urodeles to regenerate amputated limbs in contrast to the failure of regeneration in mammals has led to an increasing awareness of the role of nerve in developmental processes. The elegant studies of Singer and his co-workers have clearly demonstrated the requirement of an adequate nerve supply for the regenerative process in amphibia (Singer, 1946, 1952, 1959) and reptiles (Singer, 1961; Simpson, 1961). In the adult newt *Diemictylus viridescens*, interruption of the sensory nerves (spinal nerves 3, 4, and 5) at the time of amputation will completely prevent limb regeneration. Moreover, when these nerves are resected after partial regeneration has occurred, further growth is prevented and the partial regenerate may regress to the original amputation site (Singer and Craven, 1948). Singer (1952) has postulated that the role of nerves in regeneration is mediated by a chemical trophic factor. Attempts to identify the factor have been unsuccessful.

Studies to date have been primarily morphological, and as a starting point for a more quantitative biochemical approach to the reacting system, we have sought to delineate the effect of denervation on the synthesis of important macromolecules of the regeneration blastema.

This paper reports the changes in synthesis of DNA, RNA, and protein in blastemal tissue *in vivo* and *in vitro* consequent to denervation.

[1] This is publication No. 466 of the Robert W. Lovett Memorial Group for the Study of Diseases Causing Deformities, Department of Medicine, Harvard Medical School at the Massachusetts General Hospital. This work was supported by a grant from the National Foundation, the National Institutes of Health (AM 3564) and by a fellowship from the U. S. Atomic Energy Commission.

[2] Present address: Department of Biochemistry, Baylor University College of Medicine, Houston, Texas 77025.

311

MATERIALS AND METHODS

Adult newts, *Diemictylus viridescens*, obtained from Connecticut Valley Biological Supply Company, were kept in large tanks at room temperature (23 ± 1°C) and fed shredded beef liver once a week. Animals were amputated with sharp scissors through both front limbs proximal to the elbow. After the initial amputation the protruding bone was trimmed and the animals were kept six to a bowl in distilled water changed thrice weekly.

Three to four weeks later regeneration had progressed to the early palette stage (Singer, 1952). In most cases both arms of a single animal appeared to regenerate at similar rates. In the experiments reported below only animals with both limbs at approximately the same regeneration stage (no later than the palette stage) were used.

Denervation was performed under anesthesia (MS-222, Sandoz Chemical Company, Hanover, New Jersey). The operation consisted of resection of approximately 5 mm of the third and fourth spinal nerves in the scapular region (this also removes the fifth spinal nerve since it usually combines with the fourth nerve high in this region).

The animals were injected with radioactive precursors intraperitoneally under MS-222 anesthesia using a sterile disposable syringe with a No. 25-gauge needle. They were then kept in a moist finger bowl at room temperature for 4–6 hours, at which time they were anesthetized and the limbs were removed with a scalpel.

For measurement on the *in vitro* synthetic activity of regenerating limb tips, the animals were sterilized by overnight treatment with penicillin (10^6 units) and streptomycin (0.4 gm) per liter of distilled water. The limbs were removed, washed in Tyrode solution (Nagai *et al.*, 1966) and placed in Leighton tubes (Bellco Glass Co., Vineland, New Jersey) containing 0.25 ml of Dulbecco's medium (Grand Island Biological Co., Grand Island, New York) under sterile conditions. Incubation was at 30°C (Stephenson, 1966) in a 5% CO_2–95% O_2 atmosphere.

Incubation *in vitro* and *in vivo* were terminated by placing the limb tissues in 1 ml of 1 N NaOH and incubating at 37°C until the tissue had dissolved (approximately 30 minutes). One milliliter of 50% trichloroacetic acid was added, and the precipitate was allowed to settle overnight in the cold. The precipitate was harvested by centrifugation at 25,000 g in a Servall RC-2 refrigerated centrifuge and dissolved in 0.5 ml of 0.2 N NaOH. An 0.2-ml aliquot was added to Bray's scintillation fluid (Bray, 1960) containing 4% Ca-bo-sil (Cabot Corp., Boston, Massachusetts) and counted in a Packard Model 3375 TriCarb scintillation spectrometer. The efficiency for ³H was approximately 30 % and for ¹⁴C about 70 %. In double-label experiments (Fig. 1) each isotope was counted in a separate channel and corrections were made for crossover between the channels. In most experiments the results were normalized to protein content due to very low DNA concentrations in the blastema. In control experiments it was found that normalization to DNA or protein content yielded essentially the same results.

When uridine-¹⁴C was used as tracer for RNA synthesis, the initial 1 N NaOH step was omitted (since RNA is hydrolyzed under these conditions) and the tissue was placed directly in 4 ml of 10 % trichloroacetic acid. After centrifugation and removal of the supernatant, the tissue was solubilized with 0.5 ml of 0.2 N NaOH.

Protein concentrations were measured by the Lowry method (Lowry et al., 1951). DNA was assayed by the method of Burton (1956).

Radioactive precursors (thymidine-³H, leucine-¹⁴C, uridine-¹⁴C) were purchased from commercial sources in sterile solutions and diluted with sterile Tyrode medium.

EXPERIMENTS AND RESULTS

Temporal Relationship of Denervation and Biosynthesis in Blastema

Animals having attained the early palette stage of regeneration (20–25 days) were denervated on one side (the other limb serving as a control) and after various periods of time post denervation were injected with radioactive precursors. Four to six hours after isotope injection the limbs were collected and the radioactivity incorporated by each blastema was determined.

In Fig. 1 we have plotted the radioactivity incorporated per unit of protein as a ratio of the control (innervated regenerate) versus the experimental (denervated) limb tips as a function of time after denervation. In order to maximize reproducibility, each point represents the average of 4 animals; each animal was injected with a mixture of two radioactive precursors (either thymidine-³H and leucine-¹⁴C, or thymidine-³H and uridine-¹⁴C).

To ascertain that the resultant changes in DNA and protein synthesis were due to denervation, sham operations were performed on several animals. This operation did not affect DNA and protein synthesis.

Temporal Relationship of Denervation and Biosynthesis in Nonregenerating Limb Tissues

To determine whether this effect of denervation on macromolecular synthesis is specific for regenerating tissues, we denervated normal ani-

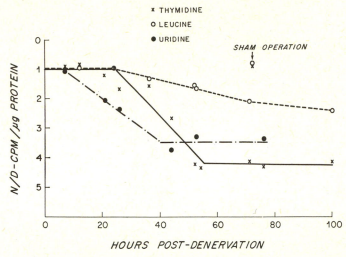

FIG. 1. Temporal relation of denervation and biosynthesis in blastema. Groups of four animals in the palette stage of regeneration were denervated unilaterally. After various time periods after denervation, the animals were anesthetized and injected intraperitoneally with a mixture of thymidine-^{3}H (10 μC/gm of body weight; 360 μC/μmole) and either leucine-^{14}C (1 μC/gm; 2.5 μC/μmole) or uridine-^{14}C (1 μC/gm; 50 μc/μmole). The animals were revived and placed in individual finger bowls with a small amount of water. Four to five hours after the injection the animals were killed, the blastema were removed, and aliquots were assayed for radioactivity and protein content. The results are plotted as a ratio of control blastema (N) to denervated blastema (D) and represent the average of the four animals.

mals and 55 hours after the operation assayed their capacity to incorporate thymidine-^{3}H.

In contrast to the considerably decreased capacity of blastema tips to incorporate thymidine after denervation, nonregenerating arms showed relatively small changes in the incorporation of thymidine-^{3}H (Table 1).

We further investigated the effect of denervation on regenerating and nonregenerating tissues by comparing the incorporation of thymidine-^{3}H into the blastema and the proximal tissues of regenerating limbs.

Animals in the early palette stage were denervated unilaterally and were injected with thymidine-^{3}H 17, 48, and 96 hours later. Regenerating blastema and proximal tissue fragments (close to the shoulder) from both denervated and control limbs were removed and assayed for radioactivity.

Table 2 shows that the time-dependent decrease in thymidine in-

TABLE 1

TEMPORAL RELATIONSHIP BETWEEN DENERVATION AND
BIOSYNTHESIS IN NORMAL LIMBS[a]

	Animal	Cpm	Protein (µg)	Cpm/µg protein	N/D[b]
1	N	197	114	1.73	1.075
	D	159	99	1.61	
2	N	474	165	2.91	1.05
	D	403	145	2.78	
3	N	176	141	1.25	1.19
	D	143	136	1.05	
4	N	520	180	2.89	1.33
	D	378	173	2.18	

[a] Four normal animals were denervated unilaterally. Fifty hours later the animals were injected with thymidine-³H (10 µc/gm body weight; 360 µc/µmole) under anesthesia. The animals were revived and kept in individual finger bowls for 5 hours; at this time the forearms were removed and assayed for radioactivity and for protein content. The results are expressed as a ratio of the radioactivity incorporated by the control (N) arm versus the denervated (D) arm.

[b] Average ratio N/D = 1.16.

TABLE 2

RELATION OF THYMIDINE-³H INCORPORATION AND DENERVATION IN BLASTEMA
AND PROXIMAL TISSUES[a]

Hours after denervation	N/D	
	Blastema	Proximal
100	4.14 (2.98–5.63)	1.58 (1.49–1.63)
52	3.20 (2.14–4.71)	1.61 (0.85–2.25)
21	1.50 (1.04–1.78)	1.75 (0.77–2.97)

[a] Groups of 4 animals having regenerated to the palette stage were denervated unilaterally. At 96, 48, and 17 hours after denervation the animals were injected with thymidine-³H (10 µc/gm body weight; 360 µc/µmole) under anesthesia. The animals were revived and kept in individual finger bowls for 4 hours at room temperature; the arms were then removed. The blastema and a proximal slice of the arm (approximately the same size as the blastema and derived from an area midway between the elbow and shoulder) were removed from each arm and assayed for radioactivity and protein content. The results are presented as the ratio (calculated as cpm/µg protein) between the control (N) and the denervated (D) limbs. The average of the four animals and the range within each group are shown.

corporation of the blastema is not paralleled by an equal decrease in the nonregenerating parts of the limb proximal to the blastema. The relatively high N/D found in this (but not other) experiments may have been due to selection of proximal tissue fragments excessively close to the area of the operation.

Effect of Regeneration Distally on Thymidine-³H Incorporation in Proximal Tissue Fragments

The same experiment was performed on fully innervated limbs to determine the difference in DNA synthesis between proximal tissues of a regenerating amputation stump and that of equivalent regions of a non-regenerating limb. Adult newts were amputated on one side, and, after regeneration had progressed to the early palette stage, fragments of tissue from the shoulder to elbow region of the regenerating and of the normal limbs were assayed for their capacity to incorporate thymidine-³H. Table 3 shows little difference in DNA synthesis between the two tissues.

DNA Biosynthesis in Vitro of Denervated and Intact Blastema

In order to ascertain that the decrease in thymidine incorporation upon denervation was not a function of changes in circulation in the de-

TABLE 3

EFFECT OF REGENERATION DISTALLY ON INCORPORATION OF THYMIDINE-³H[a]

	Animal	Cpm	Protein (μg)	Cpm /μg protein	R/NR[b]
1	R	182	93.2	1.95	1.10
	NR	206	117.3	1.77	
2	R	33	71.9	0.459	1.00
	NR	41	88.9	0.461	
3	R	85	96.0	0.855	1.06
	NR	100	124.0	0.804	
4	R	137	107.4	1.27	1.37
	NR	114	123.5	0.923	

[a] Four animals were amputated unilaterally. When the amputated arm had regenerated to the palette stage, the animals were injected intraperitoneally with thymidine-³H (10 μc/gm body weight; 360 μc/μmole). Four hours later the animals were sacrificed, and a slice of tissue midway between the elbow and the shoulder was removed from each arm and assayed for radioactivity and for protein content. The results are expressed as a ratio of radioactivity in the regenerating arm (R) versus radioactivity in the nonregenerating arm (NR).

[b] Average R/NR = 1.13.

nervated arm, we investigated the synthetic capacity of blastema in organ culture. Limbs of regenerating animals were denervated for various periods of time, the animals were treated overnight with penicillin and streptomycin, and the whole blastemas were incubated at 30°C (Stephenson, 1966) in the culture medium containing thymidine-^{3}H.

The blastema derived from animals 47 hours after denervation showed a decreased capacity to incorporate thymidine *in vitro* over the control nondenervated limbs (Table 4). In contrast, blastema from animals denervated 17 hours previously did not show a decreased synthetic capacity.

Contribution of DNA Degradation to Experimental Results

The changes in the incorporation of radioactive precursors accompanying denervation could be due to changes in the pools of the metabolites.

TABLE 4

INCORPORATION OF THYMIDINE-^{3}H IN BLASTEMA *in Vitro*[a]

Group	Animal		Cpm	Cpm/μg DNA	N/D	Average N/D
Post-denervation, 47 hours	1	N	254	12.6	1.63	
		D	137	7.75		
	2	N	144	8.1	4.03	2.74
		D	37	2.01		
	3	N	240	11.7	2.55	
		D	87	4.6		
Post-denervation, 17 hours	4	N	152	7.92	0.63	
		D	263	12.5		
	5	N	120	5.67	1.13	1.09
		D	105	5.0		
	6	N	196	9.97	1.51	
		D	142	6.47		

[a] Groups of 3 animals in the palette stage were denervated unilaterally 47 and 17 hours before removal of the blastema. The animals were kept in penicillin (10^6 u/liter) and streptomycin (0.4 gm/liter) overnight. Both forearms were removed with scissors, and the blastemas were rinsed in 100 ml of Tyrode's solution. They were then placed in Leighton tubes containing 0.25 ml Dulbecco's medium and 0.25 μC thymidine-^{3}H (360 μc/μmole). Incubation was at 30°C in a 5% CO$_2$–95% air atmosphere. After 20 hours of incubation, the blastemas were removed from the incubation medium and assayed for radioactivity and DNA content. The results are expressed as cpm/μg DNA in control (N) and denervated (D) blastema.

TABLE 5

DÉNERVATION AND DNA BREAKDOWN[a]

Group	Average N/D
67 hours post denervation	1.15
	(1.33, 1.06)
48 hours post denervation	0.98
	(0.79, 0.99, 1.16)
24 hours post denervation	1.05
	(1.07, 0.92, 1.16)
Nondenervated	0.93
	(0.97, 0.89)

[a] Ten animals in the palette stage of regeneration were injected intraperitoneally with thymidine-^{3}H (10 μc/gm body weight; 360 μC/μmole). Groups of 2 and 3 animals were denervated unilaterally 4, 5, and 6 days later. Two animals were not denervated. Seven days after the injection of thymidine the animals were sacrificed and the blastemas were assayed for acid-precipitable radioactivity and for protein content. The results are presented as a ratio of cpm/μg protein in control blastema (N) and denervated blastema (D).

Since direct measure of the pools of thymidine, uridine, and leucine would be difficult due to the small size of the blastema (approximately 0.1–0.2 mg of protein per blastema), we measured the effect of denervation on the stability of DNA; changes in pool size might be expected to result from the breakdown of preexisting macromolecules.

The DNA of the regenerating blastema were labeled by a single intraperitoneal injection of thymidine-^{3}H and after 5 days groups of animals were denervated for various periods of time, and the denervated and control arms were then assayed for radioactivity. Whereas a 4-fold decrease in capacity to incorporate thymidine-^{3}H had occurred 48 hours after denervation (Fig. 1), the content of labeled acid-insoluble DNA had decreased by only 15%. (Table 5).

DISCUSSION

The object of these studies was to analyze the early effects of denervation on the regeneration of limbs in adult newts. The blastemal cells of limbs in the early palette stage of regeneration show a high rate of mitosis (Hay and Fischman, 1961) and thus offered a convenient system for the quantitative study of changes in macromolecular syntheses upon denervation. Singer and Craven (1948) previously demonstrated that denervation of such limbs led to a decrease in mitosis (although regeneration is not totally suppressed), and more recently Robbins (1967) has reported that denervation led to a decreased capacity for thymidine incorporation, using radioautography of frog taste buds.

The data presented in this report show quantitatively that significant changes in macromolecular syntheses of the regenerating limb tissues of adult newts occur upon denervation of the limb. Within 7 hours after denervation the capacity of the blastema to incorporate labeled uridine begins to fall, while thymidine and leucine incorporation begin to decrease within 24 hours after denervation. These changes are not seen in sham-operated animals. Robbins (1967) in his experiments on frog taste buds demonstrated that the effect on thymidine incorporation was not attributable to overall changes in blood flow, and our experiments with sham-operated animals and with blastema in culture confirm this.

While the nerve thus functions to maintain a high level of RNA, DNA, and protein synthesis in regenerating tissue, the macromolecular synthetic capacities of nonregenerating tissues (of normal arms or of areas proximal to the blastema) are much less affected by the presence or absence of the nerve supply (Tables 2 and 3). The explanation may be a quantitative one, since some regenerating tissues contain more nerve components than do corresponding nonregenerating tissues (Singer, 1949). On the other hand, blastemal cells may react qualitatively differently to the nerve influence.

The effects of denervation on the incorporation of thymidine-^{3}H *in vivo* can be mimicked *in vitro*, enhancing the significance of the *in vivo* findings and lending support to the notion that the nerve has a specific biochemical effect on regeneration processes. These data do not tell us the nature of the nerve trophic agent, but they may provide a convenient system for the assay of this factor. Further investigations in this area are under way.

Denervation seems to lead to the inhibition of one facet in the control of macromolecular synthesis, for this operation reduces the synthetic capacity of the regenerating blastema to that of nonregenerating tissues.

SUMMARY

Denervation of the regenerating limbs of adult newts results in a time-dependent loss in the capacity to incorporate radioactive precursors of RNA, DNA, and protein in the regenerating tissues, but not in neighboring nonregenerative tissue. The capacity to incorporate radioactive uridine decreases within 7 hours after denervation, whereas thymidine and leucine incorporation are not affected until 24 hours after denervation.

The decrease in synthetic capacity of regenerating tissue can be reproduced in *in vitro* organ culture.

It is suggested that the nerve may exert a direct control over the synthesis of DNA, RNA, and protein in regenerating tissues.

The author wishes to express his gratitude to Dr. Jerome Gross for his support and intuition. The technical assistance of Miss G. Ingram is gratefully acknowledged.

REFERENCES

BRAY, G. A. (1960). A simple efficient liquid scintillator for counting aqueous solutions in a liquid scintillation counter. *Anal. Biochem.* **1,** 279.

BURTON, K. (1956). A study of the condition and mechanism of the diphenylamine reaction for the colorimetric estimation of deoxyribonucleic acid. *Biochem. J.* **62,** 315.

HAY, E. D., and FISCHMAN, D. A. (1961). Origin of the blastema in regenerating forelimbs of the newt *Triturus viridescens.* An autoradiographic study using tritiated thymidine to follow cell proliferation and migration. *Develop. Biol.* **3,** 26.

LOWRY, O. H., ROSEBROUGH, N. J., FARR, A. L., and RANDALL, R. J. (1951). Protein measurements with the Folin phenol reagent. *J. Biol. Chem.* **193,** 265.

NAGAI, Y., LAPIERE, C. M., and GROSS, J. (1966). Tadpole collagenase. Preparation and purification. *Biochemistry* **5,** 3123.

ROBBINS, N. (1967). The role of the nerve in maintenance of frog taste buds. *Exptl. Neurol.* **17,** 364.

SIMPSON, S. B., JR. (1961). Induction of limb regeneration in the lizard, *Lygosoma laberale,* by augmentation of the nerve supply. *Proc. Soc. Exptl. Biol. Med.* **107,** 108.

SINGER, M. (1946). The nervous system and regeneration of the forelimbs of adult *Triturus.* V. The influence of number of nerve fibers, including a quantitative study of limb innervation. *J. Exptl. Zool.* **101,** 299.

SINGER, M. (1949). The invasion of the epidermis of the regenerating forelimb of the urodele, *Triturus,* by nerve fibers. *J. Exptl. Zool.* **111,** 189.

SINGER, M. (1952). The influence of the nerve in regeneration of the amphibian extremity. *Quart. Rev. Biol.* **27,** 169.

SINGER, M. (1959). *In* "Regeneration in Vertebrates" (C. S. Thornton, ed.), p. 59. Univ. of Chicago Press, Chicago, Illinois.

SINGER, M. (1961). Induction of regeneration of body parts in the lizard, *Anolis. Proc. Soc. Exptl. Biol. Med.* **107,** 106.

SINGER, M., and CRAVEN, L. (1948). The growth and morphogenesis of the regenerating forelimb of adult *Triturus* following denervation at various stages of development. *J. Exptl. Zool.* **108,** 279.

STEPHENSON, N. G. (1966). Effects of temperature on reptilian and other cells. *J. Embryol. Exptl. Morphol.* **16,** 455.

Reprinted from *J. Morphol.*, **114**, 425–435 (1964)

Analysis of Tail Regeneration in the Lizard *Lygosoma laterale*

I. INITIATION OF REGENERATION AND CARTILAGE DIFFERENTIATION: THE ROLE OF EPENDYMA[1,2,3]

26

SIDNEY B. SIMPSON, JR.[4]
Department of Zoology, Tulane University, New Orleans, Louisiana

Lizards represent the highest group of vertebrates that have retained a considerable measure of regenerative ability, and even they are limited to an imperfect regeneration of the tail.

The lizard tail is an unique system in which to study regeneration. First of all, lizards are the only vertebrates that regenerate an appendage from a discrete preformed break (autotomy) plane. Secondly, the regenerate that forms after normal autotomy of the tail has two very distinct features: (1) the regenerate does not contain vertebrae; instead the vertebral column is replaced by an unsegmented cartilage tube (Duges, '29); (2) more notably, the regenerated spinal cord is imperfect, consisting only of descending fiber tracts from the old cord, the ependymal lining of the central canal and scattered glial cells (Hooker, '12).

Few experimental studies of lizard tail regeneration appear in the literature. Woodland ('20) was probably the first to perform significant experiments. He demonstrated that separation of the tail at the autotomy plane was not a prerequisite to regeneration. He also described the experimental production of bifid regenerates, some of which failed to differentiate cartilage tubes. Kamrin and Singer ('55) demonstrated the dependence of tail regeneration upon the presence of the spinal cord in the lizard *Anolis*. This observation was later confirmed for *Lygosoma* by Simpson ('62) and was extended to demonstrate that the spinal cord determines the position and orientation of the tail regenerate.

The present investigation was undertaken to determine: (1) which component

of the spinal cord is essential for the initiation of tail regeneration in the lizard and (2) which component of the regenerating spinal cord is responsible for the induction of cartilage in the lizard tail regenerate.

MATERIALS AND METHODS

Care of animals. The brown skink, *Lygosoma laterale*, was used exclusively in this study. This particular lizard was selected for several reasons: (1) it can be collected in large numbers throughout the year in the New Orleans area; (2) it is easily maintained for long periods of time in the laboratory and (3) the tails of over 50% of the animals collected from nature bear regenerated portions. All animals were collected from the Bonnet Carre spillway near Norco, Louisiana. A total of 152 lizards was used in these experiments. The lizards were maintained in individual glass stacking dishes provided with a circle of filter paper to absorb the feces. A small stendor dish of water was also provided. They were maintained at room temperature (70°–75°F) under controlled conditions of 12 hours light and 12 hours dark. They were fed a diet of meal worms which they readily accepted.

Histological techniques. The lizards were sacrificed at prescribed intervals and the tissues to be studied were fixed in

[1] A portion of a dissertation submitted to the Department of Zoology, Tulane University, New Orleans, Louisiana, in partial fulfillment of the requirements for the degree of Doctor of Philosophy in Zoology.

[2] This study was completed during the tenure of a National Institutes of Health Predoctoral Fellowship.

[3] This study was supported in part by a Public Health Service Research Grant (no. CA04511CB) from the National Cancer Institute to Dr. Merle Mizell, Zoology Department, Tulane University.

[4] Present address: Anatomy Department, Western Reserve Medical School, Cleveland, Ohio.

Bouin's fixative. Following a 24 hour period of fixation, the tissues were decalcified in "DECAL" and washed in running tap water for four hours. The tissues were then dehydrated, cleared in cedarwood oil and embedded *in vacuo* in 58°mp. Tissuemat. The embedded tissues were sectioned at 10 µ, mounted on slides and stained with Mayer's hemalum and eosin, Casson's modified Mallory or the Bodian stain for nerve fibers.

Operative techniques. Operative techniques used in the experimental series will be described in context.

I. *The role of the ependyma in the initiation of tail regeneration*

Spinal cord ablation. In order to test the importance of the nerve fibers contributed by the spinal cord, an attempt was made to reduce the nerve complement of the cord to a point where regeneration would not ensue.

Using operative techniques described by Simpson ('61) each lizard was curarized and pinned to the operating mat. Via a dorsal incision, the vertebral column was exposed for a distance of three vertebra lengths back from the autotomy surface. The roofs of the vertebrae were removed, thus producing a vertebral trough containing the spinal cord. Various portions of the exposed spinal cord were removed using glass needles. Following the removal of a portion of the cord, the remaining tissues were dusted with sulfadiazine and the incision closed with sutures.

Two experimental groups were prepared. The first group consisted of eight lizards in which the right half of the cord was removed for a distance of three vertebra lengths. This distance was chosen since preliminary experiments involving ten animals had shown that the cord must be removed for a distance of at least two and one-half vertebra lengths in order to prevent regrowth of the nerves and ependyma into the area of regeneration. The second group consisted of five lizards in which the dorsal half of the spinal cord was removed for the same distance. A control series was also prepared consisting of eight lizards in which the entire spinal cord was removed for a distance of three vertebra lengths. A group of five sham-operated animals

was prepared in which the spinal cord was merely split into right and left halves, lifted from the vertebral trough and then replaced. The experimental, sham-operated and control animals were maintained for a period of approximately 30 days and then sacrificed. The results of these operations are as follows:

Tail regeneration occurred normally in all of the sham-operated animals, but was effectively blocked in all of the controls. Examination of Bodian stained serial sections of these controls revealed that none of the descending nerve fibers from the severed end of the cord reached the autotomy surface. Furthermore, the ependymal lining of the central canal had not reached the autotomy surface. Histologically, the controls presented a typical picture of a non-regenerating structure. A cartilage callus had formed across the end of the terminal vertebra and a dense pad of scar tissue was interposed between the callus and the already differentiated epidermis. The failure of these controls to regenerate further substantiates the previous findings of Kamrin and Singer ('55) and Simpson ('62). Of more immediate importance however, is the fact that these results show that the descending nerve fibers of the cord cannot regenerate across a distance of three vertebra lengths in time to influence regeneration. Therefore, the removal of half of the cord in the experimental series effectively removed the nerve fiber contribution of that portion of the cord. In these experimentals, normal regenerates formed in every case. A study of the Bodian stained serial sections of the two experimental groups revealed that in 11 of the 13 cases, the number of nerve fibers contributed by the spinal cord was drastically reduced. Although only the right or dorsal half of the cord was surgically removed in each case, the actual amount of spinal cord remaining after degeneration of the wounded cord was considerably less than half. In most cases the motor perikaria within the remaining half of the cord degenerated, as did many of its nerve fibers. Although very little of the remaining half of the spinal cord persisted, a significant portion of the ependymal lining of the central canal was always present.

Thus we see that although the number of nerve fibers contributed by the spinal cord was drastically reduced in 11 of the cases, all developed normal tail regenerates. These results can be interpreted in at least two ways. First, if the nerve fibers of the spinal cord are of prime importance in regeneration, the threshold number for the lizard tail must indeed be very low. A second interpretation is that the nerves are not the critical factor in lizard tail regeneration and that some other factor within the spinal cord is important. It should be remembered that although the nerve fiber number was reduced in each of the previous cases, the ependymal lining of the central canal was always present.

Ependymal implantation. A second experiment was performed to determine whether or not the ependymal lining is capable of initiating tail regeneration. I was unable to remove the ependyma from the spinal cord of a *normal* lizard tail. However, the regenerated lizard tail contains an "imperfect spinal cord" in which the "dissection has been performed",by the process of regeneration. It is therefore possible to obtain, from the regenerated tail, ependyma that is free of nerve fibers and perikaria. By merely amputating a regenerated tail, the nerve fibers of the regenerated cord, in the amputated portion, are separated from their perikaria in the old spinal cord and therefore degenerate.

Thus portions of the regenerated cord obtained from the amputated portion consist of only the ependyma and scattered glial cells. A group of 13 lizards whose tails bore mature regenerated portions was collected and brought into the laboratory. In each case, the lizard was lightly curarized before the operation. The terminal half of the mature regenerated tail was removed (amputated), figure 1a. The cartilage tube containing the regenerated cord was dissected out and placed in cold Ringer's. The tube was then cut up into three 3 mm lengths, figure 1b. Next, using iridectomy scissors, three wounds were made in the dorsal epidermis of the tail. With the aid of a blunt probe the wounds were extended into the dorsal muscle masses, figure 1c. These wounds had the form of small, anteriorly directed channels in the muscle. Particular care was taken to insure that the vertebral column was not disturbed in any way. The previously prepared lengths of cartilage tube containing ependyma were then inserted into two of the formed channels, figure 1d; the remaining channel received a control implant which consisted of a length of cartilage tube from which the ependyma had been removed. Thus each animal received two implants containing ependyma and one without ependyma. It should be noted again that *no* intact nerve fibers were carried over with the ependyma-bearing implants since the nerve fibers associated

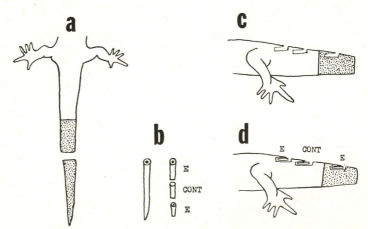

Fig. 1 Diagram of operation in which segments of cartilage tube containing ependyma (e) were transplanted into the dorsal muscle masses of the same animal (autografts). A control graft (cont) of cartilage tube alone was also transplanted. The stippled portion indicates the mature regenerated portion of the tail; see text for explanation.

with the ependyma had been severed from their perikaria in the old spinal cord.

A photograph of one of the lizards, taken 30 days after transplantation, is shown in figure 3. In all cases small tail regenerates were present in association with each of the implants *that contained ependyma. No* regenerates formed in association with the control implants of cartilage tube alone.

Examination of Bodian stained serial sections of these induced regenerates showed that, despite their reduced size, they were typical tail regenerates, figure. 4. They possessed well formed cartilage tubes and segmental muscle, figure 5. Their small size was probably due to the fact that the cells contributing to their formation came from the *de*differentiation of, at most, two dorsal muscle masses. Study of the Bodian stained serial sections also revealed that these regenerates contained very few nerve fibers. Obviously the implant could not contribute nerve fibers. The few nerve fibers that were present were peripheral fibers from the damaged muscle mass that had grown into the regenerate; and in only one case were these nerve fibers in juxtaposition with the ependyma of the regenerate. In this case only three nerve fibers were present. In no instance was the central axis of the lizard tail injured or involved in the induction of these regenerates.

Examination of Bodian stained serial sections of the control grafts of cartilage tube alone showed that no regenerative response had been induced by their presence. A cartilage callus was forming over the end of the implant and a dense pad of scar tissue had formed between the callus and the epidermis, figure 6. A number of nerve fibers from the wounded muscle mass had grown into the area, but no regeneration ensued.

The results just described, together with the results from the first experiment, strongly suggest that the ependymal lining of the central canal, and not the nerve fibers, is responsible for inducing regeneration of the tail in *Lygosoma*. However, a possible ancillary effect of the few nerve fibers in the graft areas cannot yet be completely ruled out. Experiments are in progress to clear up this point.

II. *The role of the regenerating ependyma in cartilage induction*

The first positive evidence that the regenerating ependyma was involved in the induction of the cartilage tube came from a study of five experimental animals from the two previous experimental groups in which half of the spinal cord was removed. In each of the five cases, either the dorsal or right half of the spinal cord had been removed for a distance of three vertebra lengths back from the autotomy surface.

These five animals not only regenerated a normally directed tail from the autotomy surface but also regenerated a second, dorsally directed tail from the juncture between the autotomy surface and the dorsal wound surface. Histological examination of these double regenerates revealed a very striking feature. In each case the regenerating ependyma had been deflected into the dorsally directed regenerate, thus leaving the normally directed regenerate without an ependyma. The regenerate receiving the ependyma was a normally constituted regenerate. However, the regenerate lacking an ependyma did not possess a cartilage tube. Except for the obvious absence of the cartilage tube this regenerate was, in all other aspects, normally constituted.

Study of the Bodian stained serial sections of these animals revealed that in three of the five cases, many descending nerve fibers from the old cord had grown out into both regenerates. This observation demonstrates that these nerve fibers, *per se*, are not capable of inducing cartilage in the regenerate. This observation further demonstrates the importance of the ependyma in the induction of cartilage.

The fact that one of the regenerates of each pair regenerated without an ependyma would seem to contradict the importance of the ependyma in the initiation of regeneration. However, one must remember that both regenerates developed from the same general area of regeneration. Also, the normally directed member (without ependyma) was always more advanced than the dorsally directed member (containing ependyma). The latter suggests that the deflection of the ependyma dorsally, occurred after the initiation of the normally directed member. It would ap-

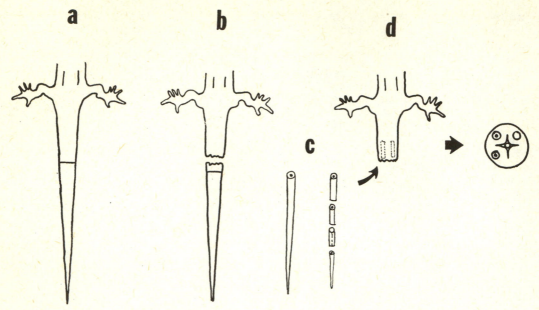

Fig. 2 Diagram of operation in which autografts of cartilage tube containing ependyma were transplanted into the dorsal, lateral or ventral muscle masses of the freshly autotomized tail; see text for explanation.

pear that once induced or initiated, the tail regenerate is no longer dependent on the ependyma for further growth.

The fact that the regenerates lacking cartilage tubes had normally constituted segmental muscle is also of interest. This suggests that differentiation of the cartilage and muscle axes are independent events governed by separate factors.

Since it was shown that the absence of the cartilage tube was associated with the absence of the ependyma, an experiment was designed to determine the effect of autoplastic ependyma transplants into freshly autotomized tails. Operations were performed in such a manner that additional lengths of ependyma were inserted into the dorsal, ventral and lateral muscle masses of freshly autotomized lizard tails. This was accomplished in the following manner:

Lizards whose tails bore a regenerated portion were collected from nature. In each case, the lizard was caused to autotomize its tail at a joint just anterior to the juncture of the native and regenerated portions, figure 2b. The cartilage tube, containing the ependyma (see page 427), was dissected from the regenerate,

placed in cold Ringer's solution, and cut into three 5 mm lengths, figure 2c. The pieces were lined up side by side in the dish so that their anterior-posterior axes would not be confused. Usually three of the pieces were placed into preformed channels in the muscle masses of the same animal (autografts), figure 2d. Two of the implants contained the ependyma, whereas the ependyma was removed from the third implant. The latter served as a control. The positions of the implants were rotated so that among the 14 lizards prepared in this way, experimental and control implants were placed in every combination of dorsal, lateral or ventral regions. It should be noted here that the experimental implants contained only the ependymal lining of the central canal and its associated glial component. The nerve fibers emanating from the old cord and associated with the ependyma degenerated within 24 hours, since their connections with the perikaria in the old cord were severed. The lizards were maintained for 30 days.

Study of Bodian stained serial sections of these regenerates revealed that in every case, accessory cartilage tubes were induced by the transplants of cartilage tube containing ependyma. No cartilage was induced in control transplants of cartilage tube alone. A typical example is shown in figures 7, 8 and 9. Study of the Bodian stained material also showed that no nerve fibers were associated with the ependyma transplants in 12 of the 14 cases. The regenerate in figure 10 illustrates the lack of nerve fibers, but the presence of ependyma and a resulting induced cartilage tube.

In two cases, a small number of peripheral nerve fibers from the muscle invaded the anterior end of the cartilage tube-ependyma implant. These fibers were closely associated with the ependyma and extended out into the regenerate along with the ependyma. However, in some cases nerve fibers also invaded the control implant of cartilage tube, continued through the tube and extended into the regenerate, yet no cartilage was induced in these cases.

The regenerating enpendyma from the implants, especially those positioned in the medio-lateral muscle masses, occupied an area of the regenerate that would normally have formed segmental muscle. However, under the influence of the ependyma, the blastema cells in this area differentiated into cartilage. The influence of the ependyma was so strong in most cases that only small, scattered strands of segmental muscle formed between the induced cartilage tube and the dermis. In some cases the ependyma grew out in such close contact with the wound epithelium that no muscle formed between the induced cartilage tube and the dermis. Moreover, the surface of the induced cartilage tube, bordering on the dermis, was itself often incomplete in several areas.

These results demonstrate that the regenerating ependyma and/or its associated glial cells are responsible for the induction of the cartilage tube in the lizard tail regenerate.

DISCUSSION

Role of the ependyma in the initiation of tail regeneration

The importance of the spinal cord in regeneration of the salamander tail (S. Holtzer, '56) and lizards (Kamrin and Singer, '55; Simpson, '62) has been well documented. However the exact nature of its influence has not been critically tested. The failure of lizards tails to regenerate after removal of the spinal cord has been interpreted to be due to the removal of the nerve complement supplied by the cord (Kamrin and Singer, '55; and Simpson, '62). However, experiments of this nature demonstrate only that some portion of the spinal cord is necessary if regeneration is to ensue.

Results obtained in the present study tend to deemphasize the importance of the nerves and at the same time suggest that the ependymal lining of the central canal is the influential factor. The fact that tail regeneration can proceed normaly in *Lygosoma* after removal of as much as three-quarters of the spinal cord casts serious doubt on the idea of the importance of nerves in this system. Recent experiments dealing with the role of the spinal cord in larval salamander tail regeneration (Kiortsis and Droin, '61) suggest that the inductive action exerted by the spinal cord is specific. They found that neither transplants of spinal ganglia nor the nerve fiber complement of two deviated sciatic nerves could take the place of the spinal cord in inducing tail regeneration.

Evidence obtained from experiments in the present study strongly suggests that the ependymal lining of the central canal and/or its associated glial cells is responsible for the initiation of tail regeneration in *Lygosoma*. In all cases where tail regeneration proceeded after removal of the majority of the spinal cord, the ependyma was present. More critical proof of the importance of the ependyma in the initiation of tail regeneration was demonstrated in the transplantation experiments. Implantation of pieces of cartilage tube containing ependyma into the dorsum of the tail resulted in the induction of small but complete tail regenerates in every case, whereas no regenerates were induced by implantation of the cartilage tube alone.

Role of the ependyma in cartilage induction

The involvement of the spinal cord in axial cartilage differentiation has been well documented in: (1) salamander embryos (Holtzer and Detwiler, '53) and regenerating salamander tails (S. Holtzer, '56); (2) the developing chick embryo (Strudel, '53; Watterson et al., '54); and (3) the developing mouse embryo (Grobstein and Parker '54; Grobstein and Holtzer, '55). In all of the preceding cases, the primary inductive activity has been shown to reside in the ventral half of the developing or regenerating spinal cord.

The present study demonstrates the importance of the regenerating spinal cord in the induction of the cartilage tube of the lizard tail regenerate and furthermore demonstrates that the inductive activity resides in the ependymal lining of the central canal and/or its associated glial cell component. In the case of five experimentally produced *bifid* tail regenerates, the regenerating ependymal lining of the central canal of the spinal cord extended into only one of the regenerates. In each case the member of the pair receiving the ependyma possessed a cartilage tube, while the other member lacking an ependyma possessed no cartilage tube. Also in each case, although the cartilage tube was missing, the arrangement of the segmental muscle was quite normal, suggesting that differentiation of the two systems is controlled by different factors.

Woodland ('20) described the experimental production of three bifid tail regenerates in the lizard *Hemidactylus*. In each case he found that one of the regenerates was missing the cartilage tube and its contained ependyma.

The results obtained in the transplantation studies demonstrate in an even more dramatic way the importance of the ependyma and/or its glial cell component in the induction of cartilage. An explanation of the specific nature of the influence exerted by the ependyma must await further experimentation.

SUMMARY

1. Reduction of the nerve fiber contribution of the spinal cord, via destruction of approximately three-quarters of the cord, does not prevent regeneration of the tail. Moreover, supernumerary tail regenerates were induced by "nerve-free" ependyma autografts. These inductions took place in host areas which contained few nerve fibers. Therefore this study deemphasizes the importance of nerve fibers in the initiation of lizard tail regeneration. At the same time this study suggests that it is the ependymal lining of the central canal which is the primary initiator of tail regeneration in this lizard.

2. Failure of a regenerate to differentiate a cartilage tube is always associated with the absence of a regenerated ependymal lining of the central canal.

3. Autoplastic transplantation of additional lengths of ependyma into a freshly autotomized lizard tail results in the induction of accessory cartilage tubes in the ensuing regenerate.

4. This study clearly demonstrates that the ependymal lining of the central canal and/or its associated glial cells is responsible for cartilage differentiation in lizard tail regeneration.

ACKNOWLEDGMENT

I wish to express my appreciation to Dr. Merle Mizell, for his continuing interest in this study and for the many stimulating discussions from which came many helpful suggestions and criticisms.

LITERATURE CITED

Duges, A. 1829 Mémoire sur les espèces du genre *Lacerta*. Ann. Sci. Natl., 16: 337–389.

Grobstein, C., and G. Parker 1954 *In vitro* induction of cartilage in mouse somite mesoderm by embryonic spinal cord. Proc. Soc. Exp. Biol. and Med., 85: 447–481.

Grobstein, C., and H. Holtzer 1955 *In vitro* studies of cartilage induction in mouse somite mesoderm. J. Exp. Zool., 128: 333–357.

Holtzer, H., and S. Detwiler 1953 An experimental analysis of the development of the spinal column. III. Induction of skeletogenous cells. J. Exp. Zool., 123: 335–369.

Holtzer, S. 1956 The inductive activity of the spinal cord in urodele tail regeneration. J. Morph., 99: 1–40.

Hooker, D. 1912 Die Nerven im regenerieten Schwanz der Eidechsen. Arch. Mikr. Anat., 80: 217–222.

Kamrin, R. P., and M. Singer 1955 The influence of the spinal cord in regeneration of the tail of the lizard, *Anolis carolinensis*. J. Exp. Zool., 128: 611–627.

Kiortsis, V., and A. Droin 1961 La régénération caudale des urodeles (induction et réactivité du territoire). J. Embryol. Exp. Morph., 9: 77–96.

Simpson, S. B. 1961 Induction of limb regeneration in the lizard, *Lygosoma laterale*, by the augmentation of the nerve supply, Proc. Soc. Exp. Biol. and Med., *107:* 108–111.

——— 1962 Regeneration of lizard extremities. Masters Thesis, Tulane University, New Orleans, La.

Strudel, G. 1953 L'influence morphogene du tube nerveux sur la differenciation de la coloone vertebrale. Comp. rend. Soc. Biol., *147:* 132–133.

Watterson, R. et al. 1954 The role of the neural tube and notochord in development of the axial skeleton of the chick. Amer. J. Anat., *95:* 337–400.

Woodland, W. N. F. 1920 Some observations on caudal anatomy and regeneration in the gecko (*Hemidactylus flaviviridis,* Ruppel), with notes on the tails of *Sphenodon and Pygopus.* Quart. J. Micr. Science, *65:* 64–100.

PLATE 1

EXPLANATION OF FIGURES

3　Photograph of a liazrd 30 days after receiving two dorsal transplants of cartilage tube with ependyma and one dorsal graft of cartilage tube alone. Regenerates (1 and 2) developed in conjunction with the grafts of cartilage tube containing ependyma. No regenerate developed in connection with the control graft (arrow) of cartilage tube alone.

4　Sagittal section through the induced regenerate, labeled (2) in figure 3. This regenerate was induced in the regenerated portion of the tail. The ependyma (ep) of the implanted piece of cartilage tube (im) can be seen extending into the cartilage tube of the induced tail. Bodian stain, 35 ×.

5　Longitudinal section through the same induced regenerate as shown in figure 4, showing the differentiating cartilage tube (dct) and segmental muscle (m) of the induced regenerate. 125 ×.

6　Sagittal section through the control implant of the same tail as shown in figure 3. No regeneration was induced by the control implant of cartilage tube alone. Cartilage callus (cc) is forming over the end of the cartilage tube, and a pad of scar tissue (st) has formed between the callus and the epidermis (e). Bodian stain, 125 ×.

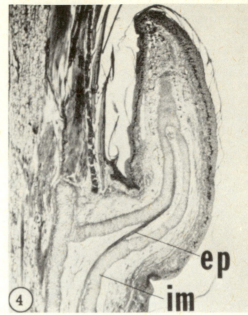

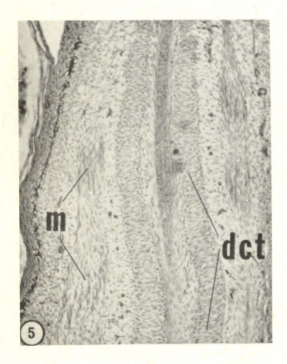

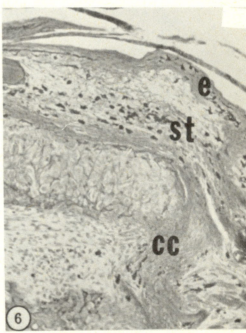

433

PLATE 2

7 Longitudinal section through experimental and control implants of cartilage tube-ependyma and cartilage tube alone. A differentiating cartilage tube has formed in conjunction with the ependyma (ep)-bearing implant. An arrow marks the juncture between the cartilage of the implant and the induced cartilage of the regenerate. No cartilage tube was induced in conjunction with the control implant of cartilage tube alone (c). Bodian stain, 125 ×.

8 Higher magnification of the experimental implant shown in figure 7, showing the ependyma (ep) and the absence of nerve fibers. 250 ×.

9 Higher magnification of the control implant shown in figure 7, showing the absence of the ependyma as well as the absence of an induced cartilage tube. 250 ×.

10 Longitudinal section through a regenerated tail showing tangential cuts through an induced cartilage tube (ict), and the cartilage tube of the normal axis (ct). Note that the induced cartilage tube contains only the ependyma (ep). No nerve fibers are present. Nerve fibers (arrows) as well as ependyma can be seen inside the cartilage tube of the normal axis. Bodian stain, 125 ×.

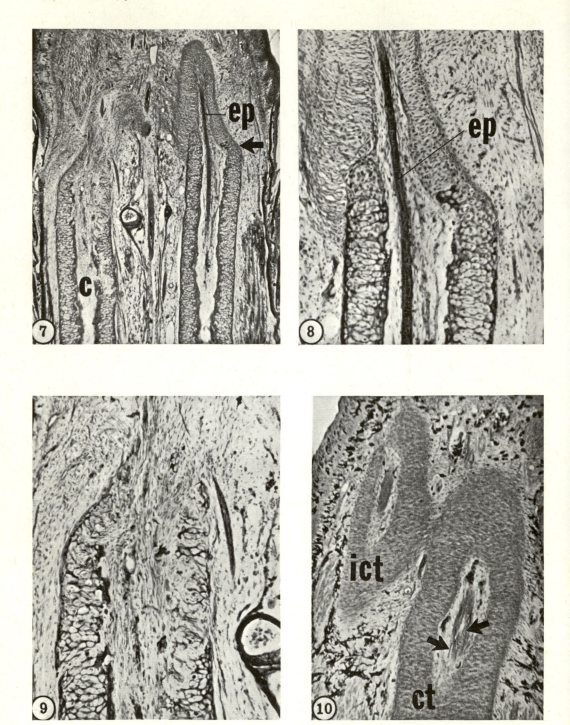

Mammalian Regeneration

IV

Editors' Comments on Papers 27, 28, and 29

27 Goss: *Experimental Investigations of Morphogenesis in the Growing Antler*

28 Carlson: *Relationship Between the Tissue and Epimorphic Regeneration of Muscles*

29 Mizell: *Limb Regeneration: Induction in the Newborn Opossum*

This aspect of regeneration is the most provocative of all, both to most workers in the field (particularly when not speaking for publication) and certainly to our patrons in general. It is somewhat annoying to us, as members of the class under discussion, to realize that our evolutionary paths have led us to a body whose regenerative powers are feeble indeed when compared to some "lower" vertebrates. Mammalian skin, bone, and tendons can repair themselves in an effective manner although the result is usually rather unesthetic. Those processes seem better described by the term "healing" than by "regeneration." Liver, too, can "regenerate," but again this involves more the accretion of cells of the same basic type than the demonstration of the complexities involved in the production of a new and wholly functional limb. Even so, regeneration of internal organs has received relatively little attention. Possibly the upset in homeostatic mechanisms resulting from pervasive loss of function of a vital organ has usually been so severe that whatever value the upset may have had as a signal to regenerate has been exceeded to the point where the organism dies from secondary effects.

Antlers are shed and regrown, although the loss of them is made good the next year. In this case, the "amputation plane" activates itself and is always at the same place, like deciduous leaves. The experiments discussed in the paper by Richard Goss indicates that morphogenetic fields are present, as they are in most other developing systems. The causative agents of such fields have been discovered only in a few relatively simple organisms, and their dissection in more complex organisms is one of the central questions of the day.

The extent of healing in mammalian muscle has been explored by Bruce Carlson, himself an M.D.–Ph.D. This "tissue regeneration" in muscle is demonstrated by excising a muscle and implanting in its place a wad of minced muscle. The resulting structure, while ill-formed and weak, is obviously a functional muscle, with attachments to bone. Carlson cites work indicating that cartilage can be induced to form in mammals, and indeed, as the ability to regenerate limbs is progressively lost in maturing clawed frogs, cartilage spikes grow on amputation surfaces long past the time that muscular elements have ceased to reappear. It is possible to speculate that restitution of lost limbs of mammals may finally come about by a combination of stimulation of cartilagenous elements, along with variations stemming from Carlson's work on tissue regeneration.

A most intriguing demonstration is M. Mizell's nerve-mediated production of a very good new limb from an amputation surface on the hind limb of a neonatal opossum. While the tissue was very young at the time the amputations were performed, the resulting limbs do lend obvious encouragement to those who are committed to continuing research in this area.

Reprinted from *J. Embryol. Exptl. Morphol.*, **9**, 342–354 (1961)

Experimental Investigations of Morphogenesis in the Growing Antler

27

by R. J. GOSS[1]

From The Department of Biology, Brown University, Providence, Rhode Island

WITH FIVE PLATES

INTRODUCTION

D'ARCY THOMPSON (1942) described the form of antlers as having developed in a two-dimensional pattern which, during growth, may have become more or less distorted depending on the species. In some (e.g. moose, fallow deer), the antlers may exhibit a palmate configuration; in most deer, however, they are branched structures formed by the repeated two-dimensional bifurcation of the original outgrowth. This process gives rise to a series of tines which vary in number according to age and species. In the sika deer, the first set of antlers, produced in yearlings, are unbranched spikes which may grow as much as 6 inches in length. The following year, these are replaced by branched antlers usually having three points each. Mature bucks ordinarily possess 4 points per antler.

The annual growth cycle of antlers has been thoroughly documented by Waldo & Wislocki (1951), and Wislocki (1956) in the Virginia deer. In the sika deer it is briefly as follows. Shedding of the old antlers occurs in the spring, having been preceded by the local swelling of the skin of the pedicle immediately below the base of the antler. There is reason to believe that the incipient growth of the new antler is somehow instrumental in bringing about the loss of the old. Following autotomy, a scab forms on the stump of the pedicle, and wound healing ensues by the migration of epidermis and subjacent connective tissue from the margins. It is from this undifferentiated tissue derived from the dermis of the pedicle skin that the new antler bud appears to develop. Subsequent invasion and proliferation of such cells results in the formation, within a few weeks, of a rounded knob growing on top of the pedicle. This structure branches dichotomously as it grows throughout the spring and summer, attaining its full size by late August. Growth occurs almost exclusively at the apical end of each branch, where there exists a mass of undifferentiated cells resembling the blastema of regenerating structures in lower vertebrates. In older parts of the growing antler, located progressively more proximal, are zones of chondrification

[1] *Author's address*: Department of Biology, Brown University, Providence 12, Rhode Island, U.S.A.

[J. Embryol. exp. Morph. Vol. 9, Part 2, pp. 342–354, June 1961]

and ossification which are honeycombed by numerous small blood-vessels. The whole antler is enveloped in a richly vascularized layer of skin, and is said to be 'in velvet' because of the numerous hair follicles which are formed *de novo* throughout the growth period (Billingham, Mangold, & Silvers, 1959). Once the ultimate size of the antler is attained and further growth ceases, the entire inner parts of the antler become converted to bone which is solid except for the many small blood-vessels which permeate it. At the end of the summer, apparently due to vascular restrictions, the entire antler dies and the investing layer of desiccated skin is rubbed off by the animal, exposing the bare bony antlers which persist until the following spring when a new cycle is initiated.

It is natural to regard the growing antler as a developing system analogous to certain embryological structures, or comparable to the regenerating appendages encountered among the cold-blooded vertebrates. Inasmuch as the latter systems, during their formative periods, tend to exhibit varying degrees of regulatory capabilities, it was of interest to determine to what extent the antler can adjust its morphogenesis to compensate for various defects surgically inflicted at different periods in the growth cycle. Specifically, the experiments to be described below were devised to provide information concerning (*a*) the capacity of the antler to exhibit morphogenetic regulation, (*b*) the degree to which such abilities may be restricted to certain phases of the growth cycle, and (*c*) the extent of the antler 'territory', i.e. the distribution of tissues potentially capable of participating in antler development.

MATERIALS AND METHODS

All of the experiments in the present account were performed on the sika deer (*Cervus nippon*), a species which has been bred in captivity for many years and is easier to work with and to obtain than the native Virginia deer. Animals were maintained either in individual enclosures or in small groups confined in larger pens.

In order to anesthetize animals in preparation for operations, equipment manufactured by the Palmer Chemical and Equipment Company, of Atlanta, Georgia, was used. This consisted of a 'Cap-chur' rifle, powered by compressed carbon dioxide, which shoots a projectile designed to inject a drug of choice automatically into the muscle of the deer. In these cases, Anectine (generously donated by Burroughs, Wellcome & Co.) was used in doses of approximately 6 to 8 mg./100 lb. body-weight. Such an intramuscular dose took effect in about 10 minutes and kept the animal unconscious for approximately half an hour. In addition to this general anesthesia, the local area of the pedicle and antler was injected subcutaneously with 1 ml. of 2 per cent. Xylocaine hydrochloride to render the areas to be operated insensitive to pain. Although operations were performed with non-sterile, but clean, instruments and without benefit of antibiotics, no infections occurred. Surgery was accomplished with scalpels in most instances, and with a hacksaw where it was necessary to cut through bone.

Bleeding was often profuse at first, but always subsided within minutes. Wounds resulting from such operations became covered by scabs and healed promptly. All operations were performed on the left antlers only in order to reserve the right antlers as normal controls. Further details of specific operations are presented in the appropriate context below.

The progress of experiments on individual animals was recorded periodically by careful examination of anesthetized deer and photographs of the regions under observation. Operations were performed in the springs of 1959 and 1960, and the final results obtained late in the respective summers.

RESULTS

Because of the difficulty and expense involved in acquiring and maintaining deer, it is not always feasible to repeat experiments on numbers of animals. Since most of the various operations described below were performed on single animals, the results obtained cannot be regarded as being as reliable as one might prefer. However, in those cases in which similar experiments were carried out on more than one deer the results were highly consistent, arguing in favor of the validity of other single experiments.

Total transverse amputation

Shedding of the dead antler in the spring may be regarded as an example of autotomy which is followed by the regeneration of a new antler from the pedicle. Obviously, therefore, the greatest antler-producing potential resides in the tissues of the pedicle immediately proximal to the level at which the old antler breaks off. Since this natural case of amputation is followed by the production of a new antler, it is of some interest to explore the antler-producing capacities of other regions, namely, the more proximal parts of the pedicle, as well as the growing antler itself.

Inasmuch as the developing antler appears to elongate by virtue of the growth of an apically located mass of undifferentiated cells, experiments were performed to determine the extent to which the antler can compensate for the loss of this growing tip. In two deer, this part of the antler was amputated transversely about 1 cm. below the apex. In one animal (deer 36), the end of a 6-inch bifurcate antler was removed on June 7 (Plate 1, fig. A). This operation effectively interfered with further development, except for the production of a tapered 2-inch outgrowth formed from the stump (Plate 1, fig. B). The opposite antler became a normal 3-pointed structure nearly twice as long as the operated one. In a younger animal (deer 9) the tip of a growing antler about 1·5 inches long was amputated on July 23 (Plate 1, fig. C). This antler was able to produce only a pointed outgrowth less than 2 inches in length (Plate 1, fig. D), in contrast to the unoperated contralateral antler which had two points and was three times the length of the experimental one. These experiments testify to the importance of the growing tip in ensuring normal elongation and morphogenesis in the antler. Yet because some growth did occur in both cases despite the removal of the terminal end of the antler, it is apparent that the antler is capable of compensating to a very limited extent for the loss of its growing tip. These results are best explained on the basis of the distribution of undifferentiated, proliferating cells in the growing region of the antler. Largely concentrated at the very apex, significant amounts of such tissue also extend down the sides of the antler. The level of amputation was sufficiently distal to leave behind some of this marginal tissue, which was apparently capable of subsequent reorganization and limited resumption of growth. Indeed, injury to this side of an antler regularly brings about the formation of exostoses, and Bubenik (1959) described the growth of a new side branch in a reindeer antler after the original one had been broken off.

Amputation of the pedicle results in less serious disturbances to the growth of antlers. Normal antler production following loss of both antler and pedicle has been reported in the moose (Jaczewski, 1954) and in the red deer (Jaczewski, 1955). In the present investigation, two deer were subjected to pedicle amputation. One animal (deer 19) was nearly 2 years old and had short, 1-inch spike antlers which had not yet been shed. Amputation was performed by sawing through the middle of the left pedicle on 18 March, which was 77 days before the right antler was normally shed (June 3). The operated pedicle failed to heal until after the opposite antler was lost. Thus, for nearly 3 months the severed pedicle bone protruded as a bare stump, although the skin on the sides of the pedicle exhibited the usual tumescence characteristic of the preparatory changes in that region prior to actual antler growth (Plate 1, fig. E). After shedding of the old antler on the right side, healing of the pedicle occurred normally on both sides, and identical normal antlers developed from both pedicles (Plate 1, fig. F). In deer 38 the left pedicle was completely removed at the level of its basal attachment to the skull on 7 June (Plate 1, fig. G), which was less than 1 week after the old antlers had been shed. The wound healed normally and then proceeded to give rise to an antler which appeared normal except that its development lagged behind that of the normal right antler. By 26 August, when the right antler had attained its full size and form (3 points), the left antler had achieved a length equal to half that of the control, and exhibited every indication of still being in an active state of growth (Plate 1, fig. H). At its tip it was rounded and soft, and was beginning to branch. Further growth resulted in a terminal bifurcation. On 24 September it was in the process of shedding its skin, several weeks after the control antler had done so.

Partial excision of antler tissues

In order to investigate the possibility of morphogenetic regulation in the antler following surgically inflicted defects, a series of experiments was performed in which portions of pedicles or of antler buds were removed at various

times before or after shedding and initiation of antler growth. Future development was observed to determine the extent to which the morphology of the
resultant antler reflected the effects of the original operation.

In one group of animals the posterior half of the antler bud or pedicle was
extirpated. When the posterior half of a young antler about 1 inch in height was
removed on 23 April (deer 4: Plate 2, fig. A), growth continued anteriorly to
produce a normal brow tine, but the development of the rest of the antler was
seriously disturbed. By 21 May, when the main branch of the control antler was
elongating rapidly (having attained a length of about 6 inches), the operated
antler had produced an abortive outgrowth directed posteriorly (Plate 2, fig. B).
This structure continued to grow throughout the summer at a very slow rate,
eventually becoming about 3 inches long (Plate 2, fig. C). On the opposite side,
a large normal 3-point antler was formed. A similar operation on another
younger animal (deer 32) was performed at the same stage of development. The
posterior part of the young bud was excised on 3 June. In this case, less tissue
was removed than in the previous example (cf. Plate 2, figs. A and D). Two
weeks later the brow tine had elongated considerably, but there were no longer
indications of growth of the rest of the antler from the posterior region (Plate 2,
fig. E). At the end of the summer, only a very small (1-inch) spur had formed
posteriorly, while the anterior brow tine had grown to a length of some 6 inches
(Plate 2, fig. F), nearly twice as long as the corresponding tine on the opposite
normal 3-point antler.

Removal of the posterior half of an even younger antler primordium was
performed on deer 22. The left antler had been shed less than 1 week previously
and the pedicle stump had healed over but was still flattened on the end. Only
the soft tissues of the incipient antler plus part of the adjacent pedicle skin were
removed on 20 April (Plate 3, fig. A). Healing of the wound was followed by
the posteriorly directed growth of the remaining anterior part of the antler
bud (Plate 3, fig. B). After 1 month, a large rounded mass of tissue had formed
which proceeded to become elevated posteriorly (Plate 3, fig. C). There were no
indications of the formation of a brow tine, but this was also lacking in the
opposite control antler, except for an extremely diminutive protuberance. The
posterior outgrowth from the left antler had become nearly 2 inches long 6
weeks after the operation (Plate 3, fig. D) and continued to grow throughout
the summer. It eventually grew to a length of approximately 5 inches (Plate 3,
fig. E), which was about one-third the length of the control antler.

A final experiment in this series involved the removal of the distal, posterior
half of the left pedicle of deer 25 on 29 April, about 3 weeks prior to shedding.
Wound healing occurred normally after the operation (Plate 3, fig. F) as well as
after shedding. The antler bud formed on the experimental side (Plate 3, fig. G)
was originally semicircular in shape but gradually assumed a more symmetrical
outline as tissue invaded what had once been the posterior side of the pedicle. It
then grew into a normal 3-point antler during the course of the summer (Plate 3,

fig. H). Thus, with respect to the posterior half of the antler, it would appear that the earlier the defect is inflicted, the more complete is the morphogenetic regulation.

Additional studies of the regulatory capacities of the pedicle were conducted on 3 deer. In one animal (deer 31), the anterior half of the pedicle was removed on the day (21 May) that the old antler was shed (Plate 4, fig. A). The right antler had been lost the day before. After 2 weeks, wound healing had taken place and a rudimentary antler had formed. Four weeks after the operation the young antler had become about 2 inches long and had partly filled in the missing anterior half of the pedicle (Plate 4, fig. B). Subsequent growth resulted in the production of a nearly normal antler (Plate 4, fig. C). Although it lacked a brow tine, and was slightly shorter than the control at its branched end, it is clear that the absence of the anterior half of the pedicle causes only minor disturbances in the development of the antler. In deer 28 the median half of the pedicle was ablated (21 May) a few days after shedding of the old antler (Plate 4, fig. D). From this deficient pedicle there developed a 3-point antler of normal size and proportions (Plate 4, fig. F). The brow tine was absent, but in this particular animal the control antler also lacked this branch.

Although excision of the anterior, posterior, and median halves of the pedicle had little or no effect on the final outcome of antler development, removal of the lateral half of the pedicle was responsible for the complete failure of antler renewal. In deer 30, which had lost its left antler less than a week before, the lateral half of the incipient antler bud, along with the pedicle, was removed on 21 May (Plate 4, fig. G). Although healing occurred as in previously described cases, the residual tissues of the early antler bud failed to grow any farther, and at the end of the summer only a rounded remnant of the bud persisted on the pedicle (Plate 4, fig. H). From this result it would appear that the lateral portion of the antler bud and/or pedicle is particularly important for the production of a normal antler, a conclusion which is substantiated by the experiment described below.

To learn to what extent a young antler bud is morphogenetically determined, an anterior-posterior wedge of tissue was excised from the middle of a bud 1 inch high (deer 5, operated 23 April). Although this antler bud was thus split into median and lateral halves (Plate 5, fig. E), an antler grew only from the lateral half. The other side of the original bud remained only as a small excrescence on the median basal region of the otherwise normal 4-point antler (Plate 5, fig. F). This experiment demonstrates that the lateral half of the young antler bud is capable of giving rise to a complete antler, but that the median half is not. It had been expected that if the bud were split into two halves, there would grow a set of double antlers. Since this did not occur the first time, the experiment was repeated on another animal (deer 27) the following year. This time the anterior-posterior cleft in the 1-inch high antler bud was made just lateral to the mid-line on 6 May (Plate 5, fig. A). A week later healing was complete and the median and lateral halves of the divided antler were developing independently (Plate 5, fig. B). Already, however, bifurcation was evident in the lateral half, while the median portion was deficient in this regard. After 4 weeks (Plate 5, fig. C), the two antlers had grown to a height of approximately 6 inches, and by the end of the summer (Plate 5, fig. D) they had become almost as long as the normal 4-point antler produced on the right side. Neither the left median nor the left lateral antler had become branched at their upper ends, a result possibly attributable to the mutual reduction in the total amount of formative material available to each half. Although both sets of antlers were abnormal, it is noteworthy that the lateral one was the more complete inasmuch as it had a brow tine. This again emphasizes the dominant nature of the lateral portion of the young antler bud.

A final experiment along similar lines involved the cleavage of an antler bud in the median-lateral direction (Plate 5, fig. G). This operation was performed on deer 35 on June 7, and served to emphasize precociously the natural bifurcation which normally gives rise to the brow tine anteriorly and the rest of the antler posteriorly. Thus, this kind of a defect had no observable effect on the future development of the antler. In this deer, both left and right sides produced normal and equal 3-point antlers (Plate 5, fig. H). Similar experiments on roe bucks by Bubenik & Pavlansky (1959), however, yielded duplicate, albeit abnormal, antlers.

DISCUSSION

The experimental results herein related demonstrate that the developing antler is capable of morphogenetic regulation following injury, but that this is profoundly affected by the stage of antler development at the time of operation as well as the location of the defect. Well-developed antlers in which growth is still in progress are especially susceptible to injury and incapable of compensating for loss of the growing tip. Nevertheless, if part of the undifferentiated apex of an antler remains (as was the case in the present studies), it is able to continue growth but to a very limited extent. In this respect, it would be of interest to remove just half of the tip of an advanced antler to determine if this would affect subsequent morphogenesis in the same way that excision of parts of the young antler bud does. In the latter instance, removal of the posterior half of the early antler bud (1 inch long) resulted in serious deficiencies in the future growth of the antler. Regulation at this stage is practically absent. The comparable experiment performed on an even younger bud which had not yet become rounded, however, was followed by a significant degree of morphogenetic adjustment. Indeed, when the posterior half of the pedicle itself is excised before loss of the old antler, there develops a completely normal antler. Similarly, removal of the anterior or median halves of pedicles around the time of antler shedding does not have serious consequences on the future course of antler development.

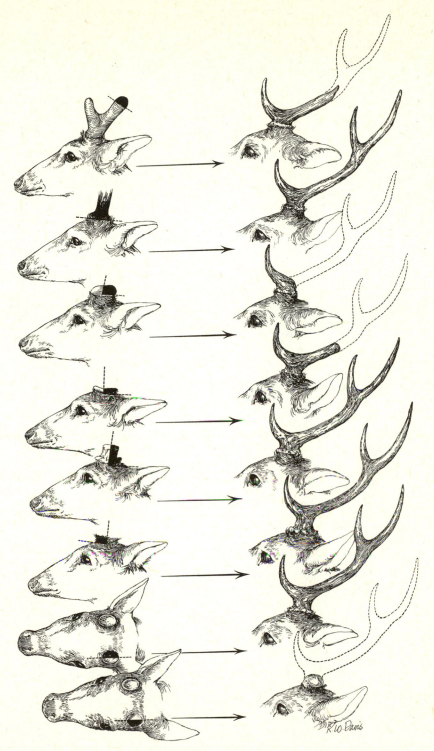

TEXT-FIG. 1. Illustrative summary of certain of the experiments and their results described more fully in text. Blackened areas on left designate portions of antlers and/or pedicles removed. On the right are diagrammatically illustrated the resulting extents of antler growth and development in relation to normal control antler structures (broken lines).

Thus, with the exception to be noted below, the antler pedicle is capable of considerable regulatory potentialities before shedding or during the very early phases of antler growth. How long after this it retains such capacities has not been determined as yet. The incipient antler can likewise compensate for experimental injuries, but as growth proceeds this capacity rapidly diminishes.

The lateral side of the pedicle appears to be particularly important in antler production. Perhaps this is related to the fact that antlers normally grow at a laterally directed angle from the head of the deer. Not only can the lateral half of the pedicle alone give rise to a complete and normal antler, as already mentioned above, but also the lateral half of the 1-inch antler bud can develop into an entire structure, something which the anterior half cannot do. Conversely, if the lateral half of the pedicle is removed, no antler develops. Removal of the posterior, anterior, or median halves of pedicles permits typical antler growth. Judging from these results, the greatest morphogenetic potential of the pedicle and the antler bud is concentrated in the lateral region. Nevertheless, this area is capable of being subdivided, for double antlers can be produced experimentally by longitudinally cleaving the antler bud lateral to the centre. Naturally occurring instances of antler duplication have been reported in the literature. Bland-Sutton (1890) mentioned a case of dichotomy in moose antlers, and Kitchener (1954) described an antler of a Malayan Sambar deer which possessed a double brow tine. When duplication of a structure is possible, the opposite phenomenon of unification may also occur. Thus, Dove (1936) was able to produce an artificial unicorn in an Ayrshire bull by fusing horn buds in the calf. Tegner (1954), Fooks (1955), and Whitehead (1955a) described partially coalesced antlers in roe deer. Inasmuch as duplication and fusion have been shown to occur, it would appear that antlers (and horns) develop under the influence of morphogenetic fields.

The primary site of antler production resides in the pedicle. Under normal circumstances the distal region of the pedicle provides the requisite materials for antler formation. This usually follows loss of the old antler, but antler renewal will occur even if the shedding of the old one is delayed. In this eventuality, growth occurs to the side of the persistent old antler, usually in a lateral direction. Accessory antler production from the side of the pedicle has occasionally been reported (Gadow, 1902; Rhumbler, 1916; Whitehead, 1955b), and is probably the result of injury to the pedicle. Even removal of the entire pedicle is not sufficient to preclude antler production. The limits of the 'antler territory' have not yet been determined.

The regeneration of the antler is somewhat unique since the structure formed is not always an exact replica of that which is replaced. As the deer matures, successive sets of antlers may develop to increasingly elaborate extents, from a single spike in the young buck to repeatedly branched racks in mature stags. It is of interest to correlate several sets of facts related to this phenomenon. The larger the animal, for example, the more highly branched are its antlers. This is correlated with the greater diameter of the pedicle supporting such antlers. These large antlers are shed earlier than are smaller antlers in young deer. This early loss of the larger antlers is followed by the precocious initiation of new antler development and a longer growing season which is required for the production of such a massive amount of tissue. From these observations it is tempting to speculate that the degree of morphological development of the antler (i.e. the number of branches) may be solely related to the amount of tissue initially available in the young, undifferentiated antler bud. This, in turn, is obviously related to the diameter of the pedicle and the age of the animal. As development progresses and bifurcation occurs, the sizes of the growing apices diminish and further branching is correspondingly reduced. When the young antler bud is bisected (as in deer 27) double antlers are produced which fail to branch terminally. Perhaps this is due to the mutual reduction in the amount of formative material in the two halves. Further experiments designed to test the possible relation between mass of available tissue and degree of morphogenetic expression are in order.

SUMMARY

1. Investigations of the morphogenesis of antler growth have been undertaken in the sika deer (*Cervus nippon*). Operations were performed on the left antlers, while the intact right antlers served as controls.

2. Amputation of the terminal centimetre of growing antlers seriously impedes further development except for the production of an abortive outgrowth from the stump. By contrast, transverse amputation through the middle of the pedicle is followed by normal antler formation. Removal of the entire pedicle permits antler development to occur, although at a retarded rate. Therefore, antler-forming potentiality resides in the growing tips of the elongating antler as well as in the tissues of the entire pedicle.

3. The extent of morphogenetic regulation for defects surgically inflicted on the antler bud or pedicle depends on the age of the antler and the location of the injury. Excision of the posterior halves of 1-inch antler buds results in corresponding deficiencies in the antlers produced. The same injury to an extremely rudimentary bud, however, is followed by considerable regulation, and removal of the posterior half of the pedicle prior to shedding of the old antler has no effect on the normal development of the resulting antler. Ablation of anterior or median halves of pedicles likewise causes no morphogenetic abnormality, but if the lateral half of the pedicle is removed no antler develops. The importance of the lateral half of the antler-forming structure is further illustrated by the fact that a complete and normal antler can grow from just the lateral half of a pedicle or of a 1-inch antler bud.

4. The morphogenetic field of the antler can give rise to double antlers if bisected longitudinally; transverse subdivision, however, does not alter subsequent normal production of a single antler. It is proposed that a relationship

between the initial mass of antler-forming material and the degree of morpho-genetic expression may possibly exist.

RÉSUMÉ

Recherches expérimentales sur la morphogenèse des bois du Cerf en cours de croissance

1. Des recherches sur la morphogenèse et la croissance des bois ont été entreprises chez le Cerf Sika (*Cervus nippon*). Les opérations ont été réalisées sur les bois gauches, ceux de droite, intacts, servant de témoins.

2. L'amputation du centimètre terminal des bois en cours de croissance empêche leur développement ultérieur, excepté en ce qui concerne la formation d'une excroissance abortive à partir du moignon. Par contraste, l'amputation transversale au milieu de la tige est suivie de la formation d'un bois normal. L'ablation de la tige entière permet le développement du bois, mais à un rythme ralenti. Ainsi, la potentialité formatrice du bois réside dans l'extrémité en croissance de celui-ci aussi bien que dans les tissus de la tige entière.

3. L'étendue de la régulation morphogénétique des blessures infligées chirur-gicalement au bourgeon ou à la tige du bois dépend de l'âge de ce dernier et de l'emplacement de la blessure. L'excision des moitiés postérieures de bourgeons de 25 mm. provoque des déficiences corrélatives dans les bois produits. Néan-moins, la meme blessure infligée à un bourgeon très rudimentaire est suivie d'une régulation considérable et l'ablation de la moitié postérieure de la tige, avant la chute du vieux bois, n'a pas d'effet sur le développement normal du bois suivant. De meme, l'ablation des moitiés antérieure ou médiane des tiges ne provoque pas d'anomalies morphogénétiques, mais il ne se développe pas de bois si on ôte la moitié latérale de la tige. L'importance de cette moitié latérale est encore soulignée par le fait qu'un bois complet et normal peut croître à partir de la seule moitié latérale d'une tige ou d'un bourgeon de 25 mm.

4. Le champ morphogénétique du bois peut donner naissance à des bois doubles s'il est sectionné en deux longitudinalement. Une subdivision transver-sale, néanmoins, n'altère pas la formation ultérieure normale d'un bois unique. On suggère l'existence possible d'un rapport entre la quantité initiale de matériel formateur du bois et le degré de son expression morphogénétique.

ACKNOWLEDGEMENTS

The indispensable technical assistance of Miss Marsha Rankin is gratefully acknowledged. The author is also indebted to Mr. Robert Bolinder, of the Dama Dama Game Farm, Middleborough, Massachusetts, for his co-operation in caring for the experimental animals.

These investigations have been supported by a research grant (B–923) from the National Institute of Neurological Diseases and Blindness of the National Institutes of Health.

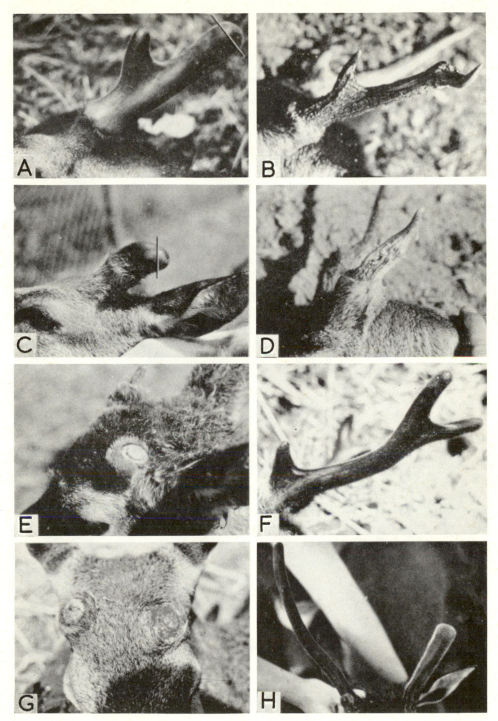

R. J. GOSS

Plate 1

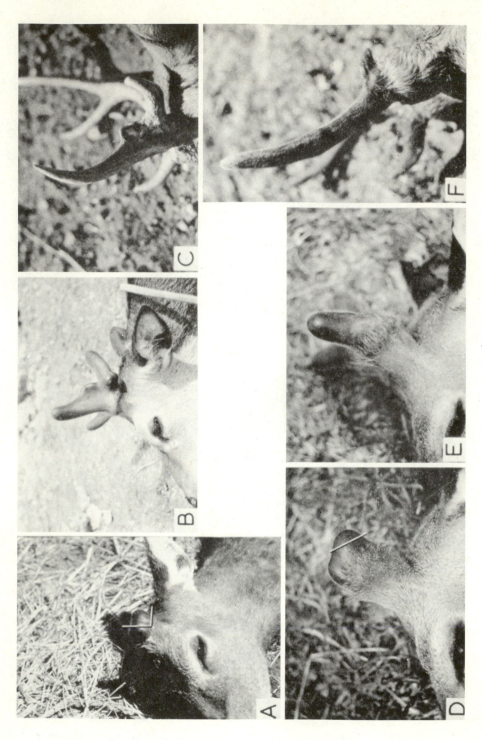

REFERENCES

BILLINGHAM, R. E., MANGOLD, R., & SILVERS, W. K. (1959). The neogenesis of skin in the antlers of deer. *Ann. N.Y. Acad. Sci.* **83**, 491–98.

BLAND-SUTTON, J. (1890). *Evolution and Disease*, London: Walter Scott.

BUBENIK, A. B. (1959). Ein weiterer Beitrag zu den Besonderheiten der Geweihtrophik beim Ren. *Z. Jagdwissensch.* **5**, 51–55.

—— & PAVLANSKY, R. (1959). Von welchem Gewebe geht der eigentliche Reiz zur Geweihentwicklung aus? III. Mitteilung: Operative Eingriffe am Bastgeweih. *Säugetierkundliche Mitteilungen*, **7**, 157–63.

DOVE, W. F. (1936). Artificial production of the fabulous unicorn. A modern interpretation of an ancient myth. *Sci. Mon.* **42**, 431–36.

FOOKS, H. A. (1955). Coalesced roe heads. *Field*, **205**, 536.

GADOW, H. (1902). The evolution of horns and antlers. *Proc. Zool. Soc. London*, **1**, 206–22.

JACZEWSKI, Z. (1954). Regeneracja rogów u tosia (*Alces alces* L.). *Kosmos, Warszawa*, **4**, 260.

—— (1955). Regeneration of antlers in red deer, *Cervus elaphus* L. *Bull. Acad. Polon. Sci.* Ser. II, **3**, 273–78.

KITCHENER, H. J. (1954). Malformation in antlers of the Malayan Sambar. *J. Bombay nat. Hist. Soc.* **52**, 588–9.

RHUMBLER, L. (1916). Der Arterienverlauf auf der Zehnerkolbenstange von *Cervus elaphus* L. und sein Einfluß auf die Geweihform. *Z. wiss. Zool.* **115**, 337–67.

TEGNER, H. (1954). Coalesced roe heads. *Field*, **203**, 453.

THOMPSON, D'A. W. (1942). *On Growth and Form*. Cambridge: University Press.

WALDO, C. M., & WISLOCKI, G. B. (1951). Observations on the shedding of the antlers of Virginia deer (*Odocoileus virginianus borealis*). *Amer. J. Anat.* **88**, 351–96.

WHITEHEAD, G. K. (1955a). Coalesced roe heads. *Field*, **205**, 353.

—— (1955b). Deformities of antlers. *Country Life, London*, **117**, 991.

WISLOCKI, G. B. (1956). The growth cycle of deer antlers. In *Ageing in Transient Tissues*, Wolstenholme, G. E. W., and Millar, E. C. P., eds., *Ciba Foundation Colloquia on Ageing*, **2**, 176–87.

EXPLANATIONS OF PLATES

PLATE 1

FIG. A. Photograph of left antler of deer 36 on 7 June, the day the apex was amputated at the leve indicated by the transverse line.

FIG. B. Same antler on 26 August, showing the abortive terminal point.

FIG. C. Left antler of deer 9, showing level of amputation of its tip on 23 July.

FIG. D. Same antler on 25 September. A short, tapered point has grown from the stump.

FIG. E. Deer 19, on 29 April, 42 days after the distal half of the left pedicle had been amputated, and 35 days before shedding of the right antler. In the absence of wound healing, the pedicle bone has remained exposed.

FIG. F. Normal antler (26 August) produced from the left pedicle of deer 19.

FIG. G. Appearance of deer 38 on 7 June, when its left pedicle was totally removed at the level of the skull.

FIG. H. Same animal on 26 August. The left antler, though retarded, is otherwise normal.

PLATE 2

FIG. A. Deer 4, on 23 April, showing the amount of tissue removed from the posterior portion of the young antler bud.

FIG. B. Four weeks later, this antler has formed a typical brow tine, but is abnormal posteriorly.

FIG. C. Same antler on 24 September. The final configuration of the antler indicates that the development of the main branch of the antler was seriously impeded by excision of the posterior half of the young bud.

FIG. D. Deer 32, on 3 June when part of the posterior region of the left antler bud was amputated at the level indicated.

FIG. E. Same antler on 18 June. The brow tine is growing normally, but no growth has occurred from the posterior part of the antler.

FIG. F. The final form of the left antler of deer 32 (26 August) is represented by an elongate brow tine.

PLATE 3

FIG. A. Deer 22 immediately after operation on 20 April. All soft tissues of the incipient antler bud were removed posterior to the level indicated by the arrow.

FIG. B. Same animal 16 days later (6 May). Wound healing has occurred and growth of the remaining anterior part of the antler is in progress.

FIG. C. Same antler on 21 May. The originally defective antler bud has become reconstituted and is beginning to elongate.

FIG. D. Same antler on 3 June. Though considerably retarded with respect to the opposite control antler, growth in the normal direction is occurring.

FIG. E. Final appearance of the left antler of deer 22 (26 August). It reached a length about one-third that of the control.

FIG. F. Appearance of left pedicle of deer 25 on 14 May, 2 weeks after excision of its posterior half. The base of the old antler, most of which had previously been cut off, remained attached to the pedicle for about another week.

FIG. G. Two-week-old antler bud (3 June) formed on anterior half of the same pedicle. Antler-forming tissue is invading the posterior region of the pedicle.

FIG. H. Same antler on 26 August, illustrating the normal appearance of the antler produced from the anterior half of the pedicle.

PLATE 4

FIG. A. Deer 31 on 21 May after removal of the anterior half of the left pedicle (arrow). The old antler was shed on the same day.

FIG. B. Same animal on 18 June. The anterior region of the pedicle has been partly filled in, and elongation of the antler is proceeding normally.

FIG. C. Appearance of the same antler on 26 August when the full extent of its growth had been achieved. It departed from the control only in its slightly shorter length and in the absence of the brow tine.

FIG. D. Deer 28 on 21 May after excision of the median half of the left pedicle (several days after shedding of the old antler).

FIG. E. Same animal 4 weeks after operation. A normal antler is forming.

FIG. F. Appearance of deer 28 on 26 August. The animal's left experimental antler is identical with the control.

FIG. G. Front view of deer 30 on 21 May when the lateral half of the left pedicle was removed less than a week after shedding of the old antler.

FIG. H. Same animal on 26 August, showing the presence of only a rudimentary mass of tissue on the operated side (arrow) in contrast with the normal antler produced on the right side.

PLATE 5

FIG. A. Appearance of deer 27 following removal of a segment of tissue slightly lateral to the mid-line of the left antler bud (May 6).

FIG. B. Front view of same antler on May 14, showing continued but separate growth of the two parts of the antler. The lateral portion has bifurcated.

FIG. C. Same antler on June 3. Further growth of the double antler.

FIG. D. Final form of the same antler in lateral view (August 26). The two components of the double antler have elongated considerable, but have failed to branch at their ends.

FIG. E. Deer 5 after excision of an antero-posterior wedge of tissue dividing the antler bud into median and lateral halves.

FIG. F. Median view of same antler after the final normal form had been attained. The antler grew entirely from the lateral half, while the median half gave rise only to a diminutive outgrowth (arrow). Remnants of the original scar can still be seen.

FIG. G. Deer 35 following bisection of the young antler into anterior and posterior halves (June 7).

FIG. H. The same animal, on August 26, had developed a normal pair of identical antlers, indicating the inefficacy of the original operation.

(Manuscript received 1 : xii : 1960)

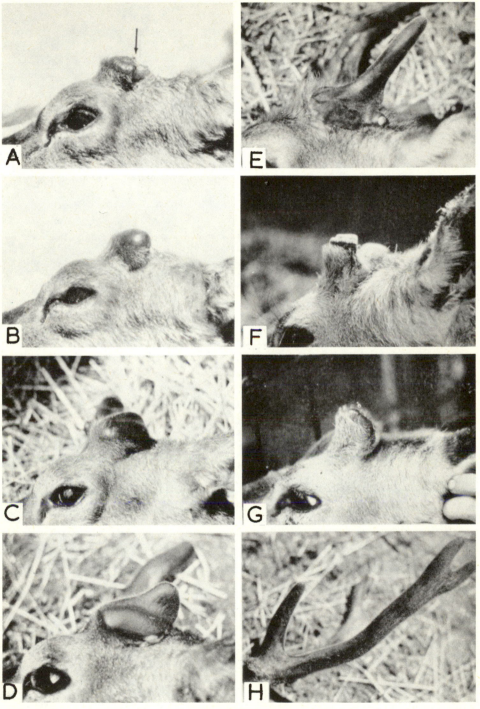

R. J. GOSS

Plate 3

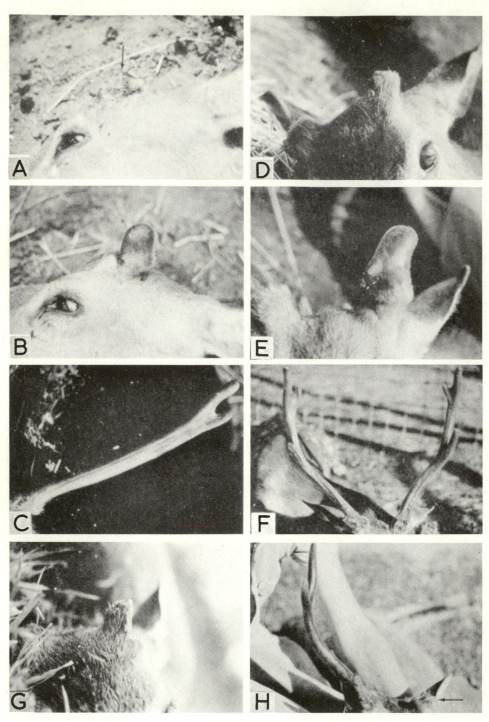

R. J. GOSS

Plate 4

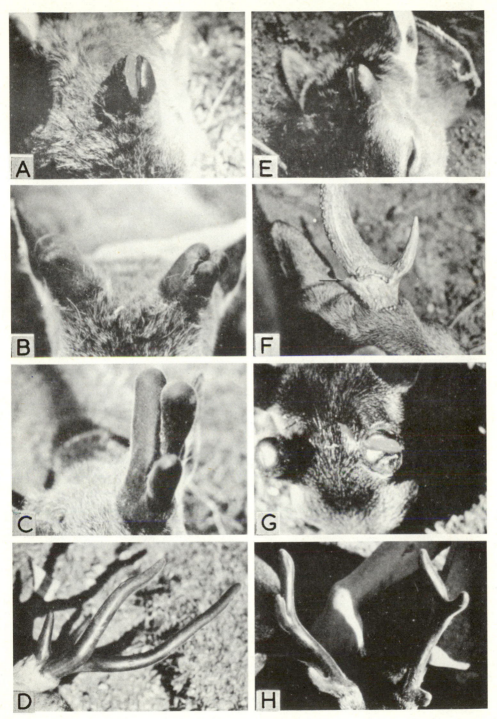

R. J. GOSS

Plate 5

Copyright © 1970 by the American Society of Zoologists

Reprinted from *Amer. Zoologist*, **10**, 175–186 (1970)

Relationship Between the Tissue and Epimorphic Regeneration of Muscles

Bruce M. Carlson

*Department of Anatomy, University of Michigan,
Ann Arbor, Michigan 48104*

28

Synopsis. Entire muscles in vertebrate extremities can regenerate from minced muscle fragments implanted into the bed of the removed muscle. The histological course of regeneration of muscle fibers in this system is the same as that described in damaged mammalian muscle. This type of regeneration has been called the tissue regeneration of muscle, and entire muscles have been regenerated by this mode in urodelan and anuran amphibians, birds, and mammals. Entire muscles are also formed in a regenerating limb (an epimorphic regenerative process). It is postulated that there are two fundamentally different modes of regeneration of muscles. Epimorphic regeneration is relatively slow, and the blastema seems to play an important role. The end product is morphologically and functionally perfect. Morphogenesis seems to be controlled in a manner analogous to that in the embryonic limb. The tissue regeneration of muscles is rapid, occurring without the mediation of a blastema. The end product is smaller and less well organized than a normal muscle, and its morphogenesis seems to require the function of the limb. A possible parallel reaction of skeletal elements is discussed.

The regeneration of skeletal muscle in mammals has long been investigated (Neumann, 1869; Volkmann, 1893; Waldeyer, 1865; reviewed by Field, 1960; Betz, et al., 1966; McMinn, 1969), but the majority of studies have treated it in the context of mammalian wound healing. As a result, there have been few attempts to relate the process of muscle regeneration in the mammal to the broad spectrum of regenerative phenomena appearing throughout the animal kingdom. A recent series of investigations of Studitsky's (1952, 1959) method of producing entire muscle regenerates from minced fragments has afforded me the opportunity to study the regeneration of mammalian muscle at the tissue and organ level and to compare certain aspects of this process with muscle regeneration in lower vertebrates. In this paper I shall review the major developmental events in regeneration from minced muscle. Following this I shall discuss the relationship between that process and other manifestations of muscle regeneration which have been studied experimentally.

This work was supported in part by grants from the University of Michigan Cancer Research Institute and the Muscular Dystrophy Associations of America.

REGENERATION OF MINCED MUSCLES IN THE RAT

Regeneration of an entire muscle in the rat can be stimulated by removing a muscle, mincing it into one-mm³ fragments and replacing the minced fragments into the site from which the muscle was removed. Details of the operative technique have been described by Studitsky, et al. (1956) and Carlson (1968). Both the gross and micro-anatomical aspects of regeneration of minced muscle in the rat have been treated in detail by Studitsky (1959, 1963), Zhenevskaya (1962) and Carlson (1968, 1970c), and they will only be summarized here.

The gross form of the newly implanted mass of muscle fragments initially conforms to the space left by the muscle which was removed. For the first day or two following implantation there is not much change in the appearance of the fragments except for a greater degree of adhesion, one to another, of fragments in the peripheral parts of the implant and the earliest evidence of revascularization in the most peripheral regions. Toward the end of the first week, the distal third of the implanted fragments has become thinner and has established fairly strong con-

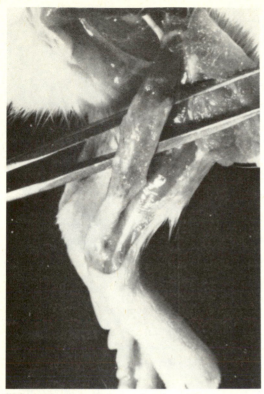

FIG. 1. Typical 68-day minced-muscle regenerate in the rat.

nections with the stump of the Achilles tendon. In the proximal two-thirds, individual muscle fragments can no longer be distinguished at the periphery of the implant, but at the center the appearance of the individual muscle fragments has not greatly changed from that at the time of implantation. By the middle of the second week all traces of the originally implanted muscle fragments have disappeared, and the regenerate has a readily recognizable tendon distally, whereas proximally it begins to assume a muscular appearance. In subsequent weeks the proximal part of the regenerate takes on more typical muscular characteristics and the regenerating Achilles tendon becomes more compact and a glistening white. Of practical significance is the formation of strong connective tissue adhesions between the lateral edges of the regenerate and the underlying tissues. The adhesions are extensive enough to

limit considerably the functional expression of a regenerated muscle. A typical 68-day regenerate is shown in Figure 1. The weight of most regenerates is 15-25% of the contralateral muscles. Studitsky (1963) has published kymographic tracings of contractions by regenerated muscles.

During the first two days after implantation of the minced fragments, sarcoplasmic degeneration begins. The muscle fragments are loosely held together by fibrinous material, and in scattered areas small pockets of acute inflammatory cells are seen. Starting with the second or third day, blood vessels begin growing into the minced mass, and the first evidence of muscle regeneration is seen at the periphery. A day or two after the first vascular invasion, a characteristic pattern of three developmental zones is established within the regenerate (Fig. 2). The central zone (C) contains the originally implanted muscle fibers. It is not vascularized and little histological evidence of regenerative activity is seen. The transitional zone (T) is characterized by sarcoplasmic fragmentation of the old muscle fibers and the appearance of early stages of muscle regeneration under the basement membranes surrounding the degenerated muscles. Ingrowing blood vessels reach only the peripheral part of

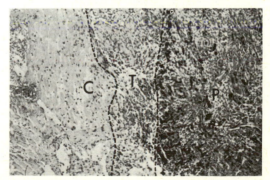

FIG. 2. Five-day minced-muscle regenerate in the rat. This specimen was injected with ink to demonstrate the pattern of vascularization. C, central non-vascularized, non-regenerating region. T, transitional zone in which histological signs of muscle regeneration are first seen. P, peripheral zone of regeneration containing myotubes and newly forming connective tissue. × 105.

the transitional zone. The inner area of this zone, in which the first histologically evident stages of muscle regeneration occur, is not directly vascularized. The peripheral zone (P) contains no traces of the originally implanted muscle fibers. Here are found the more advanced stages of regenerating muscle fibers and the connective tissue stroma. A more complete histological characterization of these zones is given in Carlson (1970*b*).

During the interval from the fourth day to the middle of the second week after implantation the three zones described above remain intact, but the peripheral zone steadily thickens and the central zone correspondingly decreases in size. The transitional zone maintains the same position relative to the other two zones, but as regeneration proceeds, this zone moves closer to the center of the regenerating muscle. Later development of minced-muscle regenerates consists primarily of maturation of cellular and tissue elements as well as alignment of the individual regenerating muscle and connective tissue fibers into a highly regular orientation.

An obvious result of the zonal pattern of development is that in any regenerate less than two weeks old there is a radial gradient of maturity of the regenerated muscle fibers ranging from the greatest maturity at the periphery to the least at the center. This is a striking contrast to the linear pattern of maturation in outgrowing regenerating systems, and it is an important factor to consider when posing questions related to the overall control of morphogenesis of regenerating muscles.

The regeneration of individual muscle fibers within this developing system does not seem to differ significantly from that described from other experimentally induced systems of muscle regeneration. Considerable uncertainty surrounds the origin of the mononuclear myoblastic cell. Current opinion is divided between the release of nucleated fragments from the muscle fiber itself (Hay, 1959, 1962; Lentz, 1969; Reznik, 1969) and the relatively undiffer-

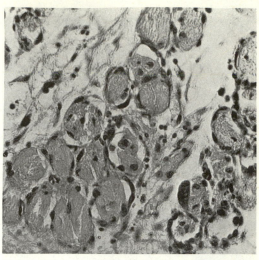

FIG. 3. Three-day minced-muscle regenerate in the rat. In the transitional zone of early minced-muscle regenerates, the sarcoplasm of the implanted muscle fibers is invaded by macrophages and is breaking up. Beneath the basement membrane of the original muscle fibers, early basophilic cuffs are forming. × 400.

entiated satellite cells which lie adjacent to but under the basement membrane of the muscle fiber (Muir, *et al.*, 1965; Shafiq and Gorycki, 1965; Church, *et al.*, 1966; Shafiq, *et al.*, 1967). Whatever their source, the mononuclear myoblasts accumulate beneath the basement membrane of the old muscle fiber and fuse with one another, forming syncytial basophilic cuffs of cells (Fig. 3). The basophilic cuffs initially surround the remnants of the degenerating sarcoplasm as well as the macrophages which are eliminating the sarcoplasmic debris. Following this rather transitory stage of development, the basophilic cuffs appear to consolidate and are recognizable as strap-like elements with long rows of regularly arranged nuclei in their center and with very basophilic cytoplasm (Fig. 4). The regenerating muscle fiber is now called a myotube or sarcoblast. With continuing maturation, there is an increasing cytoplasmic eosinophilia, the appearance of longitudinal myofibrils followed by cross striations and finally the migration of the nuclei from the central rows to the periphery of the muscle fiber. Further

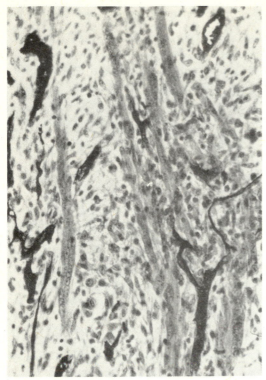

FIG. 4. Four-day minced-muscle regenerate in the rat. Multinucleated myotubes with central nuclear chains are found in the peripheral zone. The vasculature of this regenerate was perfused with ink. × 400.

structural details are given by Godman (1957), Allbrook (1962), Price *et al.* (1964), and Betz, *et al.* (1966).

REGENERATION OF MUSCLE IN THE AXOLOTL

The regeneration of entire muscles from minced fragments has been described in rats (Studitsky, 1954, 1959, 1963; Zhenevskaya, 1962; Carlson, 1968), mice (Yeasting, personal communication), guinea pigs (Studitsky, 1963), chickens and pigeons (Studitsky, 1954), and frogs (Samsonenko, 1956, Carlson, 1968. Anatomically, the development of regenerates from minced muscle follows a very similar course in all these species, and the regeneration of individual muscle fibers differs little, if at all, from that which occurs after various other types of artifically applied local trauma (Betz and Reznik, 1964;

Price, *et al.*, 1964; Allbrook, *et al.*, 1966; Shafiq, *et al.*, 1967). All of the above animals share the common characteristic of normally not being able to regenerate a limb after amputation.

I became intrigued by the question of what would happen if one minced a muscle in the limb of an animal which possesses a well developed capacity for complete limb regeneration (an epimorphic regenerative process). Would the minced muscle regenerate into a new muscle by following a developmental pathway similar to that described above for mammals or would the tremendous amount of tissue destruction trigger the formation of a supernumerary limb (an epimorphic type of regenerative response)?

To answer this question, I investigated the regeneration of the pubo-ischio-tibialis (p.i.t.) muscle in the axolotl (*Ambystoma mexicanum*). After the p.i.t. muscle was minced and replaced, the subsequent regenerative process was histologically similar to that which had been observed following mincing in the frog and rat (Carlson, 1970a). The first recognizable myoblastic elements were seen at five or six days. By 10 days the individual myoblasts had begun to fuse, and within two weeks well defined myotubes were scattered throughout the regenerate. Striated muscle fibers were found in abundance toward the end of the first month. Regenerated muscle fibers did not attain normal diameters until four or five months after implantation.

The formation of muscles in a regenerating axolotl limb differs from the process described above. Following amputation of the limb in a one- or two-year-old animal, most of the activities within the mesodermal components appear to be geared toward establishing a distal blastema. The musculature of the limb stump undergoes a progressive distal-proximal morphological dedifferentiation. The cells left in the wake of this process migrate distally and accumulate under the thickened apical epidermis to form a blastema. Occasional regenerating muscle fibers can be seen proximal to the amputation-surface during the first month after amputation, but the bulk of activity is regressive. Early in the second month the cartilaginous primordia of the skeleton differentiate in the central part of the blastema. The regenerating muscles can be first distinguished as accumulations of basophilic spindle-shaped cells lying close to one another lateral to the developing skeletal elements. These cells fuse into multinucleated fibers which from the beginning are oriented parallel to one another and collectively assume the gross form of the original muscle. The development of an individual muscle fiber from recognizable myoblastic elements does not differ histologically from that seen after mincing. The regeneration of muscle following mincing and following limb amputation is summarized in Figure 5.

THE RELATIONSHIP BETWEEN TISSUE AND EPIMORPHIC REGENERATION OF MUSCLE

The results of the muscle-mincing experiments, particularly in axolotls and newts, have led to the hypothesis that there are two morphologically distinct processes by which an entire muscle can be regenerated. One is the tissue mode of regeneration, which is seen following mincing. The other is the epimorphic mode of regeneration in which a new muscle is formed in the context of a totally regenerating limb. To determine the validity of this hypothesis, it might be instructive to compare what is known and what is not known about certain critical events in each process.

At the cellular level very little is known. The origin of the myoblastic cell has not been agreed upon in any regenerating system. Most workers in the field of amphibian limb regeneration maintain that the damaged muscle fibers of the limb stump break up and that the nucleated fragments which are released migrate distally to form the blastema (Towle, 1901; Thorn-

ton, 1938; Hay, 1959, 1962; Lentz, 1969). Simpson (1965) and Mufti (1969) have provided quite convincing evidence that the muscle in the stump of autotomized tails of lizards and salamanders is not a major contributor to the muscle arising in the tail blastema. A good summary of differing viewpoints is given by Schmidt (1968, pp. 30-37).

The majority of those studying the regeneration of damaged mammalian muscle (tissue regeneration) have been inclined to suspect the satellite cell as the precursor of the myoblast (Muir, et al., 1965; Shafiq and Gorycki, 1965; Church, et al., 1966; Shafiq, et al., 1967), but recently Reznik (1969) has concluded that in damaged mammalian muscle, mononucleated cells are pinched off the damaged cytoplasm and pass through a transitory satellite-cell stage before becoming myoblasts. At present there is no information which would help to identify the nature of the precursor cell of the muscle fibers regenerated after mincing in the axolotl.

Once begun, the differentiation of muscle fibers seems to follow the same morphological course in both minced muscle regeneration and in limb regeneration. Up to now direct comparison has only been made in urodeles and only at the light microscopic level.

Differences between minced muscle regeneration and the epimorphic regeneration of muscle are more readily apparent at a level of organization more complex than that of the single cell, i.e., at the tissue and organ level of organization. The most obvious difference is the absence of a regenerative blastema in any system of minced-muscle regeneration examined thus far. Regenerating muscle fibers arise throughout the mass of minced muscle, but no cells with the histological appearance of blastemal cells have been found in minced-muscle regenerates of newts, axolotls, frogs, or rats. There is no evidence that interactions between epidermis, nerve, or adjacent muscle and bone are necessary for the initiation of the regenerative re-

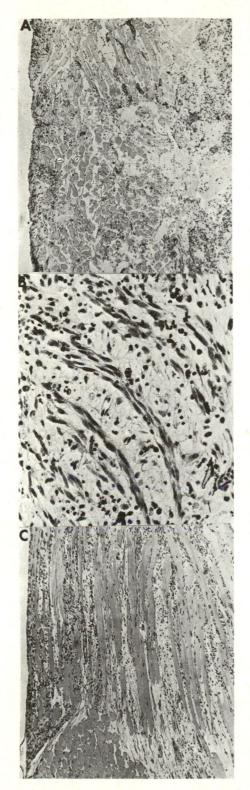

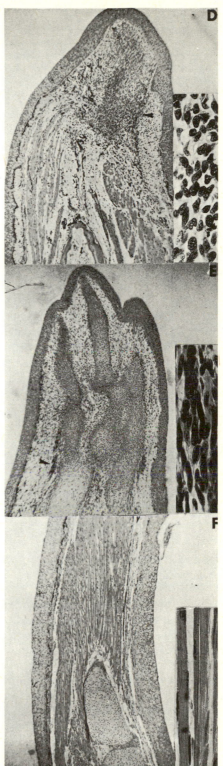

FIG. 5. Comparison between tissue- and epimorphic regeneration of muscle in the 200-mm axolotl. A. Five-day minced-muscle regenerate. Originally implanted minced-muscle fragments fill the entire field. Scattered among the muscle fragments are erythrocytes and leukocytes. Little evidence of regeneration is seen at this stage. × 53. B. Ten-day minced-muscle regenerate. Myoblasts are beginning to fuse into myotubes. Basophilic cuffs are not so evident as in higher forms. × 210. C. Minced-muscle regenerate after 150 days. The regenerated muscle fibers have attained normal diameters, but there is considerable connective tissue among them. The cross-sectioned muscle fibers at the bottom of the figure are from an adjacent normal muscle. × 53. D. Limb regenerate after 34 days. × 22. Cells in the region of future muscle formation (arrow) appear in inset (× 400). E. Limb regenerate after 44 days. × 22. An area of early myogenesis (arrow) appears in inset (× 400). F. Limb regenerate after 91 days, showing fully regenerated muscle. × 22. Inset (× 400). Figures A, B, C, reproduced by courtesy of the Wistar Press.

sponse. Studies based on vascular injection of the rat (Carlson, 1970c) indicate a relationship between the ingrowth of blood vessels and the onset of muscle regeneration in minced muscle, but it is likely that the relationship is facilitatory rather than causal. This viewpoint is supported by our recent studies on regeneration of frog muscle in diffusion chambers, and by the observations of O'Steen (1962, 1963), who recorded the regeneration of both mouse and human muscle in diffusion chambers implanted into the peritoneal cavity of mice.

In contrast to the pattern described above, a blastema of dedifferentiated cells is one of the most prominent morphological features in an epimorphic regenerative response. Although it has not been unequivocally proven that the blastema proper contains potentially myogenic cells, the bulk of available evidence points to this conclusion (Thornton, 1938; DeHaan, 1956; Hay, 1962; Simpson, 1965; Trampusch and Harrebomée, 1965). Whether myogenic cells are an integral part of the blastema or not, the fact remains that the period of the undifferentiated blastema represents a phase during which no identifiable muscle cells have been detected in the regenerate. Immunological studies (DeHaan, 1956; Laufer, 1959) have shown that during this period, cells reacting with anti-myosin antibody can not be demonstrated distal to the surface of amputation. Proximal to the blastema, the muscles of the limb stump in the axolotl show occasional newly regenerating fibers, but not infrequently in the newt the stump mus-

cles continue to dedifferentiate as the blastema is being elaborated, and very little evidence of muscle regeneration is seen during the period of the early blastema.

Another outstanding difference between minced-muscle regeneration and the epimorphic regeneration of muscle is in the rates of the two processes. The reappearance of regenerating muscle fibers after the implantation of muscle fragments is quite rapid in all species studied, and in the rat it corresponds with the rates obtained by others for the repair of muscle damage following relatively minor trauma such as transection. Thus, after mincing, myotubes are seen within three days in the rat, 7–9 days in the frog, and 8–10 days in the axolotl. The redifferentiation of muscle in epimorphic regeneration is a much more leisurely process. In the undifferentiated limb blastema of the axolotl (30 days after amputation) there are no cells which can be called myogenic cells on the basis of their morphology alone. Multinucleated muscle fibers do not appear until cartilage has differentiated in the regenerate early in the second month after amputation. The only other animal in which the rates of minced-muscle regeneration and the epimorphic regeneration of muscle have been compared is the adult newt (*Triturus viridescens*), and they are comparable to those noted in the adult axolotl.

A third instructive difference between the two proposed types of regeneration is the degree of morphological perfection of the end product. Although minced-muscle regenerates normally establish morphologically correct tendinous origins and insertions, they differ from normal muscles in several ways. Their total size is considerably less in most species studied; the proportion of connective tissue to muscle is higher, and in some cases the internal architecture is less regular than normal. Some of the muscles regenerated from minced fragments resemble their normal precursors quite closely, but the morphology of other regenerated muscles strongly indicates that other factors than the original form of the muscle determine the shape of the regenerate. A specific example of this occurs after removal of the gastrocnemius, soleus, and plantaris muscles in the rat. When these are minced and replaced into the limb, only one muscle regenerates, but the tendinous origins of all the original muscles are usually connected to the regenerate by connective tissue. It appears that the standard form of minced muscle regenerates is a muscular proximal portion which tapers to a strongly tendinous distal insertion. Local variations in tendinous attachments depend upon the individual muscles concerned. In contrast, the morphology of the muscles in regenerated limbs is almost always perfect with respect to both gross form and internal architecture. Even during the initial stages of morphogenesis, the regenerating muscle fibers are situated parallel to one another. This is in marked contrast to early minced-muscle regenerates in which the early myotubes are scattered in all directions.

Although the primarily descriptive material given above strongly suggests that the epimorphic regeneration of muscle is a distinctly different process from that which occurs after mincing, additional experimental material is needed to confirm the validity of this hypothesis. Work is presently being done in my laboratory to define the origin of the myogenic cells in the two systems, to determine the role of function in the morphogenesis of regenerating muscles, and to compare the inhibitory effects of certain procedures, such as denervation, x-irradiation and hypophysectomy upon the two processes.

TISSUE AND EPIMORPHIC REGENERATION OF SKELETAL ELEMENTS

In order to determine whether the ability to regenerate by two apparently different modes is unique to muscle or if it may be indicative of a more general phenomenon, I compared other components of epimorphic regenerating systems to see if a similar duplication of potential regenerative responses might occur. There does seem to be a parallel situation in the reactions of skeletal tissue in the amphibian limb to traumatic stimuli.

Cartilage, like muscle, can be formed by two morphologically distinct processes, each of which follows a different time schedule. Following limb amputation, the epimorphic regeneration of cartilage is accomplished by the formation of a blastema and the differentiation of competent blastemal cells into chondrocytes. The differentiation of cartilage in implant-induced supernumerary limbs is also preceded by the appearance of a blastema (Ruben, 1955; Carlson, 1967a), and the rate of this process is even slower than that occurring after the amputation of a limb. However, cartilage can be induced to form much more quickly and without the mediation of a blastema (tissue regeneration) after such diverse insults to the limb as fractures (Pritchard and Ruzicka, 1950), an injection of beryllium nitrate (Carlson, 1970d) or multiple injections of saline (Carlson and Morgan, 1967). Cartilage of the same type is formed around the cut ends of bone in non-regenerating limbs of frogs (Goode, 1967; Polezhaev, 1968; confirmed by me) and mammals (Schotté and Smith, 1959; Scharf, 1961). It is also seen at the cut end of bone in regenerating urodelan limbs (Mettetal, 1939; Schmidt, 1958, 1968, p. 94; Trampusch and Harrebomée, 1965; Carlson, 1967b). In most of these cases the cartilage arises in the inner relatively undifferentiated layer of the periosteum by direct differentiation. In my experience the appearance of such cartilage has always been accompanied or preceded by the osteoclastic destruction of bone. It is

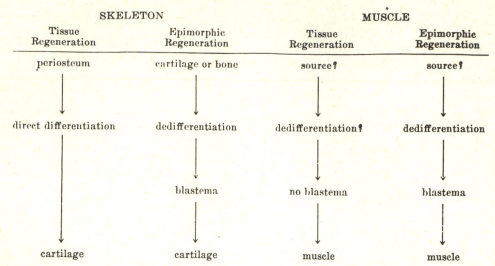

| SKELETON | | MUSCLE | |
Tissue Regeneration	Epimorphic Regeneration	Tissue Regeneration	Epimorphic Regeneration
periosteum	cartilage or bone	source?	source?
↓	↓	↓	↓
direct differentiation	dedifferentiation	dedifferentiation?	dedifferentiation
	↓	↓	↓
	blastema	no blastema	blastema
↓	↓	↓	↓
cartilage	cartilage	muscle	muscle

FIG. 6. A comparison between the major events of tissue- and epimorphic regeneration in muscular and skeletal tissues. As stated in the text, dedifferentiation of muscle fibers themselves is the likely source of myogenic cells in the epimorphic regeneration of muscle, but a contribution by satellite cells has not been ruled out. If, in the tissue regeneration of muscle, satellite cells prove to be the sole source of the myogenic cell, the process would closely parallel the tissue regeneration of skeletal tissue. However, if the myogenic cells arose from nucleated fragments of muscle fibers themselves, a form of dedifferentiation would occur, but the process would not lead to a blastema as in epimorphic regeneration.

also significant that the differentiation of this type of cartilage has a higher threshold of sensitivity to actinomycin D than does blastemally derived cartilage (Carlson, 1966, 1967b).

Thus in the amphibian limb there seems to exist a parallel range of regenerative responses in the muscular and cartilaginous elements (Fig. 6). Some types of trauma to these tissues elicit a rather prompt restorative response in which cellular and tissue differentiation occur in the absence of a blastema (tissue regeneration). Other stimuli, however, lead to the formation of a regeneration blastema from which both skeletal elements and muscles differentiate (epimorphic regeneration). A major unsolved question is whether these regenerative responses result from qualitatively different initial stimuli or whether after a given stimulus, the group of affected cells is committed to one or the other regenerative response at some later stage in the series of reactive events to the initial stimulus. Even in an overall epimorphic regenerative response a certain

amount of tissue regeneration occurs. For example, in the amputated urodelan limb, a cap of cartilage forms around the cut end of the bone before differentiation from the blastema has begun. A limited amount of muscle regeneration also occurs in the limb stump before differentiation of the blastema has begun (DeHaan, 1956; Laufer, 1959). This is best demonstrated in the mature axolotl. It is present to a much lesser extent in the adult newt.

CONCLUSION

What are the implications of the observations and the hypothesis presented above? An examination of the phylogenetic distribution of tissue and epimorphic regeneration is instructive. Some degree of tissue regeneration has been found in all classes of vertebrates examined, and the entire muscles have been regenerated by this mode in urodelan and anuran amphibians, birds, and mammals. Fishes have been woefully neglected in studies of tissue regeneration. Natural epimorphic regeneration of extremities has been described in cyclostomes, bony fishes, urodelan and larval anuran amphibians, and lizards. In addition, epimorphic regeneration of normally non-regenerating structures has been stimulated in adult frogs, in lizards, and in the limb of the newborn opossum.

Tissue regeneration is a relatively rapid process which seems to require the functional activity of the extremity for complete differentiation and morphogenesis of the damaged tissue. Tissue interactions of the embryonic or inductive type do not appear to be necessary. Although a reasonable restoration of the damaged tissue is produced, it is normally neither morphologically nor functionally perfect. In contrast, epimorphic regeneration, a relatively leisurely process, leads to the perfect morphological and functional restoration of an entire extremity as well as its individual tissue components. This process seems to require a number of specific tissue interactions in order to proceed, and the early phases of development and morphogenesis appear to be independent of on-going function. Epimorphic regeneration is confined primarily to aquatic forms or to species which do not absolutely depend upon the function of all extremities for individual survival. The lack of a given extremity, however, may seriously hinder certain activities, such as reproductive behavior, which are required for survival of the species.

Tissue regeneration is a property of muscle and bone in all vertebrates, and its phylogenetic survival may be related to its ability to occur within a functioning extremity. The imperfect form and function of the regenerated structure may represent the price paid for the relative rapidity with which the repair is made. Epimorphic regeneration, on the other hand, appears to be a developmental luxury confined to those forms for which the loss of an extremity does not constitute an immediate threat to life and which can afford the expenditure of time and morphological alterations of the remaining tissues to reproduce exactly the damaged structure.

REFERENCES

Allbrook, D. 1962. An electron microscopic study of regenerating skeletal muscle. J. Anat. 96:137-152.

Allbrook, D. B., W. deC. Baker, and W. H. Kirkaldy-Willis. 1966. Muscle regeneration in experimental animals and in man. J. Bone Joint Surg. 48B:153-169.

Betz, E. H., H. Firket, and M. Reznik. 1966. Some aspects of muscle regeneration. Int. Rev. Cytol. 19:203-227.

Betz, E. H., and M. Reznik. 1964. La régénération des fibres musculaires striées dans différentes conditions expérimentales. Arch. Biol. 75:567-594.

Carlson, B. M. 1966. Inhibition of limb regeneration in adult *Triturus viridescens* treated with actinomycin D (in Russian). Doklady 'Akad. Nauk SSSR 171: 229-232.

Carlson, B. M. 1967a. Studies on the mechanism of implant-induced supernumerary limb formation in urodeles. I. The histology of supernumerary limb formation in the adult newt. J. Exp. Zool. 164:227-242.

Carlson, B. M. 1967b. The histology of inhibition of limb regeneration in the newt, *Triturus*, by actinomycin D. J. Morphol. 122:249-263.

Carlson, B. M. 1968. Regeneration of the completely excised gastrocnemius muscle in the frog and rat from minced muscle fragments. J. Morphol. 125:447-472.

Carlson, B. M. 1970a. The regeneration of a limb muscle in the axolotl from minced fragments. Anat. Rec. 166:423-436.

Carlson, B. M. 1970b. Regeneration of the rat gastrocnemius muscle from sibling and non-sibling muscle fragments. Amer. J. Anat. (In press)

Carlson, B. M. 1970c. The regeneration of entire muscles from minced fragments. *In* A. Mauro and A. T. Milhorat, [ed.], Myogenesis and regeneration of the skeletal muscle fiber. Excerpta Medica, N.Y. (In press)

Carlson, B. M. 1970d. The effect of x-irradiation and beryllium nitrate upon implant-induced supernumerary limb formation in the newt. Oncology 24:31-47.

Carlson, B. M., and C. F. Morgan. 1967. Studies on the mechanism of implant-induced supernumerary limb formation in urodeles. II. The effect of heat treatment, lyophilization and homogenization on the inductive capacity of frog kidney. J. Exp. Zool. 164:243-249.

Church, J. C. T., R. F. X. Noronka, and D. B. Allbrook. 1966. Satellite cells and skeletal muscle regeneration. Brit. J. Surg. 53:638-642.

DeHaan, R. L. 1956. The serological determination of developing muscle protein in the regenerating limb of *Amblystoma mexicanum*. J. Exp. Zool. 133:73-85.

Field, E. J. 1960. Muscle regeneration and repair, p. 139-170. *In* G. H. Bourne, [ed.], Structure and function of muscle. III. Academic Press, N. Y.

Godman, G. C. 1957. On the regeneration and

rediflerentiation of mammalian striated muscle. J. Morphol. 100:27-82.

Goode, R. P. 1967. The regeneration of limbs in adult anurans. J. Embryol. Exp. Morphol. 18:259-267.

Hay, E. D. 1959. Electron microscopic observations of muscle dedifferentiation in regenerating *Amblystoma* limbs. Develop. Biol. 1:555-585.

Hay, E. D. 1962. Cytological studies of dedifferentiation and differentiation in regenerating amphibian limbs, p. 177-210. *In* D. Rudnick, [ed.], Regeneration. Ronald Press, N. Y.

Laufer, H. 1959. Immunochemical studies of muscle proteins in mature and regenerating limbs of the adult newt, *Triturus viridescens*. J. Embryol. Exp. Morphol. 7:431-458.

Lentz, T. L. 1969. Cytological studies of muscle dedifferentiation and differentiation during limb regeneration of the newt *Triturus*. Amer. J. Anat. 124:447-480.

McMinn, R. M. H. 1969. Tissue repair, Chap. 3. Academic Press, N. Y.

Mettetal, C. 1939. La régénération des membres chez la salamandre et le Triton. Arch. Anat. Histol. Embryol. 28:1-214.

Mufti, S. A. 1969. Tail regeneration in an adult salamander, *Desmognathus fuscus*. (Abstr.) Amer. Zoologist 9:613.

Muir, A. R., A. H. M. Kanji, and D. Allbrook. 1965. The structure of the satellite cells in skeletal muscle. J. Anat. 99:435-444.

Neumann, E. 1868. Ueber den Heilungsprozess nach Muskelverletzungen. Arch. Mikroskop. Anat. Entwicklungsmech. 4:323-333.

O'Steen, W. K. 1962. Growth activity of normal and dystrophic muscle implants in normal and dystrophic hosts. Lab. Invest. 11:412-419.

O'Steen, W. K. 1963. The growth of human dystrophic skeletal muscle in diffusion chambers. Texas Rep. Biol. Med. 21:369-379.

Polezhaev, L. V. 1968. Loss and restoration of regenerative capacity of organs and tissues in animals (in Russian). Izdatel. Nauka, Moscow, 326 p.

Price, H. M., E. L. Howes, and J. M. Blumberg. 1964. Ultrastructural alterations in skeletal muscle fibers injured by cold. II. Cells of the sarcolemmal tube: Observations on "discontinuous" regeneration and myofibril formation. Lab. Invest. 13:1279-1302.

Pritchard, J. J., and A. J. Ruzicka. 1950. Comparison of fracture repair in the frog, lizard and rat. J. Anat. 84:236-261.

Reznik, M. 1969. Origin of myoblasts during skeletal muscle regeneration. Lab. Invest. 20:353-363.

Ruben, L. N. 1955. The effect of implanting anuran cancer into non-regenerating and regenerating larval urodele limbs. J. Exp. Zool. 128:29-51.

Samsonenko, R. V. 1956. Development of muscular tissue in implanted minced muscle tissue in the site of completely removed muscular organs in frogs (in Russian). Arkh. Anat. Gist. Embryol. 33 (2):56-64.

Scharf, A. 1961. Experiments on regenerating rat digits. Growth 25:7-23.

Schmidt, A. J. 1958. Forelimb regeneration of thyroidectomized adult newts. II. Histology. J. Exp. Zool. 139:95-136.

Schmidt, A. J. 1968. Cellular biology of vertebrate regeneration and repair. Univ. Chicago Press, Chicago, 420 p.

Schotté, O. E., and C. B. Smith. 1959. Wound healing processes in amputated mouse digits. Biol. Bull. 117:546-561.

Shafiq, S. A., and M. A. Gorycki. 1965. Regeneration in skeletal muscle of mouse: some electron-microscope observations. J. Pathol. Bacteriol. 90:123-127.

Shafiq, S. A., M. A. Goryski, and A. T. Milhorat. 1967. An electron microscopic study of regeneration and satellite cells in human muscle. Neurology 17:567-575.

Simpson, S. B. 1965. Regeneration of the lizard tail, p. 431-443. *In* V. Kiortsis and H. A. L. Trampusch, [ed.], Regeneration in animals and related problems. North-Holland Publ. Co., Amsterdam.

Studitsky, A. N. 1952. Restoration of muscle by means of transplantation of minced muscle tissue (in Russian). Doklady Akad. Nauk SSSR 84 (2):389-392.

Studitsky, A. N. 1954. Principles of restoration of muscle in higher vertebrates (in Russian). Trudy Inst. Morphol. Zhivot. im. A. N. Severtsova 11:225-264.

Studitsky, A. N. 1959. Experimental surgery of muscles (in Russian). Izdatel. Akad. Nauk SSSR, Moscow, 338 p.

Studitsky, A. N. 1963. Dynamics of the development of myogenic tissue under conditions of explantation and transplantation, p. 171-200. *In* G. G. Rose, [ed.], Cinemicrography in cell biology, Academic Press, N. Y.

Studitsky, A. N., R. P. Zhenevskaya, and O. N. Rumyantseva. 1956. Fundamentals of the technique of restoration of muscle by means of transplantation of minced muscle tissue (in Russian). Cesk. Morfol. 4:331-340.

Thornton, C. S. 1938. The histogenesis of muscle in the regenerating fore limb of larval *Amblystoma punctatum*. J. Morphol. 62:17-47.

Towle, E. W. 1901. On muscle regeneration in the limbs of *Plethodon*. Biol. Bull. 2:289-299.

Trampusch, H. A. L., and A. E. Harrebomée. 1965. Dedifferentiation a prerequisite of regeneration, p. 341-374. *In* V. Kiortsis and H. A. L. Trampusch, [ed], Regeneration in animals and related problems. North-Holland Publ. Co., Amsterdam.

Volkmann, R. 1893. Ueber die Regeneration des quergestreiften Muskel-gewebes beim Menschen und Säugethier. Beitr. Pathol. Anat. Allg. Pathol. 12:233-332.

Waldeyer, W. 1865. Ueber die Veränderungen der quergestreiften Muskeln bei der Entzündung und dem Typhusprozess, sowie über die Regeneration derselben nach Substanzdefekten. Arch. Pathol. Anat. Physiol. 34:473-514.

Zhenevskaya, R. P. 1962. Experimental histologic investigation of striated muscle tissue. Rev. Can. Biol. 21:457-470.

Reprinted from *Science*, **161**, 283–285 (1968)

29

Limb Regeneration:
Induction in the Newborn Opossum
Merle Mizell

Abstract. *The marsupial* Didelphys virginiana *(the North American opossum) is uniquely suited for studies of mammalian limb replacement. By transplanting nervous tissue to the limb, regeneration has been successfully induced in this mammal.*

In lower vertebrates at least two factors, nervous (1) and hormonal (2), are decisive in determining whether or not regeneration will occur after limb amputation. Investigators have used this knowledge to induce regeneration in typically nonregenerating appendages of adult anurans (frogs) (3) and reptiles (lizards) (4). However, previous attempts to induce mammalian limb regeneration have been disappointing (5).

With increasing age the rate of regeneration and even the ability to regenerate diminish; for example, the developing hindlimbs of metamorphosing frogs can regenerate during early stages of development but lose this ability before the adult condition is attained (6). (This does not mean that adults undergo irreversible modifications in limb tissue which preclude regeneration, for the successful induction of adult anuran and reptilian limb regeneration indicates that this is not the case. But, as a general rule, "younger" tissues are more

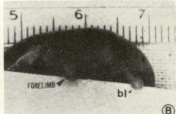
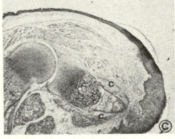
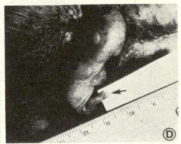

Fig. 2. Controls. (A) Newborn opossum (1 cm long) in pouch showing dichotomy of limb development: forelimbs well-developed with digits and claws (arrow); hindlimb with margins of the digits just becoming evident (× 7.8). (B) Twelve-day old opossum. Simple amputation of left forelimb and left hindlimb was performed 6 days after birth (thus, photo shows animal after 6 days "regeneration"). No external indication of regenerative response in forelimb, but note blastema-like appearance of hindlimb (*bl*) (scale in centimeters). (C) Photomicrograph of sagital section of forelimb in Fig. 2B. Note thickened apical epidermis, but also note the premature redifferentiation of subjacent tissues and the cartilagenous callus (*c*) forming around the radius (× 21). (D) Control hindlimb showing the best regenerative response after simple amputation. Note digit-like protuberance (arrow). Animal 42 days old, amputated above ankle 5 days after birth (therefore pictured after 35 days "regeneration") (scale in centimeters).

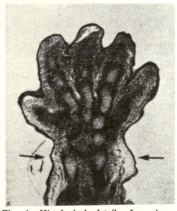

Fig. 1. Histological detail of newborn opossum hindlimb. Arrows indicate level of amputation (× 36).

responsive.) Some developing mammalian tissue can adjust to environmental changes and insults. Nevertheless, during mammalian ontogeny this ability wanes, and at birth mammals are incapable of regenerating limbs.

Nicholas had shown that amputated forelimbs of 14-day-old, or older, rat embryos *in utero* do not regenerate (7). And pilot experiments with mouse digits indicated that our present knowledge was still too rudimentary for induction of regeneration in newborn rodents to be attempted. Marsupials are unique in that they are born at a very early state of development. The newborn opossum thus presents a rare opportunity to perform chronic experiments on extremely young mammalian limbs.

Each series of experiments began in January; by early summer, work with the living pouch young was completed, and the parents were released. The opossums were trapped in areas adjoining New Orleans during the January to May breeding season. Some of the captured females contained pouch litters young enough to be used in the following experiments [developmental charts prepared by earlier investigators were used to determine the approximate age of the pups (8, 9)]. However, most animals were bred in captivity, and newborns were obtained from females that gave birth in the laboratory.

Females were anesthetized with an inhalant anesthetic, Penthrane (Abbott Laboratories), and all amputations were performed under a dissecting scope with iridectomy scissors while the young remained attached to the teats of the anesthetized mother. Limbs were amputated directly above the wrist or ankle. Small loss of blood made the use of ligatures unnecessary. Although aseptic techniques were employed, no elaborate sterilization procedures were utilized, nor was it necessary to employ antibiotics, for in over 220 amputations infection was never noted.

Wound healing and all changes in the limb stumps after amputation were followed by repeated gross observations; at various intervals, when noteworthy changes occurred, the limbs were photographed. Periodically some limbs were fixed for histological examination; limbs were embedded in paraffin and sectioned at 8 μm. Serial sections were stained with hematoxylin and eosin or Mallory's polychromatic stain.

Throughout the course of these experiments, the animals remained attached to the mother within the protective and nourishing environment of the marsupial pouch. Adult opossums had free access to water and Gaines Meal (General Foods Corp.).

Although McCrady (10) had shown that newborn opossum limbs (Figs. 1 and 2A) were not capable of regenerating after simple amputation, our first experiments were performed to determine the nature of their response to amputation. At birth, the embryo-like opossums display a striking dichotomy in limb development (Fig. 2A). Forelimbs have

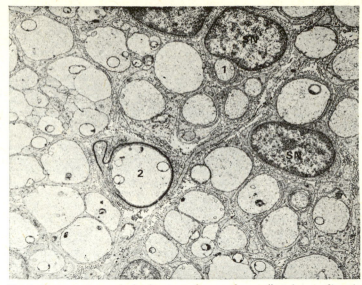

Fig. 4. Electron micrograph showing preponderance of unmyelinated nerve fibers in hindlimb of 3-week-old opossum. Schwann cell nucleus (*SN*); axis cylinder with the earliest indications of myelin envelopment (1); slightly later stage in myelinization (2); all other nerve fibers are essentially naked ($\times$ 9,000).

well-developed digits complete with claws, which permit the newborn animal to crawl from the birth canal to the pouch; but hindlimbs are rather rudimentary structures, with merely the early external indications of digits. However, histological examination has shown that cartilagenous models of all the phalanges are already present in the hindlimb of the newborn opossum (Fig. 1).

Gross observations verified that forelimbs were unable to regenerate after simple amputation (Fig. 2B). But histological examination disclosed that even the newborn forelimb exhibited a limited regenerative response (Fig. 2C); the epidermis underwent apical thickening, and the underlying tissues did undergo some dedifferentiation. However, although an appreciable amount of dedifferentiation did occur in the stump tissues, this was immediately followed by premature redifferentiation; the typical pattern of callus formation can be seen in Fig. 2C.

In contrast to the forelimb, the less differentiated tissues of the hindlimb displayed a rather remarkable regenerative response and in one case even gave rise to a blastema-like structure (Fig. 2B). Nevertheless, this structure also underwent premature redifferentiation and resulted in very limited replacement of the amputated portion of the limb. Simple amputation of the hindlimb does not lead to regeneration, and our results confirm McCrady's (10) finding that newborn opossums cannot regenerate hindlimbs after simple amputation. Of

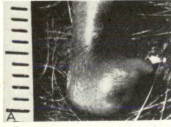

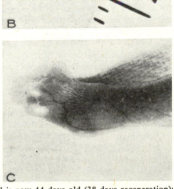

Fig. 3. Nerve implant regenerates. (A) Xenoplastic: Hindlimb of 51-day-old opossum. Nerve tissue from *Rana pipiens* tadpole cerebrum was implanted at 1 day, and the limb was amputated 6 days later. A heteromorphic, club-shaped regenerate formed in the ankle region and possessed a single digit-like protuberance (arrow). Limb pictured after 44 days regeneration (scale in centimeters). (B) Homoplastic: Hindlimb of animal 24 days old. Nerve tissue from young opossum cerebrum was transplanted when the recipient was 2 days old; limb was amputated 4 days later. After 18 days regeneration the early indication of digit differentiation can be seen (arrows) (scale in centimeters). (C) A later stage of the hindlimb regenerate in Fig. 3B. Animal is now 44 days old (38 days regeneration); the foot and basal portions of three toes are clearly visible ($\times$ 3.5).

94 hindlimb controls, Fig. 2D depicts the best regenerative response obtained after simple amputation. In this case, a flattened structure developed, and 37 days after amputation the very earliest indication of a digit-like protuberance was visible (Fig. 2D). This remarkable response of the hindlimb, albeit insufficient to produce a regenerate, lent encouragement for future attempts at inducing limb regeneration in the hindlimb.

Since preliminary experiments indicated that hindlimbs amputated later than 1 week after birth exhibited a pattern of healing similar to that of the forelimb, subsequent experiments were performed on hindlimbs of animals less than 1 week old. Earlier investigations have shown that during the first week of life, the opossum's immune mechanism is inoperative (11). In fact, at birth the thymus itself is merely an epithelial anlage (9). This lack of an immune mechanism permitted homologous and even heterologous transplants to persist in these young opossums.

Developing forebrain (cerebral cortex) was chosen as the source of nervous tissue for the transplantation experiments. The brain was removed from the donor and placed in chilled normal saline. The cerebrum was cut into several small pieces which were then picked up by a previously prepared fine-drawn glass pipette (inner diameter approximately 0.5 mm) whose surfaces had been lightly coated with silicone (Siliclad, Clay-Adams, Inc.) to facilitate transfer of the tissue. The slender point of a watchmaker's forceps was inserted into the proximal thigh region of the recipient hindlimb and extended distad, thereby creating a channel in the 1- or 2-day-old opossum limb. The pipette was inserted into this channel, and the small cylinder of nervous tissue (approximately 0.5 by 1.5 mm) was transferred to the hindlimb so that the long axis of the implant was parallel to the long axis of the limb.

Two to four days later the limb was amputated so that the amputation plane transected the implant. All amputations were attempted through the distal portions of the tibia and fibula (Fig. 1); but in some cases, because of the small size, amputation was inadvertently made through the ankle. When the operation was successful, a regenerative response ensued.

When nervous tissue from forebrain of young *Rana pipiens* tadpoles (Taylor-Kollros stage VII) was used, a positive response was noted in 3 of 14 cases. The best regenerate resulting from these heterologous nerve transplants was a curiously shaped outgrowth (Fig. 3A) consisting of a distal club-shaped structure which emerged from the ankle region and possessed a single digit-like protuberance on its medial surface. This response surpassed that of controls with simple amputations, but did not approach the extent of development achieved by the homologous nerve transplants.

Opossum nerve tissue evoked a positive response in 8 of 30 cases. The best regenerate resulting from homologous transplants of young opossum cerebrum is shown in Fig. 3, B and C. Figure 3B shows the animal at 24 days of age (nerve tissue implanted at 2 days and limb amputated 4 days later); after 18 days of regeneration a recognizable foot-like structure possessing the first indications of the fourth and fifth toes (Fig. 3B) could be seen. Development continued, and 20 days later (38 days regeneration) a heteromorphic foot containing the basal portions of three toes was evident (Fig. 3C). Since hindlimb development at birth (6 days before amputation) had already reached a state where all of the bones were present as cartilagenous models, the replacement of the foot and three toes must be interpreted as regeneration and not embryonic regulation.

The manner in which regeneration proceeds in the opossum hindlimb is very similar to regeneration of the metamorphosing frog hindlimb. As in hindlimb regeneration in the metamorphosing frog, histological landmarks which indicate the original plane of amputation are soon lost. Initially there exists a slight difference in tissue densities in the regenerating opossum limb, but it soon becomes impossible to detect the level of amputation. The histological aspects of regenerative phenomena in these two developing limb systems have been compared (12).

Two unique characteristics of the newborn opossum undoubtedly contributed to the development of regenerates: (i) absence of an immune mechanism, which prevented rejection of nerve implants; and (ii) short gestation period—a mere 12.75 days [at birth the opossum is equivalent in development to a 12-day rat embryo or a 2-month human fetus (13)]. Another feature of the opossum hindlimb which may participate in its ability to regenerate is the relatively "immature" state of the nerve fibers after they make their appearance in the limb. As late as 3 weeks after birth,

nerves of the hindlimb are essentially in an unmyelinated condition (Fig. 4).

My studies demonstrate that young opossum limbs can regenerate when additional nervous tissue is supplied. Results of control experiments indicate that neither the trauma of simple amputation, the trauma of implantation, nor the implantation of other homoplastic tissues (for example, liver or kidney) can evoke the regenerative response which results after implantation of brain tissue. Although the opossum has afforded the opportunity to induce regeneration in young mammals, it should be pointed out that we are no closer to an understanding of the mechanism of nerve action in regeneration than before. However, we are now in a position to compare this mammalian limb regeneration with regeneration in lower vertebrates; as their similarities and differences become apparent, additional insight into the phenomenon of mammalian cellular differentiation should be gained. Hopefully, once these factors are ascertained in young opossum regenerates, the newly acquired knowledge can then be successfully applied to other mammals.

MERLE MIZELL

Department of Biology,
Tulane University,
New Orleans, Louisiana 70118

References and Notes

1. M. Singer, *Quart. Rev. Biol.* **27**, 169 (1952); M. Singer, in *Regeneration in Vertebrates*, C. S. Thornton, Ed. (Univ. of Chicago Press, Chicago, 1959), p. 59.
2. O. E. Schotté and A. B. Hall, *J. Exp. Zool.* **121**, 521 (1952); O. E. Schotté and R. Bierman, *Rev. Suisse Zool.* **63**, 353 (1956).
3. O. E. Schotté and J. F. Wilber, *J. Embryol. Exp. Morphol.* **6**, 247 (1958); M. Singer, *Proc. Soc. Exp. Biol. Med.* **76**, 413 (1951); M. Singer, *J. Exp. Zool.* **126**, 419 (1954).
4. S. B. Simpson, Jr., *Proc. Soc. Exp. Biol. Med.* **107**, 108 (1961); M. Singer, *ibid.*, p. 106.
5. O. E. Schotté and C. B. Smith, *J. Exp. Zool.* **146**, 209 (1961); J. A. Bar-Moar and G. Gitlin, *Transplantation Bull.* **27**, 460 (1961); A. Scharf, *Growth* **27**, 255 (1963).
6. E. Marcucci, *Arch. Zool.* **8**, 89 (1916); O. E. Schotté and M. Harland, *J. Morphol.* **73**, 329 (1943); J. W. Forsyth, *ibid.* **79**, 287 (1946); J. M. Van Stone, *ibid.* **97**, 345 (1955).
7. J. S. Nicholas, *Proc. Soc. Exp. Biol. Med.* **23**, 436 (1926).
8. C. G. Hartman, *J. Morphol. Physiol.* **46**, 143 (1928); C. R. Moore and D. Bodian, *Anat. Rec.* **76**, 319 (1940).
9. M. Block, *Ergeb. Anat. Entwickl.* **37**, 237 (1964).
10. E. McCrady, Jr., *The Embryology of the Opossum* (Wistar Institute, Philadelphia, 1938), p. 203.
11. D. T. Rowlands, Jr., M. F. La Via, M. Block, *J. Immunol.* **93**, 157 (1964).
12. M. Mizell, in preparation.
13. M. Block, *Nature* **187**, 340 (1960).
14. Supported by NSF grant GB 7101 and by a grant from the Cancer Association of Greater New Orleans, Inc. I thank Miss Joyce Isaacs for technical assistance, J. P. Butler and S. Ziesenis for care and maintenance of our opossum colony, B. O. Spurlock and Dr. J. C. Harkin for the electron micrograph, and Dr. T. A. Rawls for veterinary assistance.

8 May 1968

Author Citation Index

Abbott, J., 101
Adensamer, W., 199
Adolph, E. F., 198
Albright, J. F., 265, 296
Allbrook, D. B., 499, 500
Amatatu, H., 295
Arey, L. B., 139
Ashbaugh, A., 358
Ashley, C. A., 77
Avery, G., 47, 319

Baker, R. F., 78
Baker, W. deC., 499
Ball, J. N., 443, 444
Banerjee, S. D., 5
Barber, L. W., 209
Barfurth, D., 209
Bar-Moar, J. A., 504
Barr, H. J., 101, 366
Barth, L. G., 198
Bassett, C. A. L., 46
Bassina, J. A., 46, 209
Bates, R. W., 443
Bélanger, L. F., 157
Bennett, H. S., 77
Berg, G. C., 140, 158
Berman, R., 331, 443
Bern, H. A., 331, 443, 444
Bernfield, M. R., 5
Betz, E. H., 499
Bier, A., 209
Bierman, R., 421, 445
Billingham, R. E., 485
Bischler, V., 101
Blacher, L. J., 209
Bland-Sutton, J., 485
Block, M., 504
Blocker, T. G., Jr., 158
Bloom, W., 77
Blum, H. F., 139
Blum, J., 78

Blumberg, J. M., 500
Bodemer, C. W., 77, 139, 157, 161, 395, 406
Bodian, D., 379, 504
Bogoroch, R., 158
Böhmel, W., 199
Bond, V. P., 139, 140
Bonneville, M., 421
Borssuk, R. A., 209
Brachet, J., 77
Bragdon, D. E., 444
Bray, G. A., 175, 457
Brecher, G., 140
Briggs, R., 102
Bromley, N. W., 209
Bromley, S. C., 5
Brønsted, H. V., 198
Brunst, V. V., 23, 46, 139
Bubenik, A. B., 485
Burnett, A. L., 101, 227
Burton, K., 457
Butler, E. G., 5, 23, 47, 101, 139, 175, 198
    209, 251, 252, 345, 379, 395, 406
Byrnes, E., 209

Cahn, M. B., 101
Cahn, R. D., 101
Cameron, G. R., 421
Campbell, D. H., 46
Capriata, A., 320
Carlson, B. M., 499
Caro, L. G., 101
Carpenter, R., 319
Cason, J. E., 101
Caufield, J. B., 78
Chadwick, C. S., 444, 445
Chaikoff, I. L., 157
Chalkley, D. T., 47, 77, 101, 139, 175, 227,
    444
Chamberlain, J., 421, 445
Chérémetiéva, E. A., 23, 46, 47, 139
Chèvremont, M., 319

505

Subject Index

511

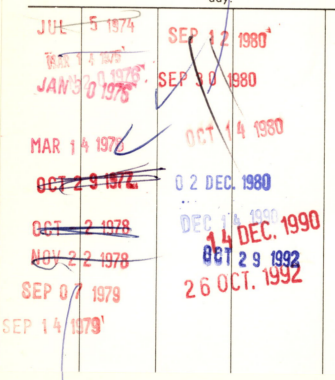